Linear Algebra II

Frederick P. Greenleaf
Courant Institute

Sophie Marques
Stellenbosch University

30

Linear Algebra II

Courant Institute of Mathematical Sciences
New York University
New York, New York

American Mathematical Society
Providence, Rhode Island

2010 *Mathematics Subject Classification.* Primary 97H60, 15-01, 15A04, 15A15, 15A18, 15A21, 40A05, 40A25, 20B30.

For additional information and updates on this book, visit
www.ams.org/bookpages/cln-30

Library of Congress Cataloging-in-Publication Data

Names: Greenleaf, Frederick P., author. | Marques, Sophie, 1986- author.
Title: Linear algebra II / Frederick P. Greenleaf, Courant Institute, New York University, Sophie Marques, Stellenbosch University.
Description: Providence, RI : American Mathematical Society, [2020] | Series: Courant lecture notes in mathematics, 1529-9031 ; 30 | Includes bibliographical references.
Identifiers: LCCN 2019059330 | ISBN 9781470454258 (paperback) | ISBN 9781470456429 (ebook)
Subjects: LCSH: Algebras, Linear. | AMS: Mathematics education – Algebra – Linear algebra. | Linear and multilinear algebra; matrix theory – Instructional exposition (textbooks, tutorial papers, etc.). | Linear and multilinear algebra; matrix theory – Basic linear algebra – Linear transformations, semilinear transformations. | Linear and multilinear algebra; matrix theory – Basic linear algebra – Determinants, permanents, other special matrix functions [See also 19B10, 19B14]. | Linear and multilinear algebra; matrix theory – Basic linear algebra – Matrix exponential and similar functions of matrices. | Linear and multilinear algebra; matrix theory – Basic linear algebra – Eigenvalues, singular values, and eigenvectors. | Linear and multilinear algebra; matrix theory – Basic linear algebra – Canonical forms, reductions, classification. | Sequences, series, summability – Convergence and divergence of infinite limiting processes – Convergence and divergence of series and sequences. | Sequences, series, summability – Convergence and divergence of infinite limiting processes – Approximation to limiting values (summation of series, etc.) For the Euler-Maclaurin summation formula, | Group theory and generalizations – Permutation groups – Symmetric groups.
Classification: LCC QA184.2 .G7255 2020 — DDC 512/.5–dc23
LC record available at https://lccn.loc.gov/2019059330

∞ The paper used in this book is acid-free and falls within the guidelines established to ensure permanence and durability.
Visit the AMS home page at `https://www.ams.org/`

10 9 8 7 6 5 4 3 2 1 25 24 23 22 21 20

Contents

Preface ix
Organization of the Text xiii
Acknowledgments xiv

Chapter 1. Generalized Eigenspaces and the Jordan Decomposition 1
An Overview of This Chapter 1
1.1. Nilpotent Operators and Examples 3
"Stable Range" and "Stable Kernel" of a Linear Map 5
Basic Properties of Nilpotent Operators 6
Cyclic Vectors and Cyclic Subspaces 8
1.2. Fine Structure of Nilpotent Operators 9
The General Inductive Step in Theorem 1.16 14
Cyclic Subspace Decompositions 16
Uniqueness of Cyclic Subspace Decomposition 16
1.3. Generalized Eigenspaces 21
Jordan Decomposition of a Generalized Eigenspace 24
1.4. The Generalized Eigenspace Decomposition 25
Further Properties of Characteristic Polynomials 27
Proof of Generalized Eigenspace Decomposition 29
The Jordan Canonical Form 31
1.5. Consequences of the Jordan Canonical Form (JCF) 37
The General Spectral Mapping Theorem 37
Cayley-Hamilton Theorem 39
Alternative Proof of Cayley-Hamilton 40
Further Examples of Jordan Form 41
Jordan Form $J(T)$ as a Similarity Invariant 43
The Minimal Polynomial $m_T(x)$ for $T : V \to V$ 44
Additional Exercises 46
1.6. Appendix A: Brief Review of Diagonalization 55
Algebraic vs. Geometric Multiplicity 58

Chapter 2. Further Applications of Jordan Form 63
An Overview of This Chapter 63
2.1. The Jordan Form and Differential Equations 64
Solving Higher-Order Differential Equations 66
General Comments on Constant Coefficient ODE 68

Transforming an ODE to a First-Order System $\{dX/dt = A \cdot X(t)$ 68
The Case of a Double Root 71
2.2. Normal Forms for Linear Operators over $\mathbb{R}$ 75
Complexification: A Case Study 76
Complexification of Real Vector Spaces and Operators 77
Relations Between $\mathbb{R}[t]$ and $\mathbb{C}[t]$ 78
Complexification of Linear Operators over $\mathbb{R}$ 80
Subspaces of Real Type in $V_{\mathbb{C}}$ 83
Concordance Between Eigenspaces of T in V and of $T_{\mathbb{C}}$ in $V_{\mathbb{C}}$ 85
The Normal Form of a Linear Operator Over $\mathbb{R}$ 88
When $\boldsymbol{T}_{\mathbb{C}}$ Is Not Diagonalizable 89
Additional Exercises 94

Chapter 3. Bilinear, Quadratic, and Multilinear Forms 99
An Overview of This Chapter 99
3.1. Basic Definitions and Examples 100
Variants: Sesquilinear Forms 102
Dealing with Degeneracy of B 106
3.2. Canonical Models for Bilinear Forms 106
The Automorphism Group of a Bilinear Form B. 107
Canonical Forms. Case 1: B Symmetric, $\mathbb{K} = \mathbb{R}$ 109
Sylvester's Theorem: Invariance of the Signature on $\mathrm{O}(p, q)$ 113
A Diagonalization Algorithm 115
The Gauss-Seidel Algorithm 117
Canonical Forms. Case 2: B Symmetric, $\mathbb{K} = \mathbb{C}$ 118
Canonical Forms: Case 3: B Antisymmetric, $\mathbb{K} = \mathbb{R}$ or $\mathbb{C}$ 120
3.3. Sesquilinear Forms ($\mathbb{K} = \mathbb{C}$) 122
Sesquilinear Change of Basis Formula 123
Additional Exercises 127

Chapter 4. Tensor Fields, Manifolds, and Vector Calculus 131
An Overview of This Chapter 131
4.1. Tangent Vectors, Cotangent Vectors, and Tensors 134
Smooth Functions and Mappings on Manifold M 138
Tangent Vectors and the Tangent Spaces TM_p 141
Some Objections: Finding the Right Definition 142
Change of Coordinates 147
Vector Fields as Differential Operators on M 149
The Differential of a Map Between Manifolds 151
4.2. Cotangent Vectors and Differential Forms 154
The Canonical d-Operator on Functions 155
Exterior Derivatives (an Overview) 157
Primitives of 1-Forms 158
Interpretations of 1-Forms and Their Primitives 160

4.3. Differential Forms on M and Their Exterior Derivatives 161
Action of Permutation Groups S_n on Tensors 165
Products of Tensors and Tensor Fields 167
Wedge Product of Antisymmetric Tensors 167
Calculating Wedge Products on a Chart 170
Exterior Derivative of Higher Rank k-Forms 172
Primitives of k-Forms. 175
Poincaré's Lemma 176
Transferring Calculations Between M and $\mathbb{R}^m$ 177
Proof of Poincaré's Lemma 178
4.4. Div, Grad, Curl, and All That 181
Vector Operations in Two Dimensions 185
Differential Forms and the Cross Product 186
Additional Exercises 189

Chapter 5. Matrix Lie Groups 203
An Overview of This Chapter 203
5.1. Matrix Groups and the Implicit Function Theorem 204
Smooth Mappings and Their Differentials 205
Smooth Hypersurfaces and the Implicit Function Theorem (IFT) 207
Rank vs Dimension of Level Sets 211
The "Maximal Rank" Case 212
The Inverse Mapping Theorem (IMT) 213
Differentiable Manifolds (General Definition) 214
Differentiable Manifolds and the IFT 215
5.2. Matrix Lie Groups 221
Real vs Complex Matrix Groups 222
Examples of Lie Groups 225
Translations and Automorphisms on a Lie Group 231
Tangent Spaces TM_p of a Manifold 233
The Differential of a Map 234
Smooth Vector Fields on a Manifold 235
Smooth Vector Fields as Differential Operators on M 238
5.3. Lie Algebra Structure in Tangent Spaces of Lie Groups 242
Lie Algebra of a Matrix Lie Group 243
Adjoint Action of G on its Lie Algebra $\mathfrak{g}$ 246
Lie Algebras of the Classical Groups 247
Some Lie Algebras Determined by Commutation Relations 252
5.4. The Exponential Map for Matrix Lie Groups 254
One-Parameter Subgroups in Matrix Lie Groups 255
The Logarithm $Log(A)$ of a Matrix 256
Singularities of the Exponential Map on $M(n, \mathbb{C})$ 258
$exp : \mathfrak{g} \rightarrow G$ 260
Case Study: Nilpotent Lie Groups 261
The Maps Exp, Ad, ad Revisited 264

Connected Lie Groups 270
The Campbell-Baker-Hausdorff Formula: Recovering G from $\mathfrak{g}$ 272
5.5. The Lie Correspondence: Lie Groups vs Lie Algebras 274
Further Development of Lie Theory 276
Additional Exercises 277

Bibliography 289

Preface

The previous volume, Linear Algebra I (LA I), provided a self-contained account of basic topics that might be covered in one semester at the introductory graduate level, ending with:

- In Chapter 5, a general discussion of diagonalization including: norm convergence of linear operators, matrix-valued power series, and use of the exponential map $\text{Exp}(A) = e^A$ to solve systems of differential equations $dX/dt = A \cdot X(t)$ in which A is diagonalizable.
- In Chapter 6 we discuss orthogonal diagonalization of operators on inner product spaces, including spectral decomposition, spectral mapping theorem, polar decomposition $T = U \cdot |T|$, and singular value decomposition.

Linear Algebra II begins with more challenging basic topics, presented in Chapters 1–3, whose contents are described below. The final Chapters 4 and 5 are different. They are intended as more-or-less independent and self-contained surveys of two special topics:

- Chapter 4: Tensor Fields, Differentiable Manifolds and Vector Calculus
- Chapter 5: Matrix Lie Groups

Both are vast subjects, so presentations in these chapters will not be as fully developed as those in preceding chapters or LA I. Our intent was to present some advanced topics that illustrate uses of linear algebra in realms beyond a standard second course in linear algebra. In practice, every instructor could choose which to present, according to his or her interests and those of the class. Each chapter in LA II begins with a detailed overview of the topics that will be covered. Here is brief description of the chapter contents.

CHAPTER 1. GENERALIZED EIGENSPACES AND THE JORDAN DECOMPOSITION. The first serious obstacle to diagonalization is dealing with nilpotent operators, which is addressed by working out a detailed procedure for computing their cyclic subspace decompositions. Although there are elegant existence proofs regarding such decompositions, the desired cyclic subspaces are not unique (unlike eigenspaces in diagonalization), and any procedure for finding them is inevitably complicated by the need to make some arbitrary choices.

For any linear operator $T : V \to V$ the space V is uniquely a direct sum (the Fitting decomposition) $V = K_\infty(T) \oplus R_\infty(T)$ of T-invariant subspaces, the "stable kernel" and "stable range" of T, on which $T|_{K_\infty}$ is nilpotent and $T|_{R_\infty}$ is bijective. If λ is an eigenvalue of T, $(T - \lambda I)$ is obviously nilpotent on its stable kernel $M_\lambda(T) = K_\infty(T - \lambda I)$, the *generalized λ-eigenspace* of T. These spaces are T-invariant and linearly independent. Whenever the characteristic polynomial $p_T(x) = \det(T - \lambda I)$ splits in $\mathbb{K}[x]$, a proof by induction on $\dim(V)$ employs the Fitting decomposition to show that V is the direct sum $V = \bigoplus_\lambda M_\lambda(T)$ of its generalized eigenspaces. Since each component $M_\lambda(T)$ has its own cyclic subspace decomposition, this "generalized eigenspace decomposition" leads directly to the Jordan canonical form $J(T)$ for T, which has many applications.

Chapter 1 ends with a brief appendix reviewing key theorems about diagonalization that were covered in Linear Algebra I, illustrating a few techniques of proof that will be relevant in discussing what to do when diagonalization fails.

CHAPTER 2. FURTHER APPLICATIONS OF THE JORDAN FORM. In the first half of this chapter we employ the Jordan form to solve higher-order ODEs by converting them into linear systems $dX/dt = A \cdot X(t)$ of constant coefficient first-order equations. We then recast the coeffcent matrix A in Jordan canonical form $J(A)$ made up of diagonal blocks $B_k = \lambda_k I + E_k$ in which E_k is an elementary nilpotent matrix. Owing to nilpotence, the one-parameter groups e^{tB_k} can be written explicitly, and the system solved by taking $X(t) = e^{tJ(A)} \cdot X(0)$. The solution of the original n^{th}-order ODE is easily read from this.

These techniques work when the characteristic polynomial $p_A(x)$ splits into linear factors in the space of polynomials $\mathbb{K}[x]$. The second half of Chapter 2 concerns itself with complexifying real vector spaces and linear operators $T : V \rightarrow V$ to get complex spaces $V_\mathbb{C}$ and operators $T_\mathbb{C}$ to which the preceding methods apply. By reverse-engineering the construction of the Jordan form $J(T_\mathbb{C})$, the original $\mathbb{R}$-linear operator T can be recast in a "real-normal form" that reveals its structure. There are actually two such real forms, depending on whether $T_\mathbb{C}$ is diagonalizable or not (in which case one applies the Jordan form).

CHAPTER 3. BILINEAR, QUADRATIC, AND MULTILINEAR FORMS. Bilinear forms $B(x, y)$ on a vector space V arise often in many areas of mathematics, inner products on real vector spaces being just one example. We restrict attention to the *symmetric* and *antisymetric* B, the ones most frequently encountered. Bilinear forms can be described by $n \times n$ matrices $[B]_\mathfrak{X} = [B(e_i, e_j)]$ once a basis $\mathfrak{X} = \{e_i\}$ in V is specified. Modulo a change of basis, bilinear forms over $\mathbb{R}$ or $\mathbb{C}$ can be classified according to their normal forms, for which there are only three possibilites. We also work out the transformation law $[B]_\mathfrak{X} \rightarrow [B]_\mathfrak{Y}$ between coordinate descriptions of B.

Every bilinear form $B : V \times V \rightarrow \mathbb{K}$ determines a group $\text{Aut}(B)$ of *automorphisms*, linear operators that leave B invariant, so $B(Tx, Ty) = B(x, y)$. As examples we have the *orthogonal group* of linear rigid motions $\text{O}(n)$ on $\mathbb{R}^n$, and the *unitary operators* $\text{U}(n)$ on $\mathbb{C}^n$, that preserve the usual inner products on these spaces; there is also a third family, the *symplectic groups* $\text{Sp}(2n, \mathbb{K})$. These *classical groups* are prevalent in physics and geometry, or wherever one has to deal with symmetries. As an example we discuss the *Lorentz group*, the automorphism group $\text{Aut}(B) = \text{SO}(3,1)$ for the form

$$B(\mathbf{x}, \mathbf{y}) = x_1y_1 + x_2y_2 + x_3y_3 - x_4y_4 \quad \text{for } \mathbf{x}.\mathbf{y} \in \mathbb{R}^4.$$

This is the symmetry group underlying Einstein's theory of special relativity (taking x_1, x_2, x_3 to be space coordinates and x_4 = time). It tells us how to transform measurements made in one space-time coordinate frame to those made in a different frame moving at constant velocity. (The outcome is decidedly nonintuitive!)

Beyond all this, Chapter 3 makes an initial foray into the world of *tensors* and *multilinear forms*, which have become central topics in differential geometry, and physics too — general relativity is all about tensors, as is much of electromagnetic theory and the analysis of stress in solid media. This part of Chapter 3 covers a few basic facts that set the stage for a much more extensive discussion of tensor algebra and differential geometry in Chapter 4, should the instructor choose to pursue this special topic upon completion of Chapters 1–3.

CHAPTER 4. TENSOR FIELDS, MANIFOLDS, AND VECTOR CALCULUS. This chapter explores aspects of linear and multilinear algebra that lie at the heart of modern differential geometry and ends with a reinterpretation of the main results of traditional multivariate calculus. After a brief review of the classical vector operators *div*, *grad*, and *curl* of calculus, attention shifts to the concept of a differentiable manifold and the present-day interpretation of the vector fields that live on them, which are now recognized to be of many types — fields of tangent vectors, cotangent vectors, and tensors of various ranks.

Early on, manifolds were viewed as smooth hypersurfaces embedded in some Euclidean space; tangent vectors, normal vectors, and tensor were realized as objects within that external space. But everyone knew this universal Euclidean space was a fiction. Furthermore, in trying to do calculus on, say, a 4-dimensional hypersurface M embedded in a 10-dimensional space, one is confronted with some vexing questions, for instance:

> *What do we mean by a "tangent vector to M" if we are forbidden to speak of anything external to M?*

More generally, how can one get rid of all reference to some mythical "all-encompassing Euclidean space" and deal with such questions in an intrinsic and coordinate-independent way?

A satisfactory answer emerged in the first quarter of the twentieth century, based on the concept of a *differentiable manifold* which circumvented these philosophical issues. Along with it came the realization that many different types of vector fields could be associated with a manifold. We illustrate the possibilities by examining a few examples from the physical sciences, in which we explain why:

- *velocity fields*, as in fluid flow, must be regarded as fields of tangent vectors,

- *force fields* such as electric fields, gravitational fields, etc., must be interpreted as fields of cotangent vectors.
- On the other hand, *magnetic fields* are usually described in introductory courses as vector fields, but the vectors involved are something quite different, neither tangent nor cotangent vectors. They are in fact represented by fields of rank-2 antisymmetric tensors.

Our goals in Chapter 4 are to:

- Develop enough of the theory of differentiable manifolds, and the multilinear forms (tensor fields) that live on them, to reformulate traditional multivariate calculus in modern terms.
- Work through basic concepts of multilinear algebra needed to understand the concepts upon which modern differential geometry is based.
- Provide enough exposure to these concepts so that one can carry out meaningful computations.

The last section of this chapter is focused on tying the concepts discussed in this chapter, to what you might have been taught in lower-level multivariate calculus.

CHAPTER 5. MATRIX LIE GROUPS. The subject of *Lie groups* is the nexus where three major areas of mathematics come into play — calculus-style analysis, modern algebra, and differential geometry are all intertwined in this field, underpinned by the tools linear algebra provides. That interplay among so many disciplines can be daunting, but it is also what makes Lie groups such an interesting subject.

This brief chapter cannot be a treatise on the subject. It is intended to be an introduction to the players involved, a few of the most important concepts, and examples illustrating the techniques involved in working with them. Some knowledge of differential geometry will be assumed, but the necessary topics reprised early in the chapter to make the discussion self-contained. For additional information see Section 4.1.

Many Lie groups of interest in physics and geometry were originally modeled as curvilinear smooth hypersurfaces M embedded in Euclidean spaces $\mathbb{R}^n$ (and sometimes $\mathbb{C}^n$), for example the "classical matrix groups" — the orthogonal groups $\mathrm{O}(n)$, unitary groups $\mathrm{U}(n)$, and symplectic groups $\mathrm{Sp}(2n)$ introduced in Chapter 3 as the symmetry groups associated with various bilinear forms that arise throughout mathematics — are realized as subsets of matrix space $\mathrm{M}(n, \mathbb{K})$ with $\mathbb{K} = \mathbb{R}$ or $\mathbb{C}$.

Beginning in the twentieth century Lie theory (and most of differential geometry as well) moved away from the notion of embedded hypersurfaces and developed an intrinsic theory of such spaces — now called *differentiable manifolds* — in which Lie groups and manifolds were regarded as self-contained universes in their own right. However, to keep things simple we focus on *matrix Lie groups*, which can be modeled as subsets of matrix space over $\mathbb{R}$ or $\mathbb{C}$. Besides, almost everything said for matrix Lie groups carries over to the general theory of Lie groups, although some proofs get harder in that setting.

Main topics of Chapter 5 are:

- To show how differentiable structure can be imposed on a matrix group G using the implicit function theorem to define Euclidean coordinates on a family of open sets U_α that cover G. Equipped with these *coordinate charts*, G becomes a *Lie group*, in which the group multiplication operation $G \times G \to G$ becomes a differentiable map.
- To discuss what *tangent vectors*, *tangent spaces*, and *derivatives of functions* $f : G \to \mathbb{R}$" might mean on a group G equipped with differentiable structure, which might be realized as curvilinear hypersurface of high dimension. This is the realm of differential geometry.
- The result is a construct G that has both geometric and algebraic aspects. Lie groups can be quite complicated, but deal with them we must because they turn up at the heart of modern physics (and other fields), for instance as the symmetry groups that govern: the interactions of subatomic particles; the behavior of the cosmos according to special and general relativity; and the seemingly paradoxical rules of quantum mechanics. Even the periodic table of chemistry can in the end be deduced as a consequence of the mathematics associated with that classic example of a Lie group — the *special orthogonal group* SO(3) of rotations in 3-dimensional space.
- Finally, we will show that the tangent space to a Lie group G at its identity element, which at first sight is just a vector space of the same dimension as the coordinate charts that cover G, acquires algebraic structure induced by the multiplication law in G and becomes what is known as a *Lie algebra*.

  The important point is that this Lie algebra is a *linear* structure, much easier to deal with than G, and yet it encodes almost all the information to completely reconstruct and understand the Lie group from which it was derived.

Organization of the Text

The handling of exercises is somewhat unconventional. Many are placed within the main text, as the topics they address first occur. Each chapter ends with an extensive set of section-by-section Additional Exercises. Some recap the main topics of of each section; others are longer and intended to be more challenging; each begins with a block of true/false questions, which students often find more challenging than you might expect.

There are a few unconventional notations, explained as they are introduced:

- We often write $|V|$ for the dimension $\dim(V)$, and $R(T), K(T)$ for the range and kernel of T, respectively, when this is convenient.
- A special symbol ł is used to distinguish the constant function (or polynomial) everywhere equal to 1, from the scalar 1.

The symbol $\mathbb{K}$ indicates a generic ground field.

Acknowledgments

The authors are particularly indebted to Paul Monsour, managing editor of the AMS/Courant Lecture Notes Series, and the in-house production editor, Neelang Parghi, for their tireless efforts in creating this volume. Thanks also to Ina Mette, AMS acquisitions editor, and to Holli Chopra, our copy-editor, who was able to catch many errors in the original LaTeX file — those that remain, for which we apologize in advance, are entirely the authors' responsibility.

CHAPTER 1

Generalized Eigenspaces and the Jordan Decomposition

What to do when diagonalization fails.

An Overview of This Chapter. In the previous volume, *Linear Algebra I*, we analyzed operators that were diagonalizable and the techniques by which they can be cast into diagonal form. The current volume, *Linear Algebra II*, explores what can be done to understand the structure of a linear operator when diagonalization fails. We continue the conventions employed in Linear Algebra I, referred to as "LA I" in the present *Notes*. For instance, we often write $\dim(V) = |V|$ where appropriate; $\mathbb{K}$, $\mathbb{C}$, $\mathbb{R}$, $\mathbb{Q}$, for number fields; and $\mathbb{1}$ for the constant polynomial in $\mathbb{K}[x]$.

Chapter 1 ends with a brief appendix that recapitulates key concepts developed in LA I about diagonalizing linear operators. This appendix also includes some details of proof illustrating techniques that will be relevant to our efforts in the current chapter.

Once we pass beyond the topics of LA I, the analysis gets harder and the revealed structures more complicated. With that in mind, and as a guide to readers of this chapter, we shall briefly outline the major steps needed to move beyond the realm of diagonalizability.

- The first serious obstacle is dealing with *nilpotent operators*, those for which $T^n = 0$ (the zero operator) for all sufficiently large $n \in \mathbb{N}$. These have $\lambda = 0$ as their only eigenvalue for any ground field $\mathbb{K}$, and the only eigenspace $E_{\lambda=0}(T)$ can be $\subsetneqq V$. Decomposing V into eigenspaces gets you nowhere.
- The "Fitting decomposition" of V provides a path toward understanding an arbitrary linear operator $T : V \to V$. In it, we extend the notions of kernel $K(T)$ and range $R(T)$ of an operator by defining the *stable kernel* $K_\infty(T)$ and *stable range* $R_\infty(T)$:

$$K_\infty(T) = \bigcup_{n\in\mathbb{N}} K(T^n) \quad \text{and} \quad R_\infty(T) = \bigcap_{n\in\mathbb{N}} R(T^n).$$

  The subspaces $K(T^n)$ are strictly increasing until they become constant, with $K(T^n) = K(T^{n+1}) = \cdots = K_\infty(T)$; the ranges are decreasing and stabilize at the same exponent n, with $R(T^n) = R_\infty(T)$. These stable subspaces are T-invariant and yield a direct sum decomposition

$V = K_\infty(T) \oplus R_\infty(T)$; this Fitting decomposition is the basis for all further progress. Notice that the restricted operator $\widetilde{T} = T|_{K_\infty(T)}$ is nilpotent, with $\widetilde{T}^{n-1} \neq 0$, $\widetilde{T}^n = 0$, and nilpotence degree $n = \dim(K_\infty)$.

- The structure of a nilpotent operator is revealed by decomposing V as a direct sum $V = C_1 \oplus \cdots \oplus C_r$ of *cyclic subspaces*, where each C_k contains a vector v_0 whose iterates

$$v_0, T(v_0), \ldots, T^{d_k - 1}(v_0)$$

span V and are linearly independent, with $T^{d_k}(v_0) = 0$. Thus,

$$\mathfrak{X} = \{T^k(v_0) : 0 \leq j \leq d_k - 1\}$$

is a basis, $\dim(C_k) = d_k$, and the restriction $T|_{C_k}$ is nilpotent with nilpotence degree $d(T|_{C_k}) = d_k$. If we list these basis vectors in reverse order, the restriction has a particularly simple matrix realization as an *elementary nilpotent matrix*

$$[T|_{C_k}]_{\mathfrak{X}\mathfrak{X}} = \begin{pmatrix} 0 & 1 & & & 0 \\ & 0 & 1 & & \\ & & \ddots & \ddots & \\ & & & 0 & 1 \\ 0 & & & & 0 \end{pmatrix}_{d_k \times d_k} .$$

- The *cyclic subspace decomposition* $V = C_1 \oplus \cdots \oplus C_r$ is, unfortunately, not unique. The subspaces C_k are not uniquely determined, unlike the eigenspaces of a diagonalizable operator T; "algorithms" for finding suitable subspaces C_k will necessarily involve some arbitrary choices that cannot be determined in advance. On the other hand, this decomposition *is* unique in the sense that the number of subspaces C_k, their dimensions d_k, and the elementary nilpotent matrix realization of $T|_{C_k}$ for a suitably chosen basis are unique and can be computed explicity. The trouble comes if you find it necessary to exhibit an actual basis $\mathfrak{X}$ that does the job.
- After working out a detailed procedure for handling nilpotent operators, we turn to analysis of general operators whose characteristic polynomial $p_T(x) = \det(T - \lambda I)$ splits into linear factors in $\mathbb{K}[x]$. This is always true when $\mathbb{K} = \mathbb{C}$, and in any case, this obstacle can always be surmounted by "enlarging" the ground field $\mathbb{K}$, a process that will be illustrated for $\mathbb{K} = \mathbb{R}$ vs $\mathbb{K} = \mathbb{C}$ in Chapter 2.
- This brings us to the central result of Chapter 1, the *generalized eigenspace decomposition* of a linear operator T and the associated *Jordan canonical form* (JCF) for matrices, which apply whenever $p_T(x)$ splits over $\mathbb{K}$. For any $\lambda \in \mathbb{K}$, we define the *generalized eigenspace*

$$\begin{aligned} M_\lambda(T) &= K_\infty(T - \lambda I) \\ &= \{x \in V : (T - \lambda I)^k x = 0 \text{ for some } k \in \mathbb{N}\}, \end{aligned}$$

and λ is a *generalized eigenvalue* for T if $M_\lambda(T) \neq (0)$. Clearly, λ is an eigenvalue $\Leftrightarrow$ it is a generalized eigenvalue, but $E_\lambda(T) \subsetneqq M_\lambda(T)$ in general and that makes all the difference. The $M_\lambda(T)$ are T-invariant, and when $p_T(x)$ splits over $\mathbb{K}$, we get a direct sum decomposition

$$V = \bigoplus_\lambda M_\lambda(T)$$

indexed by the distinct eigenvalues of T in $\mathbb{K}$.

- That analysis begins by considering a fixed eigenvalue λ and invoking both the Fitting decomposition and the generalized eigenspace decomposition. Observe that the stable kernel of the operator $T - \lambda I$ is invariant under both $(T - \lambda I)$ and T, and that $(T - \lambda I)$ is nilpotent on its stable kernel

$$K_\infty(T - \lambda) = \{x \in V : (T - \lambda I)^k x = 0 \text{ for some } k\} = M_\lambda(T).$$

Therefore, the restriction of $(T - \lambda I)$ to this stable kernel has a cyclic subspace decomposition $M_\lambda(T) = \bigoplus_{j=1}^{s} C_{\lambda,j}$. Each cyclic subspace has a basis such that the restriction of $(T-\lambda I)$ to $C_{\lambda,j}$ is represented by an elementary nilpotent matrix. Since T itself leaves these subspaces invariant, the matrix of $T = (T - \lambda I) + \lambda I$ with respect to the same basis is a typical *Jordan block*

$$B_k = [T|_{C_{\lambda,j}}]_{\mathfrak{X}\mathfrak{X}} = \begin{pmatrix} \lambda & 1 & & & 0 \\ & \lambda & 1 & & \\ & & \ddots & & \\ & & & \lambda & 1 \\ 0 & & & & \lambda \end{pmatrix}_{d_k \times d_k}.$$

- In the endgame (only sketched here, leaving details to the main text), we proceed by returning to the Fitting decomposition, which becomes $V = M_\lambda(T) \oplus R_\infty$ (abbreviating $R_\infty = R_\infty(T - \lambda I)$). We will show the following:
  (i) R_∞ is T-invariant.
  (ii) The characteristic polynomial $p_{T'}$ of $T' = T|_{R_\infty}$ splits over $\mathbb{K}$ if p_T does.

  That means the restriction of T to R_∞ satisfies all the hypotheses of the theorem we are trying to prove, and since $\dim(R_\infty) < \dim(V)$, we may argue by induction on $\dim(V)$ to conclude that we have similar "fine structure" in each of the generalized eigenspaces.

That concludes our outline of things to come. Now for the details.

1.1. Nilpotent Operators and Examples

Nilpotent operators present the first obstacle in attempts to diagonalize a linear operator.

DEFINITION 1.1. We say $T : V \to V$ is *nilpotent* if $T^k = 0$ for some $k \in \mathbb{N}$ and *unipotent* if $T = I + N$ with N nilpotent. The smallest exponent such that

$T^d = 0$ is the *nilpotent degree* $\deg(T)$. Obviously, T is unipotent $\Leftrightarrow T - I$ is nilpotent. Nilpotent and unipotent matrices $A \in \mathrm{M}(n, \mathbb{K})$ are defined the same way as for operators.

Nilpotent operators cannot be diagonalized unless T is the zero operator (or $T = I$, if unipotent). Any analysis of normal forms must therefore examine these operators in detail. As examples, all strictly upper triangular matrices (with zeros on the diagonal) as well as those that are strictly lower triangular are nilpotent in view of the following observations.

EXERCISE 1.2. If A has upper triangular form with zeros on and below the diagonal, prove that

$$A^2 = \begin{pmatrix} 0 & 0 & & * \\ & 0 & \ddots & \\ & & \ddots & 0 \\ 0 & & & 0 \end{pmatrix}, \quad A^3 = \begin{pmatrix} 0 & 0 & 0 & & * \\ & \ddots & \ddots & \ddots & \\ & & \ddots & \ddots & 0 \\ & & & \ddots & 0 \\ 0 & & & & 0 \end{pmatrix},$$

etc., so that $A^n = 0$.

Matrices of the same form but with 1's on the diagonal all correspond to unipotent operators.

We will see that if $N : V \to V$ is nilpotent, there is a basis $\mathfrak{X}$ such that

$$[N]_{\mathfrak{X}\mathfrak{X}} = \begin{pmatrix} 0 & & * \\ & \ddots & \\ & & \ddots \\ 0 & & 0 \end{pmatrix},$$

but this is not true for all bases. Furthermore, a lot more can be said about the off-diagonal terms $(*)$ for suitably chosen bases.

EXERCISE 1.3. In $\mathrm{M}(n, \mathbb{K})$, show that the following sets of upper triangular matrices are both subgroups in $\mathrm{GL}(n, \mathbb{K})$.

(a) The STRICTLY UPPER TRIANGULAR GROUP $\mathcal{N} = \left\{ \begin{pmatrix} 1 & & * \\ & \ddots & \\ & & \ddots \\ 0 & & 1 \end{pmatrix} \right\}$

with entries in $\mathbb{K}$.

(b) The FULL UPPER TRIANGULAR GROUP $\mathcal{P} = \left\{ \begin{pmatrix} a_{1,1} & & * \\ & \ddots & \\ & & \ddots \\ 0 & & a_{n,n} \end{pmatrix} \right\}$

with entries in $\mathbb{K}$ such that $\prod_{i=1}^n a_{ii} \neq 0$.

Verify that $\mathcal{N}$ and $\mathcal{P}$ are closed under taking matrix products and inverses.

EXAMPLE 1.4. Let $A = \begin{pmatrix} 0 & 1 \\ 0 & 0 \end{pmatrix}$ in $\mathrm{M}(2, \mathbb{K})$. This is a nilpotent matrix, and in any ground field $\lambda = 0$ is the only root of its characteristic polynomial

$$p_A(\lambda) = \det(A - \lambda I) = \lambda^2.$$

There is a nontrivial eigenvector $e_1 = (1,0)$ corresponding to eigenvalue $\lambda = 0$ because $\ker(A) = \mathbb{K} \cdot e_1$ is nontrivial (as it must be for any nilpotent operator). But you can easily verify that scalar multiples of e_1 are the only eigenvectors, so there is no basis of eigenvectors in $\mathbb{K}^2$ and A cannot be diagonalized by any similarity transformation, regardless of the ground field $\mathbb{K}$. We now turn to an important tool for advancing beyond the limitations of diagonalization, the Fitting decomposition.

"Stable Range" and "Stable Kernel" of a Linear Map. Given a linear operator $T : V \to V$ on a finite-dimensional vector space (arbitrary ground field), let $K_i = K(T^i) = \ker(T^i)$ and $R_i = R(T^i) = \text{range}(T^i)$ for $i = 0, 1, 2, \ldots$. These spaces are nested

$$\begin{gathered}(0) \subseteq K_1 \subseteq K_2 \subseteq \cdots \subseteq K_i \subseteq K_{i+1} \subseteq \cdots \\ V \supseteq R_1 \supseteq R_2 \supseteq \cdots \supseteq R_i \supseteq R_{i+1} \supseteq \cdots,\end{gathered}$$

and if $\dim(V) < \infty$, they must each stabilize at some point, say with $K_r = K_{r+1} = \cdots$ and $R_s = R_{s+1} = \cdots$ for some integers r and s. In fact, if r is the first (smallest) index such that $K_r = K_{r+1} = \cdots$ the sequence of ranges must also stabilize at the same point because $|V| = |K_i| + |R_i|$ at each step. With this in mind, we define

$$\text{STABLE RANGE OF } T\text{:} \quad R_\infty = \bigcap_{i=1}^{\infty} R_i = R_r = R_{r+1} = \cdots$$

$$\text{STABLE KERNEL OF } T\text{:} \quad K_\infty = \bigcup_{i=1}^{\infty} K_i = K_r = K_{r+1} = \cdots$$

PROPOSITION 1.5. *$V = R_\infty \oplus K_\infty$ and the spaces R_∞, K_∞ are T-invariant. Furthermore, $R_{i+1} \neq R_i$ and $K_{i+1} \neq K_i$ for $i < r$.*

PROOF. To see there is a nontrivial jump $R_i \supsetneqq R_{i+1}$ at every step until $i = r$, it suffices to show that $R_{i+1} = R_i$ at the ith step implies $R_{i+2} = R_{i+1}$ (a similar result for kernels then follows automatically). Obviously, $R_{i+2} \subseteq R_{i+1}$ for all $i \geq r$ since the ranges get smaller; to prove the reverse inclusion, let $v \in R_{i+1}$. Then there is some $w_1 \in V$ such that $v = T^{i+1}(w_1) = T\big(T^i(w_1)\big)$. By hypothesis, $T^{i+1}(V) = R_{i+1} = R_i = T^i(V)$, so there is some $w_2 \in V$ such that $T^i(w_1) = T^{i+1}(w_2)$. Thus,

$$v = T^{i+1}(w_1) = T(T^i(w_1)) = T(T^{i+1}(w_2)) = T^{i+2}(w_2) \in R_{i+2},$$

so $R_{i+1} \subseteq R_{i+2}$. Hence $R_i = R_{i+1} = R_{i+2}$, and by induction $R_r = R_{r+1} = \cdots = R_\infty$.

For T-invariance, we have $T(R_i) = R_{i+1}$ for $i \geq r$, with $|R_i| = |R_{i+1}|$ by definition of r. Thus $T : R_i \to R_{i+1}$ is a bijection for all large i, and hence so is $T : R_\infty \to R_\infty$. Similarly, $T(K_i) = K_{i+1}$ with $|K_i| = |K_{i+1}|$ for all large i and $T : K_\infty \to K_\infty$ is also bijective.

To see $V = K_\infty \oplus R_\infty$ we show (i) $R_\infty + K_\infty = V$ and (ii) $R_\infty \cap K_\infty = \{0\}$. For (ii), if $v \in R_\infty = R_r$, there is some $w \in V$ such that $T^r(w) = v$; but if $v \in K_\infty = K_r$, then $T^r(v) = 0$. Consequently $T^{2r}(w) = T^r(v) = 0$. But

$T : R_i \to R_{i+1}$ is a bijection for $i \geq r$. Since $v = T^r(w) \in R_r$, and since $T^r : R_r \to R_{2r} = R_r$ is a bijection, we get

$$0 = T^{2r}(w) = T^r(T^r(w)).$$

Hence $v = T^r(w) = 0$, so $K_\infty \cap R_\infty = \{0\}$.

Then *(ii)* $\Rightarrow$ *(i)* because

$$\begin{aligned}|R_\infty + K_\infty| &= |R_r + K_r| = |R_r| + |K_r| - |K_r \cap R_r| \\ &= |K_\infty| + |R_\infty| = |K_r| + |R_r| = |V|\end{aligned}$$

(by a variant LA I-Exercise 1.67 of the dimension theorem, LA I-Theorem 1.54). We conclude that $R_\infty + K_\infty = V$, proving *(i)*. □

LEMMA 1.6 (Fitting Decomposition). *$T|_{K_\infty}$ is a nilpotent operator on K_∞ and $T|_{R_\infty}$ is a bijective linear map of $R_\infty \to R_\infty$. Hence, every linear operator T on a finite-dimensional space V over any field has a direct sum decomposition*

$$T = (T|_{R_\infty}) \oplus (T|_{K_\infty})$$

such that $T|_{K_\infty}$ is nilpotent and $T|_{R_\infty}$ bijective on R_∞.

PROOF. $T^r(K_\infty) = T^r(\ker(T^r)) = \{0\}$ so $(T|_{K_\infty})^r = 0$ and $T|_{K_\infty}$ is nilpotent of degree $\leq r$, the index at which the ranges stabilize at R_∞. □

Basic Properties of Nilpotent Operators.

LEMMA 1.7. *If $N : V \to V$ is nilpotent, the unipotent operator $I + N$ is invertible.*

PROOF. If $N^k = 0$, the geometric series $I + N + N^2 + \cdots + N^{k-1} + \cdots = \sum_{k=0}^{\infty} N^k$ is finite and a simple calculation shows that

$$(I - N)(I + N + \cdots + N^{k-1}) = I - N^k = I.$$

Since $N \neq I$ (by nilpotence), we get

$$(I - N)^{-1} = I + N + \cdots + N^{k-1}. \tag{1.1}$$

□

LEMMA 1.8. *If $T : V \to V$ is nilpotent, then $p_T(\lambda) = \det(T - \lambda I)$ is equal to $(-1)^n \lambda^n$ ($n = \dim V$), so $\lambda = 0$ is the only eigenvalue (over any field $\mathbb{K}$). It is an eigenvalue since $\ker(T) \neq \{0\}$ and this kernel is the full subspace of $\lambda = 0$ eigenvectors $E_{\lambda=0}(T)$.*

PROOF. Take a basis $\mathfrak{X} = \{e_1, \cdots, e_n\}$ that runs first through $K(T) = K_1 = \ker(T)$ and is then augmented to a basis in $K_2 = \ker(T^2)$, etc. With respect to such a basis, $[T]_{\mathfrak{X}\mathfrak{X}}$ is an upper triangular matrix with blocks of zero matrices on the diagonal (see Exercise 1.10 below). Obviously, $[T - \lambda I]_{\mathfrak{X}\mathfrak{X}}$ has diagonal values $-\lambda$, so $\det(T - \lambda I) = (-1)^n \lambda^n$ as claimed. □

Similarly, a unipotent operator T has $\lambda = 1$ as its only eigenvalue (over any field), and its characteristic polynomial is $p_T(x) = (1-x)^n$. The sole eigenspace $E_{\lambda=1}(T)$ is the set of *fixed points* $\mathrm{Fix}(T) = \{v : T(v) = v\}$ in V.

EXERCISE 1.9. Prove the following:

(a) A nilpotent operator T is diagonalizable if and only if $T = 0$.
(b) A unipotent operator T is diagonalizable if and only if T is the identity operator $I = \mathrm{id}_V$.

EXERCISE 1.10. If $T : V \to V$ is a nilpotent linear operator on a finite-dimensional space and $\{e_1, \dots, e_n\}$ is a basis that passes through successive kernels $K_i = \ker(T^i)$, $1 \le i \le d = \deg(T)$, prove that $[T]_{\mathfrak{X}\mathfrak{X}}$ is upper triangular with $m_i \times m_i$ zero-blocks on the diagonal, $m_i = \dim(K_i/K_{i-1})$.

HINT. The problem is to devise efficient notation to handle this question. Partition the indices $1, 2, \dots, n$ into consecutive intervals $J_1, \dots, J_d$ ($d = \deg(T)$) with $\{e_j : j \in J_1\}$ a basis for K_1, $\{e_i : i \in J_1 \cup J_2\}$ a basis for K_2, etc. Matrix entries T_{ij} are determined by the system of vector equations

$$T(e_i) = \sum_{j=1}^{n} T_{ji} e_j \quad (1 \le i \le n = \dim V).$$

What do the inclusions $T(K_i) \subseteq K_{i-1}$ tell you about the coefficients T_{ij}?

If $T : V \to V$ is nilpotent, the powers T^k eventually kill every vector $v \neq 0$, so there is an $m = m(v)$ in $\mathbb{N}$ such that $\{v, T(v), \dots, T^{m-1}(v)\}$ are nonzero and $T^m(v) = 0$. The *nilpotence degree* $\deg(T)$ is the smallest exponent $d = 0, 1, 2, \dots$ such that $T^d = 0$.

PROPOSITION 1.11. *Let $T : V \to V$ be nilpotent and $v_0 \neq 0$. If*

$$v_0, T(v_0), \dots, T^{m-1}(v_0) \text{ are nonzero and } T^m(v_0) = 0,$$

the subspace $W(v_0) = \mathbb{K}\text{-span}\{v_0, T(v_0), \dots, T^{m-1}(v_0)\}$ is T-invariant and the vectors $\{v_0, T(v_0), \dots, T^{m-1}(v_0)\}$ are independent. Hence they are a basis for this "cyclic subspace" $W(v_0)$ determined by v_0 and the action of T.

PROOF. The $\{T^k(v_0) : 0 \le k \le m-1\}$ span $W(v_0)$ by definition. They are independent because if $0 = c_0 v_0 + c_1 T(v_0) + \cdots + c_{m-1} T^{m-1}(v_0)$ for some choice of $c_k \in \mathbb{K}$, then

$$\begin{aligned} 0 = T^{m-1}(0) &= T^{m-1}\big(c_0 v_0 + c_1 T(v_0) + \cdots + c_{m-1} T^{m-1}(v_0)\big) \\ &= c_0 T^{m-1}(v_0) + c_1 \cdot 0 + \cdots + c_{m-1} \cdot 0, \end{aligned}$$

which implies $c_0 = 0$ since $T^{m-1}(v_0) \neq 0$ by minimality of the exponent m. Next, apply T^{m-2} to the original sum, which now has the form $c_1 T(v_0) + \cdots + c_{m-1} T^{m-1}(v_0)$; we get

$$\begin{aligned} T^{m-2}(0) &= T^{m-2}\big(c_1 T(v_0) + \cdots + c_{m-1} T^{m-1}(v_0)\big) \\ &= c_1 T^{m-1}(v_0) + 0 + \cdots + 0 \end{aligned}$$

and then $c_1 = 0$. Continuing this process, we eventually obtain $c_0 = c_1 = c_2 = \cdots = c_{m-1} = 0$. □

Obviously, $W(v_0)$ is T-invariant and $T_0 = T|_{W(v_0)}$ is nilpotent (with degree $m = \deg(T_0) \le \deg(T)$) because for each basis vector $T^k(v_0)$ we have

$$T_0^m(T^k(v_0)) = T^k(T^m(v_0)) = 0,$$

but in fact $\deg(T_0) = m$ because $T_0^{m-1}(v_0) \neq \{0\}$. Now consider the ordered basis

$$\mathfrak{X} = \{e_1 = T^{m-1}(v_0), e_2 = T^{m-2}(v_0), \dots, e_m = v_0\} \text{ in } W(v_0).$$

Since $T(e_{k+1}) = e_k$ for each $k \ge 1$ and $T(e_1) = 0$, the matrix $[T]_{\mathfrak{X}\mathfrak{X}}$ has the form

$$[T]_{\mathfrak{X}\mathfrak{X}} = \begin{pmatrix} 0 & 1 & 0 & \cdots & \cdots & 0 \\ 0 & 0 & 1 & \ddots & & \vdots \\ 0 & 0 & 0 & \ddots & \ddots & \vdots \\ \vdots & & \ddots & \ddots & \ddots & 0 \\ \vdots & & & \ddots & \ddots & 1 \\ 0 & \cdots & \cdots & 0 & 0 & 0 \end{pmatrix},$$

and the action on these ordered basis vectors is

$$0 \xleftarrow{T} e_1 \xleftarrow{T} e_2 \xleftarrow{T} \cdots \xleftarrow{T} e_{m-1} \xleftarrow{T} e_m = v_0.$$

The "top vector" $e_m = v_0$ is referred to as a *cyclic vector* for the invariant subspace $W(v_0)$. Any matrix having the form

$$\begin{pmatrix} 0 & 1 & 0 & \cdots & \cdots & 0 \\ \vdots & \ddots & \ddots & \ddots & & \vdots \\ \vdots & & \ddots & \ddots & \ddots & \vdots \\ \vdots & & & \ddots & \ddots & 0 \\ \vdots & & & & \ddots & 1 \\ 0 & \cdots & \cdots & \cdots & \cdots & 0 \end{pmatrix}$$

is called an *elementary nilpotent matrix.*

Cyclic Vectors and Cyclic Subspaces. To put this in its proper context, we leave the world of nilpotent operators for a moment.

DEFINITION 1.12. If $\dim(V) < \infty$, $T : V \to V$ is any linear operator, and $W \subseteq V$ is a nonzero T-invariant subspace, we say W is a *cyclic subspace* if it contains a "cyclic vector" $v_0 \in W$ such that $W = \mathbb{K}\text{-span}\{v_0, T(v_0), T^2(v_0), \dots\}$.

Only finitely many iterates $T^i(v_0)$ under the action of T can be linearly independent, so there will be a first (smallest) exponent $k = k(v_0)$ such that $\{v_0, T(v_0), \dots, T^{k-1}(v_0)\}$ are linearly independent and $T^k(v_0)$ is a linear combination of the previous vectors.

LEMMA 1.13. *Let $T : V \to V$ be an arbitrary linear operator on a finite-dimensional vector space. If $v_0 \in V$ is nonzero, there is a unique exponent $k = k(v_0) \ge 1$ such that $\{v_0, T(v_0), \dots, T^{k-1}(v_0)\}$ are linearly independent and $T^k(v_0)$ is a linear combination of these vectors. Obviously,*

$$W = \mathbb{K}\text{-span}\{T^j(v_0) : j = 0, 1, 2, \dots\} = \mathbb{K}\text{-span}\{v_0, T(v_0), \dots, T^{k-1}(v_0)\}$$

and $\dim(W) = k$. *Furthermore, $T(W) \subseteq W$ and W is a cyclic subspace in V.*

PROOF. By definition of $k = k(v_0)$, $T^k(v_0)$ is a linear combination $T^k(v_0) = \sum_{j=0}^{k-1} c_j T^j(v_0)$. Arguing recursively,

$$\begin{aligned} T^{k+1}(v_0) = T(T^k(v_0)) &= \sum_{j=0}^{k-1} c_j T^{j+1}(v_0) \\ &\in c_{k-1}T^k(v_0) + \mathbb{K}\text{-span}\{v_0, T(v_0), \dots, T^{k-1}(v_0)\} \end{aligned}$$

Since we already know $T^k(v_0)$ lies in $\mathbb{K}$-span$\{v_0, T(v_0), \dots, T^{k-1}(v_0)\}$, so does $T^{k+1}(v_0)$. Continuing this process, we find that all iterates $T^i(v_0)$ $(i \geq k)$ lie in W. By definition $v_0, T(v_0), \dots, T^{k-1}(v_0)$ are linearly independent and span W, so $\dim(W) = k$. □

When T is nilpotent, there is a simpler description of the cyclic subspace W generated by the action of T on $v_0 \neq 0$. Since $T^d = 0$ on all of V when $d = \deg(T)$, there is a smallest exponent l with $\{v_0, T(v_0), \dots, T^{\ell-1}(v_0)\}$ nonzero and $T^\ell(v_0) = T^{\ell+i}(v_0) = 0$ for all $i \geq 0$. These vectors are independent and $T^\ell(v_0) = 0$ lies in $\mathbb{K}$-span$\{v_0, T(v_0), \dots, T^{\ell-1}(v_0)\}$, so ℓ is precisely the exponent of the previous lemma and $W = \mathbb{K}$-span$\{v_0, T(v_0), \dots, T^{\ell-1}(v_0)\}$ is the cyclic subspace generated by v_0.

1.2. Fine Structure of Nilpotent Operators

Continuing the discussion of nilpotent operators, observe that if $T : V \to V$ is nilpotent and nonzero, the chain of kernels $K_i = \ker(T^i)$,

$$\{0\} = K_0 \subsetneqq K_1 = \ker(T) \subsetneqq K_2 \subsetneqq \cdots \subsetneqq K_d = V \quad (d = \deg(T)),$$

terminates at V in finitely many steps. The difference sets partition $V \sim (0)$ into disjoint "layers"

$$V \sim (0) = (K_d \sim K_{d-1}) \cup \cdots \cup (K_i \sim K_{i-1}) \cup \cdots \cup (K_1 \sim K_0)$$

where $K_0 = (0)$. The layers $K_i \sim K_{i-1}$ correspond to the quotient spaces K_i/K_{i-1}, and by examining the action of T on these quotients, we will determine the structure of the operator T.

EXERCISE 1.14. If v_0 is in the "top layer" $V \sim K_{d-1}$, prove that the subspace W spanned by $\{T^j(v_0) : j \geq 0\}$ has dimension d and that *every* v_0 in this layer is a cyclic vector under the action of T on W.

REMARK 1.15. Since $|K_{d-1}| < |K_d| = |V|$, the subspace K_{d-1} is a very thin subset of V and has "measure zero" in V when $\mathbb{K} = \mathbb{R}$ or $\mathbb{C}$. If you could pick a vector $v_0 \in V$ at random, you would have $v_0 \in V \sim K_{d-1}$ "with probability 1," and every such choice of v_0 would generate a cyclic subspace of dimension d. "Unsuccessful" choices, which occur with "probability zero," yield cyclic subspaces $W(v_0)$ of dimension $< d$. ○

We now state the main structure theorem for nilpotent operators.

THEOREM 1.16 (Cyclic Subspace Decomposition). *Given a nilpotent linear operator $T : V \to V$ on a finite-dimensional vector space V, there is a decomposition $V = C_1 \oplus \cdots \oplus C_r$ into cyclic T-invariant subspaces. Obviously the restrictions $T_i = T|_{C_i}$ are nilpotent with nilpotence degrees*

$$m_i = \dim(C_i) = (\text{smallest } m \text{ such that } T^m \text{ kills the cyclic generator } v_i \in E_i).$$

Listed in descending order $m_1 \geq m_2 \geq \cdots \geq m_r > 0$ (repeats allowed), these degrees are unique and $\sum_{i=1}^r m_i = \dim(V)$.

While it is nice to know such structure exists, it is equally important to develop a constructive procedure for finding suitable cyclic subspaces $C_1, \ldots, C_r$. This is complicated by the fact that C_k are not necessarily unique, unlike the eigenspaces $E_\lambda(T)$ in the diagonalization problem. Any algorithm for constructing suitable C_j will necessarily involve some arbitrary choices.

The rest of this section provides a proof of Theorem 1.16 that yields an explicit construction of the desired subspaces. There are some shorter, more elegant proofs of Theorem 1.16, but they are existential rather than constructive and so are less informative.

COROLLARY 1.17. *If $T : V \to V$ is nilpotent, there is a decomposition into cyclic spaces $V = C_1 \oplus \cdots \oplus C_r$, so there is a basis $\mathfrak{X}$ such that $[T]_{\mathfrak{X}\mathfrak{X}}$ consists of elementary nilpotent blocks on the diagonal:*

$$[T]_{\mathfrak{X}\mathfrak{X}} = \begin{pmatrix} \boxed{B_1} & 0 & & \cdots & & 0 \\ 0 & \boxed{B_2} & 0 & & & \\ & 0 & \ddots & \ddots & & \vdots \\ \vdots & & \ddots & \ddots & \ddots & \\ & & & \ddots & \ddots & 0 \\ 0 & & \cdots & & 0 & \boxed{B_r} \end{pmatrix}$$

with

$$B_i = \begin{pmatrix} 0 & 1 & 0 & \cdots & \cdots & 0 \\ 0 & 0 & 1 & \ddots & & \vdots \\ 0 & 0 & 0 & \ddots & \ddots & \vdots \\ \vdots & & \ddots & \ddots & \ddots & 0 \\ \vdots & & & \ddots & \ddots & 1 \\ 0 & & \cdots & & 0 & 0 \end{pmatrix}.$$

We start with the special case in which T has the largest possible degree of nilpotence.

LEMMA 1.18. *If T is nilpotent and $\deg(T) = \dim(V)$, there is a cyclic vector in V and a basis $\mathfrak{X}$ such that $[T]_{\mathfrak{X}\mathfrak{X}}$ is an elementary nilpotent matrix.*

PROOF. If $\deg(T) = d$ is equal to $\dim(V)$ the spaces $K_i = \ker(T^i)$ increase, with $|K_{i+1}| \geq 1 + |K_i|$ at each step in the chain $\{0\} \subsetneqq K_1 \subsetneqq \cdots \subsetneqq K_{d-1} \subsetneqq K_d = V$, but there are only $d = \dim(V)$ steps, so we must have $|K_{i+1}| = 1 + |K_i|$. If $v_0 \in V \sim K_{d-1}$, then $T^d(v_0) = 0$ and by definition of K_{d-1} the vectors $v_0, T(v_0), \ldots, T^{d-1}(v_0)$ are all nonzero, so v_0 is a cyclic vector for the iterated action of T. □

If $T : V \to V$ is nilpotent of degree d, the idea behind the proof of Theorem 1.16 is to look at the kernels $K_i = \ker(T^i)$:

$$V = K_d \supsetneqq K_{d-1} \supsetneqq \cdots \supsetneqq K_2 \supsetneqq K_1 = \ker(T) \supsetneqq \{0\}.$$

As the kernels get smaller, more of V is "uncovered": the differences $V \sim K_s$ and the quotients V/K_s get bigger, and $T|_{K_s}$ reveals more details about the full action of T on V.

It will be important to note that $T(K_i) \subseteq K_{i-1}$ (since $0 = T^i(x) = T^{i-1}(T(x))$ and $T(x) \in K_{i-1}$). Furthermore, $x \notin K_i$ implies that $0 \neq T^i(x) = T^{i-1}(T(x))$ so that $T(x) \notin K_{i-1}$. Therefore,

$$T \text{ maps } (K_{i+1} \sim K_i) \text{ into } (K_i \sim K_{i-1}) \quad \text{for all } i. \tag{1.2}$$

However, although $T(K_i) \subseteq K_{i-1}$, it is not generally true that $T(K_i) = K_{i-1}$, so the induced map on the quotient space $\widetilde{T} : K_i/k_{i-1} \to K_i/K_{i-1}$ need not be surjective.

DEFINITION 1.19. Let $T : V \to V$ be an arbitrary linear map and W a T-invariant subspace. We say that vectors $e_1, \dots, e_m$ in V

1. ARE INDEPENDENT (MOD W) if their images $\overline{e}_1, \dots, \overline{e}_m$ in V/W are linearly independent. Since $\sum_{i=1}^m c_i\overline{e}_i = 0$ in V/W if and only if $\sum_{i=1}^m c_ie_i$ is in W, that is the same as saying
$$\sum_{i=1}^m c_ie_i \in W \Rightarrow c_1 = \cdots = c_m = 0 \qquad (c_i \in \mathbb{K}).$$
2. SPAN V (MOD W) if $\mathbb{K}\text{-span}\{\overline{e}_i\} = V/W$, which means that given v in V, there are $c_i \in \mathbb{K}$ such that $\left(v - \sum_i c_ie_i\right) \in W$, or $\overline{v} = \sum_{i=1}^m c_i\overline{e}_i$ in V/W.
3. ARE A BASIS FOR V (MOD W) if the images $\{\overline{e}_i\}$ are a basis in V/W, which happens if and only if (1) and (2) hold.

EXERCISE 1.20. Let $W \subseteq \mathbb{R}^5$ be the solution set of system

$$\begin{cases} x_1 + x_3 + 2x_4 = 0 \\ x_1 - x_4 + 2x_5 = 0, \end{cases}$$

and let $\{e_i\}$ be the standard basis in $V = \mathbb{R}^5$.

(a) Find vectors v_1, v_2 that are a basis for V (mod W).

(b) If v_1, v_2 are the vectors in (a), is $\mathfrak{X} = \{e_1, e_2, e_3, v_1, v_2\}$ a basis for V?

(c) Find a basis $\{f_1, f_2, f_3\}$ for the subspace W.

EXERCISE 1.21. Let $T : V \to V$ be an arbitrary linear map and let W be a T-invariant subspace. Explain why independence of vectors $f_1, \dots, f_r$ mod a T-invariant subspace $W \subseteq V$ implies their independence (mod W') for any smaller T-invariant subspace $W' \subseteq W \subseteq V$.

PROOF. (Proof of Theorem 1.16.) Below we will construct two related sets of vectors $\mathcal{F}_1, \mathcal{F}_2, \mathcal{F}_3, \dots$ and $\mathcal{E}_1 = \mathcal{F}_1 \subseteq \mathcal{E}_2 \subseteq \mathcal{E}_3 \subseteq \cdots \subseteq \mathcal{E}_r$ such that $\mathcal{E}_r$ is a basis for V aligned with the kernels $K_d = V \supseteq K_{d-1} \supseteq \cdots \supseteq K_1 = \ker(T) \supseteq \{0\}$.

When the construction terminates, the vectors in $\mathcal{E}_r$ will be a basis for all of V that provides the desired decomposition into cyclic subspaces.

(INITIAL) STEP 1: Let $\mathcal{F}_1 = \mathcal{E}_1$ be a set of vectors $\{e_i : i \in I_1\} \subseteq V \sim K_{d-1}$ that are a basis for $V \pmod{K_{d-1}}$, so their images $\{\overline{e}_i\}$ are a basis in V/K_{d-1}. Obviously the index set I_1 has cardinality $|I_1| = |V/K_{d-1}| = |V| - |K_{d-1}|$, the dimension of the quotient space.

You might feel more comfortable indicating the index sets $I_1, I_2, \ldots$ being constructed here as consecutive blocks of integers, say $I_1 = \{1, 2, \ldots, s_1\}$, $I_2 = \{s_1 + 1, \ldots, s_2\}$, etc., but this notation becomes really cumbersome after the first two steps. And in fact there is no need to explicitly name the indices in each block. From here on, you should refer to the chart shown in Figure 1.1, which lists all the players that will emerge in our discussion.

STEP 2: The T-images $T(\mathcal{F}_1)$ lie in the layer $T(V \sim K_{d-1}) \subseteq K_{d-1} \sim K_{d-2}$, as noted in (1.2). In this step we shall verify two assertions.

CLAIM (i): Vectors $T(\mathcal{F}_1) = \{T(e_i) : i \in I_1\} \subseteq K_{d-1} \sim K_{d-2}$ are independent $\pmod{K_{d-2}}$.

If these vectors are not already representatives of a basis for K_{d-1}/K_{d-2}, we can adjoin additional vectors $\mathcal{F}_2 = \{e_i : i \in I_2\} \subseteq K_{d-1} \sim K_{d-2}$ chosen so that $T(\mathcal{F}_1) \cup \mathcal{F}_2$ corresponds to a basis for K_{d-1}/K_{d-2}; otherwise we take $\mathcal{F}_2 = \emptyset$.

CLAIM (ii): The vectors

$$\mathcal{E}_2 = \mathcal{F}_2 \cup \big[\mathcal{E}_1 \cup T(\mathcal{F}_1)\big] = \mathcal{E}_1 \cup \big[T(\mathcal{F}_1) \cup \mathcal{F}_2\,\big]$$

are a basis for all of $V \pmod{K_{d-2}}$.

REMARK 1.22. In Chapter 1 of LA I we saw that if $\{e_1, \ldots, e_r\}$ is a basis for W, we can adjoin successive "outside vectors" $e_{r+1}, \cdots, e_n$ to get a basis for V. (These can even be found by deleting some of the vectors in a preordained basis in V.) Then the images $\{\overline{e}_{r+1}, \ldots, \overline{e}_n\}$ are a basis for the quotient space V/W, as in the proof of the dimension formula $|V| = |W| + |V/W|$ for finite-dimensional V (Theorem 1.79 in LA I). ○

PROOF. (Proof of Claim (i)). If there is a relation

$$\sum_{i \in I_1} a_i T(e_i) = T\Big(\sum_{i \in I_1} a_i e_i\Big) \equiv 0 \pmod{K_{d-2}},$$

then $\sum_{i \in I_1} a_i e_i$ is in K_{d-2} and also lies in the larger space $K_{d-1} \supseteq K_{d-2}$. But by definition, vectors in $\mathcal{F}_1 = \{e_i : i \in I_1\}$ are independent $\pmod{K_{d-1}}$, so we must have $a_i = 0$ for $i \in I_1$, proving independence $\pmod{K_{d-1}}$ of the vectors in $T(\mathcal{F}_1)$. □

PROOF. (Proof of Claim (ii)). Suppose $a_i^{(1)}, a_i^{(2)}, b_i \in \mathbb{K}$ are coefficients such that

$$\sum_{i \in I_1} a_i^{(1)} e_i + \sum_{i \in I_1} b_i T(e_i) + \sum_{i \in I_2} a_i^{(2)} e_i \equiv 0 \ (mod\ K_{d-2}). \tag{1.3}$$

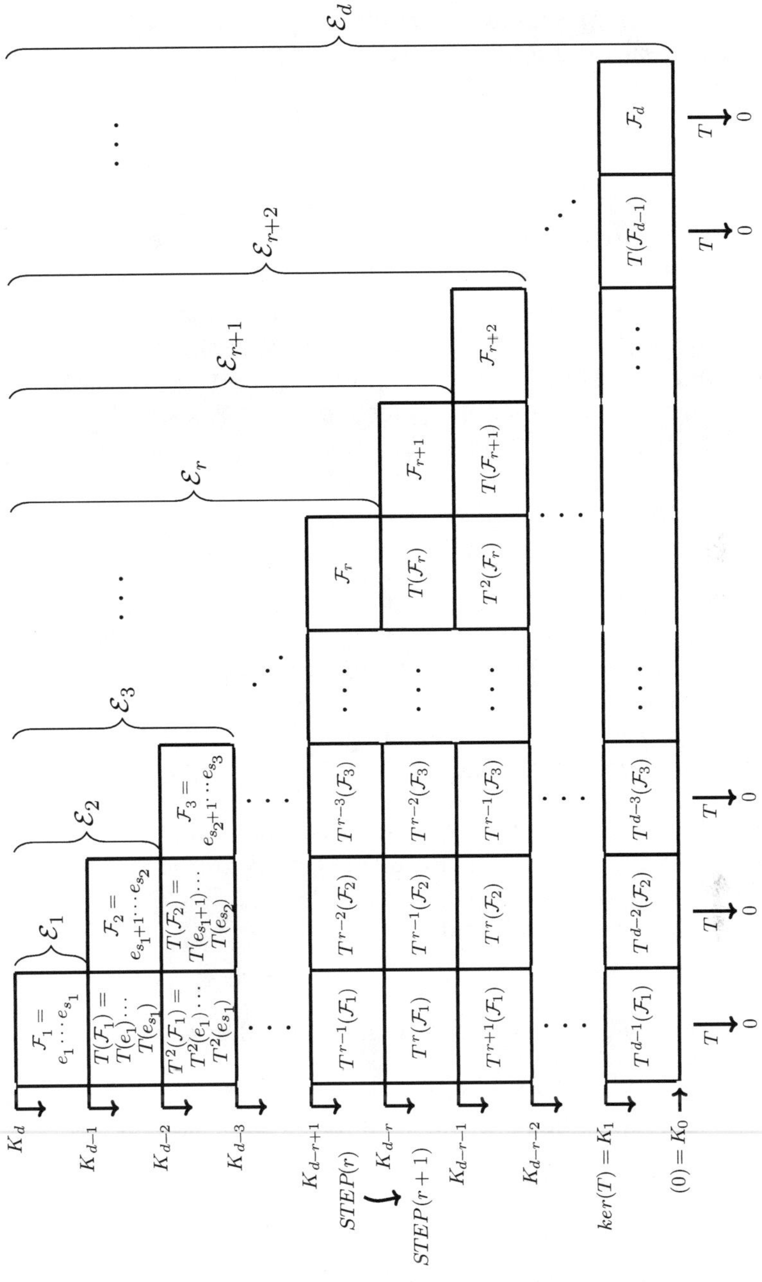

FIGURE 1.1. Steps in the construction of a basis that decomposes vector space V into cyclic subspaces under the action of a nilpotent linear operator $T : V \to V$. The subspaces K_i are the kernels of the powers T^i for $1 \le i \le d = \deg(T)$, with $K_d = V$ and $K_0 = (0)$.

This sum lies in K_{d-2}, hence also in the larger subspace K_{d-1}, and the last two terms are already in K_{d-1} because $\mathcal{F}_2 \cup T(\mathcal{F}_1) \subseteq K_{d-1} \sim K_{d-2}$. Thus

$$\sum_{i \in I_1} a_i^{(1)} e_i \equiv 0 \ (\mathrm{mod}\ \mathrm{K}_{d-1}),$$

and since $\{e_i : i \in I_1\}$ are independent (mod K_{d-1}) we must have $a_i^{(1)} = 0$ for all $i \in I_1$. Now the sum (1.3) reduces to its last two terms, which all lie in K_{d-1}. But by construction, $\mathcal{F}_2 \cup T(\mathcal{F}_1)$ is a basis for K_{d-1} (mod K_{d-2}), which implies $a_i^{(2)} = 0$ for $i \in I_2$ and $b_i = 0$ for $i \in I_1$. Thus $\mathcal{E}_2 = \mathcal{F}_1 \cup [T(\mathcal{F}_1) \cup \mathcal{F}_2] = [\mathcal{F}_1 \cup T(\mathcal{F}_1)] \cup \mathcal{F}_2 \subseteq V$ is an independent set of vectors (mod K_{d-2}).

It remains to show $\mathcal{E}_2$ spans V (mod K_{d-2}). If $v \in V$ is not contained in K_{d-1}, then $v - v_1 \equiv 0$ (mod K_{d-1}) for some $v_1 \in \mathbb{K}$-span$\{\mathcal{F}_1\}$, so $v - v_1 \in K_{d-1}$. If this difference lies outside of K_{d-2}, we can find some $v_2 \in T(\mathcal{F}_1) \cup \mathcal{F}_2$ such that $v = (v_1 + v_2) \in K_{d-2}$ because $T(\mathcal{F}_1) \cup \mathcal{F}_2$ is a basis for K_{d-1}/K_{d_2}. Thus $v = v_1 + v_2$ (mod K_{d-2}), and since $v_1 + v_2 \in \mathbb{K}$-span$\{\mathcal{F}_1 \cup T(\mathcal{F}_1) \cup \mathcal{F}_2\}$, statement (ii) is proved. □

That completes Step 2, which fills in the second row of Figure 1.1. Further inductive steps that fill in the remaining rows involve no new ideas, but things can get out of hand unless the notation is managed carefully. Below we spell out the general inductive step in this process, which could be skipped on first reading. It is followed by a final paragraph proving uniqueness of the multiplicities m_i (which you should read). □

The General Inductive Step in Theorem 1.16. With the chart from Figure 1.1 in hand to keep track of the players, we continue the inductive construction of basis vectors. At step r, we have defined sets $\mathcal{F}_i \subseteq K_{d-i+1} \sim K_{d-i}$ for $1 \le i \le r$ such that $\mathcal{E}_1 = \mathcal{F}_1$ and

$$\mathcal{E}_r = \mathcal{E}_{r-1} \cup \left[\, T^{r-1}(\mathcal{F}_1) \cup \cdots \cup T(\mathcal{F}_{r-1}) \cup \mathcal{F}_r \right]$$

is a basis for V/K_{d-r}. At the next step, we take the vectors

$$T^{r-1}(\mathcal{F}_1) \cup T^{r-2}(\mathcal{F}_2) \cup \cdots \cup \mathcal{F}_r \subseteq K_{d-r+1} \sim K_{d-r}$$

created in the previous steps and form their T-images

$$T^r(\mathcal{F}_1) \cup \cdots \cup T(\mathcal{F}_r) \subseteq K_{d-r} \sim K_{d-r-1}.$$

To complete the inductive step we show the following:

1. These vectors in K_{d-r} are independent (mod K_{d-r-1}).
2. When we adjoin additional vectors $\mathcal{F}_{r+1} \subseteq K_{d-r} \sim K_{d-r-1}$ as needed to produce a basis for K_{d-r}/K_{d-r-1}, taking $\mathcal{F}_{r+1} = \varnothing$ if the vectors $T^r(\mathcal{F}_1) \cup \cdots \cup T(\mathcal{F}_r)$ are already basis representatives for K_{d-r}/K_{d-r-1}, then

$$\mathcal{E}_{r+1} = \mathcal{E}_r \cup \left[T^r(\mathcal{F}_1) \cup \cdots \cup T(\mathcal{F}_r) \cup \mathcal{F}_{r+1}\right]$$

is a basis for V (mod K_{d-r-1}).

When this process terminates, the final set $\mathcal{E}_d$ is a basis for all of V.

PROOF. STEP 1. If the vectors $T^r(\mathcal{F}_1) \cup \cdots \cup T(\mathcal{F}_r)$ are not representatives for independent vectors in K_{d-r}/K_{d-r-1}, there would be sets of coefficients

$$\{c_i^{(1)} : i \in I_1\}, \dots, \{c_i^{(r)} : i \in I_r\}$$

such that

$$\sum_{i \in I_1} c_i^{(1)} T^r(e_i) + \cdots + \sum_{i \in I_r} c_i^{(r)} T(e_i) \equiv 0 \qquad (\text{mod } \mathrm{K}_{\mathrm{d-r-1}}). \tag{1.4}$$

This sum is in K_{d-r-1} and also in the larger space K_{d-r}. But the vectors

$$T^{r-1}\{e_i : i \in I_1\} \cup \cdots \cup \{e_i : i \in I_r\}$$

are independent (mod K_{d-r}) by hypothesis and are representatives for a basis in K_{d-r+1}/K_{d-r}. We may rewrite the congruence in (1.4) as

$$T\Big(\sum_{i \in I_1} c_i^{(1)} T^{r-1}(e_i) + \cdots + \sum_{i \in I_r} c_i^{(r)} e_i\Big) \equiv 0 \quad (\text{mod } K_{d-r-1}).$$

Thus $T(\cdots) \in K_{d-r-1}$, and $(\cdots) \in K_{d-r}$ too. By (mod K_{d-r}) independence of the e_i, we must have $c_i^{(j)} = 0$ in $\mathbb{K}$ for all i, j so the vectors $T^r(\mathcal{F}_1) \cup \cdots \cup T(\mathcal{F}_r)$ are independent (mod K_{d-r-1}) as claimed.

STEP 2. To verify independence of the updated set of vectors

$$\mathcal{E}_{r+1} = \mathcal{E}_r \cup \big[T^r(\mathcal{F}_1) \cup \cdots \cup T(\mathcal{F}_r) \cup \mathcal{F}_{r+1}\big]$$

in V/K_{d-r-1}, suppose some linear combination $S = S' + S''$ is zero (mod K_{d-r-1}) where S' is a sum over vectors in $\mathcal{E}_r$ and S'' a sum over vectors in $T^r(\mathcal{F}_1) \cup \cdots \cup \mathcal{F}_{r+1}$. Then $S \equiv 0$ (mod K_{d-r-1}) implies $S \equiv 0$ (mod K_{d-r}), and then by independence of vectors in $\mathcal{E}_r$ (mod K_{d-r}), all coefficients appearing in S' are zero. The remaining term S'' in the reduced sum lies in $K_{d-r} \sim K_{d-r-1}$, and by independence of the set of vectors $T^r(\mathcal{F}_1) \cup \cdots \cup \mathcal{F}_{r+1}$ in K_{d-r}/K_{d-r-1}, all coefficients in S'' are also zero. Thus $\mathcal{E}_{r+1} \subseteq V$ corresponds to an independent set in K_{d-r}/K_{d-r-1}.

Dimension counting reveals that

$$\begin{aligned}
|V/K_{d-1}| &= |\mathcal{F}_1|, \\
|K_{d-1}/K_{d-2}| &= |T(\mathcal{F}_1)| + |\mathcal{F}_2| = |\mathcal{F}_1| + |\mathcal{F}_2|, \\
&\vdots \\
|K_{d-r}/K_{d-r-1}| &= |\mathcal{F}_1| + \cdots + |\mathcal{F}_{r+1}|.
\end{aligned}$$

Thus $|V/K_{d-r-1}| = |V/K_{d-1}| + \cdots + |K_{d-r}/K_{d-r-1}|$ is precisely the number $|\mathcal{E}_{r+1}|$ of basis vectors appearing in the first $r+1$ rows from the top of the chart in Figure 1.1. But this is also equal to $\dim(V/K_{d-r-1})$, so $\mathcal{E}_{r+1}$ is a basis for V/K_{d-r-1} and Step $(r+1)$ of the induction is complete. That also completes the proof of Theorem 1.16. □

Cyclic Subspace Decompositions. A direct sum decomposition of V into cyclic subspaces can now be read out of the chart in Figure 1.1, in which basis vectors have been constructed row-by-row. For this, consider what happens when we partition into columns. For each $e_i \in \mathcal{F}_1$, $(i \in I_1)$, we have $e_i, T(e_i), T^2(e_i), \dots, T^{d-1}(e_i) \neq 0$ and $T^d(e_i) = 0$, so these vectors span a cyclic subspace $E(e_i)$ such that $T|_{E(e_i)}$ has nilpotence degree d with e_i as its cyclic vector. Since the vectors that span $E(e_i)$ are part of a basis $\mathcal{E}_d$ for all of V, we obtain a direct sum of cyclic T-invariant subspaces $\bigoplus_{i\in I_1} E(e_i) \subseteq V$ with $|I_1| = |\mathcal{F}_1|$ summands.

Vectors $e_i \in \mathcal{F}_2$ $(i \in I_2)$ generate cyclic subspaces $E(e_i)$ with $\dim\big(E(e_i)\big) = \deg(T|_{E(e_i)}) = d - 1$; these become part of

$$\bigoplus_{i\in I_1} E(e_i) \oplus \bigoplus_{i\in I_2} E(e_i),$$

etc. (Of course $\bigoplus_{i\in I_2} E(e_i) = (0)$ when $I_2 = \varnothing$.) In the last step, the vectors $e_i \in \mathcal{F}_d$ $(i \in I_d)$ determine T-invariant one-dimensional cyclic spaces such that $T(\mathbb{K}e_i) = (0)$, with nilpotence degree $= 1$—i.e., for $i \in I_d$, the spaces $E(e_i) = \mathbb{K}e_i$ all lie within $\ker(T)$. The end result is a cyclic subspace decomposition

$$\Big(\bigoplus_{i_1\in I_1} E(e_{i_1})\Big) \oplus \Big(\bigoplus_{i_2\in I_2} E(e_{i_2})\Big) \oplus \cdots \oplus \Big(\bigoplus_{i_d\in I_d} E(e_{i_d})\Big) \tag{1.5}$$

of the entire space V, since all basis vectors in $\mathcal{E}_r$ are accounted for. (Various summands in (1.5) may of course be trivial.)

Uniqueness of Cyclic Subspace Decomposition. A direct sum decomposition $V = \bigoplus_{j=1}^{s} E_j$ into T-invariant cyclic subspaces can be refined by gathering together those E_i of the same dimension, writing

$$V = \bigoplus_{k=1}^{d} \mathcal{H}_k \quad \text{where} \quad \mathcal{H}_k = \bigoplus \big\{E_i : \dim(E_i) = \deg(T|_{E_i}) = k\big\}$$

for $1 \leq k \leq d = \deg(T)$. The action of T is the same on each summand in $\mathcal{H}_k$.

PROPOSITION 1.23. *In any direct sum decomposition $V = \bigoplus_{j=1}^{s} E_j$ into cyclic T-invariant subspaces, the number of spaces of dimension $\dim(E_i) = k$, $1 \leq k \leq d = \deg(T)$ can be computed in terms of the dimensions of the quotients K_i/K_{i-1}. These numbers are then the same for all cyclic decompositions.*

PROOF. Let us regard Figure 1.1 as a $d \times d$ array of "cells" with C_{ij} the cell in Row_i (from the top) and Col_j (from the left); the "size" $|C_{ij}|$ of a cell is the number of basis vectors it contains. Note that

1. $|C_{ij}| = 0$ if the cell lies above the diagonal, with $j > i$, because those cells are empty (others may be empty too).
2. $|C_{ij}| = |\mathcal{F}_j|$ for all cells on and below the diagonal in Col_j of the array. In particular, $|C_{j1}| = |\mathcal{F}_1|$ for all nonempty cells in Col_1, $|C_{j2}| = |\mathcal{F}_2|$ for those in Col_2, etc.

By our construction, it is evident that vectors in the nonempty cells in Row_r of Figure 1.1 correspond to a basis for the quotient space K_{d-r}/K_{d-r-1}. Counting the total number of basis vectors in Row_r, we find that

$$\dim(K_{d-r}/K_{d-r-1}) = |C_{r,1}| + \cdots + |C_{r,r}| = |\mathcal{F}_1| + \cdots + |\mathcal{F}_{r+1}|.$$

We may now recursively compute the values of $|C_{rj}|$ and $|\mathcal{F}_j|$ from the dimensions of the quotent spaces K_i/K_{i-1}. But as noted above, each $e_i \in \mathcal{F}_k$ lies in the diagonal cell C_{kk} and generates a distinct cyclic space in the decomposition. That completes the proof of Proposition 1.23 $\square$

REMARK 1.24. To summarize,

1. We define $K_i = \ker(T^i)$ for $1 \leq d = i$ nilpotence degree of T.
2. The following relations hold:

$$\begin{aligned}
\mathcal{E}_1 &= \mathcal{F}_1 \subseteq V \sim K_{d-1} \text{ is a basis for } V/K_{d-1},\\
\mathcal{E}_2 &= \mathcal{E}_1 \cup \left[\, T(\mathcal{F}_1) \cup \mathcal{F}_2 \,\right] \subseteq V \sim K_{d-2} \quad \text{is a basis for } V/K_{d-2},\\
&\ \vdots\\
\mathcal{E}_{r+1} &= \mathcal{E}_r \cup \left[\, T^r(\mathcal{F}_1) \cup T^{r-1}(\mathcal{F}_2) \cup \cdots \cup \mathcal{F}_{r+1} \right] \subseteq V \sim K_{d-r} \text{ is a basis for } V/K_{d-r-1},\\
&\ \vdots\\
\mathcal{E}_d &= \mathcal{E}_{d-1} \cup \left[\, T^{d-1}(\mathcal{F}_1) \cup \cdots \cup T(\mathcal{F}_{d-1}) \cup \mathcal{F}_d \,\right] \text{ is a basis for all of } V.
\end{aligned}$$

$\bigcirc$

In working examples, it usually helps to start by determining a basis $\mathcal{B}^{(0)} = \mathcal{B}^{(1)} \cup \cdots \cup \mathcal{B}^{(d)}$ for V aligned with the kernels so that $\mathcal{B}^{(1)}$ is a basis for K_1, $\mathcal{B}^{(2)}$ determines a basis for K_2/K_1, etc. This yields a convenient basis in V to start the construction.

EXAMPLE 1.25. Let $V = \mathbb{K}^5$ and $T : V \to V$ the operator

$$T(x_1, \ldots, x_5) = (0, x_3 + x_4, 0, x_3, x_1 + x_4)$$

whose matrix with respect to the standard basis $\mathfrak{X} = \{e_1, \ldots, e_5\}$ in $\mathbb{K}^5$ is

$$A = [T]_{\mathfrak{X}\mathfrak{X}} = \begin{pmatrix} 0 & 0 & 0 & 0 & 0 \\ 0 & 0 & 1 & 1 & 0 \\ 0 & 0 & 0 & 0 & 0 \\ 0 & 0 & 1 & 0 & 0 \\ 1 & 0 & 0 & 1 & 0 \end{pmatrix}.$$

Show that T is nilpotent, and then determine $\deg(T)$ and the kernels

$$V = K_d \supseteq K_{d-1} \supseteq \cdots \supseteq K_1 \supseteq \{0\}.$$

Find a basis $\mathfrak{Y}$ such that $[T]_{\mathfrak{Y}\mathfrak{Y}}$ has block diagonal form, with each block B_i an elementary nilpotent matrix. This is the *Jordan canonical form* for a nilpotent linear operator.

DISCUSSION. First find bases for the kernels $K_i = \ker(T^i)$. We have

$$\begin{aligned} K_1 = \ker(T) &= \{\mathbf{x} : x_3 + x_4 = 0,\ x_3 = 0,\ x_1 + x_4 = 0\} \\ &= \{\mathbf{x} : x_4 = x_3 = 0, x_1 + x_4 = 0\} = \{\mathbf{x} : x_1 = x_3 = x_4 = 0\} \\ &= \{(0, x_2, 0, 0, x_5) : x_2, x_5 \in \mathbb{K}\} = \mathbb{K}\text{-span}\{e_2, e_5\}. \end{aligned}$$

Iteration of T sends $\mathbf{x}$ to

$$\begin{aligned} T(\mathbf{x}) &= (0, x_3 + x_4, 0, x_3, x_1 + x_4) \\ T^2(\mathbf{x}) &= T(T(\mathbf{x})) = (0, x_3, 0, 0, x_3) \\ T^3(\mathbf{x}) &= (0, \dots, 0) \end{aligned}$$

for $\mathbf{x} \in \mathbb{K}^5$. Clearly T is nilpotent with $\deg(T) = 3$, and

$$\begin{aligned} &|K_1| = 2 : K_1 = \mathbb{K}\text{-span}\{e_2, e_5\} = \{\mathbf{x} : x_1 = x_3 = x_4 = 0\}, \\ &|K_2| = 4 : K_2 = \ker(T^2) = \{\mathbf{x} : x_3 = 0\} = \mathbb{K}\text{-span}\{e_1, e_2, e_4, e_5\}, \text{and} \\ &|K_3| = 5 : K_3 = \mathbb{K}^5. \end{aligned}$$

In this example, $\mathfrak{X} = \{e_2, e_5\,; e_1, e_4\,; e_3\} = \mathcal{B}^{(1)} \cup \mathcal{B}^{(2)} \cup \mathcal{B}^{(3)}$ is an ordered basis for V aligned with the kernels, running through $(0) \subseteq K_1 \subseteq K_2 \subseteq K_3 = V$. From this, we can determine the families $\mathcal{F}_1, \mathcal{F}_2, \mathcal{F}_3$ of Theorem 1.16.

STEP 1: Since $|K_3/K_2| = 1$, any $\mathbf{x} \neq \mathbf{0}$ in the layer $K_3 \sim K_2 = \{\mathbf{x} : x_3 \neq 0\}$ yields a basis vector for K_3/K_2. We shall take $\mathcal{F}_1 = \{e_3\}$ chosen from the standard basis $\mathfrak{X}$, and then $\mathcal{E}_1 = \{e_3\}$ too. (Any $\mathbf{x}$ with $x_3 \neq 0$ would also work.)

STEP 2: The image set $T(\mathcal{F}_1)$ consisting of the single vector $T(e_3) = e_2 + e_4$ lies in the next layer

$$\begin{aligned} K_2 \sim K_1 &= \{\mathbf{x} : x_3 = 0\} \sim \mathbb{K}\text{-span}\{e_2, e_5\} \\ &= \{\mathbf{x} : x_3 = 0 \text{ and } x_1, x_4 \text{ are not both } = 0\}. \end{aligned}$$

Since $|T(\mathcal{F}_1)| = 1$ and $\dim(K_2/K_1) = |K_2| - |K_1| = 4 - 2 = 2$, we must adjoin one suitably chosen new vector $\mathbf{x}$ from layer $K_2 \sim K_1$ to $T(\mathcal{F}_1)$ to get representatives for the desired basis in K_2/K_1. Then $\mathcal{F}_2 = \{\mathbf{x}\}$, and

$$\mathcal{E}_2 = \big(\mathcal{F}_1 \cup T(\mathcal{F}_1)\big) \cup \mathcal{F}_2 = \{e_3, e_2 + e_4, \mathbf{x}\}$$

is a basis for V/K_2 as in the first inductive step of Theorem 1.16.

A suitable vector $\mathbf{x} = (x_1, \dots, x_5)$ in $K_2 \sim K_1$ must have $x_3 = 0$ so $\mathbf{x} \in K_2$ and x_1, x_3, x_4 not all zero (so $x \notin K_1$). This holds if and only if $x_3 = 0$ and x_1, x_4 are not both 0. But we must also ensure that our choice of $\mathbf{x}$ makes $\{e_3, e_2 + e_4, \mathbf{x}\}$ independent (mod K_1). The following lemma is helpful.

LEMMA 1.26. *Let $V = \mathbb{K}^n$, W a subspace, $\mathfrak{X} = \{v_1, \dots, v_r\}$ independent vectors in V, and $\mathfrak{Y} = \{w_1, \dots, w_{n-r}\}$ a basis for W. If $M = \mathbb{K}$-span$\{v_1, \dots, v_r\}$, the following assertions are equivalent:*

1. *$\mathfrak{X}$ is a set of representatives for a basis in V/W.*
2. *$\mathfrak{Y} \cup \mathfrak{X} = \{w_1, \dots, w_{n-r}, v_1, \dots, v_r\}$ is a basis for V.*
3. *$V = W \oplus M$ (direct sum of subspaces).*

In every case, we have $r = |V/W|$.

PROOF. In LA I, we showed that the images $\overline{v}_1, \dots, \overline{v}_r$ are a basis for $V/W \Leftrightarrow \{v_1, \dots, v_r\} \cup \mathfrak{Y}$ are a basis for V. Obviously (2) $\Leftrightarrow$ (3). $\square$

COROLLARY 1.27. *In the setting of Theorem 1.79 in LA I, the "outside vectors" $v_1, \dots, v_r \notin W$ are representatives for a basis $\{\overline{v}_1, \dots, \overline{v}_r\}$ in V/W if and only if the $n \times n$ matrix A whose rows are $R_1 = v_1, \dots, R_r = v_r, R_{r+1} = w_1, \dots, R_n = w_{n-r}$ has rank$(A) = n$.*

Armed with this observation (and the known basis $\{e_2, e_5\}$ for K_1), we seek a vector $\mathbf{x} = (x_1, \dots, x_5)$ such that (i) $\mathbf{x} \in K_2$ (so $x_3 = 0$), (ii) x_1, x_4 are not both equal to 0, and (iii)

$$A = \begin{pmatrix} e_3 \\ e_2 + e_4 \\ (x_1, x_2, 0, x_4, x_5) \\ e_2 \\ e_5 \end{pmatrix} = \begin{pmatrix} 0 & 0 & 1 & 0 & 0 \\ 0 & 1 & 0 & 1 & 0 \\ x_1 & x_2 & 0 & x_4 & x_5 \\ 0 & 1 & 0 & 0 & 0 \\ 0 & 0 & 0 & 0 & 1 \end{pmatrix}$$

has Rowrank$(A) = 5$. Symbolic row operations put this into the form

$$\begin{pmatrix} x_1 & x_2 & 0 & x_4 & x_5 \\ 0 & 1 & 0 & 0 & 0 \\ 0 & 0 & 1 & 0 & 0 \\ 0 & 0 & 0 & 1 & 0 \\ 0 & 0 & 0 & 0 & 1 \end{pmatrix},$$

which has rank $= 5$ if and only if $x_1 \neq 0$.

Thus we may take e_1 as the additional vector we seek, and then

$$\mathcal{F}_1 = \{e_3\}, \qquad T(\mathcal{F}_1) = \{e_2 + e_4\}, \qquad \mathcal{F}_2 = \{e_1\},$$

and $\mathcal{E}_2 = [\mathcal{F}_1 \cup T(\mathcal{F}_1)] \cup \mathcal{F}_2$. That completes Step 2. (Actually any $\mathbf{x}$ with $x_1 \neq 0$, $x_3 = 0$ would work.)

STEP 3: In the next layer $K_1 \sim K_0$, we have the vectors

$$T^2(\mathcal{F}_1) = \{T^2(e_3) = T(e_2 + e_4) = e_2 + e_5\} \quad \text{and} \quad T(\mathcal{F}_2) = \{T(e_1)\} = \{e_5\}.$$

Since $|K_1/K_0| = |K_1| = 2$, there is no need to adjoin additional vectors from this layer, so $\mathcal{F}_3 = \emptyset$. A suitable basis in V is

$$\mathcal{E}_3 = \mathcal{F}_1 \cup \left[T(\mathcal{F}_1) \cup \mathcal{F}_2\right] \cup \left[T^2(\mathcal{F}_1) \cup T(\mathcal{F}_2)\right] = \{e_3\,;\, e_2 + e_4, e_1\,;\, e_2 + e_5, e_5\}.$$

The diagram in Figure 1.1 takes the form

$\mathcal{F}_1 = e_1^{(1)} = e_3$	
$T(\mathcal{F}_1) = e_2^{(1)} = e_2 + e_4$	$\mathcal{F}_2 = e_1^{(2)} = e_1$
$T^2(\mathcal{F}_1) = e_3^{(1)} = e_2 + e_5$	$T(\mathcal{F}_2) = e_2^{(2)} = e_5$

,

and the iterated action of T sends

$$e_3 \to T(e_3) = e_2 + e_4 \to T^2(e_3) = e_2 + e_5 \to 0 \quad \text{and} \quad e_1 \to T(e_1) = e_5 \to 0.$$

The cyclic subspaces are

$$E_1 = \mathbb{K}\text{-span}\{e_3, T(e_3), T^2(e_3)\} = \mathbb{K}\text{-span}\{e_3, e_2+e_4, e_2+e_5\} \text{ and}$$
$$E_2 = \mathbb{K}\text{-span}\{e_1, T(e_1)\} = \mathbb{K}\text{-span}\{e_1, e_5\},$$

with $V = E_1 \oplus E_2$. With respect to this basis, $[T]_{\mathfrak{X}\mathfrak{X}}$ has the block diagonal form

$$[T]_{\mathfrak{X}\mathfrak{X}} = \left(\begin{array}{ccc|cc} 0 & 1 & 0 & 0 & 0 \\ 0 & 0 & 1 & 0 & 0 \\ 0 & 0 & 0 & 0 & 0 \\ \hline 0 & 0 & 0 & 0 & 1 \\ 0 & 0 & 0 & 0 & 0 \end{array}\right),$$

with each diagonal block an elementary nilpotent matrix. The number and size of such blocks are uniquely determined, but the bases are not unique, nor are the cyclic subspaces in the splitting $V = E_1 \oplus E_2$. That concludes our discussion of Example 1.25. ○

EXERCISE 1.28. Let W be the 3-dimensional subspace in $V = \mathbb{K}^5$ determined by the equations

$$\begin{cases} x_1 - 2x_2 + x_3 - x_5 = 0, \\ 3x_1 + 5x_3 - x_4 = 0, \end{cases}$$

which is equivalent to the matrix equation $A\mathbf{x} = \mathbf{0}$ with

$$A = \begin{pmatrix} 1 & -2 & 1 & 0 & -1 \\ 3 & 0 & 5 & -1 & 0 \end{pmatrix}.$$

(a) Find vectors $\{v_1, v_2, v_3\}$ that are a basis for W.

(b) Find 2 vectors $\{v_4, v_5\}$ that are representatives for a basis in V/W.

(c) Find two of the standard basis vectors $\{e_1, e_2, e_3, e_4, e_5\}$ in $\mathbb{K}^5$ that are a basis for $V \pmod W$.

EXERCISE 1.29. Do either of the vectors

$$f_1 = 2e_1 - 3e_2 + e_3 + e_4 \qquad f_2 = -e_1 + 2e_2 + 5e_3 - 2e_4$$

in $\mathbb{K}^5$ lie in the subspace W determined by the system of equations in Exercise 1.28? Do these vectors form a basis for $\mathbb{K}^5 \pmod W$?

EXERCISE 1.30. Which of the following matrices A are nilpotent?

$$\text{(a)} \begin{pmatrix} 0 & 0 & 0 \\ 1 & 0 & 0 \\ 0 & 1 & 0 \end{pmatrix} \qquad \text{(b)} \begin{pmatrix} 0 & 1 & 2 \\ 0 & 0 & 3 \\ 1 & 0 & 0 \end{pmatrix}$$

$$\text{(c)} \begin{pmatrix} 1 & 2 & -1 \\ -1 & -2 & 1 \\ -1 & -2 & 1 \end{pmatrix} \qquad \text{(d)} \begin{pmatrix} 5 & -6 & -6 \\ -1 & 4 & 2 \\ 3 & -6 & 4 \end{pmatrix}$$

If A is nilpotent, find a basis for $\mathbb{K}^3$ that puts A into block diagonal form with elementary nilpotent blocks. What is the resulting block diagonal form if the blocks are listed in order of decreasing size?

HINT. As a quick test before launching into calculations, if A is nilpotent what does that tell you about $\mathrm{Tr}(A)$ and $\det(A)$?

EXERCISE 1.31. If N_1, N_2 are nilpotent, is N_1N_2 nilpotent? What if N_1 and N_2 commute, with $N_1N_2 = N_2N_1$?

EXERCISE 1.32. Prove the following:

(a) If N_1, N_2 are nilpotent operators that commute, so $N_1N_2 = N_2N_1$, then linear combinations $c_1N_1 + c_2N_2$ are also nilpotent.
(b) If $N_1, \dots, N_r$ are nilpotent and commute pairwise, so $[N_i, N_j] = 0$ for $i \neq j$, then all operators in $\mathbb{K}\text{-span}\{N_1, \dots, N_r\}$ are nilpotent.

EXERCISE 1.33. Let $V = \mathcal{P}_n(\mathbb{K})$ be the polynomials $f = \sum_{k=0}^{n} c_k x^k \in \mathbb{K}[x]$ of degree $\leq n$.

(a) Show that the differentiation operator $D : V \to V$
$$Df = \frac{df}{dx} = c_1 + 2c_2\, x + \cdots + n \cdot c_n x^{n-1}$$
is nilpotent with $\deg(D) = n + 1 = \dim(V)$.
(b) Prove that any constant-coefficient differential operator $L : V \to V$ of the form $a_1D + a_2D^2 + \cdots + a_nD^n$ that has *no constant term* c_0I is nilpotent on V.
(c) Does this remain true if there is a nonzero constant term c_0I in (b)?

EXERCISE 1.34. In the space of polynomials $\mathbb{R}[x]$, let
$$V_1 = \{f : f(-x) = f(x), \text{ the even polynomials}\},$$
$$V_2 = \{f : f(-x) = -f(x), \text{ the odd polynomials}\}.$$
Prove that $\mathcal{P}_n$ is their direct sum $V_1 \oplus V_2$. Are these subspaces invariant under differentiation?

EXERCISE 1.35. Explain why $\mathrm{Tr}(T) = 0$ and $\det(T) = 0$ for any nilpotent linear operator $T : V \to V$ on a finite-dimensional space. Is the converse true?

1.3. Generalized Eigenspaces

We will give a general structure theorem for linear operators T over a field $\mathbb{K}$ large enough that the characteristic polynomial $p_T(x) = \det(T - xI)$ splits into linear factors $f(x) = c \cdot \prod_{i=1}^{s}(x - a_i)^{m_i}$ in $\mathbb{K}[x]$. This is always true if $\mathbb{K} = \mathbb{C}$ but need not be true for other fields; and even if $p_T(x)$ does split, that alone is not enough to guarantee T is diagonalizable. In Appendix A at the end of this chapter, we briefly review diagonalizability of linear operators over a general field $\mathbb{K}$, which means that V contains a basis of eigenvectors, so the eigenspaces $E_\lambda(T)$ span V. If you already have a good understanding of these matters, you may nevertheless wish to review Theorem A.5 in this appendix, as well as its proof. The techniques used there will be applied to develop the *generalized eigenspace decomposition* and *Jordan canonical form*, the main topics of this section.

The Fitting decomposition (Lemma 1.6), reprised below, is a first step toward decomposing linear operators $T : V \to V$ that are not diagonalizable.

> (FITTING DECOMPOSITION) *Given a linear map $T : V \to V$ on a finite-dimensional vector space over any field, there is a direct sum decomposition $V = N \oplus S$ into T-invariant subspaces such that $T|_S : S \to S$ is a bijection (invertible linear operator on S) and $T|_N : N \to N$ is nilpotent.*

The relevant subspaces are the stable kernel and stable range of T,

$$K_\infty = \bigcup_{i=1}^{\infty} K_i \;\; (K_i = \ker(T^i) \;\text{ with } \{0\} \subsetneqq K_1 \subsetneqq \cdots \subsetneqq K_r = K_{r+1} = \cdots = K_\infty),$$

$$R_\infty = \bigcap_{i=1}^{\infty} R_i \;\; (R_i = \text{range}(T^i) \;\text{ with } \{0\} \supsetneqq R_1 \supsetneqq \cdots \supsetneqq R_r = R_{r+1} = \cdots = R_\infty)$$

(see Section 1.1). Obviously, $T = (T|_{R_\infty}) \oplus (T|_{K_\infty})$ which splits T into canonically defined nilpotent and invertible parts.

EXERCISE 1.36. Prove that the Fitting decomposition is unique: If $V = N \oplus S$, both T-invariant, such that $T|_N : N \to N$ is nilpotent and $T|_S : S \to S$ invertible, show that $N = K_\infty(T)$ and $S = R_\infty(T)$.

Given a linear operator $T : V \to V$, we may apply these remarks to the operators $(T - \lambda I)$ associated with eigenvalues λ in $\text{sp}_{\mathbb{K}}(T)$. The eigenspace $E_\lambda(T) = \ker(T - \lambda I)$ is the first in an ascending chain of T-invariant subspaces:

$$\{0\} \subsetneqq \ker(T-\lambda) = E_\lambda(T) \subsetneqq \ker(T-\lambda)^2 \subsetneqq \cdots \subsetneqq \ker(T-\lambda)^r = \cdots = K_\infty(T-\lambda).$$

DEFINITION 1.37. If $\lambda \in \mathbb{K}$, the "stable kernel" of $(T - \lambda I)$

$$K_\infty(\lambda) = \bigcup_{m=1}^{\infty} \ker(T - \lambda I)^m$$

is called the *generalized λ-eigenspace*, which we will hereafter denote by $M_\lambda(T)$.

$$\begin{aligned} M_\lambda(T) &= \{v \in V : (T - \lambda I)^k v = 0 \text{ for some } k \in \mathbb{N}\} \\ &\supseteq E_\lambda(T) = \{v : (T - \lambda I)v = 0\} \end{aligned} \tag{1.6}$$

We refer to any $\lambda \in \mathbb{K}$ such that $M_\lambda(T) \neq (0)$ as a *generalized eigenvalue* for T. Since $M_\lambda(T) \neq \{0\} \Leftrightarrow E_\lambda(T) \neq \{0\} \Leftrightarrow \det(T - \lambda I) = 0$, these are just the usual eigenvalues of T in $\mathbb{K}$.

Generalized eigenspaces have the properties laid out below.

LEMMA 1.38. *The spaces $M_\lambda(T)$ are T-invariant.*

PROOF. T commutes with all the operators $(T - \lambda)^m$, which commute with each other. Thus, $v \in M_\lambda(T) \Rightarrow (T - \lambda I)^k v = 0$ for some $k \in \mathbb{N} \Rightarrow$

$$(T - \lambda I)^k T(v) = T(T - \lambda I)^k v = T(0) = 0.$$

Hence $T(v) \in M_\lambda(T)$. □

We now show that $T|_{M_\lambda}$ has upper triangular form with respect to a suitably chosen basis in M_λ.

PROPOSITION 1.39 (Block Upper Triangular Decomposition of $M_\lambda(T)$). *Each generalized eigenspace $M_\lambda(T)$, $\lambda \in \text{sp}(T)$, has a basis $\mathfrak{X}$ that puts the matrix of $T|_{M_\lambda(T)}$ into upper triangular form:*

$$[T|_{M_\lambda}]_{\mathfrak{X}\mathfrak{X}} = \begin{pmatrix} \lambda & & & * \\ & \ddots & & \\ & & \ddots & \\ 0 & & & \lambda \end{pmatrix}.$$

PROOF. We already know that any nilpotent operator N on a finite-dimensional vector space can be put into strictly upper triangular form by a suitable choice of basis:

$$[N]_{\mathfrak{X}\mathfrak{X}} = \begin{pmatrix} 0 & & & * \\ & \ddots & & \\ & & \ddots & \\ 0 & & & 0 \end{pmatrix}.$$

Now write

$$T|_{M_\lambda} = (T - \lambda I)|_{M_\lambda} + \lambda I|_{M_\lambda}$$

in which $V = M_\lambda$, $N = (T - \lambda I)|_{M_\lambda}$. Since $[I|_{M_\lambda}]_{\mathfrak{X}\mathfrak{X}} = I_{m\times m}$ for any basis, a basis that puts $(T-\lambda I)|_{M_\lambda}$ into strict upper triangular form automatically yields

$$[T|_{M_\lambda}]_{\mathfrak{X}\mathfrak{X}} = \big[(T - \lambda I)|_{M_\lambda}\big]_{\mathfrak{X}\mathfrak{X}} + \lambda[I_\lambda|_{M_\lambda}]_{\mathfrak{X}\mathfrak{X}} = \begin{pmatrix} \lambda & & * \\ & \ddots & \\ 0 & & \lambda \end{pmatrix}. \qquad \square$$

A more precise result is obtained using the cyclic subspace decomposition for nilpotent operators (Theorem 1.16) to guide the choice of basis. As a preliminary step, we might pick a basis $\mathfrak{X}$ aligned with the kernels

$$(0) \subsetneqq K_1 = \ker(T) \subsetneqq K_2 = \ker(T^2) \subsetneqq \cdots \subsetneqq K_d = V$$

where $d = \deg(T)$. As in Proposition 1.39, $[T]_{\mathfrak{X}\mathfrak{X}}$ is then upper triangular with $m_i \times m_i$ blocks of zeros on the diagonal and $m_1 + \cdots + m_d = \deg(T)$. If we apply this to the action of $(T - \lambda I)$ on $M_\lambda(T)$, the matrix of the nilpotent operator $T - \lambda I$ becomes upper triangular with zero blocks on the diagonal. Writing $T = (T - \lambda I) + \lambda I$, the matrix of T with respect to any basis $\mathfrak{X}$ running through successive kernels $K_i = \ker(T - \lambda I)^i$ must have the form

$$\big[T|_{M_\lambda}\big]_{\mathfrak{X}\mathfrak{X}} = \lambda \cdot I_{n\times n} + [\,T - \lambda I\,]_{\mathfrak{X}\mathfrak{X}}$$

$$(1.7) \qquad = \begin{pmatrix} \lambda \cdot \boxed{I_{m_1\times m_1}} & & & * \\ & \lambda \cdot \boxed{I_{m_2\times m_2}} & & \\ & & \ddots & \\ 0 & & & \lambda \cdot \boxed{I_{m_r\times m_r}} \end{pmatrix}$$

with $m_i = \dim(K_i/K_{i-1}) = \dim(K_i) - \dim(K_{i-1})$ and $n = \dim(V) = \sum_i m_i$. The shape of the "block upper triangular form" (1.7) is completely determined by the dimensions of the kernels $K_i = K_i(T - \lambda I)$.

In short, (1.7) says that $T|_{M_\lambda} = \lambda I_\lambda + N_\lambda$ where $I_\lambda = \mathrm{id}_{M_\lambda}$, λI_λ is a scalar operator on M_λ, and $N_\lambda = (T - \lambda I)|_{M_\lambda}$ is a nilpotent operator whose matrix with respect to the basis $\mathfrak{X}$ is the matrix in (1.7) but with $m_i \times m_i$ zero-blocks on the diagonal. We conclude that the restriction $T|_{M_\lambda}$ has an "additive decomposition" $T|_{M_\lambda} = (scalar) + (nilpotent)$ into commuting scalar and nilpotent parts,

$$T|_{M_\lambda} = \lambda I + N_\lambda = \begin{pmatrix} \lambda & & * \\ & \ddots & \\ 0 & & \lambda \end{pmatrix} = \begin{pmatrix} \lambda & & 0 \\ & \ddots & \\ 0 & & \lambda \end{pmatrix} + \begin{pmatrix} 0 & & * \\ & \ddots & \\ 0 & & 0 \end{pmatrix}.$$

Both terms in this decomposition commute with $T|_{M_\lambda}$.

There is also a "multiplicative decomposition"

$$T|_{M_\lambda} = (scalar) \cdot (unipotent) = (\lambda I) \cdot U_\lambda,$$

where U_λ is the unipotent operator $(I + N_\lambda)$; the corresponding matrices are

$$\begin{pmatrix} \lambda & & * \\ & \ddots & \\ 0 & & \lambda \end{pmatrix} = \begin{pmatrix} \lambda & & 0 \\ & \ddots & \\ 0 & & \lambda \end{pmatrix} \cdot \begin{pmatrix} 1 & & * \\ & \ddots & \\ 0 & & 1 \end{pmatrix}.$$

The off-diagonal entries $(*)$ in N_λ and U_λ need not be the same in these two decompositions.

Jordan Decomposition of a Generalized Eigenspace. We now show that the block-upper triangular description of the action of T on $M_\lambda(T)$ can be refined to provide much more information about the off-diagonal terms $(*)$. We will also see that for many purposes the block upper-triangular form will suffice and is easier to compute than the Jordan decomposition since we only need to determine the kernels K_i.

Now consider what happens if we take a basis $\mathfrak{Y}$ in M_λ corresponding to a cyclic subspace decomposition of the nilpotent operator

$$N_\lambda = (T - \lambda I)|_{M_\lambda} = (T|_{M_\lambda}) - \lambda I_\lambda \quad (I_\lambda = I|_{M_\lambda}).$$

Then $[\lambda I_\lambda]_{\mathfrak{Y}\mathfrak{Y}}$ is λ times the identity matrix (as it is for any basis in M_λ) while $[N_\lambda]_{\mathfrak{Y}\mathfrak{Y}}$ consists of diagonal blocks, each an elementary nilpotent matrix N_i.

$$[N_\lambda]_{\mathfrak{Y}\mathfrak{Y}} = \big[(T - \lambda I)|_{M_\lambda}\big]_{\mathfrak{Y}\mathfrak{Y}} = \begin{pmatrix} \boxed{N_1} & & & 0 \\ & \ddots & & \\ & & \ddots & \\ 0 & & & \boxed{N_r} \end{pmatrix}$$

with

$$N_i = \begin{pmatrix} 0 & 1 & & 0 \\ & \ddots & \ddots & \\ & & \ddots & 1 \\ 0 & & & 0 \end{pmatrix},$$

of size $d_i \times d_i$, but with N_i a 1×1 zero matrix when $d_i = 1$. This yields the *Jordan Block Decomposition* of T on a typical generalized eigenspace M_λ.

THEOREM 1.40 (Jordan Decomposition of $T|_{M_\lambda}$). *If $T : V \to V$ is a linear operator on a finite-dimensional space, then in each generalized eigenspace M_λ, there is a basis $\mathfrak{Y}$ that puts the restriction $T|_{M_\lambda}$ into block upper triangular form*

$$(1.8) \quad [T|_{M_\lambda}]_{\mathfrak{YY}} = \lambda \cdot [I|_{M_\lambda}]_{\mathfrak{YY}} + [N_\lambda]_{\mathfrak{YY}} = \begin{pmatrix} \boxed{B_1} & & & & 0 \\ & \ddots & & & \\ & & \ddots & & \\ & & & \boxed{B_m} & \\ 0 & & & & \boxed{\lambda I_{r\times r}} \end{pmatrix}.$$

When $d_1 > 1$, where each $B_i = \lambda I_{d_i \times d_i} +$ (elementary nilpotent) is referred to as a "Jordan block,"

$$B_i = \begin{pmatrix} \lambda & 1 & & 0 \\ & \ddots & \ddots & \\ & & \ddots & \ddots & \\ & & & \ddots & 1 \\ 0 & & & & \lambda \end{pmatrix}.$$

REMARK 1.41. The last block in (1.8) is exceptional. The others correspond to restrictions $T|_{C_i(\lambda)}$ to cyclic subspaces of dimension $d_i > 1$ in a cyclic subspace decomposition

$$M_\lambda(T) = C_1(\lambda) \oplus \cdots \oplus C_m(\lambda).$$

However, some cyclic subspaces might be one-dimensional, and any such $C_i(\lambda)$ is contained in the ordinary eigenspace $E_\lambda(T)$. If there are r such degenerate cyclic subspaces, we may lump them together into a single subspace

$$E = \bigoplus \{C_i(\lambda) : \dim(C_i(\lambda)) = 1\} \subseteq E_\lambda(T) \subseteq M_\lambda(T).$$

If we write $\dim(E) = r$ and $T|_E = \lambda \cdot I_E$, it should also be evident that

$$r + d_1 + \cdots + d_m = \dim\big(M_\lambda(T)\big). \qquad \bigcirc$$

1.4. The Generalized Eigenspace Decomposition

So far we have only determined the structure of T restricted to a single generalized eigenspace $M_\lambda(T)$. Several issues must be resolved to arrive at the overall structure of T on V.

- If the generalized eigenspaces $M_{\lambda_i}(T)$ fail to span V (for instance if $p_T(x)$ fails to split in $\mathbb{K}[x]$), knowing the behavior of T only on their span

$$M = \sum_{\lambda \in \mathrm{sp}(T)} M_\lambda(T)$$

  leaves the global behavior of T beyond reach.

- It is equally important to prove (as we did in LA I, Proposition 2.29, for ordinary eigenspaces) that the span of the generalized eigenspaces is in fact a DIRECT SUM,
$$M = \bigoplus_{\lambda \in \mathrm{sp}(T)} M_\lambda(T).$$
That means the actions of T on different M_λ are *independent* and can be examined separately to get $T|_M = \bigoplus_{\lambda \in \mathrm{sp}(T)} (T|_{M_\lambda})$.

Both issues will be resolved in our favor if $p_T(x)$ splits into linear factors in $\mathbb{K}[x]$, which is always true if $\mathbb{K} = \mathbb{C}$. We are not so lucky for linear operators over $\mathbb{K} = \mathbb{R}$ or over finite fields such as $\mathbb{K} = \mathbb{Z}_p$. When this program succeeds, the result is the *generalized eigenspace decomposition.*

THEOREM 1.42 (Generalized Eigenspace Decomposition). *If T is a linear operator on a finite-dimensional vector space V and its characteristic polynomial $p_T(x) = \det(T - xI)$ splits over $\mathbb{K}$, then V is a direct sum of the generalized eigenspaces for T,*
$$V = \bigoplus_{\lambda \in \mathrm{sp}(T)} M_\lambda(T).$$
Since the spaces M_λ are T-invariant, we obtain a decomposition of T itself

$$T = \bigoplus_{\lambda \in \mathrm{sp}(T)} T|_{M_\lambda(T)} \tag{1.9}$$

into operators, each of which can be put into Jordan form (1.8) *by choosing bases compatible with a decomposition of M_λ into T-invariant subspaces that are cyclic under the action of $(T - \lambda I)$.*

PROOF. The following proof that the generalized eigenspaces are independent components in a direct sum follows the same lines as a similar result for ordinary eigenspaces (Proposition 2.29 of LA I) but with additional technical complications. Proof that they span V will require some new ideas based on the Fitting decomposition.

PROPOSITION 1.43 (Independence of the M_λ). *If $T : V \to V$ is a linear operator on a finite-dimensional space, the span $M = \sum_{\lambda \in \mathrm{sp}(T)} M_\lambda(T)$ of the generalized eigenspaces (which might be a proper subspace in V) is always a direct sum, $M = \bigoplus_{\lambda \in \mathrm{sp}(T)} M_\lambda(T)$.*

PROOF. Let $\lambda_1, \dots, \lambda_r$ be the distinct eigenvalues in $\mathbb{K}$. By definition of "direct sum" we must show the components M_λ are independent so that

$$0 = v_1 + \cdots + v_r \text{ with } v_i \in M_{\lambda_i} \Rightarrow \text{ each term } v_i \text{ is zero.} \tag{1.10}$$

Fix an index k. For each $1 \le i \le r$, let
$$m_j = \deg(T - \lambda_j I)|_{M_{\lambda_j}}.$$
If $v_k = 0$, we're done, and if $v_k \neq 0$ let $m = m(v_k) \le m_k$ be the smallest exponent such that $(T - \lambda_k I)^m v_k = 0$ and $(T - \lambda_k I)^{m-1} v_k \neq 0$. Then $w = (T - \lambda_k I)^{m-1} v_k$ is a nonzero eigenvector in E_{λ_k}.

Define

$$A = \prod_{i \neq k} (T - \lambda_i I)^{m_i} \cdot (T - \lambda_k)^{m-1},$$

which is keyed to the particular index λ_k and base point w as above. Then

$$\begin{aligned}
0 = A(0) &= 0 + Av_k \quad \text{(since } Av_i = 0 \text{ for } i \neq k\text{)} \\
&= \prod_{i \neq k} (T - \lambda_i)^{m_i} ((T - \lambda_k)^{m-1} v_k) = \prod_{i \neq k} (T - \lambda_i)^{m_i} w \\
&= \prod_{i \neq k} \big((T - \lambda_k) + (\lambda_k - \lambda_i)\big)^{m_i} w \quad \text{(a familiar algebraic trick)} \\
&= \prod_{i \neq k} \sum_{s=0}^{m_i} \binom{m_i}{s} (T - \lambda_k)^{m_i - s} (\lambda_k - \lambda_i)^s w \quad \text{(binomial expansion of } (\cdots)^{m_i}\text{)}.
\end{aligned}$$

All terms in the binomial sum are zero except when $s = m_i$, so we get

$$0 = \Big(\prod_{i \neq k} (\lambda_k - \lambda_i)^{m_i} \Big) \cdot w.$$

The factor $(\cdots)$ is nonzero because the λ_i are the distinct eigenvalues of T in $\mathbb{K}$, so w must be zero. This is a contradiction because $w \neq 0$ by definition. We conclude that *every* term v_k in (1.10) is zero, so the span M is a direct sum of the M_λ. □

EXERCISE 1.44. If a linear operator $T : V \to V$ has characteristic polynomial

(a) $p_T(x) = x(x-2)^2$,
(b) $p_T(x) = (x-2)^2(x-5)^2$,

make diagrams showing the possible Jordan canonical forms T can have. Normalize your diagrams by listing Jordan blocks for each eigenvalue in order of decreasing size.

Further Properties of Characteristic Polynomials. Before proving Theorem 1.42, we digress to develop a few more facts about characteristic polynomials needed to work out the relation between the eigenvalues of T and those of the restriction $T|_{R_\infty}$, where $R_\infty = R_\infty(T - \lambda_1 I)$. These will be needed in the induction proof of Theorem 1.42

LEMMA 1.45. *If* $A = \left(\begin{array}{c|c} B & * \\ \hline 0 & C \end{array}\right)$, *where* B *is* $r \times r$ *and* C *is* $(n-r) \times (n-r)$, *then* $\det(A) = \det(B) \cdot \det(C)$.

COROLLARY 1.46. *If* $A \in \mathrm{M}(n, \mathbb{K})$ *is upper triangular with values* $c_1, \ldots, c_n$ *on the diagonal, then* $\det(A) = \prod_{i=1}^{n} c_i$.

PROOF. (Proof of Lemma 1.45) Consider an $n \times n$ template corresponding to some $\sigma \in S_n$. If any of the marked spots in columns $C_1, \dots, C_r$ occur in a row R_i with $r+1 \le i \le n$, then $a_{ij} = a_{i,\sigma(i)} = 0$ and so is the corresponding term in $\sum_{\sigma \in S_n}(\cdots)$. Thus, columns $C_j, 1 \le j \le r$, can only be marked in rows $R_1, \dots, R_r$ if the template is to yield a nonzero term in $\det(A)$. It follows immediately that all columns C_j with $r+1 \le j \le n$ must be marked in rows R_i with $r+1 \le i \le n$ if σ is to contribute to

$$\det(A) = \sum_{\sigma \in S_n} \operatorname{sgn}(\sigma) \cdot \prod_{i=1}^{n} a_{i,\sigma(i)}.$$

Therefore, only permutations σ that leave invariant the blocks of indices $[1, r]$, $[r+1, n]$ can contribute. These σ are composites of permutations $\mu \in S_r$ and $\tau \in S_{n-r}$ (identified with a permutation $\tilde{\tau}$ of $\{r+1, \dots, n\}$),

$$\sigma(k) = \mu \times \tilde{\tau}(l) = \begin{cases} \mu(k) & \text{if } 1 \le k \le r, \\ \tilde{\tau}(k) = r + \tau(k-r) & \text{if } r+1 \le k \le n. \end{cases}$$

Furthermore, we have $\operatorname{sgn}(\sigma) = \operatorname{sgn}(\mu \times \tau) = \operatorname{sgn}(\mu) \cdot \operatorname{sgn}(\tau)$ by definition of the signature sgn because μ, τ operate on disjoint subsets of indices in $[1, n]$.

In the matrix A, we have

$$\begin{aligned} B_{k,\ell} &= A_{k,\ell} && \text{for } 1 \le k,\ \ell \le r, \\ C_{k,\ell} &= A_{k+r,\ell+r} && \text{for } r+1 \le k,\ \ell \le n, \end{aligned}$$

so we get

$$\begin{aligned} \det(A) &= \sum_{(\mu,\tau) \in S_r \times S_{n-r}} \operatorname{sgn}(\mu \times \tilde{\tau}) \cdot \Big(\prod_{i=1}^{r} B_{i,\mu(i)}\Big) \cdot \Big(\prod_{j=1}^{n-r} C_{j,\tilde{\tau}(j)}\Big) \\ &= \Big(\sum_{\mu \in S_r} \operatorname{sgn}(\mu) \cdot \prod_{i=1}^{r} B_{i,\mu(i)}\Big) \cdot \Big(\sum_{\tau \in S_{n-r}} \operatorname{sgn}(\tilde{\tau}) \cdot \prod_{j=1}^{n-r} C_{j,\tilde{\tau}(j)}\Big) \\ &= \det(B) \cdot \det(C). \end{aligned}$$

□

COROLLARY 1.47. *If $T : V \to V$ is a linear operator on a finite-dimensional vector space and $M \subseteq V$ is a T-invariant subspace, the characteristic polynomial $p_{T|_M}(x)$ of the restriction $T|_M$ divides $p_T(x) = \det(T - xI)$ in $\mathbb{K}[x]$.*

PROOF. If $M \subseteq V$ is T-invariant and we take a basis $\mathfrak{X} = \{e_i\}$ that first spans M and then picks up additional vectors to get a basis for V, the matrix $[T]_{\mathfrak{X}\mathfrak{X}}$ has block upper triangular form $\left(\begin{array}{c|c} A & * \\ \hline 0 & B \end{array}\right)$, and then

$$[T - xI]_{\mathfrak{X}\mathfrak{X}} = \left(\begin{array}{c|c} A - xI & * \\ \hline 0 & B - xI \end{array}\right).$$

But it is trivial to check that $A - xI = \big[(T - xI)|_M\big]_{\mathfrak{X}'\mathfrak{X}'}$ where $\mathfrak{X}' = \{e_1, \cdots, e_r\}$ are the initial vectors that span M. Thus, $\det(A - xI) = \det((T - xI)|_M) = p_{T|M}(x)$ divides $p_T(x) = \det(A - xI) \cdot \det(B - xI)$. □

EXERCISE 1.48. Given (V, M, T) as in the previous corollary, T induces a linear map $\widetilde{T} : V/M \to V/M$ such that $\widetilde{T}(v + M) = T(v) + M$ for $v \in V$. Prove that the characteristic polynomial $p_{\widetilde{T}(x)} = \det_{V/M}(\widetilde{T} - xI)$ also divides $p_T(x) = \det(A - xI) \cdot \det(B - xI)$.

LEMMA 1.49. *If f and P are nonconstant polynomials in $\mathbb{K}[x]$ and P divides f, so $f(x) = P(x)Q(x)$ for some other $Q \in \mathbb{K}[x]$, then $P(x)$ must split over $\mathbb{K}$ if $f(x)$ does.*

PROOF. If Q is constant, there is nothing to prove. Nonconstant polynomials $f \neq 0$ in $\mathbb{K}[x]$ have unique factorizations into irreducible polynomials $f = \prod_{i=1}^{r} F_i$ where F_i cannot be written as a product of nonconstant polynomials of lower degree. Each polynomial f, P, Q has such a factorization:

$$P = \prod_{k=1}^{m} P_k\,,\; Q = \prod_{j=1}^{m} Q_j, \qquad \text{so} \qquad f = PQ = \prod_k P_k \cdot \prod_j Q_j.$$

Since f splits over $\mathbb{K}$, it can also be written as a product of linear factors $f(x) = \prod_{i=1}^{n}(x - \alpha_i)$, where $\{\alpha_i\}$ are the roots of $f(x)$ in $\mathbb{K}$ counted according to their multiplicities. Linear factors $(x - \alpha)$ are obviously irreducible, and the two irreducible decomposition of $f(x)$ must agree. Thus P (and Q) are products of linear factors, and $P(x)$ splits over $\mathbb{K}$. □

This lemma has a useful corollary.

COROLLARY 1.50. *If the characteristic polynomial $p_T(x)$ of a linear operator $T : V \to V$ splits over $\mathbb{K}$, so does $p_{T|_W}$ for any T-invariant subspace $W \subseteq V$.*

Proof of Generalized Eigenspace Decomposition. We now take up the proof of Theorem 1.42. The operator T has eigenvalues in $\mathbb{K}$ because p_T splits, and its eigenvalues $\{\lambda_1, \dots, \lambda_r\}$ are the distinct roots of p_T in $\mathbb{K}$. Recall that $E_\lambda \neq \{0\} \Leftrightarrow M_\lambda \neq \{0\}$.

Pick an eigenvalue λ_1 and let $V = K_\infty \oplus R_\infty$ be the Fitting decomposition of the operator $(T - \lambda_1 I)$, so K_∞ is the generalized eigenspace $M_{\lambda_1}(T)$ while R_∞ is the stable range of $(T - \lambda_1 I)$. Both spaces are invariant under $T - \lambda_1 I$ and also under T since $\lambda_1 I$ commutes with T. It will be important to note that

$$\lambda_1 \textit{ cannot be an eigenvalue of } T|_{R_\infty}$$

because if $v \in R_\infty$ is nonzero, then $(T - \lambda_1 I)v = 0 \Rightarrow v \in K_\infty \cap R_\infty = \{0\}$. Hence $\mathrm{sp}(T|_{R_\infty}) \subseteq \{\lambda_2, \dots, \lambda_r\}$.

We now argue by induction on $n = \dim(V)$. There is little to prove if $n = 1$. (There is an eigenvalue, so $E_\lambda = V$ and $T = \lambda I$.) So assume $n > 1$ and that the theorem has been proved for all spaces V' of dimension $< n$ and for all operators $T' : V' \to V'$ such that $\det(T' - \lambda I)$ splits over $\mathbb{K}$. The natural move is to apply this inductive hypothesis to $T' = T|_{R_\infty} = T|_{R_\infty(T - \lambda_1 I)}$ since $\dim(R_\infty) = \dim(V) - \dim(M_{\lambda_1}) < \dim(V) = n$. But to do so, we must show $p_{T'}$ splits over $\mathbb{K}$, as required in the induction hypothesis.

By Lemma 1.49, the characteristic polynomial of $T' = T|_{R_\infty}$ splits over $\mathbb{K}$ and, by induction on dimension $R_\infty(T')$, is a direct sum of generalized eigenspaces for the restricted operator T',

$$R_\infty(T') = \bigoplus_{\mu \in \mathrm{sp}(T')} M_\mu(T'),$$

where $\mathrm{sp}(T') =$ the distinct roots of $p|_{T'}$ in $\mathbb{K}$. To compare the roots of p_T and $p_{T'}$, we invoke our observation: $p_{T'}$ divides p_T. The roots $\mathrm{sp}(T') = \mathrm{sp}(T|_{R_\infty})$ are a subset of the roots $\mathrm{sp}(T)$ of $p_T(x)$, so every eigenvalue μ for T' is an eigenvalue for T. Now label the distinct eigenvalues of T so that

$$\mathrm{sp}(T') = \{\lambda_s, \lambda_{s+1}, \dots, \lambda_r\} \subseteq \mathrm{sp}(T) = \{\lambda_1, \dots, \lambda_r\}$$

(with $s > 1$ because $\lambda_1 \notin \mathrm{sp}(T|_{R_\infty})$, as we observed earlier).

Furthermore, for each $\mu \in \mathrm{sp}(T')$, the generalized eigenspace $M_\mu(T')$ is a subspace of $R_\infty(T - \mu I) \subseteq V$ and must be contained in $M_\mu(T)$ because

$$(T' - \mu I)^k v = 0 \Rightarrow (T - \mu I)^k v = 0 \quad \text{for all } v \in M_\mu(T').$$

It is obvious from the definitions that $R_\infty(T') = R_\infty(T)$, so we shall abbreviate $R_\infty(T') = R_\infty$. Then

$$R_\infty = \bigoplus_{\mu \in \mathrm{sp}(T')} M_\mu(T') \subseteq \sum_{\mu \in \mathrm{sp}(T')} M_\mu(T) \subseteq \sum_{\lambda \in \mathrm{sp}(T)} M_\lambda(T)$$

(since $R_\infty = R_\infty(T') = \bigoplus_{\mu \in \mathrm{sp}(T')} M_\mu(T')$ by the induction hypothesis). Therefore, the generalized eigenspaces $M_\lambda(T)$, $\lambda \in \mathrm{sp}(T)$, must span V because

$$V = K_\infty \oplus R_\infty = M_{\lambda_1}(T) \oplus R_\infty \subseteq M_{\lambda_1}(T) \oplus \Big(\bigoplus_{\mu \in \mathrm{sp}(T')} M_\mu(T') \Big)$$

$$\text{(1.11)} \qquad \subseteq M_{\lambda_1}(T) + \Big(\sum_{\mu \in \mathrm{sp}(T')} M_\mu(T) \Big) \quad \text{(because } M_\mu(T') \subseteq M_\mu(T)\text{)}$$

$$\subseteq M_{\lambda_1}(T) + \Big(\sum_{\lambda \in \mathrm{sp}(T)} M_\lambda(T) \Big) \subseteq V \quad \text{(because } \mathrm{sp}(T') \subseteq \mathrm{sp}(T)\text{)}.$$

We draw the following conclusion: for $\lambda \in \mathrm{sp}(T)$, the M_λ span V, so V is a direct sum of its generalized eigenspaces by Proposition 1.43. That finishes the proof of Theorem 1.42. □

It is worth noting that

$$\mathrm{sp}(T') = \{\lambda_2, \dots, \lambda_r\} \quad \text{and} \quad M_{\lambda_i}(T') = M_{\lambda_i}(T) \quad \text{for } 2 \le i \le r.$$

Since $M_\mu(T') \subseteq M_\mu(T)$ for all $\mu \in \mathrm{sp}(T')$ and $M_{\lambda_1}(T) \cap V' = (0)$, λ_1 cannot appear in $\mathrm{sp}(T')$; on the other hand, every $\mu \in \mathrm{sp}(T')$ must lie in $\mathrm{sp}(T)$.

The Jordan Canonical Form. The preceding discussion has several corollaries that lead to a fully developed version of the Jordan canonical form (JCF). However, keep in mind that some things can be proved using just the simpler block upper-triangular form presented below.

COROLLARY 1.51 (Block Upper Triangular Form). *If the characteristic polynomial of $T : V \to V$ splits over $\mathbb{K}$, and in particular if $\mathbb{K} = \mathbb{C}$, there is a basis $\mathfrak{X}$ such that $[T]_{\mathfrak{X}\mathfrak{X}}$ has block upper triangular form*

$$[T]_{\mathfrak{X}\mathfrak{X}} = \begin{pmatrix} \boxed{T_1} & & & 0 \\ & \ddots & & \\ & & \ddots & \\ & & & \ddots \\ 0 & & & \boxed{T_r} \end{pmatrix} \tag{1.12}$$

with blocks on the diagonal

$$T_i = \begin{pmatrix} \lambda_i & * & & * \\ & \ddots & \ddots & \\ & & \ddots & \ddots \\ & & \ddots & * \\ 0 & & & \lambda_i \end{pmatrix}$$

of size $m_i \times m_i$ such that:

1. *$\lambda_1, \dots, \lambda_r$ are the distinct eigenvalues of T (which may appear in several different blocks).*
2. *The block sizes are the algebraic multiplicities m_i of the λ_i in the splitting of the characteristic polynomial $p_T(t)$ (see the next corollary for details).*
3. *$p_T(x) = (-1)^n \cdot \prod_{j=1}^{r}(x - \lambda_j)^{m_j}$ with $n = m_1 + \cdots + m_r$.*

The blocks T_i may or may not have off-diagonal terms.

COROLLARY 1.52. *If the characteristic polynomial of an $n \times n$ matrix A splits over $\mathbb{K}$, there is a similarity transform $A \mapsto SAS^{-1}$, $S \in \mathrm{GL}(n, \mathbb{K})$, such that SAS^{-1} has the block upper-triangular form shown above.*

COROLLARY 1.53. *If the characteristic polynomial of $T : V \to V$ splits over $\mathbb{K}$, and in particular if $\mathbb{K} = \mathbb{C}$, then for every $\lambda \in \mathrm{sp}(T)$ we have*

$$(\textit{algebraic multiplicity of } \lambda) = \dim(M_\lambda) = m_j,$$

where m_j is the block size in (1.12).

PROOF. Taking a basis that casts $[T]_{\mathfrak{X}\mathfrak{X}}$ in the form (1.12), $[\,T - xI\,]_{\mathfrak{X}\mathfrak{X}}$ will have the same form but with diagonal entries λ_j replaced by $(\lambda_j - x)$. Then

$$\det[T - xI]_{\mathfrak{X}\mathfrak{X}} = \prod_{j=1}^{r} (\lambda_j - x)^{\dim(M_{\lambda_j})} = p_T(x)$$

since the blocks T_j correspond to $T|_{M_{\lambda_j}}$. Obviously, the exponent on $(\lambda_j - x)$ is also the multiplicity of λ_j in the splitting of p_T. □

COROLLARY 1.54. *If the characteristic polynomial of an $n \times n$ matrix A with distinct eigenvalues $\mathrm{sp}_{\mathbb{K}}(A) = \{\lambda_1, \dots, \lambda_r\}$ splits over $\mathbb{K}$, then the following hold:*

1. $\det(A) = \prod_{i=1}^r \lambda_i^{m_i}$, *the product of eigenvalues counted according to their algebraic multiplicities m_i.*
2. $\mathrm{Tr}(A) = \sum_{i=1}^r m_i\lambda_i$, *the sum of eigenvalues counted according to their algebraic multiplicities m_i.*
3. *More generally, if $\mathbb{K} = \mathbb{C}$, there are explicit formulas for all coefficients of the characteristic polynomial when we write it in the form*

$$p_A(x) = \det(A - xI) = \sum_{k=0}^{n} (-1)^k c_k(A)\, x^k.$$

If eigenvalues are listed according to their multiplicities $m_k = m(\lambda_k)$, say as $\mu_1, \dots, \mu_n$ with $n = \dim(V)$ and

$$\mu_1 = \cdots = \mu_{m_1} = \lambda_1, \quad \mu_{m_1+1} = \cdots = \mu_{m_1+m_2} = \lambda_2, \quad \textit{etc.},$$

then $c_n = 1$ and

$$c_{n-1} = \sum_{j=1}^{n} \mu_j = \mathrm{Tr}(A),$$

$$c_{n-2} = \sum_{j_1<j_2}^{n} \mu_{j_1}\mu_{j_2},$$

$$\vdots$$

$$c_{n-k} = \sum_{j_1<\cdots<j_k} \mu_{j_1}\cdots\mu_{j_k},$$

$$\vdots$$

$$c_0 = \mu_1 \cdots \mu_n = \det(A).$$

These formulas need not be true if $\mathbb{K} = \mathbb{R}$ and $p_T(x)$ has nonreal roots in $\mathbb{C}$.

Given a matrix $A \in \mathrm{M}(n, \mathbb{K})$, $\mathrm{Tr}(A)$ and $\det(A)$ are examples of *similarity invariants* on matrix space because they are unaffected by similarity transformations $A \to SAS^{-1}$. The rank

$$\mathrm{rank}(A) = \dim\big(\mathrm{range}(L_A)\big) \quad \text{where } L_A(\mathbf{x}) = A \cdot \mathbf{x} \text{ for } \mathbf{x} \in \mathbb{K}^n$$

is another example because $\mathrm{rank}(A) = \mathrm{rank}(SAS^{-1})$ (Why?). A similarity invariant need not be a scalar. For instance, when $\mathbb{K} = \mathbb{C}$, the spectrum $\mathrm{sp}_{\mathbb{K}}(A) = \mathrm{sp}_{\mathbb{K}}(L_A)$, the algebraic multiplicities of the eigenvalues, and the characteristic polynomial $p_A(t) = \det(A - tI)$ in $\mathbb{K}[t]$, are all constant on similarity classes within $\mathrm{M}(n, \mathbb{K})$.

By invariance of the spectrum, all the combinations of eigenvalues $c_0 = \det(A), \dots, c_{n-1} = \mathrm{Tr}(A)$ listed in Theorem 1.54 are similarity invariants of matrices. Since they are all determined by the characteristic polynomial, it is natural to ask

> *Are matrices $A, A' \in \mathrm{M}(n, \mathbb{C})$ similar when they have the same characteristic polynomial?*

The answer is “*No.*”

EXERCISE 1.55. Explain why the matrices

$$A = \begin{pmatrix} 1 & 0 \\ 0 & 1 \end{pmatrix} \quad \text{and} \quad B = \begin{pmatrix} 1 & 1 \\ 0 & 1 \end{pmatrix},$$

which have the same characteristic polynomial in $\mathbb{C}[t]$, cannot be in the same similarity class (so $B \neq SAS^{-1}$ for any S).

Clearly $p_T(A)$ is a similarity invariant, but by itself it is not enough to completely determine the similarity class of A.

COROLLARY 1.56. *If the characteristic polynomial of an $n \times n$ matrix A splits over $\mathbb{K}$, then $T : V \to V$ is diagonalizable if and only if*

$$(\textit{algebraic multiplicity}) = (\textit{geometric multiplicity})$$

for each $\lambda \in \mathrm{sp}(T)$. Both multiplicities are then equal to $\dim(E_\lambda(T))$.

PROOF. $E_\lambda \subseteq M_\lambda$ for every eigenvalue, and by the previous corollary we know that

$$(\textit{geometric multiplicity}) = \dim(E_\lambda) \leq \dim(M_\lambda) = (\textit{algebraic multiplicity})$$

Furthemore, $M_\lambda = \ker(T - \lambda I)^N$ for large $N \in \mathbb{N}$. Writing $V = M_{\lambda_1} \oplus \cdots \oplus M_{\lambda_r}$, the implication $(\Leftarrow)$ follows because

$$\begin{aligned} (\textit{algebraic multiplicity}) &= (\textit{geometric multiplicity}) \\ &\Rightarrow \dim(E_{\lambda_i}) = \dim(M_{\lambda_i}) \ \text{ for all } i \\ &\Rightarrow M_{\lambda_i} = E_{\lambda_i} \text{ since } M_{\lambda_i} \supseteq E_{\lambda_i} \\ &\Rightarrow V = \bigoplus_{i-1}^{r} E_{\lambda_i} \text{ and } T \text{ is diagonalizable.} \end{aligned}$$

For $(\Rightarrow)$, if T is diagonalizable, we have $V = \bigoplus_{i=1}^r E_{\lambda_i}$, but $E_{\lambda_i} \subseteq M_{\lambda_i}$. Compare this with the decomposition $V = \bigoplus_{i=1}^r M_{\lambda_i}$ to get $M_{\lambda_i} = E_{\lambda_i}$. □

THEOREM 1.57 (Jordan Canonical Form). *If $T : V \to V$ is a linear operator whose characteristic polynomial splits over $\mathbb{K}$, there is a “Jordan basis” $\mathfrak{X}$ such that $A = [T]_{\mathfrak{X}\mathfrak{X}}$ is made up of “Jordan blocks” along the diagonal:*

$$(1.13)\quad [T]_{\mathfrak{X}\mathfrak{X}} = \begin{pmatrix} \boxed{B_1} & & & 0 \\ & \ddots & & \\ & & \ddots & \\ 0 & & & \boxed{B_r} \end{pmatrix} \quad \textit{with} \quad B_i = \begin{pmatrix} \lambda_i & 1 & & 0 \\ & \ddots & \ddots & \\ & & \ddots & 1 \\ 0 & & & \lambda_i \end{pmatrix}.$$

Blocks with the same size or diagonal values λ_i can appear more than once.

With respect to the same basis, the powers $I, T, T^2, \ldots$ take the form

$$[T^k]_{\mathfrak{X}\mathfrak{X}} = \begin{pmatrix} \boxed{B_1^k} & & & 0 \\ & \ddots & & \\ & & \ddots & \\ 0 & & & \boxed{B_r^k} \end{pmatrix}. \tag{1.14}$$

Such a basis can be reordered to gather together all blocks involving the same λ_i. The resulting "superblocks" correspond to the generalized eigenspaces $M_\lambda(T)$.

PROOF. A generalized eigenspace $M_\lambda(T)$ is the stable kernel $K_\infty(T - \lambda I)$. Decompose $V = \bigoplus_{\lambda \in \mathrm{sp}(T)} M_\lambda$ into generalized eigenspaces, and in each M_λ, find a basis aligned with the kernels $\ker(T - \lambda I)^j$. Decomposing each $M_\lambda = \bigoplus_j C_{\lambda,j}$ into cyclic subspaces as in Theorem 1.16, we obtain a basis that puts T_{M_λ} into Jordan canonical form. As for (1.14), the powers $[T^k] = [T]^k$ are actually easy to compute. Each block $[T_i]$ has the form $[T_i] = \lambda_i I + N_i$ with N_i an elementary nilpotent matrix, so we have

$$[T_i^k] = \sum_{j=0}^{k} \binom{k}{j} \lambda_i^{k-j} N_i^j = \lambda_i^k I + \binom{k}{1} \lambda_i^{k-1} N_i + \cdots + N_i^k,$$

a binomial expansion with $N_i^j = 0$ for $j \geq \deg(N_i)$. □

EXERCISE 1.58. If N is an $r \times r$ elementary nilpotent matrix,

$$N = \begin{pmatrix} 0 & 1 & & & 0 \\ & \ddots & \ddots & & \\ & & \ddots & \ddots & \\ & & & \ddots & 1 \\ 0 & & & & 0 \end{pmatrix}, \quad \text{show that} \quad N^2 = \begin{pmatrix} 0 & 0 & 1 & & 0 \\ & \ddots & \ddots & \ddots & \\ & & \ddots & \ddots & 1 \\ & & & \ddots & 0 \\ 0 & & & & 0 \end{pmatrix}.$$

Each new multiplication by N moves the diagonal file of $1's$ one step to the upper right, with $N^r = 0$ at the last step.

EXERCISE 1.59. If $T : V \to V$ is diagonalizable, explain why its JCF is

$$A = [T]_{\mathfrak{X}\mathfrak{X}} = \begin{pmatrix} \lambda_1 \boxed{I_{m_1 \times m_1}} & & 0 \\ & \ddots & \\ 0 & & \lambda_r \boxed{I_{m_r \times m_r}} \end{pmatrix}.$$

EXERCISE 1.60. If the characteristic polynomial of $T : V \to V$ splits over $\mathbb{K}$, there is a basis that puts $[T]_{\mathfrak{X}\mathfrak{X}}$ in Jordan form $A = \mathrm{diag}(B_1, \ldots, B_r)$ consisting of diagonal blocks

$$B_k = \lambda I_{k \times k} + N_k = \begin{pmatrix} \lambda & 1 & & 0 \\ & \ddots & \ddots & \\ & & \ddots & 1 \\ 0 & & & \lambda \end{pmatrix}_{k \times k}.$$

Assuming there are just two blocks, of size 2×2 and 4×4 with diagonal entries λ, μ respectively,

(a) Compute the exponential matrix

$$\mathrm{Exp}(tA) = e^{tA} = \sum_{n=0}^{\infty} \frac{t^n}{n!} A^n$$

in closed form for $t \in \mathbb{K}$ (sum the series).

(b) Is the matrix e^{tA} a Jordan canonical form for the linear operator $e^T = \sum_{n=0}^{\infty} \frac{1}{n!} T^n$? Discuss.

Does it matter if $\mu = \lambda$?

HINT. If A and B commmute, $e^{A+B} = e^A \cdot e^B$; apply the previous exercise. Since N is nilpotent, its exponential series is a finite sum.

EXAMPLE 1.61. Let $T : V \to V$ be a linear map on $V = \mathbb{R}^4$ whose matrix with respect to the standard basis $\mathfrak{X} = \{e_1, \cdots, e_4\}$ has the form

$$A = [T]_{\mathfrak{X}\mathfrak{X}} = \begin{pmatrix} 7 & 1 & 2 & 2 \\ 1 & 4 & -1 & -1 \\ -2 & 1 & 5 & -1 \\ 1 & 1 & 2 & 8 \end{pmatrix}.$$

Taking $T = L_A : \mathbb{C}^4 \to \mathbb{C}^4$, we will do the following:

1. Find the generalized eigenspaces $M_{\lambda(T)}$ of T.
2. Compute a cyclic subspace decomposition of $T - \lambda I$ on M_λ for each $\lambda \in \mathrm{sp}(T)$.
3. Compute bases for these cyclic subspaces to obtain a basis $\mathfrak{X} \subseteq \mathbb{C}^4$ that puts $[T]_{\mathfrak{X}\mathfrak{X}}$ in Jordan canonical form.

DISCUSSION. After some computations, which we omit, we find that T has $\lambda = 6$ as its only eigenvalue and that the characteristic polynomial is

$$p_T(t) = \det(A - xI) = x^4 - 4(6x^3) + 6(6^2x^2) - 4(6^3x) + 6^4 = (x-6)^4,$$

so $M_\lambda = \mathbb{C}^5$ and $\mathrm{sp}_{\mathbb{R}}(T) = \{6\}$ with algebraic multiplicity $m = 4$. Thus $M_{\lambda=6}$ is all of V, and $(T - 6I)$ is nilpotent.

To find a cyclic subspace decomposition, we start by finding basis vectors for V aligned with the kernels $K_j = \ker\left((T-6I)^j\right)$. For $j = 1$, row reduction of $[T - 6I]_{\mathfrak{X}\mathfrak{X}} = [A - 6I\,]$ yields $\ker(T - 6I) = E_{\lambda=6}(T)$. We have

$$[A - 6I] \to \begin{pmatrix} \boxed{1} & 1 & 2 & 2 \\ 0 & -3 & -3 & -3 \\ 0 & 3 & 3 & 3 \\ 0 & 0 & 0 & 0 \end{pmatrix} \to \begin{pmatrix} \boxed{1} & 1 & 2 & 2 \\ 0 & \boxed{1} & 1 & 1 \\ 0 & 0 & 0 & 0 \\ 0 & 0 & 0 & 0 \end{pmatrix}.$$

Thus, x_3 and x_4 are free variables, and

$$\begin{aligned} K_1 &= \{v = (-s-t, -s-t, s, t) : s, t \in \mathbb{R}\} \\ &= \mathbb{R}\text{-span}\{f_1^{(1)} = -e_1 - e_2 + e_3\,,\ f_2^{(1)} = -e_1 - e_2 + e_4\} \\ &= \mathbb{R}\text{-span}\{(-1,-1,1,0)\,,\ (-1,-1,0,1)\} \end{aligned}$$

$\mathcal{F}_1 = e_1^{(1)} = e_2$	
$T(\mathcal{F}_1) = e_2^{(1)} = e_1 - 2e_2 + e_3 + e_4$	
$T^2(\mathcal{F}_1) = 3e_3^{(1)} + 3e_2 - 6e_3 + 3e_4$	$\mathcal{F}_2 = e_1^{(2)} = -e_1 - e_2 + e_4$

FIGURE 1.2. This is the version of Figure 1.1 worked out in Example 1.61. Although there are three columns, column 2 is empty and has been deleted in the present diagram. All the basis vectors $e_j^{(i)}$ are shown.

with $\dim(K_1) = 2$. Next, row reduce $\ker(A - 6I)^2$ to get

$$[A - 6I]^2 = \begin{pmatrix} 0 & 3 & 3 & 3 \\ 0 & 3 & 3 & 3 \\ 0 & -6 & -6 & -6 \\ 0 & 3 & 3 & 3 \end{pmatrix} \to \begin{pmatrix} 0 & \boxed{1} & 1 & 1 \\ 0 & 0 & 0 & 0 \\ 0 & 0 & 0 & 0 \\ 0 & 0 & 0 & 0 \end{pmatrix}.$$

The first column meets no "step corner," so x_1 is a free variable; the other free variables are x_3, x_4. Thus $K_2 = \ker(A - 6I)^2$ is

$$\begin{aligned} K_2 &= \{v = (a, -b-c, b, c) \, : \, a, b, c \in \mathbb{R}\} \\ &= \mathbb{R}\text{-span}\{f_1^{(2)} = e_1\, ,\, f_2^{(2)} = e_3 - e_2\, ,\, f_3^{(2)} = e_4 - e_2\} \\ &= \mathbb{R}\text{-span}\{e_1, e_3 - e_2, e_4 - e_2\} \end{aligned}$$

with $\dim(K_2) = 3$. Finally, $(A - 6I)^3 = 0$, so $\deg(T - 6I) = 3$ and $K_3 = V$.

Now apply the procedure described in Figure 1.1 to find the cyclic subspaces. In the current example, that figure takes the form shown in Figure 1.2.

STEP 1: Find a basis for V mod K_2. Since $\dim(V/K_2) = 1$, this is achieved by taking any $v \in V \sim K_2$. One such choice is $e_1^{(1)} = e_2 = (0, 1, 0, 0)$, which obviously is not in K_2. Then compute its images under powers of $(T - 6I)$:

$$\begin{aligned} e_2^{(1)} &= (T - 6I)e_1^{(1)} = (1, -2, 1, 1) = e_1 - 2e_2 + e_3 + e_4 \in K_2 \sim K_1, \\ e_3^{(1)} &= (T - 6I)^2 e_1^{(1)} = (3, 3, -6, 3) = 3 \cdot (e_1 + e_2 - 2e_3 + e_4) \in K_1 \sim \{0\}. \end{aligned}$$

STEP 2: There is no need to augment the vector

$$e_2^{(1)} = (T - 6I)e_1^{(1)} = (1, -2, 1, 1) \in K_2$$

to get a basis for K_2/K_1 because $\dim(K_2/K_1) = 1$.

STEP 3: In $K_1 \sim \{0\}$, we must augment

$$e_3^{(1)} = (T - 6I)^2 e_1^{(1)} = 3e_1 + 3e_2 - 6e_3 + 3e_4$$

to get a basis for $K_1/K_0 \cong K_1$. We need a new vector $e_1^{(3)} \in K_1 \sim \{0\}$ such that $e_1^{(3)}$ and $e_3^{(1)}$ are independent mod $K_0 = (0)$ — i.e., vectors that are actually independent in V. We could try $e_1^{(3)} = (-1,-1,0,1) = -e_1 - e_2 + e_4$, which is in $K_1 \sim K_0$. Independence holds if and only if the matrix M whose rows are $e_3^{(1)}$ and $e_1^{(3)}$ has rank $= 2$. Row operations yield

$$\begin{pmatrix} 1 & 1 & -2 & 1 \\ -1 & -1 & 0 & 1 \end{pmatrix} \to \begin{pmatrix} \boxed{1} & 1 & -2 & 1 \\ 0 & 0 & -2 & 2 \end{pmatrix}$$

which has row rank =2, as desired.

Thus $\{T^2(e_1^{(1)}), T(e_1^{(1)}), e_1^{(1)}\,;\, e_1^{(3)}\}$ is a basis for all of V such that

$$C_1 = \mathbb{R}\text{-span}\{T^2(e_1^{(1)}), T(e_1^{(1)}), e_1^{(1)}\},$$
$$C_2 = \mathbb{R}\text{-span}\{e_1^{(3)}\},$$

are independent cyclic subspaces generated by the vectors $e_1^{(1)}$ and $e_1^{(3)}$. The ordered basis $\mathfrak{Y} = \{T^2(e_1^{(1)}), T(e_1^{(1)}), e_1^{(1)}\,;\, e_1^{(3)}\}$ puts the matrix $[T]_{\mathfrak{Y}\mathfrak{Y}}$ in Jordan canonical form

$$[T]_{\mathfrak{Y}\mathfrak{Y}} = \left(\begin{array}{ccc|c} 6 & 1 & 0 & 0 \\ 0 & 6 & 1 & 0 \\ 0 & 0 & 6 & 0 \\ \hline 0 & 0 & 0 & 6 \end{array}\right).$$

The resulting pattern of cyclic subspaces and corresponding basis vectors is shown in Figure 1.2. The vectors

$$e_1^{(1)} = e_2 \quad \text{and} \quad e_1^{(3)} = -e_1 - e_2 + e_4$$

are the cyclic vectors; the basis vectors $T^2(e_1^{(1)})$ and $e_1^{(3)}$ in the bottom row are eigenvectors for T that span the eigenspace

$$E_{\lambda=6}(T) = \mathbb{K}\text{-span}\{e_1^{(3)}, T^2(e_1^{(3)})\},$$

which is 2-dimensional. That concludes our discussion of Example 1.61. ○

1.5. Consequences of the Jordan Canonical Form (JCF)

We now describe various results that emerge from the preceding analysis. More advanced topics involving the JCF will be examined in Chapter 2.

The General Spectral Mapping Theorem. Suppose A is an $n \times n$ matrix whose characteristic polynomial splits over $\mathbb{K}$, with

$$p_A(x) = \det(A - xI) = (-1)^n \prod_{i=1}^{r} (\lambda_i - xI)^{m_i} \quad \text{and} \quad \text{sp}(T) = \{\lambda_1, \dots, \lambda_r\}.$$

Examination of the diagonal entries in the Jordan canonical form (1.13)-(1.14), or even the upper triangular form (1.12), reveals that

$$\text{sp}(T^k) = \text{ the } \textit{distinct} \text{ entries in the list of powers } \lambda_1^k, \dots, \lambda_r^k.$$

Be aware that there might be repeated entries among the λ_i^k even though the λ_i are distinct.

Therefore, the characteristic polynomial of T^k is the product of the diagonal entries $(\lambda_i^k - x)$ in the Jordan form of $(T^k - xI)$ so that

$$p_{T^k}(x) = \det(T^k - xI) = \prod_{i=1}^{r}(\lambda_i^k - x)^{m_i} \quad \text{(where } m_i = \dim(M_{\lambda_i}(T))\text{).}$$

More can be said under these hypotheses. If

$$f(t) = a_0 + a_1 t + a_2 t^2 + \cdots + a_m t^m$$

is any polynomial in $\mathbb{K}[t]$, we can form an operator $f(T) : V \to V$ to obtain a natural correspondence Φ from $\mathbb{K}[x]$ into the space of linear operators $\mathrm{Hom}_{\mathbb{K}}(V, V)$ such that

$$\Phi(\mathbb{1}) = I, \quad \Phi(t) = T, \quad \Phi(t^k) = T^k,$$

and

$$\begin{aligned} \Phi(f_1 + f_2) &= \Phi(f_1) + \Phi(f_2) && \text{(sum of linear operators),} \\ \Phi(f_1 \cdot f_2) &= \Phi(f_1) \circ \Phi(f_2) && \text{(composition of linear operator),} \\ \Phi(cf) &= c \cdot \Phi(f) && \text{for all } c \in \mathbb{K}, \end{aligned}$$

i.e., Φ is a homomorphism between these associative algebras. With this in mind we can prove the following theorem.

THEOREM 1.62 (Spectral Mapping Theorem). *If the characteristic polynomial of a linear operator $T : V \to V$ splits over $\mathbb{K}$, and in particular if $\mathbb{K} = \mathbb{C}$, every polynomial $f(t) = \sum_{i=0}^{m} a_i t^i$ in $\mathbb{K}[t]$ determines an operator $\Phi(f)$ in $\mathrm{Hom}_{\mathbb{K}}(V, V)$,*

$$\Phi(f) = \sum_{i=0}^{m} a_i T^i.$$

The correspondence $\Phi : \mathbb{K}[t| \to \mathrm{Hom}_{\mathbb{K}}(V)$ is a homomorphism of associative algebras that sends $\mathbb{1}$ to I and has the following "spectral mapping property":

$$\mathrm{sp}(f(T)) = f\big(\mathrm{sp}(T)\big) = \{f(z) : z \in \mathrm{sp}(T)\}. \tag{1.15}$$

In particular, this applies if T is diagonalizable over $\mathbb{K}$.

PROOF. It suffices to choose a basis $\mathfrak{X}$ such that $[T]_{\mathfrak{X}\mathfrak{X}}$ has block upper-triangular form

$$[T]_{\mathfrak{X}\mathfrak{X}} = \begin{pmatrix} \boxed{T_1} & & & 0 \\ & \ddots & & \\ & & \ddots & \\ 0 & & & \boxed{T_r} \end{pmatrix} \quad \text{with} \quad T_i = \begin{pmatrix} \lambda_i & & & * \\ & \ddots & \ddots & \\ & & \ddots & \ddots \\ 0 & & & \lambda_i \end{pmatrix}_{m_i \times m_i}.$$

Since $[T^k]_{\mathfrak{X}\mathfrak{X}} = [T]_{\mathfrak{X}\mathfrak{X}}^k$, we get

$$[f(T)]_{\mathfrak{X}\mathfrak{X}} = f\big([T]_{\mathfrak{X}\mathfrak{X}}\big) = a_0 I + a_1 [T]_{\mathfrak{X}\mathfrak{X}} + \cdots + a_m [T]_{\mathfrak{X}\mathfrak{X}}^m$$

because $f(A) = a_0 I + a_1 A + \cdots + a_m A^m$.

As in the previous corollary, $[f(T)]_{\mathfrak{X}\mathfrak{X}}$ is made up of blocks on the diagonal, each of which is upper-triangular with diagonal values $f(\lambda_i)$; then the characteristic polynomial of $f(T)$ is

$$p_{f(T)}(t) = \det(f(T) - t\,I) = \prod_{i=1}^{r} (f(\lambda_i) - t\,I)^{m_i}, \quad m_i = \dim(M_{\lambda_i}).$$

This is zero if and only if $t = f(\lambda_i)$ for some i, so

$$\mathrm{sp}(f(T)) = \{f(\lambda_i) : 1 \le i \le r\} = f(\mathrm{sp}(T)).$$

Obviously, the characteristic polynomial of $f(T)$ splits over $\mathbb{K}$ if $p_T(t)$ splits.

Here, $\lambda_i \in \mathrm{sp}(T) \Rightarrow f(\lambda_i) \in \mathrm{sp}(f(T))$, but the multiplicity of $f(\lambda_i)$ as an eigenvalue of $f(T)$ might be greater than the multiplicity of λ_i as an eigenvalue of T because we might have $f(\lambda_i) = f(\lambda_j)$ for some $i \neq j$, and then $\mu = f(\lambda_i)$ will have multiplicity at least $m_i + m_j$ in $\mathrm{sp}(f(T))$. □

Cayley-Hamilton Theorem. Another consequence of the Jordan canonical form is the *Cayley-Hamilton theorem*. This can be proved in other ways without developing the Jordan form, but the JCF does suggest the underlying reason that the result is true and makes its validity almost obvious. On the other hand, alternative proofs can be made to work for arbitrary $\mathbb{K}$ and T without any assumptions about the characteristic polynomial $p_T(x)$. Since the result is true in this generality, we give both proofs.

THEOREM 1.63 (Cayley-Hamilton). *For any linear operator $T : V \to V$ on a finite-dimensional vector space over any $\mathbb{K}$, we have*

$$p_T(T) = \left[\, p_T(t)\big|_{t=T} \right] = \mathbf{0} \quad (\textit{the zero operator in } \mathrm{Hom}_{\mathbb{K}}(V,V)),$$

so applying the characteristic polynomial $p_T(t) = \det(T - t\,I)$ to the linear operator T itself yields the zero operator on V.

PROOF. If $p_T(x)$ splits over $\mathbb{K}$, we have $p_T(x) = \prod_{i=1}^{r} (\lambda_i - x)^{m_i}$, where $m_i = \dim(M_{\lambda_i})$ and $\{\lambda_1, \dots, \lambda_r\}$ are the distinct eigenvalues in $\mathrm{sp}_{\mathbb{K}}(T)$. We want to show that

$$\mathbf{0} = \prod_{i=1}^{r} (T - \lambda_i I)^{m_i} = \left[p_T(x)\big|_{x=T} \right].$$

But $V = \bigoplus_{i=1}^{r} M_{\lambda_i}$ and $(T - \lambda_i I)^{m_i}(M_{\lambda_i}) = (0)$. (Given a Jordan basis in M_{λ_i}, $A = [(T - \lambda_i I)|_{M_{\lambda_i}}]_{\mathfrak{X}\mathfrak{X}}$ consists of elementary nilpotent blocks N_j on the diagonal; the size $d_j \times d_j$ of such a block cannot exceed $m_i = \dim(M_{\lambda_i})$, so $N_j^{d_j} = N_j^{m_i} = 0$ for each j.) Hence

$$\prod_{j=1}^{r} (T - \lambda_i I)^{m_i}(M_{\lambda_i}) = (0),$$

so the operator $p_T(T)$ is zero on each M_{λ_i} and on all of V. □

If p_T does not split over $\mathbb{K}$, a different argument will show that $p_T(T)v = 0$ for all $v \in V$ whether or not $p_T(x)$ splits. We defer that for a moment to post an important consequence of Cayley-Hamilton.

COROLLARY 1.64. *If $p_T(x)$ is the characteristic polynomial of a linear operator $T : V \to V$ and $m_T(x)$ is the minimal polynomial for T (the monic polynomial $m(x)$ of smallest degree such that $m(T) = \mathbf{0}$), then $m_T(x)$ divides $p_T(x)$ so that $p_T(x) = m_T(x) \cdot r(x)$ for some $r(x) \in \mathbb{K}[x]$.*

PROOF. We have $p_T(T) = \mathbf{0}$ by Cayley-Hamilton, Theorem 1.63, and by minimality of $m_T(x)$, we have $\deg(m_T) \leq \deg(p_T)$. By Euclidean long division of polynomials, we can then write

$$p_T(x) = q(x) \cdot m_T(x) + r(x),$$

where the remainder $r(x)$ is either zero or $\deg(r) < \deg(m_T)$. But $r(T) = p_T(T) - m_T(T) = \mathbf{0}$ too, which is impossible if $r(x)$ is nonconstant; we conclude that $r(T) = \mathbf{0}$ and m_T divides p_T, as claimed. □

Alternative Proof of Cayley-Hamilton. The result is obvious if $v = 0$. When $v \neq 0$, there is a largest $m \geq 1$ such that $v, T(v), T^2(v), \dots, T^{m-1}(v)$ are linearly independent. Then

$$W = \mathbb{K}\text{-span}\{T^k(v) : k \in \mathbb{N}\} = \mathbb{K}\text{-span}\{T^k(v) : 0 \leq k \leq m-1\}$$

and $\mathfrak{X} = \{v, T(v), \dots, T^{m-1}(v)\}$ is a basis for the cyclic subspace W, which is clearly T-invariant. As we saw in Corollary 1.47, $p_{(T|_W)}$ divides p_T so that $p_T(x) = q(x) \cdot p_{(T|_W)}(x)$ for some $q(x) \in \mathbb{K}[x]$. Now compute $p_{(T|_W)}(x)$. For the basis $\mathfrak{X}$, $T^m(v) = T(T^{m-1}(v))$ is a unique linear combination of the vectors $T^k(v)$ with $k < m$, say

$$T^m(v) + a_{m-1}T^{m-1}(v) + \cdots + a_1T(v) + a_0v = 0.$$

Hence, the matrix of $T|_W$ is

$$[T|_W]_{\mathfrak{X}\mathfrak{X}} = \begin{pmatrix} 0 & & & 0 & -a_0 \\ 1 & \ddots & & & -a_1 \\ & \ddots & \ddots & & \vdots \\ & & \ddots & \ddots & \vdots \\ 0 & & & 1 & -a_{m-1} \end{pmatrix},$$

and

$$[(T - xI)|_W]_{\mathfrak{X}\mathfrak{X}} = \begin{pmatrix} -x & & & 0 & -a_0 \\ 1 & \ddots & & & -a_1 \\ & \ddots & \ddots & & \vdots \\ & \ddots & 1 & -x & \vdots \\ 0 & & & 1 & -x - a_{m-1} \end{pmatrix}.$$

EXERCISE 1.65. Show that

$$\begin{aligned} p_{(T|_W)}(t) &= \det(T|_W - t\,I_W) \\ &= (-1)^m(t^m + a_{m-1}t^{m-1} + \cdots + a_1t + a_0). \ \square \end{aligned}$$

It follows that $p_{(T|_W)}(T)$ satisfies the equation

$$p_{(T|_W)}(T)v = (-1)^m\big(T^m(v) + a_{m-1}T^{m-1}(v) + \cdots + a_1T(v) + a_0v\big) = 0$$

by definition of the coefficients $\{a_j\}$. But then

$$p_T(T)v = q(T)\big(p_{(T|_W)}(T)\cdot v\big) = q(T)(0) = 0.$$

(Recall that $W = \mathbb{K}\text{-span}\{T^k(v)\}$ as in Proposition 1.11.) Since this is true for all $v \in V$, $p_T(T)$ is the zero operator in $\text{Hom}_{\mathbb{K}}(V, V)$, proving Cayley-Hamilton.

Further Examples of Jordan Form.

EXERCISE 1.66. Prove that the Jordan form of $L_A : \mathbb{C}^2 \to \mathbb{C}^2$ is

$$\text{(a)}\ \begin{pmatrix} 0 & 1 \\ 0 & 0 \end{pmatrix}\quad \text{if } A = \begin{pmatrix} 0 & 2 \\ 0 & 0 \end{pmatrix} \qquad \text{(b)}\ \begin{pmatrix} 1 & 1 \\ 0 & 1 \end{pmatrix}\quad \text{if } A = \begin{pmatrix} 1 & 2 \\ 0 & 1 \end{pmatrix}.$$

In each case, produce a basis $\mathfrak{Y}$ that puts $[L_A]_{\mathfrak{Y}\mathfrak{Y}}$ into Jordan form $J(A)$ and find an intertwiner S such that $SAS^{-1} = J(A)$.

HINT. Begin by determining the generalized eigenspaces $M_\lambda(L_A)$. Then use the procedure of Theorem 1.16 to find cyclic subspace decompositions.

EXERCISE 1.67. For any $A \in \text{M}(2, \mathbb{C})$, prove that its JCF must assume one of the following forms:

$$\text{(a)}\quad J(A) = \begin{pmatrix} a & 0 \\ 0 & b \end{pmatrix} \qquad \text{(b)}\quad J(A) = \begin{pmatrix} c & 1 \\ 0 & c \end{pmatrix}.$$

HINT. If the eigenvalues are distinct, A is diagonalizable. (Why?) If not, the characteristic polymomial must have a double root in $\mathbb{C}$, so $p_A(x) = (x - \lambda)^2$. In view of the results of this chapter, what does that tell you about the pattern of generalized eigenspaces and about $J(A)$?

EXERCISE 1.68. In some accounts, elementary nilpotent matrices are written with the off-diagonal "1's" *below* the diagonal instead of above it. To see why this doesn't really matter in discussing cyclic subspace decompositions and the JCF, show that there is a matrix S such that

$$S\begin{pmatrix} 0 & 0 \\ 1 & 0 \end{pmatrix}S^{-1} = \begin{pmatrix} 0 & 1 \\ 0 & 0 \end{pmatrix}.$$

In particular, both have the same JCF. The same idea works for elementary nilpotent matrices of any size.

EXERCISE 1.69. If

$$A = \begin{pmatrix} 1 & 1 \\ 0 & 1 \end{pmatrix} \quad \text{and} \quad B = \begin{pmatrix} 1 & 2 \\ 0 & 1 \end{pmatrix},$$

verify by direct calculations that we cannot have $SAS^{-1} = B$ for any invertible $S = [a, b; c, d]$.

EXERCISE 1.70. Identify all pairs in the following list of matrices in $M(2, \mathbb{C})$ that are similar, with $B = SAS^{-1}$ for some S, and hence have the same Jordan form. Explain your reasoning.

$$\text{(a) } A = \begin{pmatrix} 1 & 0 \\ 0 & -1 \end{pmatrix} \quad \text{(b) } B = \begin{pmatrix} 1 & 1 \\ 0 & 1 \end{pmatrix}$$

$$\text{(c) } C = \begin{pmatrix} 2 & 2 \\ 0 & 1 \end{pmatrix} \quad \text{(d) } D = \begin{pmatrix} 0 & 1 \\ -1 & 0 \end{pmatrix}$$

EXERCISE 1.71. Let $T = L_A : \mathbb{C}^3 \to \mathbb{C}^3$ be determined by the real matrix

$$A = \begin{pmatrix} 11 & -4 & -5 \\ 21 & -8 & -4 \\ 3 & -1 & 0 \end{pmatrix}.$$

(a) Verify that the characteristic polynomial is

$$p_A(x) = -x^3 + 3x^2 - 7x - 164$$

and has both real and complex roots

$$\lambda_1 = -1, \quad \lambda_\pm = 2 \pm i\sqrt{7}$$

(b) Find basis vectors in $V = \mathbb{C}^3$ for each eigenspace.

(c) What can be said about the generalized eigenspaces $M_\lambda(T)$?

(d) What is the Jordan form for T?

HINT. In (a), trial and error reveals that the cubic characteristic polynomial $p_T(x)$ has $\lambda_1 = -1$ as a real root; then long divide by $x + 1 = x - (-1)$.

EXERCISE 1.72. Find the Jordan canonical form for the operators T on $\mathbb{R}^n$ whose matrices with respect to the standard basis $\mathfrak{X} = \{e_1, \ldots, e_n\}$ are

$$\text{(a)} \quad A = \begin{pmatrix} 5 & -6 & -6 \\ -1 & 4 & 2 \\ 3 & -6 & -4 \end{pmatrix} \quad \text{(b)} \quad B = \begin{pmatrix} 1 & 0 & 0 & 0 & 1 & 0 \\ 0 & 1 & 1 & 1 & 0 & -1 \\ 0 & 0 & 1 & 0 & 0 & 0 \\ 0 & 0 & 0 & 1 & 0 & 0 \\ 0 & 0 & 1 & 0 & 1 & -1 \\ 0 & 0 & 0 & 0 & 0 & 1 \end{pmatrix}$$

EXERCISE 1.73. If $\mathfrak{X}$ is the standard basis in $V = \mathbb{C}^3$, find a basis $\mathfrak{Y}$ that puts $L_A : V \to V$ into Jordan form if

$$A = \begin{pmatrix} -3 & 3 & -2 \\ -7 & 8 & -3 \\ 1 & -1 & 2 \end{pmatrix}.$$

Then find a transition matrix Q such that

$$[L_A]_{\mathfrak{Y}\mathfrak{Y}} = QAQ^{-1} = Q\,[L_A]_{\mathfrak{X}\mathfrak{X}}Q^{-1} = QAQ^{-1}.$$

HINT. In Theorem 2.47 of LA I, we described the effect of a change of basis on a matrix realization $[T]_{\mathfrak{X}\mathfrak{X}}$ of a linear operator T on V, and we showed how to find a matrix Q that effects the transition once bases $\mathfrak{X}$ and $\mathfrak{Y}$ have been specified. It is often important to know the transition matrix that does the job as well as the diagonalized form of T.

In Section 2.2 of the next chapter, we will take a very different approach to finding these "intertwining operators" Q which suggests how one might find *all* such Q (these intertwiners are not unique).

EXERCISE 1.74. Let $V = \mathbb{R}\text{-span}\{1, t, t^2, e^t, te^t\}$ in the infinite-dimensional space of $\mathcal{C}^\infty$ functions $f : \mathbb{R} \to \mathbb{R}$, and let $D = \dfrac{d}{dt}$.

(a) Prove that D leaves V invariant.

(b) Find the Jordan form for $D : V \to V$ and a basis that puts $[D]_{\mathfrak{X}\mathfrak{X}}$ into this form.

Finally, we pose a problem that would be hard to resolve without appealing to the JCF. It has infinitely many solutions, but they break into a *finite number* of distinct similarity classes. So the problem reduces to making a finite list that contains one representative from each similarity class in the full set of solutions. Individual similarity classes are infinite, but once you know one representative, it is easy to find the others in the same class.

EXERCISE 1.75 (Finding $\sqrt{-I}$ in Matrix Space). Use the Jordan canonical form to find a list of solutions $A_1, \dots, A_r \in \mathrm{M}(3, \mathbb{C})$ to the matrix identity

$$A^2 + I = 0 \quad \left(\text{the "square roots" of } -I_{2\times 2}\right)$$

such that every solution is obtained from one of the A_i via a similarity transformation $A \to SA_kS^{-1}$.

HINT. A matrix in $\mathrm{M}(n, \mathbb{C})$ is similar to its Jordan form $J(A)$ and $SA^mS^{-1} = (SAS^{-1})^m$ for all $m \in \mathbb{N}$, so we have $A^2 = -I \Leftrightarrow$ the same is true of the Jordan form $J(A)$. Thus, it suffices to find all JCF such that $J(A)^2 = -I$.

Jordan Form *J(T)* as a Similarity Invariant. The Jordan form (JCF) of a linear operator T on a complex vector space is not unique. To construct it, we must find cyclic subspace decompositions of the geneneralized eigenspaces $M_\lambda(T)$, and as we pointed out earlier in discussing nilpotent operators, the subspaces appearing in such a decomposition $M_\lambda = \bigoplus_j Q_{\lambda,j}$ are not unique. Nevertheless, in any Jordan form $J(T) = [T]_{\mathfrak{X}\mathfrak{X}}$, the number of Jordan blocks, their sizes, and their diagonal entries (eigenvalues of T) are indeed unique, even if the particular subspaces in V that implement the decompositions are not.

Since the JCF of a linear operator T is obtained by constructing a particular basis $\mathfrak{X}$ in V, all Jordan forms $J(T), J'(T)$ are related to each other and to T by similarity transformations $J'(T) = SJ(T)S^{-1}$, so the JCF are "*equal modulo a similarity.*" Likewise, using the same terminology, all matrix realizations $[T]_{\mathfrak{X}\mathfrak{X}}$ of a given linear map T are "*equal modulo a similarity.*"

The Minimal Polynomial $m_T(x)$ for $T : V \to V$.

DEFINITION 1.76. The space of linear operators $\text{Hom}_{\mathbb{K}}(V,V) \cong \text{M}(n,\mathbb{K})$ is finite-dimensional, so there is a largest exponent m such that $I, T, T^2, \dots, T^{m-1}$ are linearly independent. Thus there are coefficients c_j such that

$$T^m + \sum_{k=0}^{m-1} c_k T^k = \mathbf{0} \quad \text{(the zero operator).}$$

The monic polynomial of minimum degree such that

$$m_T(T) = T^m + \sum_{k=0}^{m-1} c_k T^k = \mathbf{0}$$

is the *minimal polynomial* $m_T(x)$ for T, and it is unique.

EXERCISE 1.77. Why does the definition of the minimal polynomial m_T determine a *unique* polynomial in $\mathbb{K}[x]$?

HINT. The degree of any minimal polynomial is uniquely determined, and it is required to be *monic* (leading term x^m). What can then be said about the difference $p_1(x) - p_2(x)$ of two proposed minimal polynomials?

Obviously $m = \deg(m_T)$ cannot exceed $n^2 = \dim\big(\text{M}(n,\mathbb{K})\big)$, but it could be a lot smaller. The minimal polynomial $m_A(x)$ for a matrix $A \in \text{M}(n,\mathbb{K})$ is defined the same way, and the minimal polynomial $m_T(x)$ for a linear operator is the same as the minimal polynomial of the associated matrix $A = [T]_{\mathfrak{X}\mathfrak{X}}$ for every basis $\mathfrak{X}$ in V. The nilpotence degree $\deg(N)$ of a nilpotent matrix will be greater than the degree of its minimal polynomial unless there is a cyclic vector in V.

We have already seen in LA I that the characteristic polynomial is a "similarity invariant" for matrices (and for linear operators), so a similarity transformation $A \mapsto SAS^{-1}$ always yields a matrix with the same characteristic polynomial with

$$p_{SAS^{-1}}(x) = p_A(x) \quad \text{in } \mathbb{K}[x]$$

for all invertible matrices S. (See LA I, Section 2.4, for details about similarity transformations, and Section 5.1 for invariance of the characteristic polynomial.) The minimal polynomial $m_T(x)$ is also a similarity invariant.

EXERCISE 1.78. Show that the minimal polynomial $m_T(x)$ of a linear operator $T : V \to V$ is a similarity invariant.

(a) Given similar matrices A, A' with $A' = SAS^{-1}$, explain why they have the same minimal polynomials in $\mathbb{K}[x]$.
(b) Explain why the minimal polynomial $m_A(x)$ of the matrix $[T]_{\mathfrak{X}\mathfrak{X}}$ is the same for all bases $\mathfrak{X}$ in V.

HINT. In (a), we already know that $SA^kS^{-1} = \big(SAS^{-1}\big)^k$ for all k; show that $S(A_1 + \cdots + A_k)S^{-1} = \sum_j SA_jS^{-1}$ and hence $Sf(A)S^{-1} = f(SAS^{-1})$ for all $f \in \mathbb{K}[x]$. In (b), the effect of change-of-basis on a matrix $[T]_{\mathfrak{X}\mathfrak{X}}$ is mediated by a similarity transformation.

EXERCISE 1.79. Given Jordan blocks

$$B_k = \begin{pmatrix} \lambda & 1 & & 0 \\ & \ddots & \ddots & \\ & & \ddots & 1 \\ 0 & & & \lambda \end{pmatrix}_{k\times k} \qquad (\lambda \in \mathbb{K}),$$

(a) What is the characteristic polynomial $p_A(x)$ of B_k?

(b) What is the minimal polynomial $m_A(x)$ of a Jordan block $A = B_k$ when $\lambda = 0$? When $\lambda \neq 0$?

(c) Do two Jordan blocks B_k, B_ℓ have the same minimal polynomials $\Leftrightarrow$ they are equal (same size and diagonal values)?

We will use Example 1.61 to illustrate the connections between the minimal polynomial $m_T(x)$, the characteristic polynomial $p_T(x) = \det(T - xI)$, and the Jordan form when p_T splits over $\mathbb{K}$. (These results remain true, but with more complicated proofs, if p_T fails to split. We won't go into the general proof in these *Notes*.)

EXAMPLE 1.80. If the characteristic polynomial of $T : V \to V$ splits over the ground field $\mathbb{K}$, let $\lambda_1, \dots, \lambda_s$ be its roots. The generalized eigenspaces are $M_\lambda(T) = \ker(T - \lambda I)^d$ where $d = d(\lambda)$ is the nilpotence degree of $(T - \lambda I)$ restricted to its stable kernel

$$K_\infty(T - \lambda I) = \bigcup_{j=1}^{\infty} \ker\,(T - \lambda I)^k,$$

on which $(T - \lambda I)$ is nilpotent. With respect to a suitable basis $\mathfrak{X}$ aligned with the subspaces M_λ, the restriction of T assumes Jordan form

$$[T|_{M_\lambda}]_{\mathfrak{X}\mathfrak{X}} = \operatorname{diag}(B_1, \dots, B_s)$$

in which the Jordan blocks

$$B_i = (\lambda I_{d_i \times d_i} + N_i) \quad (N_i \text{ a } d_i \times d_i \text{ elementary nilpotent matrix})$$

correspond to the action of T on the invariant subspaces $Q_{\lambda,i} \subseteq M_{\lambda,i}$ in a cyclic subspace decomposition of $(T - \lambda I)|_{M_\lambda}$. The block size d_i is the nilpotence degree of $(T - \lambda I)|_{Q_{\lambda,i}}$.

In discussing Cayley-Hamilton, Theorem1.63, we showed that:

> *Any polynomial $f(x) \in \mathbb{K}[x]$ such that $f(T) = \mathbf{0}$ must divide the characteristic polynomial $p_T(x)$, so $p_T(x) = f(x)\cdot g(x)$ for some $g(x) \in \mathbb{K}[x]$ with deg(g) $<$ deg $p_T(x) = \dim(V)$.*

By hypothesis, $p_T(x)$ splits as a product of linear factors $\prod_{j=1}(x-\mu_j)^{m_j}$, where the μ_j are the distinct roots of $p_T(x)$ in $\mathbb{K}$ with multiplicities m_j. Each factor $(x - \mu_j 1)$ is an *irreducible* polynomial in the sense that it cannot be written as a product of nonconstant polynomials of lower degree. We now cite a fundamental result valid for all polynomials with coefficients in a field.

UNIQUE FACTORIZATION THEOREM. *Every nonconstant polynomial $f(x) \in \mathbb{K}[x]$ has a factorization into irreducible nonconstant polynomials $f(x) = \prod_{i=1}^{k} p_i(x)$. Furthermore, except for irrelevant constant scale factors, the irreducibles $p_i(x)$ and the number of times each appears are unique.*

By definition we have $m_T(T) = \mathbf{0}$ for the minimal polynomial of T, so by Corollary 1.64 $p_T(x) = m_T(x) \cdot r(x)$. By unique factorization, the irreducible factors in $m_T(x)$ must appear appear among the irreducible factors $(x - \lambda_i)$ of $p_T(x)$, so $m_T(x)$ can only have the form

$$m_T(x) = \prod_{i=1}^{s}(x - \lambda_i \mathrm{I})^{q_i} \quad \text{with exponents } 0 \le q_i \le m_i,$$

the $m_i = \dim(M_{\lambda_i})$ being the exponents appearing in $p_T(x)$. In fact, the q_i must also be nonzero:

LEMMA 1.81. *If $p_T(x)$ splits over $\mathbb{K}$, the roots of $p_T(x) = \prod_{j=1}^{r}(x - \lambda_j)^{m_j}$ are the same as those of the minimal polynomial $m_T(x)$.*

PROOF. By Corollary 1.64, $m_T(x)$ divides p_T, so $p_T = m_T(x) \cdot q(x)$ and the roots of p_T include those of m_T. The roots of p_T are the distinct eigenvalues $\{\lambda_1, \dots, \lambda_r\}$ of T, so we must show that every eigenvalue of T is a root of $m_T(x)$. In fact, if $v \neq 0$ is an eigenvector, say with $T(v) = \lambda_i v$, then $m_T(T) = m_T(x)|_{x=T} = \mathbf{0}$, and by the spectral mapping theorem, Theorem 1.62, we have $q(T)v = q(\lambda_i) \cdot v$ for any $q \in \mathbb{K}[x]$. Hence

$$0 = m_T(T) \cdot v = m_T(\lambda_i) \cdot v.$$

Since $v \neq 0$, we must have $m_T(\lambda_i) = 0$ so λ_i is a root of the minimal polynomial, as claimed. □

Additional Exercises

Section 1.1. Basic Definitions and Examples.

1. TRUE/FALSE QUESTIONS ("True" if the statement is *always* true.)
   (a) If $W \subseteq V$ is a cyclic subspace under the action of a linear operator T, the cyclic vector v_0 is unique up to a scale factor.
   (b) The matrix
   $$A = \begin{pmatrix} 5 & -6 & -6 \\ -1 & 4 & 2 \\ 3 & -6 & 4 \end{pmatrix}$$
   is nilpotent.
   (c) If $\mathfrak{X} = \{e_1, \dots, e_m\}$ is a basis for a vector space V and $T : V \to V$ a linear operator, then T is nilpotent if there exist exponents $n_i \in \mathbb{N}$ such that $T^{n_i}(e_i) = 0$ for each basis vector e_i.
   (d) If $A \in \mathrm{M}(n, \mathbb{K})$ and $\det(A) \neq 0$, the Fitting decomposition of $T = L_A : \mathbb{K}^n \to \mathbb{K}^n$ is trivial, with stable kernel $K_\infty(T) = (0)$ and stable range $R_\infty(T) = \mathbb{K}^n$.

(e) If $T : V \to V$ is nonsingular, its stable kernel $K_\infty(T)$ is trivial.

(f) If a scalar $\lambda \in \mathbb{C}$ is a root of the characteristic polynomial $p_T(t) = \det(T - tI)$ for a linear operator T on a finite-dimensional space V, then $(T - \lambda I)$ is nilpotent.

(g) A scalar $\lambda \in \mathbb{K}$ is a generalized eigenvalue for a linear operator T on a finite-dimensional vector space $V \Leftrightarrow \ker(T - \lambda I) \neq (0)$.

(h) For a nilpotent linear operator $T : V \to V$ on a finite-dimensional vector space, $\lambda = 0$ is the only generalized eigenvalue.

2. If a nilpotent linear operator T on a vector space has a cyclic vector v_0, we know that the nonzero iterates $\{v, T(v_0), \ldots, T^{n-1}(v_0)\}$ with $T^n(v_0) = 0$ are a basis $\mathfrak{X}$ for V. When these are listed in reverse order

$$e_1 = T^{n-1}(v_0), \ldots, e_{n-1} = T(v_0), e_n = v_0,$$

we have

$$0 \overset{T}{\leftarrow} e_1 \overset{T}{\leftarrow} e_2 \overset{T}{\leftarrow} \cdots \overset{T}{\leftarrow} e_{n-1} \overset{T}{\leftarrow} e_n,$$

so $\mathfrak{X}$ is a basis that puts T into elementary nilpotent form

$$[T]_{\mathfrak{X}\mathfrak{X}} = \begin{pmatrix} 0 & 1 & 0 & & 0 \\ 0 & 0 & 1 & & \\ & \vdots & 0 & \ddots & \\ 0 & & & \ddots & 1 \\ 0 & 0 & \ldots & & 0 \end{pmatrix}_{n\times n}.$$

(a) Can there be proper subspaces $(0) \subsetneqq W \subsetneqq V$ that are cyclic under the actions of T?

(b) What is the nilpotence degree of a matrix $A = \operatorname{diag}(N_1, \ldots, N_r)$ consisting of elementary nilpotent blocks of sizes $d_i \times d_i$?

(c) Can there be proper T-invariant cyclic subspaces $(0) \subsetneqq W_1, W_2 \subsetneqq V$ that decompose $V = W_1 \oplus W_2$?

(d) Is the matrix $[T]_{\mathfrak{X}\mathfrak{X}}$ a Jordan form for T?

Explain your answers or provide simple counterexamples.

3. Let $T : V \to V$ be a linear operator on a finite-dimensional vector space and W the cyclic subspace generated by the action of T on a vector $x \neq 0$. If $\dim(W) = k$, we have shown that $\{x, T(x), \ldots, T^{k-1}(x)\}$ is a basis for W and there are coefficients such that

$$T^k(x) = -a_0 x - a_1 T(x) - \cdots - a_{k-1} T^{k-1}(x).$$

Prove that the characteristic polynomial of the restriction $T|_W$ is

$$p_{T|_W}(t) = (-1)^k\big(a_0 + a_1 t + \cdots + a_{k-1} t^{k-1}\big).$$

HINT. Expansion by minors might work.

4. For the matrices

$$\text{(i) } A = \begin{pmatrix} 0 & 1 & 0 & 1 \\ 0 & 0 & 1 & 0 \\ 0 & 0 & 0 & 1 \\ 0 & 0 & 0 & 0 \end{pmatrix} \quad \text{(ii) } A = \begin{pmatrix} -1 & 2 & 1 \\ 1 & 2 & -1 \\ 1 & 2 & 1 \end{pmatrix}$$

(a) What are the dimensions of the stable range $R_\infty(L_A)$ and stable kernel $K_\infty(L_A)$?

(b) Find bases for each of these subspaces in the Fitting decomposition

$$L_A = (L_A|_{K_\infty}) \oplus (L_A|_{R_\infty}).$$

5. If $T : \mathbb{R}^5 \to \mathbb{R}^5$ is the linear transformation such that

$$T(\mathbf{x}) = (0, x_3 + x_4, 0, x_3, x_1 + x_4),$$

find a basis for the cyclic subspace W spanned by the iterates

$$\mathbf{x}, T(\mathbf{x}), T^2(\mathbf{x}), \dots$$

when $\mathbf{x} = \mathbf{e}_1 + \mathbf{e}_4 = (1, 0, 0, 1, 0)$.

6. Let $T : \mathbb{R}^5 \to \mathbb{R}^5$ be the linear transformation such that

$$T(x_1, x_2, x_3, x_4, x_5) = (0, 0, x_2 + x_4, x_1 + x_2, x_1),$$

whose matrix with respect to the standard basis is

$$\begin{pmatrix} 0 & 0 & 0 & 1 & 1 \\ 0 & 0 & 1 & 1 & 0 \\ 0 & 0 & 0 & 0 & 0 \\ 0 & 0 & 1 & 0 & 0 \\ 0 & 0 & 0 & 0 & 0 \end{pmatrix}.$$

Verify that T is nilpotent and find an "implicit" description of each kernel in the chain

$$V = K_d = \ker(T^d) \supsetneqq K_{d-1} = \ker(T^{d-2}) \supsetneqq \cdots \supsetneqq K_1 = \ker(T) \supsetneqq (0)$$

by specifying, for each j, a system of linear equations in $x_1, \dots, x_5$ that is satisfied $\Leftrightarrow \mathbf{x} \in \mathbb{R}^5$ is in K_j.

Finding Bases for Nilpotent T. Given a nilpotent operator on $\mathbb{R}^n$, we can construct a basis for V that is "aligned" with the kernels K_j via an inductive "bottom-up" procedure employing ideas developed in Section 1.5 of the LA I *Notes*.

- Find a basis $\mathfrak{X}_1$ for $K_1 = \ker(T)$ by any method.
- Find a basis for $K_2 = \ker(T^2)$ by adjoining suitably chosen standard basis vectors $\mathbf{e}_1, \dots, \mathbf{e}_5$ to get a basis

$$\mathfrak{X}_2 = \mathfrak{X}_1 \cup (\text{adjoined vectors}).$$

- Continue until $K_d = V$ (d = nilpotence degree of T), and we arrive at a basis $\mathfrak{X} = \mathfrak{X}_d$ for V.

If we order these basis vectors as $\mathbf{v}_1, \dots, \mathbf{v}_n$, we can partition the indices $I = \{1, 2, \dots, n\}$ into consecutive disjoint blocks $I_1 \cup \cdots \cup I_r$ such that

$$\begin{aligned} \mathfrak{X}_1 &= \{\mathbf{v}_j : j \in I_1\} \\ \mathfrak{X}_2 &= \{\mathbf{v}_j : j \in I_1 \cup I_2\} \\ &\vdots \end{aligned}$$

and $K_j = \mathbb{R}\text{-span}\{\mathfrak{X}_j\}$.

7. Applying the preceding remarks to the linear operator $L_A : \mathbb{R}^5 \to \mathbb{R}^5$ introduced in Additional Exercise 6.

(a) Find a basis $\mathfrak{X}$ for $V = \mathbb{R}^5$ aligned with the kernels of the operator L_A. Write out the corresponding matrix $[L_A]_{\mathfrak{X}\mathfrak{X}}$.

(b) Explain why $\mathfrak{Y}_i = \{\mathbf{v}_j : j \in I_i\}$ are representatives for a basis $\overline{v}_j = \pi(v_j)$ in K_i/K_{i-1} (so $|I_i| = \dim(K_i/K_{i-1})$.

Section 1.2. Structure of Nilpotent Operators

1. TRUE/FALSE QUESTIONS ("True" if the statement is *always* true.)

(a) The differentiation operator D^n with $D = \dfrac{d}{dx}$ is nilpotent on the space of polynomials $p \in \mathbb{R}[x]$ of degree $\leq n-1$.

(b) Any constant-coefficient differential operator

$$L = a_0 I + a_1 D + a_2 D^2 + \cdots + a_n D^n \quad \left(D = \frac{d}{dx}\right)$$

is nilpotent on the space of polynomials of degree $\leq n-1$.

(c) A matrix A is nilpotent if and only if its transpose A^{T} is nilpotent.

(d) If $A = \operatorname{diag}\{N_1, \dots, N_r\}$ consists of $d_1 \times d_i$ elementary nilpotent blocks, the nilpotence degree $\deg(L_A)$ of A is $d = d_1 + \cdots + d_r$.

(e) A finite sum $T_1 + \cdots + T_r$ of nilpotent linear operators is nilpotent.

(f) A product $N_1 \circ N_2$ of nilpotent linear operators is nilpotent.

2. If some power T^n of a linear operator T is zero on a finite-dimensional vector space, explain why:

(a) Its trace $\operatorname{Tr}(T)$ must be zero. (Is the converse true?)

(b) Its determinant $\det(T)$ must be zero.

(c) There is a basis $\mathfrak{X}$ such that $[T]_{\mathfrak{X}\mathfrak{X}}$ is strictly upper triangular.

3. Suppose $T : V \to V$ has characteristic polynomial $p_T(t) = (-1)^n t^n$.

(a) Show that this is true if and only if T is nilpotent.

(b) Does the nature of the ground field $\mathbb{K}$ matter in answering this question?

HINT. The polynomial $p(t) = t^n$ in $\mathbb{K}[t]$ splits over any ground field $\mathbb{K}$.

4. Verify that the matrix

$$A = \begin{pmatrix} 3 & 2 & 2 \\ -3 & -1 & 1 \\ -6 & 2 & 4 \end{pmatrix}$$

is nilpotent. Find a basis $\mathfrak{X}$ in $\mathbb{K}^3$ that gives a cyclic subspace decomposition and compute the matrix of $L_A : \mathbb{K}^3 \to \mathbb{K}^3$ with respect to this basis.

5. Let $T = L_A : \mathbb{K}^n \to \mathbb{K}^n$ be determined by the matrix

$$A = \begin{pmatrix} 0 & 0 & 0 & 1 & 0 \\ 0 & 1 & 1 & 0 & -1 \\ 0 & 0 & 0 & 0 & 0 \\ 0 & 0 & 0 & 0 & 0 \\ 0 & 1 & 0 & 0 & -1 \end{pmatrix}.$$

(a) Verify that T is nilpotent.

(b) Determine the kernels $K(T^k) = \{x \in V : T^k(v) = 0\}$ and find a basis $\mathfrak{X}$ in $V = \mathbb{K}^5$ that gives a cyclic subspace decomposition for $T = L_A : V \to V$.

(c) What is the matrix of T with respect to this basis?

Section 1.3. Generalized Eigenspaces.

1. True/False Questions: ("True" if the statement is *always* true.)

(a) Let T be a linear operator on a finite-dimensional vector space that is a direct sum $V = V_1 \oplus \cdots \oplus V_r$ of T-invariant subspaces. Then for any ground field $\mathbb{K}$, the characteristic polynomial $p_T(x) = \det(T - xI)$ is a product

$$p_T(x) = \prod_{k=1}^{r} p_{T|V_k}(x).$$

(b) Every generalized eigenvalue of a linear operator $T : V \to V$ is an actual eigenvalue.

(c) If $T : V \to V$ is a nilpotent linear operator on a complex vector space, all its generalized eigenvalues are zero.

(d) If $T : V \to V$ is a linear operator on a finite-dimensional complex vector space and all its generalized eigenvalues are zero, then T must be nilpotent.

2. If $T : V \to V$ is a linear operator on a finite-dimensional vector space over $\mathbb{C}$, explain how the Fitting decomposition $V = R_\infty \oplus K_\infty$ is used to prove the generalized eigenspace decomposition, Theorem 1.42,

$$V = \bigoplus_{\lambda \in \mathrm{sp}(T)} M_\lambda(T).$$

3. For the linear operator $T = L_A$ on $\mathbb{C}^3$ with

$$A = \begin{pmatrix} 3 & -1 & -2 \\ 3 & 1 & -1 \\ 6 & -2 & -4 \end{pmatrix}:$$

(a) Find the characteristic polynomial $p_T(x)$.

(b) Find the eigenvalues of T and their multiplicities (both geometric and algebraic).

(c) Find bases for the generalized eigenspaces $M_\lambda(T)$. What is the matrix $[T]_{\mathfrak{X}\mathfrak{X}}$ with respect to bases aligned with these subspaces?

Section 1.4. The Generalized Eigenspace Decomposition.

1. TRUE/FALSE QUESTIONS: ("True" if the statement is *always* true.)

(a) Every linear operator on a finite-dimensional vector space can be represented by a matrix $[T]_{\mathfrak{X}\mathfrak{X}}$ that has Jordan form.

(b) If a vector space over a field $\mathbb{K}$ has $\dim_{\mathbb{K}}(V) = n$ and T is a linear operator whose characteristic polynomial splits over $\mathbb{K}$, then each generalized eigenspace $M_\lambda(T)$ is equal to $\ker(T - \lambda I)^n$.

(c) If V is an n-dimensional vector space over $\mathbb{K}$ and T a linear operator whose characteristic polynomial splits over $\mathbb{K}$, then

$$\prod_{\lambda \in \mathrm{sp}_{\mathbb{K}}(T)} (T - \lambda I)^{m_\lambda} = 0 \quad \text{on } V,$$

where m_λ is the algebraic multiplicity of the eigenvalue λ.

(d) Let V be a finite-dimensional vector space over a field $\mathbb{K}$, and T a linear operator whose characteristic polynomial splits over $\mathbb{K}$. Then each generalized eigenspace M_λ is a direct sum $E_1 \oplus \cdots \oplus E_k$ of subspaces that are cyclic under the action of $(T - \lambda I)$.

(e) Same question as in (d), changing the last sentence to read:
"... cyclic under the action of T."

(f) Let T be a linear operator on a finite-dimensional vector space over a field $\mathbb{K}$. If $p_T(x) = \det(T - xI)$ splits into linear factors over $\mathbb{K}$, then

$$V = \bigoplus_{\lambda \in \mathrm{sp}_{\mathbb{K}}(T)} M_\lambda(T).$$

(g) If the characteristic polynomial of a linear operator on a finite-dimensional space has a generalized eigenvalue λ, then

$$\dim M_\lambda(T) = (\text{algebraic multiplicity of } \lambda).$$

(h) The generalized eigenspace $M_\lambda(T)$ of a linear operator is always nontrivial if $\lambda = 0$, regardless of the ground field $\mathbb{K}$.

(i) If $T : V \to V$ is a linear operator on a complex vector space with $\mathrm{sp}_{\mathbb{C}}(T) = \{1 + i0\}$, then $T = I$.

2. What is the generalized eigenspace decomposition of a nilpotent linear operator $T : V \to V$? Does the answer depend on the ground field $\mathbb{K}$?

3. If T is a linear operator on a vector space that is a direct sum $V = \bigoplus_{j=1}^{r} V_j$ of T-invariant subspaces, explain why the characteristic polynomial $p_T(t) = \det(T - tI)$ is a product

$$p_T(t) = \prod_{j=1}^{r} p_{T_j}(t) \quad \text{where } T_j = T|_{V_j}.$$

4. Let M_λ be a nontrivial generalized eigenspace for a linear operator T whose characteristic polynomial splits over the ground field $\mathbb{K}$.

(a) What constraints must be satisifed by the sizes $m_i \times m_i$ of the blocks B_i in a Jordan block decomposition of $T|_{M_\lambda}$?

(b) Tabulate all sequences $(m_1 \leq \cdots \leq m_r)$ of Jordan block sizes that can occur when dim $M_\lambda(T) = 5$.

(c) What constraints must be satisfied by the dimensions $\dim(M_\lambda(T))$ when $\dim(V) = n$?

5. What can be said about the dimensions $\dim(M_\lambda(T))$ of the generalized eigenspaces for a linear operator T if the characteristic polynomial $p_T(x)$ splits over $\mathbb{K}$ and $\dim(V) = 7$?

(a) Can different M_λ have the same dimension?

(b) What does it mean if $\dim(M_\lambda) = 1$?

(c) Can $\dim(V_\lambda) = 0$? What does it mean if $\dim(M_\lambda) = \dim(V)$ for some λ?

6. Find a Jordan decomposition for $T = L_A : \mathbb{C}^2 \to \mathbb{C}^2$ when

$$\text{(a) } A = \begin{pmatrix} 0 & 2 \\ 0 & 0 \end{pmatrix} \quad \text{(b) } A = \begin{pmatrix} 0 & 2 & 0 \\ 0 & 0 & 1 \\ 0 & 0 & 0 \end{pmatrix} \quad \text{(c) } A = \begin{pmatrix} 3 & 0 & 0 & 0 \\ 0 & 0 & 1 & 2 \\ 0 & 0 & 0 & -1 \\ 0 & 0 & 0 & 0 \end{pmatrix}.$$

Section 1.5. Consequences of the Jordan Form.

1. TRUE/FALSE QUESTIONS: ("True" if the statement is *always* true.)

(a) The Jordan form of a matrix $A = \lambda I + N$ with

$$N = \begin{pmatrix} 0 & 1 & & \cdots & 0 \\ & 0 & 1 & & \\ & & \ddots & \ddots & \\ & & & \ddots & 1 \\ 0 & & & & 0 \end{pmatrix}$$

is itself.

(b) The matrix $A = [T]_{\mathfrak{X}\mathfrak{X}} = \lambda I + N$ in the previous question is a typical Jordan block for some operator in Jordan form. Is $A^2 = [T^2]_{\mathfrak{X}\mathfrak{X}}$ also in Jordan form?

(c) If a linear operator T has the form (i) below with respect to a basis $\mathfrak{X} = \{e_1, \ldots, e_n\}$ and $\mathfrak{Y} = \{f_1, \ldots, f_n\}$ is the basis obtained by relabeling the vectors in reverse order, then $[T]_{\mathfrak{Y}\mathfrak{Y}}$ is the "upside down" Jordan block shown in (ii).

$$\text{(i)} \begin{pmatrix} \lambda & 1 & 0 & \cdots & 0 \\ & \lambda & 1 & & \vdots \\ & & \lambda & \ddots & 0 \\ & & & \ddots & 1 \\ & & & & \lambda \end{pmatrix} \quad \text{(ii)} \begin{pmatrix} \lambda & 0 & 0 & \cdots & 0 \\ 1 & \lambda & 0 & & \vdots \\ & 1 & \lambda & \ddots & \vdots \\ & & \ddots & \ddots & 0 \\ & & & 1 & \lambda \end{pmatrix}$$

NOTE. Since the change of basis involves only relabeling, the transition matrix Q such that $[T]_{\mathfrak{Y}\mathfrak{Y}} = Q[T]_{\mathfrak{X}\mathfrak{X}}Q^{-1}$ is a permutation matrix. ○

2. Let T be a linear operator on a vector space V whose characteristic polynomial splits over $\mathbb{K}$. If $\dim(V) = 5$, make lists $0 < d_1 \leq \cdots \leq d_r$ showing all possible patterns of dimensions for the blocks in a Jordan decomposition of T.

HINT. Can different blocks correspond to the same λ? Can there be 1×1 blocks?

3. Make diagrams showing the possible number and sizes of the blocks in the Jordan canonical form of a $\mathbb{C}$-linear operator T if its characteristic polynomial is $p_T(t) = t^3(t+5)^2$ and $\dim_{\mathbb{C}} V = 5$.

HINT. We are only asking about the *sizes* of the Jordan blocks, not their diagonal values.

4. For the linear operator $T : \mathbb{R}^3 \to \mathbb{R}^3$ whose matrix with respect to the standard basis standard basis in $\mathbb{R}^3$ is

$$A = \begin{pmatrix} 5 & -6 & -6 \\ -1 & 4 & 2 \\ 5 & -6 & 4 \end{pmatrix}.$$

(a) Determine $\mathrm{sp}_{\mathbb{C}}(T)$ and find bases for the spaces $M_\lambda(T)$. Do they span $V = \mathbb{C}^3$?

(b) Find a basis $\mathfrak{X}$ that puts T into Jordan canonical form.

5. Can you see an easy way to prove that the following matrix cannot be nilpotent?

$$A = \begin{pmatrix} 1 & 1 & 0 & 1 & 0 \\ 0 & 1 & 0 & 0 & 0 \\ 0 & 0 & 0 & 0 & 0 \\ 0 & 0 & 1 & 1 & 0 \\ 1 & 0 & 0 & 1 & 0 \end{pmatrix}$$

Taking $V = \mathbb{C}^5$ and $T = L_A : V \to V$:

(a) Find the characteristic of polynomial $p_T(t) = \det(T - tI)$ and list the eigenvalues with their algebraic multiplicities.

(b) Find the minimal polynomial $m_T(x)$ in $\mathbb{K}[x]$.

(c) Determine the generalized eigenspaces M_λ by specifying a basis in each M_λ that puts $T|_{M_\lambda}$ into Jordan form, so M_λ is a direct sum $C_1^{(\lambda)} \oplus \cdots \oplus C_r^{(\lambda)}$ of subspaces cyclic under the action of $(T - \lambda I)$.

(d) Explain why the generalized eigenspaces span V, so that $V = \bigoplus_{\lambda \in \mathrm{sp}_{\mathbb{C}}(T)} M_\lambda$.

By Proposition 1.43, when we combine these bases we get a basis $\mathfrak{X}$ for all of V that puts the matrix $[T]_{\mathfrak{X}\mathfrak{X}}$ into Jordan canonical form.

HINT. Computing $p_A(x) = \det(A - xI)$ by hand for a 5×5 matrix is usually a chore, but so many entries in this sparse matrix are zero that the determinant is easily evaluated using expansion by minors (Theorem 4.48 of LA I).

6. Use the JCF to show that all solutions $A \in \mathrm{M}(n, \mathbb{C})$ of the matrix identity $A^2 = -I$ are similar to matrices of the form

$$B = \begin{pmatrix} iI_{r\times r} & 0 \\ 0 & -iI_{s\times s} \end{pmatrix} \quad \text{where } r + s = n\,(r, s \geq 0) \text{ and } i = \sqrt{-1}.$$

7. Find the Jordan form for the following matrix over $\mathbb{C}$ and find a transition matrix $Q \in \mathrm{M}(4, \mathbb{C})$ such that $QAQ^{-1}J(A)$:

$$A = \begin{pmatrix} 2 & 1 & 0 & 0 \\ -1 & 4 & 0 & 0 \\ -1 & 4 & 1 & -1 \\ -1 & 3 & 2 & -1 \end{pmatrix}.$$

HINT. $\lambda = 3$ is a root of the characteristic polynomial $p_A(x)$.

8. The differential operator $L = D^2 + D + cI$ acts on on the subspace $V = \mathcal{P}_3 = \mathbb{C}\text{-span}\{1, x, x^2, x^3\}$ in $\mathbb{C}[x]$.

(a) Find the matrix $[L]_{\mathfrak{X}\mathfrak{X}}$ with respect to the basis $\mathfrak{X} = \{1, x, x^2, x^3\}$ and determine $\mathrm{sp}_{\mathbb{C}}(L)$.

(b) Find the Jordan form of L. What Jordan blocks appear and of what sizes?

(c) Find a $\mathbb{C}$-basis that puts $[L]_{\mathfrak{X}\mathfrak{X}}$ into Jordan form.

Does the value of the constant $c \in \mathbb{C}$ affect your answer?

9. The following rotation matrix in $\mathrm{M}(2, \mathbb{R})$ is not diagonalizable over $\mathbb{R}$ but has a well-defined JCF when regarded as a matrix in $\mathrm{M}(2, \mathbb{C})$:

$$A = \begin{pmatrix} \cos\theta & -\sin\theta \\ \sin\theta & \cos\theta \end{pmatrix} \qquad (\theta \in \mathbb{R}).$$

Find the following:

(a) Its complex eigenvalues and bases for the eigenspaces in $\mathbb{C}^2$.

(b) A basis in $\mathbb{C}^2$ that puts A into Jordan form.

(c) A transition matrix Q such that $QAQ^{-1} = J(A)$.

10. If V is a finite-dimensional vector space and $T : V \to V$ a linear operator whose characteristic polynomial $p_T(x)$ splits in $\mathbb{K}[x]$, let

$$T = \sum_{\lambda \in \mathrm{sp}_{\mathbb{K}}(T)} \lambda P_\lambda$$

be its spectral decomposition, as in Theorem 6.94 and equation (6.15) of LA I. The spectral mapping theorem, Theorem 1.62, applies since T has a well-defined Jordan form; use it to prove the following:

(a) Each spectral projection in this decomposition has the form $P_\lambda = f_\lambda(T)$ for some polynomial $f_\lambda(x)$ in $\mathbb{K}[x]$.

(b) Every linear operator S that commutes with T commutes with the spectral projections P_λ.

(c) Every linear operator S that commutes with T commutes with $\mathrm{Exp}(T) = e^T$.

1.6. Appendix A: Brief Review of Diagonalization

DEFINITION A.1 (Eigenvalues, Eigenspaces, Diagonalization). Given λ in $\mathbb{K}$ and a linear operator $T : V \to V$ on a complex vector space, the *λ-eigenspace* is $E_\lambda(T) = \{v \in V : (T - \lambda I)v = 0\}$ and λ is an *eigenvalue* if $E_\lambda(T) \neq \{0\}$; then $\dim_{\mathbb{K}}\big(E_\lambda(T)\big)$ is its *geometric multiplicity* and $\mathrm{sp}_{\mathbb{K}}(T) = \{\lambda \in \mathbb{K} : E_\lambda \neq \{0\}\}$ is called the *spectrum* of T over $\mathbb{K}$. However, we have already seen that when $\mathbb{K} = \mathbb{R}$, some operators (such as a 90° rotation acting on $\mathbb{R}^2$) have no eigenvalues in $\mathbb{R}$, and then $\mathrm{sp}_{\mathbb{R}}(T) = \varnothing$.

EXERCISE A.2. Show that every eigenspace E_λ is a T-invariant subspace in V. If $\mathfrak{X} = \{e_1, \dots, e_r, \dots, e_n\}$ is a basis for V that first passes through E_λ, show that the matrix of T takes the form

$$[T]_{\mathfrak{X}\mathfrak{X}} = \left(\begin{array}{ccc|cc} \lambda & & 0 & * & * \\ & \ddots & & * & * \\ 0 & & \lambda & * & * \\ \hline 0 & 0 & 0 & * & * \\ 0 & 0 & 0 & * & * \end{array}\right).$$

An operator T is diagonalizable if there is a basis $\mathfrak{X} = \{e_1, \dots, e_n\}$ consisting of eigenvectors e_i, so $T(e_i) = \lambda_i e_i$ for all i. Then $[T]_{\mathfrak{X}\mathfrak{X}}$ has the diagonal form

$$[T]_{\mathfrak{X}\mathfrak{X}} = \begin{pmatrix} \lambda_1 & & 0 \\ & \ddots & \\ 0 & & \lambda_n \end{pmatrix},$$

in which there may be repeats among the λ_i. Conversely, any basis such that $[T]_{\mathfrak{X}\mathfrak{X}}$ takes this form consists entirely of eigenvectors for T. When T cannot be diagonalized, a more sophisticated choice of basis vectors puts $[T]_{\mathfrak{X}\mathfrak{X}}$ into *block diagonal form*. First, a simple observation:

EXERCISE A.3. If $T : V \to V$ is a linear operator on a finite-dimensional space, show that the following statements are equivalent.

(a) There is a basis in V consisting of eigenvectors.

(b) The eigenspaces for T span V, so

$$V = \sum_{\lambda \in \mathrm{sp}(T)} E_\lambda(T).$$

NOTE. There is actually something to be proved here: (b) requires more care selecting the ordered basis than (a). So if T is diagonalizable and $\{\lambda_1, \dots, \lambda_r\}$ are its *distinct* eigenvalues in $\mathbb{K}$, we may choose a basis of eigenvectors e_i that first runs through E_{λ_1}, then through E_{λ_2}, etc., and this choice yields a "block diagonal" matrix

$$[T]_{\mathfrak{X}} = \begin{pmatrix} \lambda_1 \boxed{I_{m_1 \times m_1}} & & & 0 \\ 0 & \lambda_2 \boxed{I_{2 \times m_2}} & & \\ \vdots & & \ddots & \\ 0 & & & \lambda_r \boxed{I_{m_r \times m_r}} \end{pmatrix}$$

in which $m_i = \dim\left(E_{\lambda_i}(T)\right)$. ○

These observations do not quite yield the definitive characterization of diagonalizability, which says

(1) DIAGONALIZABILITY CRITERION. *A linear operator T on a finite-dimensional space is diagonalizable over $\mathbb{K}$ $\Leftrightarrow$ V is the* DIRECT SUM *of its distinct eigenspaces: $V = \bigoplus_{i=1}^{r} E_{\lambda_i}(T)$.*

The implication ($\Leftarrow$) is trivial. In the reverse direction we only know that (diagonalizable) $\Rightarrow$ V is *spanned* by its eigenspaces, so $V = \sum_{i=1}^{r} E_{\lambda_i}(T)$ and every v has at least one decomposition $v = \sum_i v_i$ with $v_i \in E_{\lambda_i}(T)$. In a direct sum $\bigoplus_i E_{\lambda_i}(T)$, the decomposition is unique, and in particular $0 = \sum_i v_i$ with $v_i \in E_{\lambda_i}(T) \Leftrightarrow$ each term $v_i = 0$.

EXERCISE A.4. Finish the proof of the diagonalizability criterion of Exercise A.3. If $V = \sum_i E_{\lambda_i}(T)$ (so the eigenspaces span V), prove that every $v \in V$ has a *unique* decomposition $v = \sum_i v_i$ such that $v_i \in E_{\lambda_i}(T)$.

THEOREM A.5. *If $\{\lambda_1, \dots, \lambda_r\}$ are the distinct eigenvalues in $\mathbb{K}$ for a linear operator $T : V \to V$ on a finite-dimensional vector space, and if the eigenspaces E_{λ_i} span V, then V is a direct sum $E_{\lambda_1} \oplus \cdots \oplus E_{\lambda_r}$. Furthermore,*

1. $\dim(V) = \sum_{i=1}^{r} \dim(E_{\lambda_i}) = \sum_{i=1}^{r}$*(geometric multiplicity of λ_i).*
2. *T is diagonalizable over $\mathbb{K}$.*

PROOF. Since $V = \sum_i E_{\lambda_i}$, every vector in V has a decomposition $v = \sum_{i=1}^{r} v_i$ with $V_i \in E_{\lambda_i}(T)$, so we need only prove uniqueness of this decomposition, which in turn reduces to proving that the v_i are "independent" in the sense that

$$0 = v_1 + \cdots + v_r \text{ with } v_i \in E_{\lambda_i} \Rightarrow v_1 = \cdots = v_r = 0.$$

Note that for $\mu, \lambda \in \mathbb{K}$, the linear operators $(T - \lambda I)$, $(T - \mu I)$ commute with each other since I commutes with everybody. Now suppose $\sum_{i=1}^{r} v_i = 0$ with $T(v_i) = \lambda_i v_i$.

Fix an index i and apply the operator $S = \prod_{j \neq i}(T - \lambda_j I)$ to the sum. We get

$$0 = S(0) = S\Big(\sum_{k=1}^{r} v_k\Big) = \sum_{k=1}^{r} S(v_k). \tag{2}$$

But if $k \neq i$, we can write

$$S(v_k) = \prod_{\ell \neq i}(T - \lambda_\ell I)v_k = \Big(\prod_{\ell \neq k,i}(T - \lambda_k I)\Big) \cdot (T - \lambda_k)v_k = 0.$$

Hence the sum in (2) reduces to

$$0 = \sum_k S(v_k) = S(v_i) + 0 + \cdots + 0 = \prod_{\ell \neq i}(T - \lambda_\ell I)\, v_i.$$

Since $(T - \lambda_\ell) = (T - \lambda_i) + (\lambda_i - \lambda_\ell)I$ for all ℓ, this product becomes

$$(3) \qquad 0 = \Big(\prod_{\ell \neq i}(T - \lambda_i) + (\lambda_i - \lambda_\ell)I\Big)\, v_i = 0 + \Big(\prod_{\ell \neq i}(\lambda_i - \lambda_\ell)\Big)\, v_i$$

(because $(T - \lambda_i)v_i = 0$). The constant $c = \prod_{\ell \neq i}(\lambda_i - \lambda_\ell)$ must be nonzero because $\lambda_\ell \neq \lambda_i$. Therefore (2) $\Rightarrow v_i = 0$. This works for every $1 \leq i \leq r$ so the v_i are independent, as required. □

EXERCISE A.6. Let V be finite dimensional, $\{\lambda_1, \dots, \lambda_r\}$ the distinct eigenvalues in $\mathbb{K}$ for a linear operator $T : V \to V$. Let $E = \sum_{i=1}^r E_{\lambda_i}$ be the span of the eigenspaces. ($E \subseteq V$). Show that:

(a) E is T-invariant .
(b) The restriction $T|_E$ is diagonalizable.

EXERCISE A.7. Let $T : V \to V$ be a linear operator on a finite-dimensional vector space over $\mathbb{K}$, with $n = \dim_{\mathbb{K}}(V)$. If T has n distinct eigenvalues, prove that

(a) $V = \bigoplus_{i=1}^n E_{\lambda_i}$.
(b) The eigenspaces are all one-dimensional.
(c) T is diagonalizable.

EXERCISE A.8. If a basis $\mathfrak{X}$ for V passes through the successive eigenspaces $E_{\lambda_1}(T), \dots, E_{\lambda_r}(T)$ and we then adjoin vectors that are outside of the subspace $E = \sum_{\lambda_i \in \mathrm{sp}(T)} E_{\lambda_i}(T)$ to get a basis for V, explain why the matrix of T has the form

$$[T]_{\mathfrak{X}\mathfrak{X}} = \left(\begin{array}{ccc|cc} \boxed{\lambda_1 I_{m_1 \times m_1}} & 0 & 0 & * & * \\ \vdots & \ddots & \vdots & * & * \\ 0 & 0 & \boxed{\lambda_r I_{m_r \times m_r}} & * & * \\ \hline 0 & 0 & 0 & * & * \\ 0 & 0 & 0 & * & * \end{array}\right)$$

where $m_i = \dim(E_{\lambda_i}(T))$.

DEFINITION A.9. If $T : V \to V$ is linear operator on a finite-dimensional vector space, every root $\alpha \in \mathbb{K}$ of the polynomial $p_T(x) = \det(T - xI)$ is an eigenvalue for T, so $p_T(x)$ is divisible (without remainder) by $(x - \alpha)$. Repeated division by $(x - \alpha)$ may be possible, and it yields a factorization

$$p_T(x) = (x - \alpha)^{m_\alpha} Q(x)$$

where $Q \in \mathbb{K}[x]$ does not have α as a root and thus is not divisible by $(x - \alpha)$. The exponent m_α is called the *algebraic multiplicity* of the eigenvalue α.

Now suppose $\mathbb{K}$ is an algebraically closed field (every nonconstant polynomial $f \in \mathbb{K}[x]$ has a root $\alpha \in \mathbb{K}$), for example $\mathbb{K} = \mathbb{C}$. It follows that every such f splits completely into linear factors $f = c \cdot \prod_{i=1}(x - \alpha_i)$ where $\alpha_1, \dots, \alpha_n$ are the roots of $f(x)$ in $\mathbb{K}$ (repeats allowed). If $T : V \to V$ is a linear operator on a finite-dimensional vector space over such a field, and $\lambda_1, \dots, \lambda_r$ are its distinct eigenvalues in $\mathbb{K}$, the characteristic polynomial splits completely

$$p_T(x) = \det(T - xI) = c \cdot \prod_{j=1}^{r} (x - \lambda_j)^{m_j}$$

where m_j = the algebraic multiplicity of λ_j and $\sum_j m_j = \dim(V)$.

COROLLARY A.10. *Let $T : V \to V$ be a linear operator on a finite-dimensional space V over $\mathbb{K} = \mathbb{C}$. If the characteristic polynomial*

$$p_T(x) = \det(T - x\,I) = c \cdot \prod_{j=1}^{r} (x - \lambda_j)^{m_j}$$

has DISTINCT *roots in $\mathbb{C}$ (so $m_j = 1$ for all j), then $r = n = \dim_{\mathbb{C}}(V)$ and T is diagonalizable.*

Algebraic vs. Geometric Multiplicity.

PROPOSITION A.11. *If $\lambda \in \mathbb{K}$ is an eigenvalue for linear operator $T : V \to V$, then the algebraic multiplicity of λ as a root of $p_T(x) = \det(T - x\,I)$ is greater than or equal to (geometric multiplicity of λ) $= \dim E_\lambda$.*

PROOF. Fix an eigenvalue λ. Then $E = E_\lambda(T)$ is T-invariant and $T|_E = \lambda \cdot \mathrm{id}_E$. So if $\{e_1, \dots, e_m\}$ is a basis in E_λ and we adjoin vectors $e_{m+1}, \dots, e_n$ to get a basis $\mathfrak{X}$ for V, we have

$$[T]_{\mathfrak{X}\mathfrak{X}} = \left(\begin{array}{c|c} \lambda I_{m\times m} & * \\ \hline 0 & * \end{array}\right) \qquad (m = \dim(E_\lambda(T)).$$

Here m is the geometric multiplicity of λ and the characteristic polynomial is

$$p_T(x) = \det \left(\begin{array}{ccc|ccc} (\lambda - x) & & 0 & & & \\ & \ddots & & & & * \\ 0 & & (\lambda - x) & & & \\ \hline & & & (a_{m+1,m+1} - x) & & * \\ & 0 & & & \ddots & \\ & & & * & & (a_{n,n} - x) \end{array}\right).$$

This determinant can be written as

(4) $$\det(T - xI) = \sum_{\pi \in S_n} \mathrm{sgn}(\pi) \cdot (T - xI)_{1,\pi(1)} \cdot \cdots \cdot (T - xI)_{n,\pi(n)}.$$

Each term in this sum involves a product of matrix entries, selecting each one from a different column in each row. If the spots occupied by the selected entries in (4) are marked with a "□," the marked spots provide a "template" for making the selection, and there is a unique template for each permutation $\pi \in S_n$ obtained by marking the spot where Row_i meets Col_j where $j = \pi(i)$.

The only $n \times n$ templates that can contribute to the determinant of a block-upper triangular matrix $(T - xI)$ are those in which the first m diagonal spots have been marked (otherwise the corresponding product of terms will include a zero selected from the lower left block). The remaining marked spots must then be selected from the lower right hand block — i.e., from Row_i and Col_j with $m + 1 \le i, j \le n$, as indicated in the following diagram:

$$\left(\begin{array}{ccc|cccc} \square & & 0 & & & & \\ & \cdot & & & * & & \\ 0 & & \square & & & & \\ \hline & & & \cdot & \cdot & \square & \cdot \\ & 0 & & \square & \cdot & \cdot & \cdot \\ & & & \cdot & \square & \cdot & \cdot \end{array}\right).$$

Thus $p_T(x) = \det(T - xI)$ has the general form $(x - \lambda)^m \cdot G(x)$, in which the factor $G(x)$ might involve additional copies of λ. We conclude that

$$\big(\text{algebraic multiplicity of } \lambda\big) \ge m = \big(\text{geometric multiplicity of } \lambda\big),$$

as claimed. □

EXAMPLE A.12. Let

$$T = L_A : \mathbb{R}^3 \to \mathbb{R}^3 \quad \text{with } A = \begin{pmatrix} 4 & 0 & 1 \\ 2 & 3 & 2 \\ 1 & 0 & 4 \end{pmatrix}.$$

If $\mathfrak{X} = \{e_1, e_2, e_3\}$ is the standard Euclidean basis, then $[T]_{\mathfrak{X}\mathfrak{X}} = A$ and the characteristic polynomial is

$$\begin{aligned} p_T(x) = \det(A - xI) &= \det\begin{pmatrix} 4-x & 0 & 1 \\ 2 & 3-x & 2 \\ 1 & 0 & 4-x \end{pmatrix} \\ &= [(4-x)(3-x)(4-x) + 0 + 0] - [(3-x) + 0 + 0] \\ &= (12 - 7x + x^2)(4-x) - 3 + x \\ &= 48 - 28x + 4x^2 - 12x + 7x^2 - x^3 - 3 + x \\ &= -x^3 + 11x^2 - 39x + 45. \end{aligned}$$

To determine $\mathrm{sp}(T)$, we need to find roots of a cubic; however, we can in this case guess a root λ and then long divide by $(x - \lambda)$. A little trial and error shows that $\lambda = 3$ is a root, with $p_T(3) = -27 + 99 - 117 + 45 = 0$ and $p_T(x) = -(x-3)(x^2 - 8x + 15) = -(x-3)^2(x-5)$.

EIGENVALUES. The eigenvalues in $\mathbb{K} = \mathbb{R}$ (or $\mathbb{K} = \mathbb{Q}$) are $\lambda_1 = 3, \lambda_2 = 5$ with algebraic multiplicities, $m_1 = 2, m_2 = 1$. For the geometric multiplicities, we must compute the eigenspaces $E_{\lambda_k}(T)$.

CASE 1: $\lambda = 3$. We solve the system $(A - 3I)\mathbf{x} = \mathbf{0}$ by row reduction.

$$[A - 3I|0] = \left(\begin{array}{ccc|c} 1 & 0 & 1 & 0 \\ 2 & 0 & 2 & 0 \\ 1 & 0 & 1 & 0 \end{array}\right) \to \left(\begin{array}{ccc|c} \boxed{1} & 0 & 1 & 0 \\ 0 & 0 & 0 & 0 \\ 0 & 0 & 0 & 0 \end{array}\right)$$

Columns in the row reduced system that *do not* meet a "step corner" $\boxed{*}$ correspond to free variables in the solution. Thus x_2, x_3 can take any value in $\mathbb{K}$ while $x_1 = -x_3$, so that

$$\begin{aligned} E_{\lambda=3} &= \ker(A - 3I) = \{(-v_3, v_2, v_3) : v_2, v_3 \in \mathbb{K}, v_1 = -v_3\} \\ &= \mathbb{K} \cdot (-1, 0, 1) \oplus \mathbb{K} \cdot (0, 1, 0) = \mathbb{K}(-e_1 + e_3) \oplus \mathbb{K}e_2. \end{aligned}$$

These vectors are a basis, and (geometric multiplicity) = (algebraic multiplicity) $= \dim(E_{\lambda=3}) = 2$.

CASE 2: $\lambda = 5$. Solving $(A - 5I)\mathbf{v} = \mathbf{0}$ by row reduction sends $[A - 5I]$ to

$$\left(\begin{array}{ccc|c} -1 & 0 & 1 & 0 \\ 2 & -2 & 2 & 0 \\ 1 & 0 & -1 & 0 \end{array}\right) \to \left(\begin{array}{ccc|c} -1 & 0 & 1 & 0 \\ 0 & -2 & 4 & 0 \\ 0 & 0 & 0 & 0 \end{array}\right) \to \left(\begin{array}{ccc|c} \boxed{1} & 0 & -1 & 0 \\ 0 & \boxed{1} & -2 & 0 \\ 0 & 0 & 0 & 0 \end{array}\right).$$

Now there is only one free variable x_3, with $x_2 = 2x_3$ and $x_1 = x_3$. Thus

$$E_{\lambda=5} = \{(x_3, 2x_3, x_3) : x_3 \in \mathbb{K}\} = \mathbb{K} \cdot (1, 2, 1)$$

and

$$1 = \dim(E_{\lambda=5}) = (\text{geometric multiplicity of } \lambda = 5) = (\text{algebraic multiplicity}).$$

DIAGONALIZATION: A basis $\mathfrak{Y}$ consisting of eigenvectors is given by

$$f_1 = -e_1 + e_3, \quad f_2 = e_2, \quad f_3 = e_1 + 2e_2 + e_3,$$

and for this basis we have

$$[T]_{\mathfrak{Y}\mathfrak{Y}} = \left(\begin{array}{cc|c} 3 & 0 & 0 \\ 0 & 3 & 0 \\ \hline 0 & 0 & 5 \end{array}\right),$$

while $[T]_{\mathfrak{X}\mathfrak{X}} = A$ with respect to the standard Euclidean basis $\{e_i\}$.

It is sometimes important to know the similarity transform $SAS^{-1} = [T]_{\mathfrak{Y}\mathfrak{Y}}$ that effects the transition between bases. The matrix S can be found by writing

$$\begin{aligned} [T]_{\mathfrak{Y}\mathfrak{Y}} &= [\mathrm{id} \circ T \circ \mathrm{id}\,]_{\mathfrak{Y}\mathfrak{Y}} = [\mathrm{id}\,]_{\mathfrak{Y}\mathfrak{X}} \cdot [T]_{\mathfrak{X}\mathfrak{X}} \cdot [\mathrm{id}]_{\mathfrak{X}\mathfrak{Y}} \\ &= [\mathrm{id}]_{\mathfrak{Y}\mathfrak{X}} \cdot A \cdot [\mathrm{id}]_{\mathfrak{X}\mathfrak{Y}}. \end{aligned}$$

Then $S = [\mathrm{id}]_{\mathfrak{Y}\mathfrak{X}}$ and $S^{-1} = [\mathrm{id}]_{\mathfrak{X}\mathfrak{Y}}$ because

$$[\mathrm{id}]_{\mathfrak{X}\mathfrak{Y}} \cdot [\mathrm{id}]_{\mathfrak{Y}\mathfrak{X}} = [\mathrm{id}]_{\mathfrak{X}\mathfrak{X}} = I_{3\times3} \quad (3 \times 3 \text{ identity matrix}).$$

(All this is discussed in LA I, Section 4 of Chapter 2.)

The easiest matrix to determine is usually $[\text{id}\,]_{\mathfrak{X}\mathfrak{Y}}$ which can be written down immediately if we can write vectors in $\mathfrak{Y}$ in terms those in $\mathfrak{X}$. In the present example we have

$$\begin{cases} f_1 = \text{id}(f_1) & = & -e_1 + 0\cdot e_2 + e_3 \\ f_2 = \text{id}(f_2) & = & 0 + e_2 + 0 \\ f_3 = \text{id}(f_3) & = & e_1 + 2e_2 + e_3 \end{cases} \Rightarrow S^{-1} = [\text{id}]_{\mathfrak{X}\mathfrak{Y}} = \begin{pmatrix} -1 & 0 & 1 \\ 0 & 1 & 2 \\ 1 & 0 & 1 \end{pmatrix}.$$

The matrix $[\text{id}]_{\mathfrak{X}\mathfrak{Y}}$ is just the transpose of the coefficient array in this system of vector identities that express the f_i in terms of the e_j.

We can now find the desired inverse $S = (S^{-1})^{-1}$ by row operations (or by Cramer's rule) to get

$$S = \begin{pmatrix} -1/2 & 0 & 1/2 \\ -1 & 1 & -1 \\ 1/2 & 0 & 1/2 \end{pmatrix}$$

and then

$$SAS^{-1} = \begin{pmatrix} 3 & 0 & 0 \\ 0 & 3 & 0 \\ 0 & 0 & 5 \end{pmatrix}$$

as expected. That concludes our analysis of Example A.12.

EXERCISE A.13. Fill in the details needed to compute S.

EXAMPLE A.14. Let $A = \begin{pmatrix} 2 & 4 \\ -1 & -2 \end{pmatrix}$ with $\mathbb{K} = \mathbb{R}$ or $\mathbb{C}$. Then

$$A^2 = \begin{pmatrix} -2 & 0 \\ 0 & 0 \end{pmatrix} \quad \text{and} \quad A^3 = 0,$$

so with respect to the standard basis in $\mathbb{K}^2$, the map $T = L_A : \mathbb{K}^2 \to \mathbb{K}^2$ has matrix $[T]_{\mathfrak{X}\mathfrak{X}} = A$ and the characteristic polynomial is

$$\det(A - xI) = \det\begin{pmatrix} 2-x & 4 \\ -1 & -2-x \end{pmatrix} = -4 + x^2 + 4 = x^2.$$

Thus, $\lambda = 0$ is a root (over $\mathbb{R}$ or $\mathbb{C}$) with (algbraic multiplicity) $= 2$, but the geometric multiplicity is $\dim\left(E_{\lambda=0}\right) = 1$. Solve the system $(A - \lambda I)\mathbf{x} = A\mathbf{x} = 0$ taking $\lambda = 0$ to determine $E_{\lambda=0} = \ker(L_A)$; row reduction yields

$$\left(\begin{array}{cc|c} 2-\lambda & 4 & 0 \\ -1 & -2-\lambda & 0 \end{array}\right) = \left(\begin{array}{cc|c} 2 & 4 & 0 \\ -1 & -2 & 0 \end{array}\right) \to \left(\begin{array}{cc|c} \boxed{1} & 2 & 0 \\ 0 & 0 & 0 \end{array}\right).$$

For this system x_2, is a free variable and $x_1 = -2x_2$, so

$$E_{\lambda=0} = \mathbb{K}\cdot(-2,1) = \mathbb{K}\cdot(2e_1 - e_2)$$

and $\dim\left(E_{\lambda=0}\right) = 1 <$ (algebraic multiplicity) $= 2$. There are no other eigenvalues, so the best we can do in trying to reduce T is to find a basis such that $[T]_{\mathfrak{Y}\mathfrak{Y}}$ has the form

$$\left(\begin{array}{c|c} 0 & * \\ \hline 0 & * \end{array}\right)$$

by taking $\mathfrak{Y} = \{f_1, f_2\}$, where $f_1 = (2,1) = 2e_1 + e_2$ and f_2 is any other vector independent of f_1.

However, T is a nilpotent operator (verify this), so we can do better with a slightly different basis $\mathfrak{Z}$ that puts A into the Jordan canonical form for nilpotent operators (as in Theorem 1.16). In the present example, this is

$$[T]_{\mathfrak{Z}\mathfrak{Z}} = \begin{pmatrix} \boxed{0} & 1 \\ 0 & \boxed{0} \end{pmatrix} \quad \text{(an elementary nilpotent matrix)}$$

with two 1×1 blocks of zeros on the diagonal. In fact, in the notation of Theorem 1.16, we have kernels $(0) \subseteq K_1 = \ker(T) \subseteq K_2 = \ker(T^2) = V$ with

$$K_1 = E_{\lambda=0} = \mathbb{K} \cdot f_1 \quad \text{and} \quad K_2 = \mathbb{K}^2.$$

So if f_2 is any vector transverse to $\ker(T)$, we have $T(f_2) \in \ker(T) = \mathbb{K} \cdot f_1$. But $T(f_2) \neq 0$ since $f_2 \notin K_1$, and by scaling f_2 appropriately, we can make $T(f_2) = f_1$. Then $\mathfrak{Z} = \{f_1, f_2\}$ is a basis that puts $[T]_{\mathfrak{Z}\mathfrak{Z}}$ into the form shown above.

Exercise A.15. Repeat the analysis of the previous exercise for the matrix

$$A = \begin{pmatrix} 4 & 4 \\ -1 & 0 \end{pmatrix}.$$

That concludes our review of diagonalization.

CHAPTER 2

Further Applications of Jordan Form

An Overview of This Chapter. In Sections 5.4 and 5.5 of LA I, we showed that a linear system of constant-coefficient first-order differential equations

$$\frac{dX}{dt} = A \cdot X(t) \quad (A \in \mathrm{M}(n, \mathbb{K}) \text{ with } \mathbb{K} = \mathbb{R} \text{ or } \mathbb{C}) \tag{2.1}$$

can be solved subject to an initial condition $X(0) = \mathbf{c} \in \mathbb{K}^n$ whenever the coefficient matrix A can be diagonalized over $\mathbb{K}$, with solutions

$$X(t) = e^{tA} \cdot X(0).$$

After reviewing these basic facts from LA I and with the Jordan form JCF in hand, we will show that a similar analysis yields solutions of the same type for the much larger class of matrices that can be put into Jordan form. As we saw in Chapter 1, this can always be done if the characteristic polynomial $p_A(x) = \det(A - xI)$ splits over the ground field $\mathbb{K}$. We will also show how the exponential matrices e^{tA} needed here and the transition matrices such that $QAQ^{-1} = J(A)$ can be effectively computed by adapting the analysis in LA I.

Next we explain why solving a single n^{th}-*order* constant-coefficient differential equation $y(t) = L(x(t))$, where

$$L(x(t)) = a_0 D^n x(t) + a_1 D^{n-1} x(t) + \cdots + a_n x(t), \tag{2.2}$$

is completely equivalent to solving a related system of first-order constant-coefficient equations of the form (2.1). But we now know how to solve those whenever the coefficient matrix A can be recast in JCF by a similarity transformation (which is always possible when $\mathbb{K} = \mathbb{C}$).

In Section 2.2 we address the following question:

> *What can be done to understand the structure of a linear operator $L_A : \mathbb{R}^n \to \mathbb{R}^n$ on a vector space over $\mathbb{R}$ when A is in $\mathrm{M}(n, \mathbb{R})$ and the characteristic polynomial $p_A(x)$ does not split in $\mathbb{R}[x]$, so A has no real-valued Jordan form? We could of course regard A as a complex matrix and find its complex-valued Jordan form $J(A) \in \mathrm{M}(n, \mathbb{C})$, but that does not clearly reveal the structure of L_A as an operator $L_A : \mathbb{R}^n \to \mathbb{R}^n$.*

The result will be a useful *real-normal form* for an arbitrary matrix $A \in \mathrm{M}(n, \mathbb{R})$, derived from the JCF for matrices $A \in \mathrm{M}(n, \mathbb{C})$. This is accomplished by *complexifying* an $\mathbb{R}$-linear operator $T : V \to V$ on a real vector space to get a $\mathbb{C}$-linear operator $T_{\mathbb{C}} : V_{\mathbb{C}} \to V_{\mathbb{C}}$ on a related complex-linear space $V_{\mathbb{C}}$. By reverse-engineering proofs that work for $T_{\mathbb{C}}$, we will show how the well-defined Jordan form for $T_{\mathbb{C}}$ can be unraveled to extract a natural normal form for the original

$\mathbb{R}$-linear operator $T : V \to V$. We obtain different normal forms depending on whether $T_{\mathbb{C}}$ is diagonalizable or not, applying the JCF for $T_{\mathbb{C}}$ in the latter case instead of diagonalization.

2.1. The Jordan Form and Differential Equations

Computing the exponential $e^A = \sum_{n=0}^{\infty} \frac{1}{n!}A^n$ of a matrix is important in many applications, some of which will be illustrated below. Summing the series is often not an easy calculation. In fact, computing a product of two $n \times n$ matrices seems to require n^3 multiplication operations on matrix entries, and computing a high power such as A^{200} directly could be a formidable task. The computational effort can be reduced somewhat through clever programming, but it can be done by hand for diagonal matrices and for elementary nilpotent matrices (see Exercises 1.58–1.60) and hence for any matrix that is already in Jordan form. For a linear operator T on V the Jordan form is obtained by choosing a suitable basis in V; for a matrix A, this amounts to finding an invertible matrix S such that the similarity transform $A \mapsto SAS^{-1}$ puts A into Jordan form. Similarity transforms are invertible operations that interact nicely with the matrix-exponential operation because

$$Se^AS^{-1} = e^{SAS^{-1}} \qquad \text{for every } A \in \mathrm{M}(n, \mathbb{C}). \tag{2.3}$$

So if we can determine the Jordan form $J(A)$ for A, we can compute e^A in four simple steps:

$$\begin{aligned} A &\to \text{ Jordan form } J(A) = SAS^{-1} \\ &\to e^{J(A)} = e^{SAS^{-1}} = Se^AS^{-1} \text{ (a calculation that can be done by hand)} \\ &\to e^A = S^{-1}e^{J(A)}S. \end{aligned}$$

Done. Note that e^A will have block-diagonal form if A has Jordan form but might not itself be in Jordan form.

This idea was illustrated in LA I for *diagonalizable* matrices and operators; the following example adapts it to deal with nondiagonalizable matrices whose Jordan form can be determined. Rather than go through such a calculation just to compute e^A, we will go the extra parsec and show how the ability to compute matrix exponentials can be used to solve more general linear systems of constant-coefficient differential equations.

EXAMPLE 2.1 (Solving Linear Systems of ODE). If A is in $\mathrm{M}(n, \mathbb{K})$ for $\mathbb{K} = \mathbb{R}$ or $\mathbb{C}$ and $X(t) = (x_1(t), \dots, x_n(t))$ is a differentiable function with values in $\mathbb{K}^n$, the vector identity

$$\frac{dX}{dt} = A \cdot X(t) \qquad X(0) = \mathbf{c} = (c_1, \dots, c_n) \tag{2.4}$$

is a system of constant-coefficient first-order linear differential equations with initial conditions $\mathbf{x}(0) = \mathbf{c}$. A unique $\mathcal{C}^\infty$ solution $X(t) = e^{tA} \cdot \mathbf{c}$ exists for all t. In fact, the operator-valued function e^{tA} is infinitely differentiable with

$$\frac{d}{dt}(e^{tA}) = A \cdot e^{tA} \quad \text{for all } t \in \mathbb{R}$$

(see Section 5.5 of LA I), from which we get

$$\frac{dX}{dt} = \frac{d}{dt}(e^{tA}) \cdot \mathbf{c} = A \cdot X(t).$$

Obviously, $X(0) = \mathbf{c}$.

Solving the differential equation (2.4), therefore, reduces to computing e^{tA}. But if A has a characteristic polynomial that splits over $\mathbb{K}$ (or if $\mathbb{K} = \mathbb{C}$), we can find a basis that puts A in Jordan form. So there is some (nonunique) intertwiner $S \in \mathrm{M}(n, \mathbb{K})$ such that $J(A) = SAS^{-1}$. One way to find such an S is to compare descriptions of the linear operator $L_A : \mathbb{C}^n \to \mathbb{C}^n$ with respect to the standard basis $\mathfrak{X}$, for which $[L_A]_{\mathfrak{X}\mathfrak{X}} = A$, and a basis $\mathfrak{Y}$ such that $[L_A]_{\mathfrak{Y}\mathfrak{Y}} = J(A)$. That means

$$A = [L_A]_{\mathfrak{X}\mathfrak{X}} = [\mathrm{id}]_{\mathfrak{X}\mathfrak{Y}} \cdot [L_A]_{\mathfrak{Y}\mathfrak{Y}} \cdot [\mathrm{id}]_{\mathfrak{Y}\mathfrak{X}} = [\mathrm{id}]_{\mathfrak{X}\mathfrak{Y}} \cdot J(A) \cdot [\mathrm{id}]_{\mathfrak{Y}\mathfrak{X}}.$$

So to get $SAS^{-1} = J(A)$, we need $A = S^{-1}J(A)S$ and can take $S^{-1} = [\mathrm{id}]_{\mathfrak{X}\mathfrak{Y}}$. (The basis $\mathfrak{Y}$ is somewhat ambiguous because there are certain ambiguities in defining the Jordan form itself – see Remark 2.4 below.) But S^{-1} can be computed directly once we know the bases; it is just the transpose of the coefficient array in the system of vector identities

$$f_i = \sum_{j=1}^{n} t_{ji} e_j \quad (1 \le i, j \le n)$$

that connect $\mathfrak{X} = \{e_i\}$ and $\mathfrak{Y} = \{f_j\}$.

Starting from the identity $SAS^{-1} = J(A)$ we may write equation (2.4) as

$$\frac{dX}{dt} = A \cdot X(t) = S^{-1}J(A)S \cdot X(t) \quad \text{or} \quad \frac{dW}{dt} = J(A) \cdot W(t)$$

where $W(t) = S(X(t))$. The initial condition $X(0) = \mathbf{c}$ takes the form $W(0) = S(X(0))$, and the solution $X(t) = e^{tA} \cdot \mathbf{c}$ of the original equation becomes

$$X(t) = S^{-1}e^{tSAS^{-1}}S(\mathbf{c}) = S^{-1}e^{tJ(A)}S(\mathbf{c})$$

or

$$W(t) = e^{tJ(A)}W(t).$$

Finding a solution now reduces to computing $e^{tJ(A)}$ applied to a new initial condition $W(0) = S(\mathbf{c})$.

The rest is straightforward. For the bases we are using, $J(A) = SAS^{-1}$ consists of Jordan blocks along its diagonal:

$$J(A) = \begin{pmatrix} \boxed{B_1} & & 0 \\ & \ddots & \\ & & \ddots \\ 0 & & \boxed{B_r} \end{pmatrix},$$

each block having the form

$$B_k = \begin{pmatrix} \lambda_k & 1 & & & 0 \\ & \ddots & \ddots & & \\ & & \ddots & \ddots & \\ & & & \ddots & 1 \\ 0 & & & & \lambda_k \end{pmatrix}_{k\times k} = \lambda_k I + N_k,$$

where $\lambda_k \in \mathrm{sp}_{\mathbb{K}}(A)$ and N_k is either a $k \times k$ elementary nilpotent matrix with $k > 1$ or a 1×1 block consisting of the scalar λ_k when the nilpotent part N_k is degenerate (recall the discussion surrounding equation (1.8)). Obviously,

$$\begin{aligned} e^{tA} &= S^{-1}\big(e^{t(SAS^{-1})}\big)S = S^{-1}e^{tJ(A)}S \\ &= S^{-1}\Big(\sum_{j=0}^{\infty}\frac{t^j}{j!}(J(A))^j\Big)\,S \\ &= S^{-1}\cdot\begin{pmatrix} e^{tB_1} & & & 0 \\ & \ddots & & \\ & & \ddots & \\ 0 & & & e^{tB_r} \end{pmatrix}\cdot S = S^{-1}e^{tB}S. \end{aligned}$$

Here, $e^{tB_k} = e^{t(\lambda_k I+N_k)} = e^{t\lambda_k}\cdot e^{tN_k}$ because $e^{A+B} = e^A \cdot e^B$ when A and B commute, so we get

$$e^{tB_k} = e^{t\lambda_k}\cdot\Big(\sum_{j=0}^{k-1}\frac{t^j}{j!}N^j\Big) = e^{t\lambda_k}\cdot\begin{pmatrix} 1 & t & \frac{t^2}{2!} & \cdots & \frac{t^{k-1}}{(k-1)!} \\ & \ddots & \ddots & & \vdots \\ & & \ddots & \ddots & \frac{t^2}{2!} \\ & & & \ddots & t \\ 0 & & & & 1 \end{pmatrix}_{k\times k}.$$

The matrices S and B_k can be computed explicitly and so can the scalars $e^{t\lambda_k}$. The solution of the original equation is then $\mathbf{x}(t) = e^{tA}\cdot\mathbf{c}$.

Solving Higher-Order Differential Equations. The simplest ordinary differential equation (ODE) has the form

$$\frac{dy}{dt} = k\cdot y(t) \quad (k \text{ a constant}, t \in \mathbb{R}),$$

and a solution is a differentiable function $y = y(t)$ defined on some interval $I \subseteq \mathbb{R}$ such that $dy/dt = k\cdot y(t)$ for all $t \in I$. This can be recast as an equation $Ly = 0$ for the linear operator

$$Ly = Dy - ky = \frac{dy}{dt} - k\,y(t)$$

defined on the infinite-dimensional vector space $\mathcal{C}^{(1)}(I)$ of continuously differentiable functions $y(t)$ on I. This is said to be a *first-order* equation because the highest order derivative appearing is $D = d/dt$. An nth-*order* ODE is obtained by substituting $x = D$ in a nonconstant polynomial $P(x) = \sum_{j=0}^{n} a_{n-j}x^j$ in $\mathbb{R}[x]$ to get a differential operator

$$L = P(D) = a_0D^n + a_1D^{n-1} + \cdots + a_{n-1}D + a_nI \quad (a_0 \neq 0). \tag{2.5}$$

This is obviously a *linear* operator on the space of functions with continuous derivatives of order $0 \leq j \leq n$. The problem is to find all solutions $y(t) \in \mathcal{C}^{(n)}(I)$ of the equations

$$\begin{aligned} \text{HOMOGENEOUS ODE:} \quad & Ly(t) \equiv 0 \text{ on } I, \\ \text{INHOMOGENEOUS ODE:} \quad & Ly(t) \equiv f(t) \text{ on } I, \end{aligned}$$

where $f(t)$ is a fixed continuous function. The leading coefficient a_0 is irrelevant in solving $Ly \equiv 0$ or $Ly = f$, so we shall hereafter assume $a_0 = 1$. One could allow variable coefficients in L, replacing the a_j by fixed smooth functions $a_j(t)$ defined on some interval $I \subseteq \mathbb{R}$, but then we must assume $a_0(t) \neq 0$ on I to ensure that the leading term $a_0(t)D^n$ dominates the lower order terms. In these *Notes*, we will mostly restrict attention to constant-coefficient ODE.

Solving a first-order linear ODE with constant-coefficients is easy: solutions have derivatives proportional to the original function, with $dy/dt = k\,y(t)$. The only such functions are the exponential functions

$$y(t) = Ce^{kt} \quad (C \text{ arbitrary, } t \in \mathbb{R}).$$

These are automatically $\mathcal{C}^\infty$ on all of $\mathbb{R}$ and are uniquely determined once an "initial value" is specified, say $y(0) = c$. For higher-order ODEs, the initial conditions are more complicated: values for the first n derivatives

$$y(0) = c_1,\ y'(0) = c_2,\ y^{(2)}(0) = c_3, \dots, y^{(n-1)}(0) = c_n$$

must be specified to pin down a unique solution to (2.2). Since $a_0 \neq 0$, the value of $y^{(n)}(0)$ is determined by the identity $Ly(0) = 0$.

Even the general first-order ODE with variable coefficients

$$p(t)\frac{dy}{dt} + q(t)y(t) \equiv 0 \quad \text{with } p(t) \neq 0 \text{ on } I$$

can be solved explicitly. If we want to have $y(t_0) = y_0$, the fundamental theorem of calculus implies that

$$y(t) = y_0 \cdot \left(e^{-\int_{t_0}^{t} \frac{q(s)}{p(s)}\,ds}\right) \quad t, t_0 \in I \tag{2.6}$$

has the desired properties. Unfortunately, many important differential equations in the physical sciences are of second order or higher and are not as easily solved even if they have constant coefficients.

EXERCISE 2.2. Verify that (2.6) is a solution of the equation

$$p(t)\frac{dy}{dt} + q(t)y(t) \equiv 0$$

on any interval I on which $p(t) \neq 0$. Check that $y(t_0) = y_0$.

General Comments on Constant Coefficient ODE. When an operator (2.5) is viewed as a linear map $L : \mathcal{C}^\infty(\mathbb{R}) \to \mathcal{C}^\infty(\mathbb{R})$, the full set of solutions $H_L(0) = \ker(L)$ of $Ly = 0$ in the infinite-dimensional space $\mathcal{C}^\infty(\mathbb{R})$ is *finite dimensional* with $\dim(\ker(L)) = n$, the degree of L.

If $f(t)$ is continuous there are two possibilities for the the set of solutions $H_L(f)$ of the inhomogeneous equation $Ly = f$, just as there were for linear operators $T : V \to V$ on finite-dimensional vector spaces.

> CASE 1. There is some "particular solution" $g(t)$ of $Lg = f$, and then $H_L(f) = g(t) + \ker(L)$.
>
> CASE 2. The particular inhomogeneous equation $Ly = f$ has *no* solutions in $\mathcal{C}^\infty(\mathbb{R})$, so $f \notin \text{range}(L)$ in $\mathcal{C}^\infty(\mathbb{R})$ and $H_L(f)$ is empty.

Then $Ly = 0$ has a solution for every initial condition

$$IC(y) : \quad \left(y^{(0)}(0), \dots, y^{(n-1)}(0)\right) = \mathbf{c},$$

and as $\mathbf{c}$ runs through the standard basis vectors $\mathbf{e}_j$ in $\mathbb{C}^n$, we get a basis for $H_L = \ker(L)$ consisting of $\mathcal{C}^\infty$ functions defined for all $t \in \mathbb{R}$. Because L is a linear operator, the combination $y(t) = \sum_{j=1}^{n} c_j f_j(t)$ is the unique solution such that $IC(y) = \mathbf{c}$.

Finally, if B_k is a Jordan block with $B_k = \lambda_k I + N_k$ (N_k a $k \times k$ elementary nilpotent matrix), we have

$$e^{tB_k} = e^{t\lambda_k} \cdot e^{tN_k} = e^{t\lambda_k} \cdot \left(\sum_{j=0}^{k-1} \frac{t^j}{j!} N^j \right)$$

$$= e^{t\lambda_k} \cdot \begin{pmatrix} 1 & t & \frac{t^2}{2!} & \cdots & \frac{t^{k-1}}{(k-1)!} \\ & \ddots & \ddots & & \vdots \\ & & \ddots & \ddots & \frac{t^2}{2!} \\ & & & 1 & t \\ 0 & & & & 1 \end{pmatrix}$$

because $\lambda_k I$ and N_k commute.

Transforming an ODE to a First-Order System $dX/dt = A \cdot X(t)$. The following transformation of a higher-order constant-coefficient ODE yields an equivalent system of linear *first-order* ODE

$$\frac{dZ}{dt} = A \cdot Z(t) \quad \text{with values} \quad Z(t) \in \mathbb{R}^n.$$

This can be solved explicitly in closed form using the Jordan form. Given an equation (2.5), relabel the variables $y = y^{(0)}, y^{(1)}, \dots, y^{(n-1)}$ letting

$$z_1 = y^{(0)}, z_2 = y^{(1)}, \dots, z_n = y^{(n-1)}.$$

When the variables are arranged in $n\times 1$ column vectors, the equation $Ly(t) \equiv 0$ is equivalent to the system of identities

$$\begin{aligned}
\frac{dz_1}{dt} &= y^{(1)} = z_2(t)\\
\frac{dz_2}{dt} &= y^{(2)} = z_3(t)\\
&\vdots\\
\frac{dz_{n-1}}{dt} &= y^{(n-1)} = z_n(t)\\
\frac{dz_n}{dt} &= y^{(n)} = -\frac{a_n}{a_0}\,z_1(t) - \frac{a_{n-1}}{a_0}\,z_2(t)\ \cdots - \frac{a_2}{a_0}\,z_{n-1}(t) - \frac{a_1}{a_0}\,z_n(t).
\end{aligned} \tag{2.7}$$

An ODE with complex coefficients, whose solutions $y(t)$ take values in $\mathbb{C}$, can be handled the same way.

If we let $Z(t) = \text{col}(z_1(t), \dots, z_n(t))$, then

$$\frac{dZ}{dt} = \text{col}\left(\frac{dz_1}{dt}, \dots, \frac{dz_n}{dt}\right)$$

and (2.7) becomes the first-order system

$$\frac{dZ}{dt} = A \cdot Z(t)$$

with coefficient matrix

$$A = \begin{pmatrix} 0 & 1 & 0 & \cdots & 0 & 0\\ 0 & 0 & 1 & & 0 & 0\\ \vdots & \vdots & \vdots & \vdots & \vdots & \vdots\\ 0 & 0 & 0 & & 0 & 1\\ -\dfrac{a_n}{a_0} & -\dfrac{a_{n-1}}{a_0} & -\dfrac{a_{n-2}}{a_0} & \cdots & -\dfrac{a_2}{a_0} & -\dfrac{a_1}{a_0} \end{pmatrix}.$$

Initial conditions $y(0) = c_0, \dots, y^{(n-1)}(0) = c_{n-1}$ for a solution of $Ly = 0$ correspond to the single vector condition

$$Z(0) = \mathbf{c} = (c_0, \dots, c_{n-1})$$

for the system. Once a solution of $dZ/dt = A \cdot Z(t)$ with $Z(0) = \mathbf{c}$ is in hand, we can read out a solution $y^{(0)}(t) = z_1(t)$ of the original ODE such that $y^{(k)}(0) = c_{k+1}$.

EXAMPLE 2.3. Assuming that $a_0 = 1$, the preceding change of variables

$$z_1 = y^{(0)},\ z_2 = y^{(1)},\ z_3 = y^{(2)} \quad (y^{(k)} = d^k/dt^k)$$

transforms the 2nd-order differential equation

$$Ly = y^{(2)} + a_1 y^{(1)} + a_2 y^{(0)} = 0 \quad (a_1, a_2 \in \mathbb{C})$$

into a system $dZ/dt = A \cdot Z(t)$ of first-order equations. Since

$$\frac{dz_2}{dt} = y^{(2)} = -a_1 y^{(1)} - a_2 y^{(0)} = -a_1 z_2 - a_2 z_1,$$

we end up with the system

$$\begin{pmatrix} dz_1/dt \\ dz_2/dt \end{pmatrix} = \begin{pmatrix} z_2 \\ -a_1z_2 - a_2z_1 \end{pmatrix} = \begin{pmatrix} 0 & 1 \\ -a_2 & -a_1 \end{pmatrix} \cdot \begin{pmatrix} z_1 \\ z_2 \end{pmatrix},$$

in which

$$A = \begin{pmatrix} 0 & 1 \\ -a_2 & -a_1 \end{pmatrix} \quad \text{and} \quad Z(t) = e^{tA} \cdot Z(0).$$

We shall determine the Jordan form and a transition matrix Q such that $QAQ^{-1} = J(A)$ in order to find $Z(t)$.

DISCUSSION. The roots of the characteristic polynomial

$$p_A(x) = \det \begin{pmatrix} -x & 1 \\ -a_2 & -a_1 - x \end{pmatrix} = x^2 + a_1x + a_2 \in \mathbb{C}[x]$$

are

$$\lambda_\pm = \frac{-a_1 \pm \sqrt{a_1^2 - 4a_2}}{2}.$$

Thus, A is diagonalizable with $QAQ^{-1} = D = \mathrm{diag}(\lambda_+, \lambda_-)$ and $J(A) = D$ unless p_A has a double root. That happens if and only if $a_1^2 - 4a_2 = 0$, and then $\lambda_\pm = \lambda = -\frac{1}{2}a_1$; when there is a double root, A could still be diagonalizable by being equal to $\lambda I_{2\times 2}$.

REMARK 2.4 (Nonuniqueness of Jordan Forms). As mentioned earlier, the idea behind the Jordan form $J(A)$ is to select a nice representative $J(A)$ from the similarity class of A in $\mathrm{M}(2, \mathbb{C})$,

$$\mathrm{GL} \cdot A = \{SAS^{-1} : \det(S) \neq 0\},$$

but such representatives are only unique "*modulo a similarity transformation.*" For instance, if $A \in \mathrm{M}(2, \mathbb{C})$ is diagonalizable with distinct entries $SAS^{-1} = \mathrm{diag}(\lambda_1, \lambda_2)$, saying that "$J(A) = \mathrm{diag}(\lambda_1, \lambda_2)$" is ambiguous: we obtain a different diagonal representative $PJ(A)P^{-1} = \mathrm{diag}(\lambda_2, \lambda_1) = J(A)^\mathrm{T}$ in the same similarity class if we permute the basis vectors, taking

$$P = \begin{pmatrix} 0 & 1 \\ 1 & 0 \end{pmatrix}.$$

So which one has the greater claim to be the Jordan form of A? There is no consistent answer. If the λ_i are real, we might try impose uniqueness by listing them according to their order on the real line (but should it be "ascending" or "descending" order?). Even that can't work if some entries are complex. Despite these ambiguities, the Jordan form has proved to be such a useful concept that we have learned to live with the ambiguity. ○

EXERCISE 2.5. Let $A \in \mathrm{M}(n, \mathbb{C})$.

(a) Is $J(A) = I_{n\times n}$ if A is invertible?

(b) If $c \in \mathbb{C}$, how are $J(cA)$ and $J(A)$ related?

(c) If $B = QAQ^{-1}$ for some invertible matrix Q, how are $J(B)$ and $J(A)$ related?

The Case of a Double Root. Now we have to determine $J(A)$ directly. In this setting, there is just one eigenvalue, $A - \lambda I$ is nilpotent, and $M_\lambda(A)$ is all of $V = \mathbb{C}^2$, so there are just two possibilities for the Jordan form:

$$J(A) = \begin{pmatrix} \lambda & 0 \\ 0 & \lambda \end{pmatrix} \quad \text{or} \quad J(A) = \begin{pmatrix} \lambda & 1 \\ 0 & \lambda \end{pmatrix}.$$

The outcome is determined by computing the kernel $K = \ker(A - \lambda I)$. Unless $K = V$ (in which case $A = \lambda I, J(A) = \lambda I$, and the transition map Q is irrelevant), we have $\dim(K) = 1$. Then it is easy to find a cyclic subspace decomposition for $V = M_\lambda$. Following the procedure outlined in Figure 1.1, there are two possibilities: either $a_1 \neq 0$ (hence $a_2 \neq 0$ since $a_1^2 - 4a_2 = 0$) or

$$a_1 = 0 \quad \text{and} \quad Ly = \frac{d^2y}{dt^2} + a_2 y = 0.$$

The exceptional case $a_1 = 0$ is of great physical interest and will be taken up separately.

If $a_1 \neq 0$ and λ is a double root of $p_A(t)$, we have $V = M_\lambda(A)$ and must find a cyclic subspace decomposition of the nilpotent operator

$$(A - \lambda I) = \begin{pmatrix} -\lambda & 1 \\ -a_2 & -a_1 - \lambda \end{pmatrix} = \begin{pmatrix} a_1/2 & 1 \\ -a_2 & -a_1/2 \end{pmatrix}.$$

Since $a_1 \neq 0$, row operations put $(A - \lambda I)$ in the form $\begin{pmatrix} 1 & 2/a_1 \\ 0 & 0 \end{pmatrix}$. Hence

$$K = \{\mathbf{x} \in \mathbb{C}^2 : (A - \lambda I)\mathbf{x} = \mathbf{0}\} \quad \text{is} \quad \mathbb{C} \cdot \begin{pmatrix} -2/a_1 \\ 1 \end{pmatrix}.$$

We claim that V is already a cyclic subspace for $(A - \lambda I)$. For a cyclic vector, take any $\mathbf{f}_1 \in V \sim K$, for instance the standard basis vector $\mathbf{e}_2 = (0, 1)$. Then

$$\mathbf{f}_2 = (A - \lambda)\,\mathbf{f}_1 = \begin{pmatrix} 1 \\ -a_1/2 \end{pmatrix} \quad \text{in } K,$$

and $(A - \lambda I)$ maps $0 \xleftarrow{A-\lambda I} \mathbf{f}_2 \xleftarrow{A-\lambda I} \mathbf{f}_1$. Taking $\mathfrak{Y} = \{\mathbf{f}_1, \mathbf{f}_2\}$, we have

$$[A - \lambda I]_{\mathfrak{Y}\mathfrak{Y}} = \begin{pmatrix} 0 & 1 \\ 0 & 0 \end{pmatrix} \quad \text{and} \quad [A]_{\mathfrak{Y}\mathfrak{Y}} = \begin{pmatrix} \lambda & 1 \\ 0 & \lambda \end{pmatrix}$$

as expected.

To find a transition matrix Q such that $J(A) = QAQ^{-1}$, recall the discussion of "change of variables" in Section 2.4 of LA I and observe that

$$\begin{aligned} [A - \lambda I]_{\mathfrak{X}\mathfrak{X}} &= [\mathrm{id}]_{\mathfrak{X}\mathfrak{Y}} \cdot [(A - \lambda I)]_{\mathfrak{Y}\mathfrak{Y}} \cdot [\mathrm{id}]_{\mathfrak{Y}\mathfrak{X}} \\ &= [\mathrm{id}]_{\mathfrak{X}\mathfrak{Y}} \begin{pmatrix} 0 & 1 \\ 0 & 0 \end{pmatrix} [\mathrm{id}]_{\mathfrak{Y}\mathfrak{X}} \\ &= Q^{-1} J(A) Q. \end{aligned}$$

Taking $Q^{-1} = [\mathrm{id}]_{\mathfrak{X}\mathfrak{Y}}$, which is the transpose of the coefficient array in the system of vector identities

$$\mathbf{f}_1 = 0 \cdot \mathbf{e}_1 + 1 \cdot \mathbf{e}_2,$$
$$\mathbf{f}_2 = \mathbf{e}_1 - \frac{1}{2} a_1 \, \mathbf{e}_2,$$

that connects the bases $\mathfrak{X} = \{\mathbf{e}_1, \mathbf{e}_2\}$ and $\mathfrak{Y} = \{\mathbf{f}_1, \mathbf{f}_2\}$, we conclude that

$$Q = \begin{pmatrix} 0 & 1 \\ 1 & -a_1/2 \end{pmatrix}^{-1} = \frac{1}{(-1)} \begin{pmatrix} -a_1/2 & -1 \\ -1 & 0 \end{pmatrix} = \begin{pmatrix} a_1/2 & 1 \\ 1 & 0 \end{pmatrix}.$$

That completes our discussion of $Ly = 0$ when $a_1 \neq 0$. ○

Now consider the exceptional case $\dfrac{d^2y}{dt^2} - ky(t) = 0$ in which $a_1 = 0$.

EXAMPLE 2.6. When $a_1 = 0$ in Example 2.3, we have $L = D^2 + kI$ with $k \neq 0$. Then

$$A = \begin{pmatrix} 0 & 1 \\ -k & 0 \end{pmatrix} \qquad (A - \lambda I) = \begin{pmatrix} -\lambda_+ & 1 \\ -k & -\lambda_- \end{pmatrix},$$

and $p_A(\lambda) = \lambda^2 + k\mathbb{1}$ has distinct roots $\lambda_\pm = \pm\sqrt{-k}$, which are complex when $k > 0$. Thus, A is diagonalizable over $\mathbb{C}$ with Jordan form

$$J(A) = \begin{pmatrix} \lambda_+ & 0 \\ 0 & \lambda_- \end{pmatrix} \qquad (\lambda_- = -\lambda_+).$$

Solutions of $dZ/dt = A \cdot Z(t)$ have the form $Z(t) = e^{tA} \cdot Z(0)$ from which we can extract the desired solution $y = y^{(0)}(t)$ of $Ly = 0$ because $Z(t) = (z_1(t), z_2(t)) = (y^{(0)}(t), y^{(1)}(t))$.

This essentially solves the problem, but to get explicit solutions $y = y^{(0)}(t)$, we must find a matrix Q such that $QAQ^{-1} = D$ where $D = \mathrm{diag}(\lambda_+, \lambda_-) = [L_A]_{\mathfrak{Y}\mathfrak{Y}}$. This will allow us to write e^{tA} as a product

$$e^{tA} = Q^{-1} \cdot e^{t(QAQ^{-1})} \cdot Q = Q^{-1} e^{tD} Q.$$

When the eigenvalues are distinct there is a more direct way to find a transition matrix Q such that $QAQ^{-1} = D = \mathrm{diag}\,(\lambda_+, \lambda_-)$. If the columns in $Q = [\mathbf{f}_+; \mathbf{f}_-]$ are eigenvectors with $A\mathbf{f}_\pm = \lambda_\pm \mathbf{f}_\pm$, then $\mathfrak{Y} = \{\mathbf{f}_+, \mathbf{f}_-\}$ is a diagonalizing basis for A. If $\mathfrak{X} = \{\mathbf{e}_+, \mathbf{e}_-\}$ is the standard basis in $\mathbb{C}^2$, we have

$$Q^{-1}AQ\mathbf{e}_\pm = Q^{-1}A\mathbf{f}_\pm = \lambda_\pm Q^{-1}\mathbf{f}_\pm = \lambda_\pm \mathbf{e}_\pm = D(\mathbf{e}_\pm),$$

so $Q^{-1}AQ = D$. Once Q is in hand we get

$$\begin{aligned} Z(t) &= e^{tA} \cdot Z(0) = Q^{-1} e^{tD} Q \cdot Z(0) \\ &= Q^{-1} \begin{pmatrix} e^{t\sqrt{-k}} & 0 \\ 0 & e^{-t\sqrt{-k}} \end{pmatrix} Q \cdot Z(0) = \begin{pmatrix} y^{(0)}(t) \\ y^{(1)}(t) \end{pmatrix}, \end{aligned}$$

from which we can read off the solution to $Ly = 0$.

The eigenspaces are easily determined. Symbolic row operations yield

$$E_{\lambda_\pm} = \ker(A - \lambda_\pm I) = \mathbb{C} \cdot \begin{pmatrix} \pm 1/\sqrt{-k} \\ 1 \end{pmatrix},$$

so that

$$\begin{aligned} \mathbf{f}_+ &= (1/\sqrt{-k})\mathbf{e}_1 + \mathbf{e}_2 \quad \text{in } E_{\lambda_+}, \\ \mathbf{f}_- &= (-1/\sqrt{-k})\mathbf{e}_1 + \mathbf{e}_2 \quad \text{in } E_{\lambda_-}, \end{aligned}$$

is a diagonalizing basis for A. Since $[\mathrm{id}]_{\mathfrak{X}\mathfrak{Y}}$ is the transpose of the coefficient array in this system, we get

$$\begin{aligned} Q &= [\mathrm{id}]_{\mathfrak{X}\mathfrak{Y}}^{-1} = \frac{1}{2}\begin{pmatrix} \sqrt{-k} & 1 \\ -\sqrt{-k} & 1 \end{pmatrix}, \\ Q^{-1} &= [\mathrm{id}]_{\mathfrak{X}\mathfrak{Y}} = \begin{pmatrix} 1/\sqrt{-k} & -1/\sqrt{-k} \\ 1 & 1 \end{pmatrix}. \end{aligned}$$

Thus, we arrive at

$$\begin{aligned} Z(t) = e^{tA}Z(0) = Q^{-1}e^{tD}Q \cdot Z(0) &= Q^{-1} \cdot \begin{pmatrix} e^{t\sqrt{-k}} & 0 \\ 0 & e^{-t\sqrt{-k}} \end{pmatrix} Q \cdot Z(0) \\ &= \begin{pmatrix} 1/\sqrt{-k} & -1/\sqrt{-k} \\ 1 & 1 \end{pmatrix} \cdot \begin{pmatrix} e^{t\sqrt{-k}} & 0 \\ 0 & e^{-t\sqrt{-k}} \end{pmatrix} \cdot \frac{1}{2}\begin{pmatrix} \sqrt{-k} & 1 \\ -\sqrt{-k} & 1 \end{pmatrix} \cdot Z(0) \\ &= \begin{pmatrix} \cosh(t\sqrt{-k}) & \frac{1}{\sqrt{-k}}\sinh(t\sqrt{-k}) \\ \sqrt{-k}\sinh(t\sqrt{-k}) & \cosh(t\sqrt{-k}) \end{pmatrix} \cdot Z(0). \end{aligned}$$

so that

$$\begin{pmatrix} y^{(0)}(t) \\ y^{(1)}(t) \end{pmatrix} = Z(t) = e^{tA} \cdot Z(0).$$

Taking initial values $Z(0) = \mathbf{e}_1 = (1,0)$ or $Z(0) = \mathbf{e}_2 = (0,1)$ and rescaling, we get linearly indepent real-valued solutions of $Ly = 0$:

$$y_1(t) = y^{(0)}(t) = \cosh(t|k|^{1/2}) \quad \text{and} \quad y_2(t) = \frac{1}{\sqrt{-k}}\sinh(t|k|^{1/2})$$

when $k < 0$, which become

$$y_1(t) = \cos(t|k|^{1/2}) \quad \text{and} \quad y_2(t) = \sin(t|k|^{1/2})$$

if $k > 0$. ○

Recall that for a complex variable $z = x + iy$ we have

$$\begin{aligned} e^z &= |z|(\cos x + i \sin y) \\ \cosh(z) &= \frac{e^z + e^{-z}}{2} \quad \text{and} \quad \cos(z) = \frac{e^{iz} + e^{-iz}}{2} \\ \sinh(z) &= \frac{e^z - e^{-z}}{2} \quad \text{and} \quad \sin(z) = \frac{e^{iz} - e^{-iz}}{2i}, \end{aligned}$$

and the preceding formulas are easy consequences.

REMARK 2.7. For $k \in \mathbb{R}$, the differential equations $Ly = 0$ and $Ly = f$ involving the operators

$$L = \frac{d^2y}{dt^2} + ky(t)$$

have been important in physics since the time of Issac Newton. We mention just a few ways they appear in classical mechanics.

- NEWTON'S LAW OF MOTION. $F = ma$ describes the instantaneous response of a body subjected to an external force $F(x, t)$ that may depend on time and position. (Think of a point mass moving in one dimension for simplicity.) Newton's crucial observation was that the force F and mass m of the object are all that matter in determining the moment-to-moment *acceleration* $a(t) = d^2x/dt^2$ of the object. Thus, Newton's law is a differential equation, and one of second order at that:

$$m\frac{d^2x}{dt^2} - F(x, t) \equiv 0 \quad \text{for all } t.$$

  To find the quantities of interest, the position function $x(t)$ and velocity function $x'(t)$, we must solve this ODE.

- MOTION OF A PENDULUM. When a pedulum (a weight attached to a string) is moved away from its resting position and released, the balance between the downward force of gravity and the restraining force exerted by the string causes the weight to feel a net force pushing it back toward its equilibrium position. From the geometry of the situation, one can deduce that for small displacement this force is essentially equal to a constant $k > 0$ times the horizontal displacement $x(t)$ of the mass. For small initial displacements, its motion is then governed by the differential equation

$$m\frac{d^2x}{dt^2} + kx(t) \equiv 0 \quad \text{for all } t \tag{2.8}$$

  since $F = -kx(t)$. Owing to the $(-)$ sign appearing here, the lateral force felt by the mass attached to the string is always opposite to the horizontal dispacement, which is why the pendulum bob oscillates instead of flying off into the distance.

  An even better example is the motion of a mass suspended on a spring. Left alone, the system will find a stable equilibrium position, but it is the nature of a spring to resist displacements from that position, exerting a force proportional to the displacement but in the opposite direction. Now, the motion is exactly governed by the differential equation (2.8), and the solutions are periodic up and down oscillations about the equilibrium position. Now the constant $k > 0$ reflects the strength of the spring.

Versions of (2.8) in which the constant k is a complex number can arise in certain models of physical situations. ○

We end this section with an exercise involving a third-order differential operator whose coefficients are real, but $p_A(x)$ has some complex eigenvalues and there may be solutions of $A(\mathbf{x}) = \mathbf{b}$ with $\mathbf{x}(t)$ in $\mathbb{C}^n$. We can always find complex vector-valued solutions to (2.4) by regarding $A \in \mathrm{M}(n, \mathbb{R})$ as a complex matrix with real entries; then the solutions will have complex values $\mathbf{z}(t) = \mathbf{x}(t) + i\mathbf{y}(t)$ in $\mathbb{C}^n$, whose real and imaginary parts $\mathbf{x}(t)$ and $\mathbf{y}(t)$ are both real-valued solutions since

$$\frac{d\mathbf{z}}{dt} = \frac{d\mathbf{x}}{dt} + i\frac{d\mathbf{y}}{dt},$$

and these will include all the real vector-valued solutions.

EXERCISE 2.8. Recast the third-order ordinary differential equation

$$Ly = \frac{d^3y}{dt^3} - \frac{d^2y}{dt^2} - \frac{dy}{dt} + y(t) = 0$$

as a first-order constant-coefficient system

$$\frac{dZ}{dt} = A \cdot Z(t) \quad \text{with } Z(0) = (c_1, \dots, c_3).$$

Find the Jordan form of the coefficient matrix A. Then give a general formula for the solutions $y(t) = e^{tA} \cdot Z(0)$, and find a basis for the full set of solutions to $Ly = 0$ in $\mathcal{C}^\infty(\mathbb{R})$.

HINT. The characteristic polynomial $p_A(x)$ of the matrix A in the associated first-order system $dZ/dt = A \cdot Z(t)$ has $\lambda = 1$ as a root.

2.2. Normal Forms for Linear Operators over $\mathbb{R}$

How can one analyze a linear operator $T : V \to V$ when the characteristic polynomial $p_T(x)$ does not split over $\mathbb{K}$? One approach is via the "*rational canonical form*," which makes no attempt to replace the ground field $\mathbb{K}$ with a larger field of scalars $\widetilde{\mathbb{K}}$ over which p_T might split. We will not pursue that topic in these *Notes*. A different approach, which we will illustrate for $\mathbb{K} = \mathbb{R}$, is to examine an enlarged field of scalars $\widetilde{\mathbb{K}} \supseteq \mathbb{K}$ (a "field extension"). Then we may in an obvious way regard $\mathbb{K}[x]$ as a subalgebra within $\widetilde{\mathbb{K}}[x]$, and since $\widetilde{\mathbb{K}} \supseteq \mathbb{K}$, there is a better chance that $f(x)$ will have roots in $\widetilde{\mathbb{K}}$ and split into linear factors in $\widetilde{\mathbb{K}}[x]$:

$$f(x) = c \cdot \prod_{j=1}^{d} (x - \mu_j)^{m_j} \quad \text{with } c \text{ and } \mu_j \text{ in } \widetilde{\mathbb{K}}. \tag{2.9}$$

It is, in fact, always possible to embed $\mathbb{K}$ in a field $\widetilde{\mathbb{K}}$ that is *algebraically closed*, which means that every polynomial $f(x) \in \widetilde{\mathbb{K}}[x]$ has roots in $\widetilde{\mathbb{K}}$ and splits as in (2.9). The fundamental theorem of algebra asserts that the complex number field is algebraically closed, but the real number system $\mathbb{R}$ is not — for instance, x^2+1 and $x^2+x+2 \in \mathbb{R}[x]$ have no roots in $\mathbb{R}$ and cannot split into linear factors in $\mathbb{R}[x]$. However, they do split when regarded as an elements of $\mathbb{C}[x]$; for example,

$$x^2+x+1 = (x-z_+)\cdot(x-z_-) = \left(x-\frac{1}{2}(-1+i\sqrt{3})\right)\cdot\left(x+\frac{1}{2}(-1-i\sqrt{3})\right)$$

where $i=\sqrt{-1}$. In this simple example one can find the complex roots $z_\pm = \frac{1}{2}(-1\pm i\sqrt{3})$ using the quadratic formula.

Complexification: A Case Study. "Complexification" is a fancy word for a process you have already encountered by other names. In its simplest form, you learned to solve quadratic equations $ax^2+bx+c=0$ having coefficients in $\mathbb{R}$ by passing from the real numbers to the larger field of complex numbers $\mathbb{C}$. This is what lets us make sense of all the solutions provided by the quadratic formula

$$x_\pm = \frac{-b\pm\sqrt{b^2-4ac}}{2a}.$$

This perplexed mathematicians when the discriminant (b^2-4ac) turned out to be negative — the simplest example being the quadratic equation $x^2+1=0$ whose solutions $x=\pm\sqrt{-1}$ were deemed "imaginary" because no real number can have $x^2=-1$ (the complex number system had not yet been invented).

Nowadays this is familiar stuff, and we have already encountered this shift from $\mathbb{R}$ to $\mathbb{C}$ — for instance, to find and interpret the complex eigenvalues that can arise when you solve the polynomial equation

$$p_A(x) = \det(A-xI) = 0,$$

even if the entries in A are real. The point of this section is to provide a deeper understanding of this "complexification" process and its use to reveal the structure of matrix multiplication operators on $\mathbb{R}^n$

$$\mathbf{y} = L_A(\mathbf{x}) = A\cdot\mathbf{x} \quad \text{with} \quad y_j = \sum_{k=1}^{n} a_{jk}x_k \tag{2.10}$$

determined by matrices $A \in \mathrm{M}(n,\mathbb{R})$. Obviously, the same formula could be applied to n-tuples $\mathbf{z} = (z_1,\dots,z_n)$ having *complex* entries $z_j = x_j + iy_j$ with $x_j, y_j \in \mathbb{R}$. This yields an operator $\tilde{L}_A : \mathbb{C}^n \to \mathbb{C}^n$ given by

$$\mathbf{w} = \tilde{L}_A(\mathbf{z}) = A\cdot\mathbf{z} = A\cdot[\mathbf{x}] + iA\cdot[\mathbf{y}]$$

with $w_j = \sum_{k=1}^n a_{jk}z_k$. While L_A was $\mathbb{R}$-linear on $\mathbb{R}^n$, the "extension" $\tilde{L}_A$ to $\mathbb{C}^n$ turns out to be a complex-linear operator on the complex vector space $\mathbb{C}^n$, whose elements will hereafter be written $\mathbf{z} = \mathbf{x}+i\mathbf{y}$ if $\mathbf{z} = (z_1,\dots,z_n)$ with $z_k = x_k + iy_k = x_k + \sqrt{-1}y_k$.

EXERCISE 2.9. Verify that $\tilde{L}_A$ is a complex-linear map on $\mathbb{C}^n$ for any A in $\mathrm{M}(n,\mathbb{R})$ by showing that

(a) $\tilde{L}_A\big((a+ib)\cdot\mathbf{z}\big) = (a+ib)\cdot\tilde{L}_A(\mathbf{z})$

(b) $\tilde{L}_A(\mathbf{z}+\mathbf{w}) = \tilde{L}_A(\mathbf{z}) + \tilde{L}_A(\mathbf{w})$

for all $\mathbf{z},\mathbf{w}\in\mathbb{C}^n$ and $(a+ib)\in\mathbb{C}$.

If we restrict $\tilde{L}_A$ to the "real points" $\mathbf{z} = \mathbf{x} + i\mathbf{0}$ in $\mathbb{C}^n$ with $\mathbf{x}\in\mathbb{R}^n$, we find that $\tilde{L}_A$ is an extension of L_A in the following sense:

$$\tilde{L}_A(\mathbf{x}+i0) = L_A(\mathbf{x}) + i0 \quad \text{for } \mathbf{x}\in\mathbb{R}^n.$$

Therefore, it is natural to think of $\mathbb{C}^n$ as a "complexification" of $\mathbb{R}^n$ because $\mathbb{C}^n = \mathbb{R}^n + i\mathbb{R}^n$, and then $T_{\mathbb{C}} = \tilde{L}_A : \mathbb{C}^n \to \mathbb{C}^n$ becomes the complexification of the $\mathbb{R}$-linear operator $T = L_A$ on $\mathbb{R}^n$.

Complexification of Real Vector Spaces and Operators. We can do the same for any vector space V over $\mathbb{R}$, defining its *complexification* $V_{\mathbb{C}}$ to be the set of the formal sums $z = x + iy$ with $x, y\in V$. The same arguments as in Exercise 2.9 show that $V_{\mathbb{C}} = V + iV$ is a vector space over $\mathbb{C}$ if we define

$$\begin{aligned}(a+ib)\cdot(x+iy) &= (ax-by) + i(ay+bx)\\ (x+iy)+(u+iv) &= (x+u) + i(y+v)\end{aligned} \tag{2.11}$$

for $a,b\in\mathbb{R}$, $x,y\in V$. If $T : V\to V$ is a linear operator between real vector spaces, its complexification $T_{\mathbb{C}} : V_{\mathbb{C}} \to V_{\mathbb{C}}$ is the $\mathbb{C}$-linear map

$$T_{\mathbb{C}}(x+iy) = T(x) + iT(y) \quad \text{in } W_{\mathbb{C}} \text{ for any } x,y\in V. \tag{2.12}$$

The characteristic polynomial $p_{T_{\mathbb{C}}}(t) = \det(T_{\mathbb{C}} - tI)\in\mathbb{C}[t]$ turns out to be the same as $p_T(t)$ when we view $p_T(t)\in\mathbb{R}[t]$ as a polynomial in $\mathbb{C}[t]$ that happens to have real coefficients. Since $p_{T_{\mathbb{C}}}(t)$ always splits over $\mathbb{C}$, the theory developed in Chapter 1 applies to $T_{\mathbb{C}}$. Our task is then to interpret that theory in terms of the original real linear operator $T:\mathbb{R}^n\to\mathbb{R}^n$.

EXERCISE 2.10. If T is a linear operator from $\mathbb{R}^n\to\mathbb{R}^n$ and $T_{\mathbb{C}} : \mathbb{C}^n\to\mathbb{C}^n$ is its complexification as defined in (2.11) and (2.12), verify the following:

(a) $T_{\mathbb{C}}$ is in fact a $\mathbb{C}$-linear operator on $\mathbb{C}^n$.

(b) The characteristic polynomials $p_T(t)$ in $\mathbb{R}[t]$ and $p_{T_{\mathbb{C}}}(t)$ in $\mathbb{C}[t]$ are the same, with

$$p_{T_{\mathbb{C}}}(t) = p_T(t)$$

when we identify $\mathbb{R}[t]\subseteq\mathbb{C}[t]$.

HINT. Polynomials in $\mathbb{K}[t]$ are equal $\Leftrightarrow$ they have the same coefficients. Here, $p_{T_{\mathbb{C}}}$ has coefficients in $\mathbb{C}$ while p_T has coefficients in $\mathbb{R}$, but we are identifying $\mathbb{R} = \mathbb{R} + i0 \subseteq \mathbb{C}$.

Relations Between $\mathbb{R}[t]$ and $\mathbb{C}[t]$. When we view a real-coefficient polynomial $f(t) = \sum_{j=0}^{m} a_j t^j \in \mathbb{R}[t]$ as an element of $\mathbb{C}[t]$, it can have complex roots as well as real roots. However,

> COMPLEX ROOTS OF A REAL POLYNOMIAL. *When we view $f(t) \in \mathbb{R}[x]$ as an element of $\mathbb{C}[x]$, nonreal roots must occur in conjugate pairs $z_{\pm} = (u \pm iv)$ with u, v real and $v \neq 0$. Such roots can have nontrivial multiplicities, resulting in paired factors*
> $$(t - z_+)^m \cdot (t - z_-)^m$$
> *in the irreducible factorization of $f(t)$ in $\mathbb{C}[t]$.*

In fact, if $z = x+iy$ with x, y real and $f(z) = 0$, the complex conjugate $\overline{z} = x-iy$ is also a root of f because

$$f(\overline{z}) = \sum_{j=0}^{m} a_j \overline{z}^j = \overline{\sum_j a_j z^j} = \overline{f(z)} = 0$$

(Recall that $\overline{z+w} = \overline{z} + \overline{w}$, $\overline{zw} = \overline{z} \cdot \overline{w}$, and $(\overline{z})^- = z$ for $z, w \in \mathbb{C}$.) Thus, the number of nonreal roots is even if any exist, while the number of real roots is unrestricted and might be zero. The splitting of f in $\mathbb{C}[t]$ can then be written as

$$\begin{aligned} f(t) &= c \cdot \prod_{j=1}^{r} (t - \mu_j)^{m_j} (t - \overline{\mu}_j)^{m_j} \cdot \prod_{k=r+1}^{s} (t - r_j)^{m_k} \\ &= c \cdot \prod_{j=1}^{r} \big((t - \mu_j)(t - \overline{\mu}_j)\big)^{m_j} \cdot \prod_{k=r+1}^{s} (t - r_j)^{m_k} \end{aligned} \tag{2.13}$$

where the μ_j are complex and nonreal ($\overline{\mu} \neq \mu$) and the r_j are the distinct real roots of f. Obviously

$$n = \deg(f) = \sum_{j=1}^{r} 2m_j + \sum_{j=r+1}^{s} m_j .$$

Since f has real coefficients, all complex numbers must disappear when the products involving nonreal roots are expanded. In particular, each nonreal conjugate pair $\mu, \overline{\mu}$ yields a quadratic with real coefficients. Hence,

$$Q_\mu(t) = (t - \mu)(t - \overline{\mu}) = t^2 - 2\mathrm{Re}(\mu)\, t + |\mu|^2 \tag{2.14}$$

and

$$f(t) = c \cdot \prod_{r+1}^{s} (t - r_j)^{m_j} \cdot \prod_{j=1}^{r} (Q_{\mu_j}(t))^{m_j}$$

is a factorization of $f(t)$ into linear and *irreducible* quadratic factors in $\mathbb{R}[t]$, and every $f \in \mathbb{R}[t]$ can be decomposed this way. This is the (unique) decomposition of $f(t)$ into irreducible factors in $\mathbb{R}[t]$. By definition, the $Q_\mu(t)$ have no real roots and cannot be products of polynomials of lower degree in $\mathbb{R}[t]$, while the linear factors $(t - r_j)$ are irreducible over $\mathbb{R}$ as they stand. The following terminology is standard.

DEFINITION 2.11. A nonconstant polynomial $f \in \mathbb{K}[t]$ is *irreducible* if it cannot be factored as $f(t) = g(t)h(t)$ with g, h nonconstant, hence of lower degree than f. A polynomial is *monic* if its leading coefficient is 1. It is well known that every monic $f \in \mathbb{K}[t]$ factors uniquely as $\prod_{j=1}^{r} h_j(t)^{m_j}$ where each h_j is monic and irreducible in $\mathbb{K}[t]$. The exponent $m_j \geq 1$ is the multiplicity of h_j, and this factorization is unique.

The simplest irreducibles (over any $\mathbb{K}$) are the linear polynomials $at + b$ (with $a \neq 0$ since "irreducibility" only applies to nonconstant polynomials). This follows from the *degree formula*:

> DEGREE FORMULA. *We have* $\deg(gh) = \deg(g) + \deg(h)$ *for all nonzero* $g, h \in \mathbb{K}[t]$.

If we could factor $at + b = g(t)h(t)$, either $g(t)$ or $h(t)$ would have degree 0, which is not allowed in a nontrivial factorization. Thus $at+b$ has no nontrivial factorization. When $\mathbb{K} = \mathbb{C}$, *all* irreducible polynomials have *degree* $= 1$, but if $\mathbb{K} = \mathbb{R}$, they can have *degree* $= 1$ or 2.

LEMMA 2.12. *The irreducible monic polynomials in* $\mathbb{R}[t]$ *have the form*

1. $(t - r)$ *with* $r \in \mathbb{R}$, *or*
2. $t^2 + bt + c$ *with* $b^2 - 4c < 0$. *These are the polynomials* $(t - \mu)(t - \overline{\mu})$ *with* μ *a nonreal element in* $\mathbb{C}$.

PROOF. Linear polynomials $at + b\,(a \neq 0)$ in $\mathbb{R}[t]$ are irreducible, as noted above. If f has the form (2.6), the quadratic formula can be applied to $f(t) = t^2 + bt + c$ in $\mathbb{C}[x]$ to find its roots

$$\mu, \overline{\mu} = \frac{-b \pm \sqrt{b^2 - 4c}}{2} = \frac{-b \pm i\sqrt{4c - b^2}}{2}.$$

There are three possible outcomes:

1. When $b^2 - 4ac = 0$, $f(x)$ has a single real root with multiplicity $m = 2$, and then $f(t) = \left(t + \frac{1}{2}b\right)^2$.
2. When $a = 1$ and $b^2 - 4ac > 0$, there are two distinct real roots
$$r_\pm = \frac{1}{2}\left(-b \pm \sqrt{b^2 - 4c}\right),$$
and then $f(t) = (t - r_+)(t - r_-)$
3. When $a = 1$ and the discriminant $b^2 - 4c$ is negative, there are two distinct conjugate nonreal roots in $\mathbb{C}$,
$$\mu = \frac{-b + i\sqrt{4c - b^2}}{2} \quad \text{and} \quad \overline{\mu} = \frac{-b - i\sqrt{4c - b^2}}{2} \qquad (i = \sqrt{-1}).$$
In this case, $f(t) = (t - \mu)(t - \overline{\mu})$ has the form (2.6) in $\mathbb{R}[t]$.

The quadratic $f(t)$ is irreducible in $\mathbb{R}[t]$ when $f(t)$ has two nonreal roots; otherwise it would have a factorization $(t - r_1)(t - r_2)$ in $\mathbb{R}[t]$ and also in $\mathbb{C}[t]$ that would disagree with $(x - \mu)(x - \overline{\mu})$, contrary to unique factorization in $\mathbb{C}[t]$. That cannot occur. □

Complexification of Linear Operators over $\mathbb{R}$. We now discuss complexification of *arbitrary* vector spaces over $\mathbb{R}$ and complexifications $T_{\mathbb{C}}$ of the $\mathbb{R}$-linear operators $T : V \to V$ that act on them.

DEFINITION 2.13 (Complexification). Given an arbitrary vector space V over $\mathbb{R}$, its *complexification* $V_{\mathbb{C}}$ is made up of the formal sums $\mathbf{z} = x + iy$ with $x, y \in V$ equipped with operations

$$\mathbf{z} + \mathbf{w} = (x + iy) + (u + iv) = (x + u) + i(y + v) \quad \text{and}$$
$$(a + ib) \cdot \mathbf{z} = (a + ib)(x + iy) = (ax - by) + i(bx + ay) \quad \text{for } a + ib \in \mathbb{C}.$$

Two symbols $\mathbf{z} = (x + iy)$ and $\mathbf{z}' = (x' + iy')$ designate the same element of $V_{\mathbb{C}} \Leftrightarrow x' = x$ and $y' = y$. The *real points* in $V_{\mathbb{C}}$ are those of the form $V + i0$. This set is a vector space over $\mathbb{R}$ (but NOT over $\mathbb{C}$) because

$$(c + i0) \cdot (x + i0) = (cx) + i0 \qquad \text{for } c \in \mathbb{R}, x \in V,$$
$$(x + i0) + (x' + i0) = (x + x') + i0 \quad \text{for } x, x' \in V.$$

Clearly the operations $(+)$ and (scaling by some real number $c + i0$) match the usual operations in V when restricted to $V + i0$.

If $T : V \to V$ is an $\mathbb{R}$-linear operator, its *complexification* $T_{\mathbb{C}} : V_{\mathbb{C}} \to V_{\mathbb{C}}$ is the map

$$T_{\mathbb{C}}(x + iy) = T(x) + iT(y) \quad \text{for } x, y \in V, \tag{2.15}$$

which turns out to be a $\mathbb{C}$-linear operator on $V_{\mathbb{C}}$.

EXERCISE 2.14. Prove that $V_{\mathbb{C}}$ is actually a vector space over $\mathbb{C}$.

HINT. Check the vector space axioms. In particular, you must check that $(z_1 z_2) \cdot w = z_1 \cdot (z_2 \cdot w)$ for $z_1, z_2 \in \mathbb{C}$ and $w \in V_{\mathbb{C}}$ and that $(c + i0) \cdot (x + i0) = (c \cdot x) + i0$ for $c \in \mathbb{R}$, so $V + i0$ is a subspace over $\mathbb{R}$ isomorphic to V under the map $x \to x + i0$.

EXAMPLE 2.15. You should check that $T_{\mathbb{C}}$ is a $\mathbb{C}$-linear operator on the complex vector space $V_{\mathbb{C}}$. Most steps are routine; the messy part is proving that $T_{\mathbb{C}}(z \cdot \mathbf{w}) = z \cdot T_{\mathbb{C}}(\mathbf{w})$ for $z \in \mathbb{C}, \mathbf{w} \in V_{\mathbb{C}}$, so we will only do that.

If $z = a + ib \in \mathbb{C}$ and $\mathbf{w} = u + iv$ in $V_{\mathbb{C}}$, we get

$$\begin{aligned} T_{\mathbb{C}}\big((a + ib)(u + iv)\big) &= T_{\mathbb{C}}\big((au - bv) + i(bu + av)\big) \\ &= T(au - bv) + iT(bu + av) \\ &= aT(u) - bT(v) + ibT(u) + iaT(v) \\ &= (a + ib) \cdot (T(u) + iT(v)) = (a + ib) \cdot T_{\mathbb{C}}(u + iv). \end{aligned}$$

EXERCISE 2.16. If $A, B \in \mathrm{M}(n, \mathbb{R})$ and we define an operator $(A + iB)$ on $\mathbb{C}^n = \mathbb{R}^n + i\mathbb{R}^n$ letting

$$(A + iB) \cdot (x + iy) = (Ax - By) + i(Bx + Ay) \quad \text{for } x, y \in \mathbb{R}^n,$$

is this a $\mathbb{C}$-linear operator on $\mathbb{C}^n$? What if $A = B$?

EXERCISE 2.17. If $M_1, \dots, M_r$ are vector spaces over $\mathbb{R}$ verify that the complexification of $V = M_1 \oplus \cdots \oplus M_r$ is $V_{\mathbb{C}} = (M_1)_{\mathbb{C}} \oplus \cdots \oplus (M_r)_{\mathbb{C}}$.

LEMMA 2.18 (Bases in V vs Bases in $V_{\mathbb{C}}$). *If $\mathfrak{X} = \{e_j\}$ is an $\mathbb{R}$-basis in V, then $\{\tilde{e}_j = e_j + i0\}$ is a $\mathbb{C}$-basis in $V_{\mathbb{C}}$. In particular, $\dim_{\mathbb{R}}(V) = \dim_{\mathbb{C}}(V_{\mathbb{C}})$.*

PROOF. If $w = v + iw$ in $V_{\mathbb{C}}$, there are real coefficients $\{c_j\}, \{d_j\}$ such that

$$w = \Big(\sum_j c_j e_j\Big) + i\Big(\sum_j d_j e_j\Big) = \sum_j (c_j + id_j)(e_j + i0),$$

so $\{\tilde{e}_j\}$ span $V_{\mathbb{C}}$. As for independence, if we have

$$0 + i0 = \sum_j z_j \tilde{e}_j = \sum_j (c_j + id_j) \cdot (e_j + i0) = \Big(\sum_j c_j e_j\Big) + i\Big(\sum_j d_j e_j\Big)$$

in $V_{\mathbb{C}}$ for coefficients $z_j = c_j + id_j$ in $\mathbb{C}$, then $\sum_j c_j e_j = 0 = \sum_j d_j e_j$, which implies $c_j = 0$ and $d_j = 0$ because $\{e_j\}$ is a basis in V. Thus $z_j = 0$ for all j. $\square$

EXAMPLE 2.19. If $V = \mathbb{R}^n$, then in an obvious sense $V_{\mathbb{C}} = \mathbb{R}^n + i\mathbb{R}^n$ is the same as $\mathbb{C}^n$. If $A \in \mathrm{M}(n, \mathbb{R})$, we get an $\mathbb{R}$-linear operator $T = L_A$ that sends v to $A \cdot v$, whose matrix with respect to the standard basis $\mathfrak{X} = \{e_j\}$ in $\mathbb{R}^n$ is $[T]_{\mathfrak{X}\mathfrak{X}} = A$. If $\mathfrak{Y} = \{\tilde{e}_j = e_j + i0\}$ is the corresponding basis in $V_{\mathbb{C}}$, it is easy to check that we have

$$[(L_A)_{\mathbb{C}}]_{\mathfrak{Y}\mathfrak{Y}} = [L_A]_{\mathfrak{X}\mathfrak{X}} = A,$$

so $(L_A)_{\mathbb{C}}$ is obtained by letting the matrix A with real entries act on complex column vectors by matrix multiplication (regarding A as a complex matrix that happens to have real entries).

DEFINITION 2.20. The *conjugation operator J* on $V_{\mathbb{C}}$ maps $x + iy \to x - iy$. It is an $\mathbb{R}$-linear operator on $V_{\mathbb{C}}$ with

$$J(c \cdot w) = c \cdot J(w) \quad \text{if } c = c + i0 \in \mathbb{R},$$

but it is *conjugate linear* over $\mathbb{C}$ with

$$J(z \cdot w) = \overline{z} \cdot J(w) \quad \text{and} \quad J(w_1 + w_2) = J(w_1) + J(w_2)$$

for $z \in \mathbb{C}, w \in V_{\mathbb{C}}$.

EXERCISE 2.21. Verify the following basic properties of conjugation.

(a) $J^2 = J \circ J = \mathrm{id}$, so $J^{-1} = J$.

(b) $w \in V_C$ is a real point if and only if $J(w) = w$.

(c) $\dfrac{w + J(w)}{2} = x + i0$ and $\dfrac{w - J(w)}{2i} = y + i0$ if $w = x + iy$ in $V_{\mathbb{C}}$.

We can use the operator J to identify the $\mathbb{C}$-linear maps $S : V_{\mathbb{C}} \to V_{\mathbb{C}}$ of *real-type*, those $S \in \mathrm{Hom}_{\mathbb{C}}(V_{\mathbb{C}}, V_{\mathbb{C}})$ such that $S = T_{\mathbb{C}}$ for some $\mathbb{R}$-linear operator T on V (so $S(x + iy) = T(x) + iT(y)$).

LEMMA 2.22. *If $T : V \to V$ is $\mathbb{R}$-linear and $T_{\mathbb{C}}$ is its complexification, then*

$$JT_{\mathbb{C}} = T_{\mathbb{C}}J, \textit{ or equivalently } JT_{\mathbb{C}}J = T_{\mathbb{C}}.$$

If $S : V_{\mathbb{C}} \to V_{\mathbb{C}}$ is any $\mathbb{C}$-linear operator, the following properties are equivalent.

1. *$S = T_{\mathbb{C}}$ for some $\mathbb{R}$-linear map $T : V \to V$.*
2. *$SJ = JS$.*
3. *S leaves invariant the space $V + i0$ of real vectors in $V_{\mathbb{C}}$.*

PROOF. Implication (1) ⇒ (2) follows because

$$JT_{\mathbb{C}}J(x+iy) = JT_{\mathbb{C}}(x-iy) = J(Tx - iTy) = Tx + iTy = T_{\mathbb{C}}(x+iy).$$

As for (2) ⇒ (3), suppose $SJ = JS$. We have $\mathbf{w} \in V + i0 \Leftrightarrow \mathbf{w} = \frac{1}{2}(\mathbf{w} + J(\mathbf{w}))$, and for such vectors we have (by (2))

$$S\mathbf{w} = \frac{1}{2}(S\mathbf{w} + SJ\mathbf{w}) = \frac{1}{2}(S\mathbf{w} + JS\mathbf{w}),$$

so $S\mathbf{w}$ is in $(V + i0)$. For (3) ⇒ (1): if S leaves $V + i0$ invariant, then $S(x + i0) = T(x) + i0$ for a unique $T(x)$ in V, and we must show $x \to Tx$ is $\mathbb{R}$-linear. In fact, if $c_1, c_2 \in \mathbb{R}$ and $x_1, x_2 \in V$, we have

$$S\big((c_1x_1 + c_2x_2) + i0\big) = T(c_1x_1 + c_2x_2) + i0,$$

while by $\mathbb{C}$-linearity S must also satisfy the identities

$$\begin{aligned} S\big((c_1x_1 + c_2x_2) + i0\big) &= S\big((c_1x_1 + i0) + (c_2x_2 + i0)\big) \\ &= S\big((c_1 + i0)(x_1 + i0)\big) + S\big((c_2 + i0)(x_2 + i0)\big) \\ &= (c_1 + i0)(Tx_1 + i0) + (c_2 + i0)(Tx_2 + i0) \\ &= (c_1Tx_1 + c_2Tx_2) + i0. \end{aligned}$$

Thus T is $\mathbb{R}$-linear. Furthermore, $T_{\mathbb{C}} = S$ because

$$\begin{aligned} T_{\mathbb{C}}(x+iy) &= Tx + iTy = (Tx + i0) + i(Ty + i0) \\ &= S(x + i0) + iS(y + i0) = S\big((x + i0) + i(y + i0)\big) \\ &= S(x + iy) \end{aligned}$$

for all $x + iy \in V_{\mathbb{C}}$. □

EXERCISE 2.23. Prove the following:

(a) $(T_{\mathbb{C}})^k = (T^k)_{\mathbb{C}}$ for all $k = 0, 1, 2, \ldots$.

(b) $e^{T_{\mathbb{C}}} = (e^T)_{\mathbb{C}}$.

LEMMA 2.24. *If $p_T(x)$ is the characteristic polynomial of an $\mathbb{R}$-linear operator $T : V \to V$, then $p_{T_{\mathbb{C}}}(x) = p_T(x)$ when we regard $\mathbb{R}[x] \subseteq \mathbb{C}[x]$.*

PROOF. By Lemma 2.18, if $\mathfrak{X} = \{e_j\}$ is an $\mathbb{R}$-basis in V, then $\mathfrak{Y} = \{\tilde{e}_j = e_j + i0\}$ is a $\mathbb{C}$-basis in $V_{\mathbb{C}}$ and $[T]_{\mathfrak{X}\mathfrak{X}} = [T_{\mathbb{C}}]_{\mathfrak{Y}\mathfrak{Y}}$ because

$$\begin{aligned} T_{\mathbb{C}}(\tilde{e}_j) &= T_{\mathbb{C}}(e_j + i0) = T(e_j) + iT(0) = \Big(\sum_k t_{kj}e_k\Big) + i0 \\ &= \sum_k (t_{kj} + i0)(e_k + i0) = \sum_k (t_{kj} + i0)\,\tilde{e}_k. \end{aligned}$$

The matrices agree and so do $\det([T]_{\mathfrak{X}\mathfrak{X}} - xI)$ and $\det([T_{\mathbb{C}}]_{\mathfrak{Y}\mathfrak{Y}} - xI)$. □

With this in mind, we write $p_T(x)$ for $p_{T_{\mathbb{C}}}(x)$, leaving the context to determine whether we are speaking of elements of $\mathbb{R}[x]$ or $\mathbb{C}[x]$.

As noted earlier, $p_T(x)$ splits over $\mathbb{C}$ but not necessarily over $\mathbb{R}$. Furthermore, the nonreal complex roots in the irreducible factorization of p_T come in conjugate pairs, so we may systematically list the distinct real eigenvalues λ_j and distinct conjugate pairs $(\mu_j, \overline{\mu}_j)$ of eigenvalues in $\mathrm{sp}_{\mathbb{C}}(T_{\mathbb{C}})$, taking $\mathrm{Im}(\mu_j) > 0$ for each pair:

$$\mu_1, \overline{\mu}_1, \dots, \mu_r, \overline{\mu}_r; \lambda_{r+1}, \dots, \lambda_s \quad (\lambda_i \text{ real; } \mu_j \neq \overline{\mu}_j \text{ with } \mathrm{Im}(\mu_j) > 0). \tag{2.16}$$

For eigenvalues and conjugate pairs we write

$$d(\mu_j) = \dim_{\mathbb{C}}\left(E_{\mu_j}(T_{\mathbb{C}})\right) \quad \text{and} \quad d(\lambda_j) = \dim_{\mathbb{C}}(E_\lambda(T_{\mathbb{C}})).$$

We first take up the case when $T_{\mathbb{C}}$ is diagonalizable, followed by the general case in which the Jordan form of $T_{\mathbb{C}}$ must be used. When $T_{\mathbb{C}}$ is diagonalizable, we will show that the eigenspaces $E_{\lambda_i}, E_{\mu_j}, E_{\overline{\mu}_j}$ in $V_{\mathbb{C}}$ can be used to construct a direct sum decomposition of $V_{\mathbb{C}}$ into $T_{\mathbb{C}}$-invariant subspaces.

REMARK 2.25. If $\mu \neq \overline{\mu}$, then $\mathbf{w} \in V_{\mathbb{C}}$ is an eigenvector for $\mu \Leftrightarrow J\mathbf{w}$ is an eigenvector for $\overline{\mu}$ because

$$T_{\mathbb{C}}(J\mathbf{w}) = J(T_{\mathbb{C}}\mathbf{w}) = J(\mu\mathbf{w}) = \overline{\mu} \cdot J\mathbf{w}.$$

Hence $J\big(E_\mu(T_{\mathbb{C}})\big) = E_{\overline{\mu}}(T_{\mathbb{C}})$ and J is an $\mathbb{R}$-linear bijection between these complex subspaces of $V_{\mathbb{C}}$. ○

Observe also that $J\big(E_\mu \oplus E_{\overline{\mu}}\big) = E_\mu \oplus E_{\overline{\mu}}$ even though neither summand is J-invariant, while $J\big(E_\lambda\big) = E_\lambda$ for *real* eigenvalues of $T_{\mathbb{C}}$.

Subspaces of Real Type in $V_{\mathbb{C}}$. The complex subspaces

$$E_{\mu,\overline{\mu}} = E_\mu(T_{\mathbb{C}}) \oplus E_{\overline{\mu}}(T_{\mathbb{C}}) \quad \text{with} \quad \dim_{\mathbb{C}}(E_{\mu.\overline{\mu}}) = 2d_\mu$$

are of a special "real-type" in $V_{\mathbb{C}}$ owing to their conjugation invariance.

DEFINITION 2.26. If W is a $\mathbb{C}$-subspace in the complexification $V_{\mathbb{C}} = V + iV$, its *real points* are those in the $\mathbb{R}$-subspace $W \cap (V + i0)$ in $V_{\mathbb{C}}$, which corresponds to an $\mathbb{R}$-subspace $W_{\mathbb{R}} \subseteq V$ such that $W \cap (V + i0) = W_{\mathbb{R}} + i0$. Hereafter, we will identify $W_{\mathbb{R}} + i0$ in $V_{\mathbb{C}}$ with $W_{\mathbb{R}}$ in V, referring to both as "$W_{\mathbb{R}}$." Complexifying this $\mathbb{R}$-subspace in V, we obtain a complex subspace

$$\left(W_{\mathbb{R}}\right)_{\mathbb{C}} = W_{\mathbb{R}} + iW_{\mathbb{R}} \subseteq W$$

in $V_{\mathbb{C}}$. In general, $W_{\mathbb{R}} + iW_{\mathbb{R}}$ can be smaller than the original complex subspace W and could even be trivial. We say that a complex subspace $W \subseteq V_{\mathbb{C}}$ is of *real-type* if

$$W = W_{\mathbb{R}} + iW_{\mathbb{R}} \quad \big(W_{\mathbb{R}} = W \cap (V + i0) \text{ the real points in } W\big).$$

In other words, a complex subspace W is of real type $\Leftrightarrow$ it is the complexification of its subspace $W \cap (V + i0)$ of real points.

The following observation makes it easy to identify subspaces W of real-type in $V_{\mathbb{C}}$.

LEMMA 2.27. *A complex subspace W in $V_{\mathbb{C}}$ is of real-type* $\Leftrightarrow J(W) = W$.

PROOF. For ($\Rightarrow$) we have $W = W_{\mathbb{R}} + iW_{\mathbb{R}}$, so $JW = W_{\mathbb{R}} - iW_{\mathbb{R}} = W_{\mathbb{R}} + iW_{\mathbb{R}} = W$ since $W_{\mathbb{R}} = -W_{\mathbb{R}}$. For ($\Leftarrow$), write $\mathbf{w} \in W$ as $\mathbf{w} = x + iy$ ($x, y \in V$). Then we have

$$\frac{1}{2}(\mathbf{w} + J\mathbf{w}) = x + i0 \quad \text{and} \quad \frac{1}{2i}(\mathbf{w} - J\mathbf{w}) = y + i0,$$

both of which lie in $W_{\mathbb{R}} = W \cap (V + i0)$ because $JW = W$. Thus,

$$\mathbf{w} = (x + i0) + i(y + i0) = x + iy \quad \text{and} \quad \mathbf{w} \in W_{\mathbb{R}} + iW_{\mathbb{R}},$$

so W is of real-type. □

It follows immediately that the complex subspaces $W = E_{\lambda+i0}(T_{\mathbb{C}})$ for λ in $\mathbb{R}$ and $W = E_{\mu}(T_{\mathbb{C}}) \oplus E_{\overline{\mu}}(T_{\mathbb{C}})$ for $\mu \neq \overline{\mu}$ are of real type in $V_{\mathbb{C}}$ and are $T_{\mathbb{C}}$-invariant. We now compare what $T_{\mathbb{C}}$ does in these invariant subspaces with the action of $T_{\mathbb{C}}$ on the real subspace $W_{\mathbb{R}} + i0$. This subspace actually is $T_{\mathbb{C}}$-invariant (even though $T_{\mathbb{C}}$ is acting on an $\mathbb{R}$-subspace of $V_{\mathbb{C}}$) because

$$\begin{aligned} T_{\mathbb{C}}(W_{\mathbb{R}}) &= T_{\mathbb{C}}(W \cap (V + i0)) \subseteq T_{\mathbb{C}}(W) \cap T_{\mathbb{C}}(V + i0) \\ &= W \cap (V + i0) \quad (W \text{ is } T_{\mathbb{C}}\text{-invariant}) \\ &= W_{\mathbb{R}}. \end{aligned}$$

Furthermore, its restriction to $W_{\mathbb{R}}$ is closely tied to the action of T on a certain $\mathbb{R}$-subspace of V. This will, with some effort, provide an $\mathbb{R}$-basis for V that reveals the structure of the original $\mathbb{R}$-linear operator $T : V \to V$.

EXERCISE 2.28. Whether $\mathbb{K} = \mathbb{R}$ or $\mathbb{C}$, a matrix in $\mathrm{M}(n, \mathbb{K})$ determines a linear operator $L_A : \mathbb{K}^n \to \mathbb{K}^n$. Verify the following relationships between an operator on $V = \mathbb{R}^n$ and its complexification $V_{\mathbb{C}} = \mathbb{C}^n = \mathbb{R}^n + i\mathbb{R}^n$.

(a) If $\mathbb{K} = \mathbb{R}$, $(L_A)_{\mathbb{C}} : \mathbb{C}^n \to \mathbb{C}^n$ is the same as the operator $L_A : \mathbb{C}^n \to \mathbb{C}^n$ we get by regarding A as a complex matrix whose entries are all real.

(b) Consider $A \in \mathrm{M}(n, \mathbb{C})$ and regard $\mathbb{C}^n$ as the complexification of $\mathbb{R}^n$. Verify that $L_A : \mathbb{C}^n \to \mathbb{C}^n$ is of real type $\Leftrightarrow$ all entries in A are real, so $A \in \mathrm{M}(n, \mathbb{R})$.

EXERCISE 2.29. If S and T are $\mathbb{R}$-linear operators on a real vector space V, is the map

$$(S + iT) : (x + iy) \to S(x) + iT(y)$$

on $V_{\mathbb{C}}$ a $\mathbb{C}$-linear operator? If so, when is it of real-type? How about the same question for the map

$$(S + iT) : (x + iy) \to (S(x) - T(y)) + i(T(x) + S(y))?$$

EXERCISE 2.30. If $T : V \to V$ is an $\mathbb{R}$-linear operator on a real vector space, prove the following:

(a) $(T_{\mathbb{C}})^k = (T^k)_{\mathbb{C}}$ for all $k \in \mathbb{N}$.

(b) $e^{(T_{\mathbb{C}})} = (e^T)_{\mathbb{C}}$ on $V_{\mathbb{C}}$.

EXERCISE 2.31. If $A \in \mathrm{M}(n, \mathbb{R})$, we say $\lambda \in \mathbb{C}$ is in the *complex spectrum* of A if

$$A \cdot \mathbf{z} = \lambda \mathbf{z}$$

for some $\lambda \in \mathbb{C}$ and nonzero vector $\mathbf{z} \in \mathbb{C}^n$. This is the same as saying λ is a complex root of the characteristic polynomial $p_A(x) = \det(A - xI)$, which has real coefficients. Prove that this happens if and only if

$$\lambda \in \mathrm{sp}_{\mathbb{C}}\big((L_A)_{\mathbb{C}}\big)\,,$$

where $(L_A)_{\mathbb{C}}$ is the complexification of L_A on $\mathbb{R}^n$.

EXERCISE 2.32. Let $V_{\mathbb{C}} = V + iV$ be the complexification of a real vector space and let $S : V_{\mathbb{C}} \to V_{\mathbb{C}}$ be a $\mathbb{C}$-linear operator of real-type with $S(x + iy) = T(x) + iT(y)$ for some $T : V \to V$. If $W = W_{\mathbb{R}} + iW_{\mathbb{R}} \subseteq V_{\mathbb{C}}$ is a complex subspace of real type that is S-invariant, verify that

$$\text{(a)}\ \ S(W_{\mathbb{R}} + i0) \subseteq (W_{\mathbb{R}} + i0) \quad \text{and} \quad \text{(b)}\ \ S|_{(W_{\mathbb{R}}+i0)} = \big(T|_{W_{\mathbb{R}}}\big) + i0.$$

Concordance Between Eigenspaces of T in V and of $T_{\mathbb{C}}$ in $V_{\mathbb{C}}$. The complexified operator $T_{\mathbb{C}}$ acts on a complex vector space $V_{\mathbb{C}}$ and therefore can be put into Jordan form (or perhaps diagonalized) by methods worked out in Chapter 1. We first examine the special case when $T_{\mathbb{C}}$ is diagonalizable before resolving the main problem of interest in this chapter:

> QUESTION. *If $T_{\mathbb{C}} : V_{\mathbb{C}} \to V_{\mathbb{C}}$ is the complexification of an $\mathbb{R}$-linear operator on V, what can be said about the structure of the original $\mathbb{R}$-linear operator $T : V \to V$?*

CASE 1: $T_{\mathbb{C}}$ IS DIAGONALIZABLE. It will suffice to focus on a single eigenvalue for $T_{\mathbb{C}}$ or a single conjugate pair.

SUBCASE 1A: A REAL EIGENVALUE $\lambda + i0$ FOR $\lambda \in \mathbb{R}$. Then, the correspondence between $E_\lambda(T)$ in V and $E_{\lambda+i0}(T_{\mathbb{C}})$ in $V_{\mathbb{C}}$ is straightforward.

LEMMA 2.33. *If $T : V \to V$ is $\mathbb{R}$-linear, then*

$$\mathrm{sp}_{\mathbb{R}}(T) + i0 = \mathrm{sp}_{\mathbb{C}}(T_{\mathbb{C}}) \cap (\mathbb{R} + i0)\,,$$

and if $\lambda \in \mathrm{sp}_{\mathbb{R}}(T)$,

$$\big(E_\lambda(T)\big)_{\mathbb{C}} = E_\lambda(T) + iE_\lambda(T) \quad \textit{is equal to} \quad E_{\lambda+i0}(T_{\mathbb{C}}) \subseteq V_{\mathbb{C}}.$$

PROOF. Let $\lambda \in \mathbb{R}$ be in $\mathrm{sp}_{\mathbb{C}}(T)$. Then there is a $\mathbf{w} = u + iv \neq 0$ such that

$$T_{\mathbb{C}}(\mathbf{w}) = Tu + iTv \quad \text{is equal to} \quad (\lambda + i0)\mathbf{w} = \lambda u + i\lambda v\,,$$

which happens if and only if $Tu = \lambda u$ and $Tv = \lambda v$. Since at least one of u, v is nonzero, $\lambda \in \mathrm{sp}_{\mathbb{R}}(T)$. Conversely, if $u \in V$ is nonzero and $Tu = \lambda u$, we get

$$T_{\mathbb{C}}(u + i0) = Tu + i0 = (\lambda \cdot u) + i0 = (\lambda + i0) \cdot (u + i0),$$

so $\lambda + i0 \in \mathrm{sp}_{\mathbb{C}}(T_{\mathbb{C}})$.

The second statement follows because

$$T_{\mathbb{C}}(u + iv) = Tu + iTv = (\lambda + i0)(u + iv) \Leftrightarrow Tu = \lambda u \quad \text{and} \quad Tv = \lambda v.$$

Thus $E_{\lambda+i0}(T_{\mathbb{C}})$ is of real-type in $V_{\mathbb{C}}$ and is the complexification of its set of real points $E_\lambda(T)+i0$. If $\mathfrak{X}=\{e_j\}$ is a basis for $E_\lambda(T)$, the vectors $\mathfrak{Y}=\{\tilde{e}_j=(e_j+i0)\}$ are a $\mathbb{C}$-basis for $E_{\lambda+i0}(T_{\mathbb{C}})$ such that

$$[T|_{E_\lambda(T)}]_{\mathfrak{X}\mathfrak{X}} = [T_{\mathbb{C}}|_{E_{\lambda+i0}(T_{\mathbb{C}})}]_{\mathfrak{Y}\mathfrak{Y}}. \tag{2.17}$$

In particular, $\dim_{\mathbb{R}}(E_\lambda(T)) = \dim_{\mathbb{C}}(E_{\lambda+i0}(T_{\mathbb{C}})) = d(\lambda)$ for real eigenvalues. □

SUBCASE 1B: A NONREAL CONJUGATE PAIR. If $\mu\neq\bar{\mu}$ in $\mathrm{sp}_{\mathbb{C}}(T_{\mathbb{C}})$, we have seen that $E_{\mu,\overline{\mu}}=E_\mu(T_{\mathbb{C}})\oplus E_{\overline{\mu}}(T_{\mathbb{C}})$ is $T_{\mathbb{C}}$-invariant and of real-type. List the distinct pairs $(\mu,\overline{\mu})$ of nonreal eigenvalues (taking $\mathrm{Im}(\mu)>0$) as in (2.16) and pick a $\mathbb{C}$-basis for $E_\mu(T_{\mathbb{C}})$,

$$f_j^{(\mu)}=x_j^{(\mu)}+iy_j^{(\mu)} \quad (x_j^{(\mu)},y_j^{(\mu)}\in V)$$

for $1\le j\le d_\mu=\dim_{\mathbb{C}}E_\mu(T_{\mathbb{C}})$. Then $E_\mu(T_{\mathbb{C}})=\bigoplus_{j=1}^{d(\mu)}\mathbb{C}f_j^{(\mu)}$, a direct sum of one-dimensional complex subspaces. Since $J(E_\mu)=E_{\overline{\mu}}$, we get a matching $\mathbb{C}$-basis in $M_{\overline{\mu}}(T_{\mathbb{C}})$:

$$Jf_j^{(\mu)}=x_j^{(\mu)}-iy_j^{(\mu)} \quad \text{in} \quad E_{\overline{\mu}}=\bigoplus_{j=1}^{d}\mathbb{C}\cdot Jf_j^{(\mu)}.$$

Then define the two-dimensional complex subspaces $V_j^{(\mu)}$ in $E_{\mu,\overline{\mu}}$:

$$V_j^{(\mu)}=\mathbb{C}f_j^{(\mu)}\oplus\mathbb{C}(Jf_j^{(\mu)}),\quad j=1,2,\dots,d(\mu)=\dim_{\mathbb{C}}(E_\mu(T_{\mathbb{C}})).$$

Each is $T_{\mathbb{C}}$-invariant since $T_{\mathbb{C}}(f_j^{(\mu)})=\mu f_j^{(\mu)}$ and $T_{\mathbb{C}}(J(f_j^{\overline{\mu}}))=\overline{\mu}f_j^{\overline{\mu}}$. Furthermore, $V_j^{(\mu)}$ is of real type since it is J-invariant. Clearly $E_{\mu,\overline{\mu}}=\bigoplus_{j=1}^{d}V_j^{(\mu)}$. We claim that

LEMMA 2.34. *In each complex subspace*

$$V_j^{(\mu)}=\mathbb{C}f_j^{(\mu)}\oplus C(Jf_j^{(\mu)}) \quad j=1,\dots,d(\mu)=\dim_{\mathbb{C}}(E_\mu(T_{\mathbb{C}}),$$

we can find a $\mathbb{C}$-basis consisting of two vectors of real-type in

$$(V_j^{(\mu)})_{\mathbb{R}}=V_j^{(\mu)}\cap(V+i0),$$

something that cannot be done for the spaces $\mathbb{C}f_j^{(\mu)}$ or the eigenspace $E_\mu(T_{\mathbb{C}})$ alone.

PROOF. If $f_j^{(\mu)}=x_j^{(\mu)}+iy_j^{(\mu)}$ and $Jf_j^{(\mu)}=x_j^{(\mu)}-iy_j^{(\mu)}$, then

$$x_j^{(\mu)}+i0=\frac{1}{2}(f_j^{(\mu)}+Jf_j^{(\mu)}) \quad \text{and} \quad y_j^{(\mu)}+i0=\frac{1}{2i}(f_j^{(\mu)}-Jf_j^{(\mu)})$$

are in $(V_j^{(\mu)})_{\mathbb{R}}$, and their $\mathbb{C}$-span is obviously all of $V_j^{(\mu)}$. Since $\dim_{\mathbb{C}}(V_j^{(\mu)})=2$, they are independent over $\mathbb{C}$ and form a $\mathbb{C}$-basis for $V_j^{(\mu)}$. But they are also an

$\mathbb{R}$-basis for the two-dimensional subspace $\big(V_j^{(\mu)}\big)_{\mathbb{R}} = \big(\mathbb{R}x_j^{(\mu)} + \mathbb{R}y_j^{(\mu)}\big) + i0$ of real points in $V_j^{(\mu)}$. As previously noted, $T_{\mathbb{C}}$ leaves $V_j^{(\mu)}$ invariant; furthermore, T leaves invariant the subspace of real vectors $\big(V_j^{(\mu)}\big)_{\mathbb{R}} \subseteq V$ because

$$T_{\mathbb{C}}\big(V_j^{(\mu)} \cap (V + i0)\big) \subseteq T_{\mathbb{C}}\big(V_j^{(\mu)}\big) \cap (T(V) + i0)$$
$$\subseteq V_j^{(\mu)} \cap (V + i0) = (V_j^{(\mu)})_{\mathbb{R}}.$$

That proves Lemma 2.34 . □

We can now determine the matrix of the restriction $T|_{(V_j^{(\mu)})_{\mathbb{R}}}$ with respect to the ordered $\mathbb{R}$-basis $\mathfrak{X} = \{x_1^{(\mu)}, y_1^{(\mu)}, x_2^{(\mu)}, y_2^{(\mu)}, \dots\}$. (From here on, we suppress most superscripts "(μ)" for clarity.) If $\mu = a + ib$ and $\overline{\mu} = a - ib$ $(b \neq 0)$, we have

$$T_{\mathbb{C}}\big(f_j^{(\mu)}\big) = T_{\mathbb{C}}(x_j + iy_j) = \mu(x_j + iy_j) = (ax_j - by_j) + i(ay_j + bx_j),$$
$$T_{\mathbb{C}}\big(Jf_j^{(\mu)}\big) = T_{\mathbb{C}}(x_j - iy_j) = \overline{\mu}(x_j - iy_j) = (ax_j - by_j) - i(ay_j + bx_j).$$

Writing μ in polar form $\mu, \overline{\mu} = a \pm ib = re^{\pm i\theta} = r\cos(\theta) \pm ir\sin(\theta)$, we get

$$\begin{aligned} Tx_j + i0 = T_{\mathbb{C}}(x_j + i0) &= \frac{1}{2}T_{\mathbb{C}}\big(f_j + Jf_j\big) = \frac{1}{2}\big(T_{\mathbb{C}}f_j + T_{\mathbb{C}}Jf_j\big) \\ &= \frac{1}{2}\big((a+ib)f_j + J\big((a+ib)f_j\big)\big) \quad (T_{\mathbb{C}} \text{ and } J \text{ commute}) \\ &= (ax_j - by_j) \quad \text{(all other terms cancel)} \\ &= r\big(\cos(\theta)\cdot x_j - \sin(\theta)\cdot y_j\big) \end{aligned}$$

and similarly

$$Ty_j + i0 = T_{\mathbb{C}}\Big(\frac{f_j - Jf_j}{2i}\Big) = ay_j + bx_j = r\big(\sin(\theta)\cdot x_j + \cos(\theta)\cdot y_j\big)$$

Thus, the matrix of T restricted to $(V_j^{(\mu)})_{\mathbb{R}}$ with respect to the $\mathbb{R}$-basis $\mathfrak{X} = \{x_j, y_j\}$ in $(V_j^{(\mu)})_{\mathbb{R}}$ is a scalar multiple of a rotation matrix (by $-\theta$ radians):

$$[T|_{(V_j^{(\mu)})_{\mathbb{R}}}]_{\mathfrak{X}\mathfrak{X}} = r\cdot\begin{pmatrix} \cos\theta & \sin\theta \\ -\sin\theta & \cos\theta \end{pmatrix} = r\cdot R(\theta).$$

(Reversing the order of basis vectors $x_j^{(\mu)}, y_j^{(\mu)}$ would yield a scalar multiple $r\cdot R(\theta)$ of the matrix describing a rotation by $+\theta$ radians.)

But $\widetilde{\mathfrak{X}} = \{x_j + i0\,,\ y_j + i0\}$ is also a $\mathbb{C}$-basis for $V_j^{(\mu)}$, and by (2.17) we have

$$[\,T|_{(V_j^{(\mu)})_{\mathbb{R}}}\,]_{\mathfrak{X}\mathfrak{X}} = [\,T_{\mathbb{C}}|_{V_j^{(\mu)}}\,]_{\widetilde{\mathfrak{X}}\widetilde{\mathfrak{X}}}\,,$$

whose entries are real despite the fact that $\widetilde{\mathfrak{X}}$ is a complex basis. Of course, with respect to the original $\mathbb{C}$-basis $\mathfrak{Z} = \{f_j^{(\mu)} = x_j^{(\mu)} + iy_j^{(\mu)}\,,\ Jf_j^{(\mu)} = x_j^{(\mu)} - iy_j^{(\mu)}\}$

in $V_j^{(\mu)}$, we get the expected diagonal matrix with complex entries

$$[\,T_{\mathbb{C}}|_{V_j^{(\mu)}}\,]_{\widetilde{3}\widetilde{3}} = \begin{pmatrix} re^{i\theta} & 0 \\ 0 & re^{-i\theta} \end{pmatrix}.$$

The Normal Form of a Linear Operator Over $\mathbb{R}$**.** Whether or not the complexified operator $T_{\mathbb{C}} : V_{\mathbb{C}} \to V_{\mathbb{C}}$ is diagonalizable, we will exhibit a *real-normal form* for the original $\mathbb{R}$-linear operator $T : V \to V$. This is closely related to the JCF for operators on complex spaces; it provides geometric insight into the way T acts on the original vector space V.

When $T_{\mathbb{C}}$ is diagonalizable, we first recall that the subspace $E_{\mu,\overline{\mu}} = E_\mu \oplus E_{\overline{\mu}}$ is of real type in $V_{\mathbb{C}}$ for each distinct complex-conjugate pair in $\mathrm{sp}(T_{\mathbb{C}})$. For each pair, we claim that $E_{\mu,\overline{\mu}}$ is a direct sum of two-dimensional $T_{\mathbb{C}}$-invariant complex subspaces, each of which is of real type, with

$$(E_\mu \oplus E_{\overline{\mu}}) \cap (V + i0) = \bigoplus_{j=1}^{d(\mu)} (V_j^{(\mu)})_{\mathbb{R}}.$$

The sum on the right is direct so that $\dim_{\mathbb{R}}(\bigoplus_{\mu,\overline{\mu}} \cdots) = 2d(\mu) = 2 \cdot \dim_{\mathbb{C}}(E_\mu)$, and since

$$\dim_{\mathbb{R}} \left(E_\mu \oplus E_{\overline{\mu}}\right)_{\mathbb{R}} = \dim_{\mathbb{C}} \left(E_\mu \oplus E_{\overline{\mu}}\right) = 2d(\mu)\,,$$

the spaces coincide. When $T_{\mathbb{C}}$ is diagonalizable, previous efforts yield:

THEOREM 2.35 (Real-Normal Form of T; $T_{\mathbb{C}}$ Diagonalizable). *Consider a linear operator $T : V \to V$ on a real vector space such that $T_{\mathbb{C}}$ is diagonalizable. If we list the distinct (pairs of) eigenvalues as in* (2.16),

$$\mu_1, \overline{\mu}_1, \dots, \mu_r, \overline{\mu}_r, \lambda_{r+1}, \dots, \lambda_s \quad (\operatorname{Im} \mu_j > 0),$$

there is an $\mathbb{R}$-basis $\mathfrak{X}$ in V such that $[T]_{\mathfrak{X}\mathfrak{X}}$ has block-diagonal form

$$[T]_{\mathfrak{X}\mathfrak{X}} = \begin{pmatrix} \boxed{R_1} & & & & & 0 \\ & \ddots & & & & \\ & & \boxed{R_r} & & & \\ & & & \boxed{D_1} & & \\ & & & & \ddots & \\ 0 & & & & & \boxed{D_s} \end{pmatrix}.$$

For $1 \le k \le r$, we have

$$R_k = \begin{pmatrix} r_k \boxed{R(\theta_k)} & & 0 \\ & \ddots & \\ 0 & & r_k \boxed{R(\theta_k)} \end{pmatrix},$$

which consists of $d(\mu_k) = \dim_{\mathbb{C}}(E_{\mu_k})$ *blocks, each a scalar multiple of the same* 2×2 *rotation matrix. For* $r+1 \le k \le s$, λ_k *is real and*

$$D_k = \begin{pmatrix} \lambda_k & & & 0 \\ & \ddots & & \\ & & \ddots & \\ 0 & & & \lambda_k \end{pmatrix}$$

with $d(\lambda_k) = \dim_{\mathbb{C}}(E_{\lambda_k})$.

When $T_{\mathbb{C}}$ Is Not Diagonalizable. When $T_{\mathbb{C}}$ cannot be diagonalized, we apply the Jordan form for $T_{\mathbb{C}}$.

LEMMA 2.36. *Let* $T : V \to V$ *be a linear operator on a vector space over* $\mathbb{R}$. *For* $\mu \in \mathbb{C}$, *the generalized eigenspace in* $V_{\mathbb{C}}$ *is*

$$M_\mu = M_\mu(T_{\mathbb{C}}) = \{w \in V_{\mathbb{C}} : (T_{\mathbb{C}} - \mu I)^k w = 0 \text{ for some } k \in \mathbb{N}\}.$$

Then $w \in M_\mu \Leftrightarrow J(w) \in M_{\overline{\mu}}$ *so that*

$$J(M_\mu) = M_{\overline{\mu}} \quad \text{and} \quad J(M_{\overline{\mu}}) = M_\mu.$$

PROOF. Since $J = J^{-1}$, the map $\Phi : S \to JSJ = JSJ^{-1}$ is an *automorphism* of the algebra of all $\mathbb{C}$-linear maps $\mathrm{Hom}_{\mathbb{C}}(V_{\mathbb{C}}, V_{\mathbb{C}})$:

- It is a bijection such that $\Phi(I) = I$, and it preserves products $\Phi(S_1S_2) = \Phi(S_1)\Phi(S_2)$ because $J^2 = I \Rightarrow JS_1S_2J = (JS_1J)(JS_2J)$.
- Furthermore, $\Phi(S_1+S_2) = \Phi(S_1)+\Phi(S_2)$ and $\Phi(cS) = c\Phi(S)$ for $c \in \mathbb{C}$, and in particular, $\Phi(S^k) = \Phi(S)^k$ for $k \in \mathbb{N}$.

Thus

$$J((T_{\mathbb{C}} - \mu I)^k)J = \big(J(T_{\mathbb{C}} - \mu I)J\big)^k = \big(JT_{\mathbb{C}}J - J(\mu I)J\big)^k = (T_{\mathbb{C}} - \overline{\mu} I)^k$$

(here $JT_{\mathbb{C}} = T_{\mathbb{C}}J$ because $T_{\mathbb{C}}$ is of real-type by definition), and

$$\big(J(\mu I)J\big)(w) = J(\mu \cdot J(w)) = \overline{\mu}J^2(w) = \overline{\mu}w \quad \text{for } w \in V_{\mathbb{C}}.$$

Finally, we have $(T_{\mathbb{C}} - \mu I)^k w = 0$ if and only if

$$\begin{aligned} J(T_{\mathbb{C}} - \mu I)^k w = 0 &\Leftrightarrow J(T_{\mathbb{C}} - \mu I)^k J(J(w)) = 0 \\ &\Leftrightarrow (T_{\mathbb{C}} - \overline{\mu})^k J(w) = 0 \Leftrightarrow J(w) \in M_{\overline{\mu}}. \end{aligned}$$

Then $J(M_\mu) = M_{\overline{\mu}}$ and $M_\mu = J(M_{\overline{\mu}})$ since $J^2 = I$. □

THEOREM 2.37 (Real-Normal Form of T; $T_{\mathbb{C}}$ Not Diagonalizable). *Consider a linear operator* $T : V \to V$ *on a real vector space and list the (pairs of) eigenvalues*

$$\mu_1, \overline{\mu}_1, \ldots, \mu_r, \overline{\mu}_r, \lambda_{r+1}, \ldots, \lambda_s \quad (\mathrm{Im}\, \mu_k > 0)$$

of $T_{\mathbb{C}} : V_{\mathbb{C}} \to V_{\mathbb{C}}$ *as in* (2.16). *Then there is an* $\mathbb{R}$*-basis* $\mathfrak{X}$ *for* V *that puts* $[T]_{\mathfrak{X}\mathfrak{X}}$ *in block diagonal form:*

$$[T]_{\mathfrak{X}\mathfrak{X}} = \begin{pmatrix} \boxed{A_1} & & & 0 \\ & \ddots & & \\ & & \ddots & \\ 0 & & & \boxed{A_m} \end{pmatrix}.$$

Each block A_j has one of two possible block-upper triangular forms: either

$$A_j = \begin{pmatrix} \lambda & 1 & & 0 \\ & \ddots & \ddots & \\ & & \ddots & 1 \\ 0 & & & \lambda \end{pmatrix} \quad \text{for real eigenvalues } \lambda \text{ of } T_{\mathbb{C}}$$

or

$$A_j = \begin{pmatrix} \boxed{R_{\mu,\overline{\mu}}} & I_2 & & 0 \\ & \ddots & \ddots & \\ & & \ddots & I_2 \\ 0 & & & \boxed{R_{\mu,\overline{\mu}}} \end{pmatrix} \quad \text{for conjugate pairs } \mu = e^{i\theta},\ \overline{\mu} = e^{-i\theta}$$

where I_2 is the 2×2 identity matrix and

$$R_{\mu,\overline{\mu}} = r \begin{pmatrix} \cos\theta & -\sin\theta \\ \sin\theta & \cos\theta \end{pmatrix}.$$

PROOF. The proof will of course employ the generalized eigenspace decomposition $V_{\mathbb{C}} = \bigoplus_{z \in \mathrm{sp}(T_{\mathbb{C}})} M_z(T_{\mathbb{C}})$. As above, we write $M_{\mu,\overline{\mu}} = M_\mu(T_{\mathbb{C}}) \oplus M_{\overline{\mu}}(T_{\mathbb{C}})$ and write $d(\mu) = 2 \cdot \dim(M_\mu(T_{\mathbb{C}}))$ for each distinct pair of nonreal eigenvalues. The summands $M_\lambda, M_{\mu,\overline{\mu}}$ are of real type and are $T_{\mathbb{C}}$-invariant, so for each μ we may write $M_{\mu,\overline{\mu}}$ as the complexification of its space of real points:

$$W_\mu = M_{\mu,\overline{\mu}} \cap (V + i0). \tag{2.18}$$

Here $M_{\mu,\overline{\mu}} = W_\mu + iW_\mu$ is $T_{\mathbb{C}}$-invariant. Since $T_{\mathbb{C}}$ is of real type (by definition), it leaves invariant the $\mathbb{R}$-subspace $V + i0$; therefore the space of real points W_μ is also $T_{\mathbb{C}}$ invariant with $T_{\mathbb{C}}|_{(W_\mu + i0)} = T|_{W_\mu}$. It follows that $T_{\mathbb{C}}|_{M_{\mu,\overline{\mu}}}$ is the complexification of $T|_{W_\mu}$ and

$$(T|_{W_\mu})_{\mathbb{C}}(z) = (T|_{W_\mu})(z) + i(T|_{W_\mu})(z) \quad \text{for } z \in M_{\mu,\overline{\mu}}.$$

If we can find an $\mathbb{R}$-basis in the space (2.18) of real points $W_\mu \subseteq M_{\mu,\overline{\mu}}$ for which $T|_{W_\mu}$ takes the form described in the theorem, then our proof is complete. For this we may obviously restrict attention to a single subspace $M_{\mu,\overline{\mu}}$ (or M_λ if $\lambda \in \mathbb{R}$) in $V_{\mathbb{C}}$.

CASE 1: A REAL EIGENVALUE λ. If λ is real, then

$$\begin{aligned}(T_{\mathbb{C}} - \lambda)^k(x + iy) &= (T_{\mathbb{C}} - \lambda)^{k-1}\big[(T - \lambda)x + i(T - \lambda)y\big] = \cdots \\ &= (T - \lambda)^k x + i(T - \lambda)^k y \quad \text{for } k \in \mathbb{N}.\end{aligned}$$

Thus, $x + iy \in M_\lambda(T_{\mathbb{C}})$ if and only if $x, y \in M_\lambda(T)$, and the subspace of real points in $M_\lambda(T_{\mathbb{C}}) = M_\lambda(T) + iM_\lambda(T)$ is precisely $M_\lambda(T) + i0$. There is a $\mathbb{C}$-basis $\mathfrak{X} = \{f_j\}$ consisting of real vectors in $M_\lambda(T_{\mathbb{C}})$ that yields the same matrix for $T_{\mathbb{C}}$ and for its restriction to this subspace of real points. We conclude that

$$T_{\mathbb{C}}|_{W_\lambda} = T|_{W_\lambda} = T|_{M_\lambda(T)} \quad \text{and} \quad T_{\mathbb{C}}|_{M_\lambda(T)} = T|_{M_\lambda(T)}.$$

CASE 2: A CONJUGATE PAIR $\mu, \overline{\mu}$. The space $M_{\mu,\overline{\mu}} = M_\mu(T_\mathbb{C}) \oplus M_{\overline{\mu}}(T_\mathbb{C})$ is of real type because $J(M_\mu) = M_{\overline{\mu}}$. In fact, if $v \in M_\mu$, then $(T_\mathbb{C} - \mu)^k v = 0$ for some k. But $T_\mathbb{C} J = J T_\mathbb{C}$, so

$$(T_\mathbb{C} - \overline{\mu})^k J(v) = (T_\mathbb{C} - \overline{\mu})^{k-1} J(T_\mathbb{C} - \mu)(v) = \cdots = J(T_\mathbb{C} - \mu)^k(v) = 0.$$

Thus, $M_{\mu,\overline{\mu}}$ is the complexification of its subspace $V_\mu = M_{\mu,\overline{\mu}}(T_\mathbb{C}) \cap (V + i0)$ of real points.

By the cyclic subspace decomposition (Theorem 1.16), a generalized eigenspace $M_\mu(T_\mathbb{C})$ is a direct sum of $T_\mathbb{C}$-invariant spaces C_j that are cyclic under the action of the nilpotent operator $(T_\mathbb{C} - \mu I)$. In C_j, take a basis $\mathfrak{X}_j = \{f_1^{(j)}, \dots, f_{d_j}^{(j)}\}$ that puts $(T_\mathbb{C} - \mu I)|_{C_j}$ into elementary nilpotent form

$$\begin{pmatrix} 0 & 1 & & 0 \\ & \ddots & \ddots & \\ & & \ddots & 1 \\ 0 & & & 0 \end{pmatrix} \quad \text{so that} \quad [(T_\mathbb{C})|_{C_j}] = \begin{pmatrix} \mu & 1 & & 0 \\ & \ddots & \ddots & \\ & & \ddots & 1 \\ 0 & & & \mu \end{pmatrix}.$$

For the basis $\mathfrak{X}_j$, we have

$$(T_\mathbb{C} - \mu I) f_1 = 0 \quad \text{and} \quad (T_\mathbb{C} - \mu I) f_j = f_{j-1} \quad \text{for } j > 1,$$

which implies that $T_\mathbb{C}(f_1) = \mu f_1$ and $T_\mathbb{C}(f_j) = \mu f_j + f_{j-1}$ for $j > 1$.

If $f_j = x_j + i y_j \in W_\mu + i W_\mu$, we have $\overline{f}_j = J(f_j) = x_j - i y_j$ and $T_\mathbb{C}(f_j) = \mu f_j + f_{j-1}$. Hence, since $T_\mathbb{C}$ is of real type,

$$T_\mathbb{C}(\overline{f}_j) = T_\mathbb{C}(J(f_j)) = J(T_\mathbb{C}(f_j)) = J(\mu f_j + f_{j-1}) = \overline{\mu} J(f_j) + J(f_{j-1}),$$

and we get $T_\mathbb{C}(\overline{f}_j) = \overline{\mu}\,\overline{f}_j + \overline{f}_{j-1}$.

Writing $\mu = a + ib$ with $b \neq 0$ (or in polar form, $\mu = re^{i\theta}$ with $\theta \notin \pi\mathbb{Z}$), we get

$$\begin{aligned} T(x_j) + iT(y_j) &= T_\mathbb{C}(x_j + i y_j) = T_\mathbb{C}(f_j) = \mu f_j + f_{j-1} \\ &= (a + ib)(x_j + i y_j) + (x_{j-1} + i y_{j-1}) \\ &= (a x_j - b y_j) + i(b x_j + a y_j) + (x_{j-1} + i y_{j-1}). \end{aligned}$$

Since $\mu = re^{i\theta}$ and $\overline{\mu} = re^{-i\theta}$,

$$\begin{aligned} T(x_j) &= a x_j - b y_j + x_{j-1} = x_j \cdot r\cos(\theta) - y_j \cdot r\sin(\theta) + x_{j-1}, \\ T(y_j) &= b x_j + a y_j + y_{j-1} = x_j \cdot r\sin(\theta) + y_j \cdot r\cos(\theta) + y_{j-1}, \end{aligned}$$

with respect to the $\mathbb{R}$-basis

$$\{x_1^{(1)}, y_1^{(1)}; \dots; x_{d_1}^{(1)}, y_{d_1}^{(1)}; x_1^{(2)}, y_1^{(2)}; \dots\}$$

in $V_\mu = M_{\mu,\overline{\mu}} \cap (V + i0) = \bigoplus_{j=1}^{d} V_J^{(\mu)}$. Thus, the matrix $[T]_{\mathfrak{X}\mathfrak{X}}$ consists of diagonal blocks of size $2d_j \times 2d_j$ that have the form

$$B = \begin{pmatrix} r\,\boxed{R(\theta)} & I_2 & \ddots & 0 \\ & r\,\boxed{R(\theta)} & I_2 & \ddots \\ & & \ddots & I_2 \\ 0 & & & r\,\boxed{R(\theta)} \end{pmatrix}$$

in which I_2 is the 2×2 identity matrix and

$$r \cdot R(\theta) = \begin{pmatrix} a & -b \\ b & a \end{pmatrix} = \begin{pmatrix} r\cos\theta & -r\sin\theta \\ r\sin\theta & r\cos\theta \end{pmatrix}$$

if $\mu = a + ib = re^{i\theta}$. That concludes the proof of Theorem 2.37. □

EXAMPLE 2.38. Find the Jordan form when the following matrix in $\mathrm{M}(3, \mathbb{R})$ is viewed as a complex matrix having real entries:

$$A = \begin{pmatrix} 4 & -1 & -1 \\ -2 & 0 & 4 \\ 2 & -2 & 2 \end{pmatrix}.$$

Then:

1. Find the real-normal form for A.
2. Find an invertible transition matrix $Q \in \mathrm{M}(3, \mathbb{R})$ such that QAQ^{-1} has the normal form in 1.
3. Exhibit an $\mathbb{R}$-basis $\mathfrak{X}$ for $\mathbb{R}^3$ such that $[L_A]_{\mathfrak{X}\mathfrak{X}}$ has this normal form.

DISCUSSION. The characteristic polynomial of A is

$$\begin{aligned} p_A(x) &= \det\begin{pmatrix} 4-x & -1 & -1 \\ -2 & -x & 4 \\ 2 & -2 & 2-x \end{pmatrix} \\ &= -x^3 + 6x^2 - 16x + 16 = (x-2)(x^2 - 4x + 9)\,, \end{aligned}$$

so the eigenvalues are

$$\lambda_1 = 2, \quad \lambda_2 = 2 - 2i, \quad \lambda_3 = 2 + 2i.$$

Since the eigenvalues are distinct, the diagonal matrix $D = \mathrm{diag}(\lambda_1, \lambda_2, \lambda_3)$ is already a JCF for A. Applying symbolic row operations, we can compute bases for the one-dimensional eigenspaces

$$\begin{aligned} E_{\lambda_1}(A) &= \ker(A - \lambda_1 I) = \mathbb{C} \cdot (\mathbf{e}_1 + \mathbf{e}_2 + \mathbf{e}_3) \\ E_{\lambda_2}(A) &= \ker(A - \lambda_2 I) = \mathbb{C} \cdot \big((3-i)\mathbf{e}_1 + (3+4i)\mathbf{e}_2 + 5\mathbf{e}_3\big) \\ E_{\lambda_3}(A) &= \ker(A - \lambda_3 I) = \mathbb{C} \cdot \big((3+i)\mathbf{e}_1 + (3-4i)\mathbf{e}_2 + 5\mathbf{e}_3\big). \end{aligned}$$

The proofs of Lemmas 2.33 and 2.34, which together yield the normal form Corollary 2.35, are constructive and provide an explicit recipe for constructing $\mathbb{C}$-bases for $T_{\mathbb{C}}$ that lie within the $\mathbb{R}$-subspaces of real points in the complex vector spaces

$$E_{\lambda_1}(T_{\mathbb{C}}) \quad \text{or} \quad M_{\lambda_2,\overline{\lambda}_2}(T_{\mathbb{C}}).$$

If we identify $\mathbb{C}^3$ as the complexification of $\mathbb{R}^3$, the conjugation map becomes

$$J(\mathbf{x} + i\mathbf{y}) = \mathbf{x} - i\mathbf{y}.$$

Label the three eigenvectors as

$$\mathbf{z}_{\lambda_k} = \mathbf{x}_k + i\mathbf{y}_k = \left(x_1^{(k)} + iy_1^{(k)}, \dots, x_3^{(k)} + iy_3^{(k)}\right) \quad \text{for } k = 1, 2, 3$$

(with $d_k = d(\lambda_k) = 1$ since all eigenvalues λ_k have multiplicity $= 1$). The polar forms of the eigenvalues are

$$\lambda_1 = 2, \quad \lambda_2 = \sqrt{8}\, e^{-i\pi/2}, \quad \lambda_3 = \sqrt{8}\, e^{i\pi/2}.$$

The first eigenvector

$$\mathbf{z}_1 = (\mathbf{e}_1 + \mathbf{e}_2 + \mathbf{e}_3) + i\mathbf{0} = x_1^{(1)} + i0$$

is already of real type. The eigenvectors $\mathbf{z}_2, \mathbf{z}_3 = J(\mathbf{z}_2)$ are a $\mathbb{C}$-basis for

$$M = M_{\lambda_2, \bar{\lambda}_2}.$$

As in the proof of Theorem 2.35, the real and imaginary parts of $\mathbf{z}_2 = x_1^{(2)} + iy_1^{(2)}$,

$$\begin{aligned} x_1^{(2)} &= \mathrm{Re}(\mathbf{z}_2) = 3\mathbf{e}_1 + 3\mathbf{e}_2 + 5\mathbf{e}_3 = (3, 3, 5) \quad \text{in } \mathbb{R}^3 \\ y_1^{(2)} &= \mathrm{Im}(\mathbf{z}_2) = -\mathbf{e}_1 + 4\mathbf{e}_2 = (-1, 4, 0) \quad \text{in } \mathbb{R}^3, \end{aligned}$$

are vectors of real-type that form a $\mathbb{C}$-basis for M (so do the real and imaginary parts of $\mathbf{z}_3 = J(\mathbf{z}_2)$. The point of Theorem 2.37 is that $\mathfrak{X} = \{x_1^{(1)}, x_1^{(2)}, y_1^{(2)}\}$ is also an $\mathbb{R}$-basis in $\mathbb{R}^3$ that puts T into the real-normal form

$$[T]_{\mathfrak{X}\mathfrak{X}} = \begin{pmatrix} 1 & 0 & 0 \\ 0 & \sqrt{8}\cos\theta & \sqrt{8}\sin\theta \\ 0 & -\sqrt{8}\sin\theta & \sqrt{8}\cos\theta \end{pmatrix}$$

where $\theta = \pi/4$ radians. Compare this with the diagonalized form

$$[T]_{\mathfrak{Y}\mathfrak{Y}} = \begin{pmatrix} 2 & 0 & 0 \\ 0 & \sqrt{8}\, e^{i\pi/4} & 0 \\ 0 & 0 & \sqrt{8}\, e^{-i\pi/4} \end{pmatrix}$$

of $T_{\mathbb{C}}$ for the complex basis $\mathfrak{Y} = \{\mathbf{z}_1, \mathbf{z}_2, \mathbf{z}_3\}$ in $\mathbb{C}^3$. ○

EXERCISE 2.39. Find the Jordan form for the matrix in $\mathrm{M}(4, \mathbb{R})$,

$$A = \begin{pmatrix} 2 & 1 & 0 & 0 \\ -1 & 4 & 0 & 0 \\ -1 & 4 & 1 & -1 \\ -1 & 3 & 2 & -1 \end{pmatrix},$$

regarded as matrix in $\mathrm{M}(4, \mathbb{C})$. Use this complex JCF to find a real normal form for A.

Additional Exercises

Section 2.1. Jordan Form and Differential Equations

1. TRUE/FALSE QUESTIONS ("True" if the statement is *always* true.)

T/F Questions (a)–(e) below are concerned with systems of constant-coefficient differential equations

$$\frac{dX}{dt} = A \cdot X(t) \quad (A \in \mathrm{M}(n, \mathbb{R}))$$

and their vector-valued solutions $X(t)$ with values in $\mathbb{R}^n$.

(a) For any matrix $A \in \mathrm{M}(n, \mathbb{R})$, the system $dX/dt = A \cdot X(t)$ has solutions that are differentiable for all $t \in \mathbb{R}$.

(b) For any $A \in \mathrm{M}(n, \mathbb{R})$ and $\mathbf{c} \in \mathbb{R}^n$, there is a unique solution to the system $dX/dt = A \cdot X(t)$ that satisfies the initial condition $X(0) = \mathbf{c}$.

(c) For some choices of $A \in \mathrm{M}(n, \mathbb{R})$ and $\mathbf{c} \in \mathbb{R}^n$, the system $dX/dt = A \cdot X(t)$ has *no* smooth solutions with $X(0) = \mathbf{c}$.

(d) If $\mathbf{e}_1, \dots, \mathbf{e}_n$ are the standard basis vectors in $\mathbb{R}^n$ and $X_j(t)$ is a solution of the system such that $X_j(0) = \mathbf{e}_j$, then $X_1(t), \dots, X_n(t)$ are linearly independent in the space $\mathcal{C}^\infty(\mathbb{R}, \mathbb{R}^n)$ of infinitely differentiable vector valued function on the real line.

(e) Every solution of the system $dX/dt = A \cdot X(t)$ in (d) lies in the finite-dimensional vector subspace spanned by the particular solutions $X_1(t), \dots, X_n(t)$.

T/F Questions (f)–(g) below are concerned with the solutions of an nth-order constant-coefficient differential equation

$$Ly(t) = D^n y(t) + a_1 D^{n-1} y(t) + \cdots + a_n y(t) = 0.$$

(f) Solutions of this ODE form an n-dimensional subspace in the space $\mathcal{C}^\infty(\mathbb{R})$ of smooth real-valued functions $y(t)$ on the real line.

(g) The scalar-valued solution $y(t)$ of this ODE exists and is unique once we specify its initial value $y(0) = c$ $(c \in \mathbb{R})$ at time $t = 0$.

2. For each of the following linear systems of constant-coefficient differential equations, rewrite the system in the form $dY/dt = A \cdot Y(t)$ for vector-valued functions $Y(t)$.

(i) $\dfrac{dy_1}{dt} = y_2(t), \quad \dfrac{dy_2}{dt} = -y_1(t)$

(ii) $\dfrac{dy_1}{dt} = y_2(t), \quad \dfrac{dy_2}{dt} = y_1(t)$

(iii) $\dfrac{dy_1}{dt} = y_1(t), \quad \dfrac{dy_2}{dt} = y_1(t) + y_2(t)$

(iv) $\dfrac{dy_1}{dt} = y_1(t) + y_2(t), \quad \dfrac{dy_2}{dt} = y_1(t) + 2y_2(t), \quad \dfrac{dy_3}{dt} = -y_1(t) + y_3(t)$

Then

(a) Find solutions $Y(t) = \big(y_1(t), y_2(t)\big)$ that satisfy the initial conditions $Y(0) = \mathbf{c}$ in $\mathbb{R}^2$ for the initial values

$$\mathbf{c}_1 = (1,0) \quad \text{and} \quad \mathbf{c}_2 = (0,1) \quad \text{in } \mathbb{R}^2.$$

(b) Explain why a solution satisfying an arbitrary initial condition $Y(0) = \mathbf{c} = (c_1, c_2)$ can always be obtained by taking a linear combination of these two basic solutions (with what coefficients?).

(c) Find the general solution of this equation, independent of initial conditions.

3. Find all solutions of the the linear system of constant-coefficient differential equations $dY/dt = A \cdot Y(t)$

$$\frac{dy_1}{dt} = y_2 + y_3, \quad \frac{dy_2}{dt} = y_1 + y_3, \quad \frac{dy_3}{dt} = y_1 + y_2$$

such that $\mathbf{y}(0) = (1, 0, -1)$.

4. Continuing with Additional Exercise 3, find a basis

$$\mathfrak{X} = \{\mathbf{y}_1(t), \mathbf{y}_2(t), \mathbf{y}_3(t)\}$$

for the three-dimensional vector space of vector-valued solutions to the system $d\mathbf{y}/dt = A \cdot \mathbf{y}(t)$.

5. Find all solutions of the differential equation

$$\frac{d^3y}{dt^3} - \frac{dy}{dt} = 0.$$

Then exhibit a solution for which $y(0) = 3, y'(0) = -1$. What is the value of $d^2y/dt^2(0)$ for this solution?

6. Find all solutuions of the second-order differential equation

$$\frac{d^2y}{dt^2} + 2\frac{dy}{dt} + y(t) = 0.$$

Section 2.2. Normal Forms for Linear Operators over $\mathbb{R}$

1. TRUE/FALSE QUESTIONS ("True" if the statement is *always* true.)

In (a)–(c) below, let $A \in \mathrm{M}(n, \mathbb{C})$, define $L_A(\mathbf{z}) = A \cdot \mathbf{z}$ on $\mathbb{C}^n$, and regard $\mathbb{C}^n$ as the complexification $\mathbb{R}^n + i\mathbb{R}^n$ of $\mathbb{R}^n$.

(a) L_A is of real type $\Leftrightarrow A(\mathbf{x} + i0) \in \mathbb{R}^n + i0$ for all $\mathbf{x} \in \mathbb{R}^n$.

(b) L_A is of real type $\Leftrightarrow$ all entries in A are real.

(c) L_A is of real type $\Leftrightarrow$ the spectrum $\mathrm{sp}_{\mathbb{C}}(L_A)$ is contained in $\mathbb{R} + i0$.

(d) If T is a linear operator on a vector space over $\mathbb{R}$, then $\lambda = 0$ is in $\mathrm{sp}_{\mathbb{R}}(T) \Leftrightarrow$ it is in $\mathrm{sp}_{\mathbb{C}}(T_{\mathbb{C}})$.

(e) If V is a linear operator on a vector space over $\mathbb{R}$, then $\det(T)$ is the product of the distinct eigenvalues of the complexification $T_{\mathbb{C}}$.

2. If T is a linear operator on a vector space V over $\mathbb{R}$, is $\mathrm{rank}(T) = \dim_{\mathbb{R}}(T(V))$ equal to $\mathrm{rank}(T_{\mathbb{C}}) = \dim_{\mathbb{C}}(T_{\mathbb{C}}(V_{\mathbb{C}}))$?

3. If $A \in M(n, \mathbb{R})$, the multiplication operator

$$L_A(\mathbf{x}) = A \cdot \mathbf{x} \quad \text{for } \mathbf{x} \in \mathbb{R}^n$$

is an $\mathbb{R}$-linear operator on $\mathbb{R}^n$, but A can also be viewed as a complex matrix whose entries happen to be real.

(a) Verify that

$$\tilde{L}_A(\mathbf{z}) = A \cdot \mathbf{z} \quad \text{for } \mathbf{z} \in \mathbb{C}^n$$

defines a complex-linear operator on $\mathbb{C}^n$.

(b) If we view $\mathbb{C}^n$ as the complexification $(\mathbb{R}^n)_\mathbb{C}$ of $\mathbb{R}^n$, the complexification $(L_A)_\mathbb{C} = L_A : V_\mathbb{C} \to V_\mathbb{C}$ is also a $\mathbb{C}$-linear operator on $\mathbb{C}^n$. Verify that these constructs agree:

$$\tilde{L}_A = (L_A)_\mathbb{C} \quad \text{on } \mathbb{C}^n.$$

4. Let B be a bilinear form on a vector space V over $\mathbb{R}$, and define a map $\widetilde{B} : V_\mathbb{C} \times V_\mathbb{C} \to \mathbb{C}$ letting

$$\begin{aligned} &\widetilde{B}(x + iy, x' + iy') \\ &\quad = \big[B(x, x') - B(y, y')\big] + i\big[B(x, y') + B(x', y)\big]. \end{aligned}$$

Is $\widetilde{B}$ a complex-bilinear form on $V_\mathbb{C}$ ($\mathbb{C}$-linear in each input when the other entry is held fixed)?

5. Let T be a linear operator on a finite-dimensional real vector space V and $T_\mathbb{C}$ on $V_\mathbb{C}$ its complexification. If $\lambda_1, \dots, \lambda_n$ are the complex eigenvalues of $T_\mathbb{C}$ listed according to their multiplicities, prove the following:

$$\text{(a) } \det(T) = \det(T_\mathbb{C}) = \prod_{j=1}^{n} \lambda_j \qquad \text{(b) } \mathrm{Tr}(T) = \mathrm{Tr}(T_\mathbb{C}) = \sum_{j=1}^{n} \lambda_j.$$

6. For each of the following matrices in $M(2, \mathbb{R})$,

$$\text{(i) } A = \begin{pmatrix} 0 & -1 \\ 1 & 0 \end{pmatrix} \quad \text{(ii) } A = \begin{pmatrix} 0 & 1 \\ 1 & 0 \end{pmatrix}$$

$$\text{(iii) } A = \begin{pmatrix} 1 & 1 \\ -1 & 1 \end{pmatrix} \quad \text{(iv) } A = \begin{pmatrix} 1 & 1 \\ 1 & -1 \end{pmatrix},$$

(a) Find the Jordan form of the complexification $L_A : \mathbb{C}^2 \to \mathbb{C}^2$.

(b) Find the real-normal form of A.

(c) Find a transition matrix Q such that QAQ^{-1} puts A into the real-normal form in (b).

7. Find the complex Jordan form of the matrix

$$A = \begin{pmatrix} 3 & 2 & -2 & -1 \\ 1 & 0 & -1 & 0 \\ 3 & 3 & -2 & -1 \\ 4 & 0 & -2 & -1 \end{pmatrix}$$

regarding A as a complex matrix. Then use this to find a real-normal form for A.

HINT. $\lambda = i$ is a complex double root of p_A.

CHAPTER 3

Bilinear, Quadratic, and Multilinear Forms

An Overview of This Chapter. Bilinear forms are maps $B : V \times V \to \mathbb{K}$ that act on ordered pairs (v, w) of vectors to produce a scalar $B(v, w)$ in the ground field $\mathbb{K}$. "Bilinearity" means that the output $B(v, w)$ is a linear function of v when the other input w is held fixed, and likewise if v is held fixed while w varies. In a certain sense, a bilinear map involves the joint action of two linear maps from $V \to \mathbb{K}$, each of which is an element of the dual vector space V^*, which was a main topic in Chapter 3 of LA I.

Bilinear forms arise often in physics and many areas of mathematics. For instance, the inner products $B(u, v) = (u, v)$ on real vector spaces are well-known examples of bilinear forms. Thus, it is of some importance to find natural "*canonical forms*" that reveal the behavior of a given form B. This is analogous to the problem of diagonalizing a linear linear operator T, or finding a basis that puts the matrix $[T]_{\mathfrak{X}\mathfrak{X}}$ into Jordan canonical form when T cannot be diagonalized. Thus, we will often speak of "diagonalizing" bilinear forms, although we will soon see that the two problems are different and have markedly different outcomes. (There are only finitely many possible canonical forms for B on a vector space with $\dim_{\mathbb{K}} V = n$.)

Important symmetry groups in physics arise as groups of linear operators that leave invariant particular bilinear forms. As an example, we have the *Lorentz group*, the family of linear operators that leave invariant the "Lorentz form"

$$B\big((\mathbf{x}, t), (\mathbf{x}, t)\big) = x_1^2 + x_2^2 + x_3^2 - c^2t^2 \quad \text{(with } c = \text{the speed of light)}\,,$$

that acts on the coordinates $(\mathbf{x}, t) \in \mathbb{R}^3 \times \mathbb{R}$ of space-time in Einstein's theory of special relativity. These transformations tell us how measurements of space and time change when we shift from the viewpoint of one observer to that of a different observer moving at constant velocity. Invariance of this form embodies the observed fact that the speed of a passing light wave is the same for all observers even if they are moving relative to one another. Many consequences of this simple fact are profoundly counterintuitive, and invariance of the Lorentz form is the key to understanding them. Detailed study of this symmetry group revealed that neither space nor time measurements have any absolute meaning; on the other hand, it does allow us to compute how appearances change as we move from one frame of reference to another.

Beyond all this, we will enter the world of *tensors* and *multilinear forms*, with which you may be less familiar. These have become central topics in differential geometry and in physics too — general relativity is all about tensors; so is much of electromagnetic theory and the analysis of stress in solid media such as the Earth's mantle and core. This chapter provides a basic introduction to tensors. It is followed in Chapter 4 by a much more detailed introduction to the role of multilinear algebra in modern differential geometry, ending with a re-examination of what was really going on in vector calculus where you first encountered *div, grad, curl* and all that.

3.1. Basic Definitions and Examples

DEFINITION 3.1. A *bilinear form* is a map $B : V \times V \to \mathbb{K}$ that is linear in each entry when the other entry is held fixed so that

$$\begin{aligned} B(\alpha x, y) &= \alpha B(x, y) = B(x, \alpha y) \\ B(x_1 + x_2, y) &= B(x_1, y) + B(x_2, y) \quad \text{for } \alpha \in \mathbb{K},\ x_k \in V,\ y_k \in V \\ B(x, y_1 + y_2) &= B(x, y_1) + B(x, y_2). \end{aligned}$$

(This of course forces $B(x, y) = 0$ if either input is zero.) We say B is *symmetric* if $B(x, y) = B(y, x)$ for all x, y and *antisymmetric* if $B(x, y) = -B(y, x)$.

Similarly, a *multilinear form* (aka: *k-linear form* or a *tensor of rank k*) is a map $B : V \times \cdots \times V \to \mathbb{K}$ that is linear in each entry when the other entries are held fixed. We write $V^{(0,k)} = V^* \otimes \cdots \otimes V^*$ for the set of k-linear forms. The reason we use the dual space V^* here rather than V, and the rationale for this "tensor product" notation, will soon become clear. You can think of a rank-k tensor on a vector space V as a black box that accepts an ordered list of input vectors chosen from V and returns a single scalar in $\mathbb{K}$ as its output:

$$(v_1, \dots, v_k) \xrightarrow{\text{input}} \boxed{B} \xrightarrow{\text{output}} B(v_1, \dots, v_k) \in \mathbb{K}.$$

The set $V^{(0,2)} = V^* \otimes V^*$ of bilinear forms on V becomes a vector space over $\mathbb{K}$ if we define the following:

1. ZERO ELEMENT. $B(x, y) = 0$ for all $x, y \in V$.
2. SCALAR MULTIPLE. $(\alpha B)(x, y) = \alpha B(x, y)$ for $\alpha \in \mathbb{K}$ and $x, y \in V$.
3. ADDITION. $(B_1 + B_2)(x, y) = B_1(x, y) + B_2(x, y)$ for $x, y \in V$.

When $k > 2$, the space of k-linear forms $V^* \otimes \cdots \otimes V^*$ is also a vector space using the same definitions. The space $V^{(0,1)}$ of 1-forms (the *tensors of rank 1 on* V) is the dual space $V^* = \mathrm{Hom}_{\mathbb{K}}(V, \mathbb{K})$ consisting of all $\mathbb{K}$-linear maps $\ell : V \to \mathbb{K}$. By convention, the space of *0-forms* is identified with the ground field: $V^{(0,0)} = \mathbb{K}$. Its elements are not mappings on V. It is also possible (and useful) to define multilinear forms of mixed type, mappings $\theta : V_1 \times \cdots \times V_k \to \mathbb{K}$ in which the components V_j are not all the same. These forms also constitute a vector space. We postpone any discussion of forms of "mixed type."

If $\ell_1, \ell_2 \in V^*$, we can create a bilinear form $\ell_1 \otimes \ell_2$ by taking a "*tensor product*" of these forms, defining

$$\ell_1 \otimes \ell_2(v_1, v_2) = \langle \ell_1, v_1 \rangle \cdot \langle \ell_2, v_2 \rangle \quad \text{for } v_1, v_2 \in V.$$

Bilinearity is easily checked. More generally, if $\ell_1, \ldots, \ell_k \in V^*$, we obtain a k-linear map from $V \times \cdots \times V \to \mathbb{K}$ if we let

$$\ell_1 \otimes \cdots \otimes \ell_k(v_1, \ldots, v_k) = \prod_{j=1}^{k} \langle \ell_j, v_j \rangle.$$

We will show that monomials of the form $\ell_1 \otimes \cdots \otimes \ell_k$ with $\ell_j \in V^*$ span the space $V^{(0,k)}$ of rank-k tensors, but they do not by themselves form a vector space except when $k = 1$.

EXERCISE 3.2. If $A : V \to V$ is any linear operator on a real inner product space, verify that

$$\phi(v_1, v_2) = (Av_1, v_2) \quad \text{for } v_1, v_2 \in V$$

is a bilinear form.

NOTE. This would not be true if $\mathbb{K} = \mathbb{C}$. Inner products on a complex vector space are conjugate-linear in their second input with $(x, z \cdot y) = \overline{z} \cdot (x, y)$ for $z \in \mathbb{C}$; for $\mathbb{C}$-linearity in the second entry, we would need $(x, z \cdot y) = z \cdot (x, y)$. However, $\overline{c} = c$ for real scalars, so an inner product on a *real* vector space is a bona fide rank-2 tensor in $V^{(0,2)}$. ○

EXAMPLE 3.3. Let $A \in \mathrm{M}(n, \mathbb{K})$ and $V = \mathbb{K}^n$. Regarding elements of $\mathbb{K}^n$ as $n \times 1$ column vectors, define

$$B(\mathbf{x}, \mathbf{y}) = \mathbf{x}^{\mathrm{T}} A \mathbf{y} = \sum_{i,j=1}^{n} x_i A_{ij} y_j$$

where $\mathbf{x}^{\mathrm{T}}$ is the $1 \times n$ transpose of the $n \times 1$ column vector $\mathbf{x}$. If we interpret the 1×1 product matrix as a scalar in $\mathbb{K}$, then B is a typical bilinear form on $V = \mathbb{K}^n$.

The analogous construction for multilinear forms is more complicated. For instance, to describe a rank-3 tensor $B(\mathbf{x}, \mathbf{y}, \mathbf{z})$ on $V \times V \times V$, we would need a three-dimensional $n \times n \times n$ array of coefficients $\{B_{i_1,i_2,i_3} : 1 \le i_k \le n\}$, from which we recover the original multilinear form via a sum with n^3 terms:

$$B(\mathbf{x}, \mathbf{y}, \mathbf{z}) = \sum_{i_1,i_2,i_3=1}^{n} x_{i_1} y_{i_2} z_{i_3} B_{i_1,i_2,i_3} \quad \text{for } \mathbf{x}, \mathbf{y}, \mathbf{z} \in \mathbb{K}^3.$$

If B has rank k, the coefficient array is k-dimensional with n^k entries, where $n = \dim(V)$; B is described by an $n \times n$ matrix only for bilinear forms ($k = 2$). For the time being, we will focus on bilinear forms, which are quite important in their own right.

Many examples involve symmetric or antisymmetric bilinear forms, and in any case, we have the following result.

LEMMA 3.4. *Every bilinear form B is uniquely the sum $B = B_+ + B_-$ of a symmetric and an antisymmetric form.*

PROOF. $B_{\pm}$ are given by

$$B_+(v_1, v_2) = \frac{B(v_1, v_2) + B(v_2, v_1)}{2} \quad \text{and} \quad B_- = \frac{B(v_1, v_2) - B(v_2, v_1)}{2}.$$

As for uniqueness, you cannot have $B = B'$ with B symmetric and B' antisymmetric without both being the zero form. □

Variants: Sesquilinear Forms. If V is an inner product space over $\mathbb{C}$, a map B from $V \times V \to \mathbb{C}$ is *sesquilinear* if it is a linear function of its first entry when the other is held fixed but conjugate-linear in the second entry so that

$$B(x_1 + x_2, y) = B(x_1, y) + B(x_2, y)$$
$$B(x, y_1 + y_2) = B(x, y_1) + B(x, y_2)$$

and

$$B(\alpha x, y) = \alpha B(x, y), \quad B(x, \alpha y) = B(x, y)\,\overline{\alpha}$$

for all $\alpha \in \mathbb{C}$. (This is the same as bilinearity when $\mathbb{K} = \mathbb{R}$.) The map is *Hermitian symmetric* if

$$B(y, x) = \overline{B(x, y)}.$$

On a vector space over $\mathbb{R}$, an *inner product* is a special type of sesquilinear form, one that is *strictly positive definite* in the sense that

$$B(x, x) \geq 0 \text{ for all } x \in V \quad \text{and} \quad B(x, x) = \|x\|^2 = 0 \Rightarrow x = 0. \tag{3.1}$$

Over $\mathbb{C}$, an inner product is a map $B : V \times V \to \mathbb{C}$ that is sesquilinear, Hermitian symmetric, and satisfies the positivity conditions (3.1).

A bilinear form $B \in V^* \otimes V^*$ is completely determined by its action on a basis $\mathfrak{X} = \{e_i\}$ via the matrix $B_{\mathfrak{X}} = [B_{ij}]$ with entries

$$B_{ij} = B(e_i, e_j) \quad \text{for } 1 \leq i, j \leq n.$$

This matrix is symmetric/antisymmetric if and only if A has these properties. Given $[B]_{\mathfrak{X}}$, we recover B by writing $x = \sum_i x_i e_i$, $y = \sum_j y_j e_j$ so that

$$\begin{aligned} B(x, y) &= B\Big(\sum_i x_i e_i\,,\ \sum_j y_j e_j\Big) = \sum_i x_i\,\Big(e_i\,,\ \sum_j y_j e_j\Big) \\ &= \sum_{i,j} x_i B_{ij} y_j = [x]_{\mathfrak{X}}^{\mathrm{T}} [B]_{\mathfrak{X}} [y]_{\mathfrak{X}}\,, \end{aligned} \tag{3.2}$$

regarding a 1×1 matrix as an element of $\mathbb{K}$. Conversely, given a basis and a matrix $A \in \mathrm{M}(n, \mathbb{K})$, the previous equality determines a bilinear form B (which is symmetric if and only if $B = B^{\mathrm{T}}$) such that $[B]_{\mathfrak{X}} = A$. Thus we have isomorphisms between vector spaces over $\mathbb{K}$:

1. The space of rank-2 tensors $V^{(0,2)} = V^* \otimes V^*$ is isomorphic to $\mathrm{M}(n, \mathbb{K})$ via the correspondence $B \to [B]_{\mathfrak{X}}$ once a basis in V is specified.
2. The space of *symmetric* bilinear forms is isomorphic to the vector space of symmetric matrices.

NOTE. We write $[B]_{\mathfrak{X}}$ for the matrix associated with a bilinear form $B : V \times V \rightarrow \mathbb{R}$ given a basis $\mathfrak{X}$ in V. In this situation, both copies of V are involved on the same footing and there is no "initial space" V or "target space" W. In defining the matrix $[T]_{\mathfrak{Y}\mathfrak{X}}$ associated with a linear operator $T : V \rightarrow W$, we had to specify bases $\mathfrak{X}$ in V and $\mathfrak{Y}$ in W (even if $V = W$ and $\mathfrak{X} = \mathfrak{Y}$). ○

We next produce a basis for the space $V^{(0,2)} = V^* \otimes V^*$ of bilinear forms and determine its dimension.

PROPOSITION 3.5. *If $\mathfrak{X} = \{e_i\}$ is a basis in a finite dimensional vector space V and $\mathfrak{X}^* = \{e_i^*\}$ is the dual basis in V^* such that $\langle e_i^*, e_j\rangle = \delta_{ij}$, the monomials $e_i^* \otimes e_j^*$ given by*

$$e_i^* \otimes e_j^*(v_1, v_2) = \langle e_i^*, v_1\rangle \cdot \langle e_j^*, v_2\rangle \quad (v_1, v_2 \in V)$$

are a basis in $V^* \otimes V^*$. Hence, $\dim(V^* \otimes V^*) = n^2 = \dim(V)^2$.

PROOF. The monomials $e_i^* \otimes e_j^*$ span $V^* \otimes V^*$, because if B is any bilinear form and $B_{ij} = B(e_i, e_j)$, then $\widetilde{B} = \sum_{i,j} B_{ij} \cdot e_i^* \otimes e_j^*$ has the same action on pairs $e_k, e_\ell \in V$ as the original tensor B.

$$\begin{aligned}\widetilde{B}(e_k, e_l) &= \Big(\sum_{i,j} B_{ij} \cdot e_i^* \otimes e_j^*\Big)\langle e_k, e_\ell\rangle = \sum_{i,j} B_{ij}\langle e_i^*, e_k\rangle \cdot \langle e_j^*, e_\ell\rangle \\ &= \sum_{i,j} B_{ij}\delta_{ik}\delta_{j\ell} = B_{k\ell} = B(e_k, e_\ell),\end{aligned}$$

so $\widetilde{B} = B \in \mathbb{K}\text{-span}\{e_i^* \otimes e_j^*\}$. As for linear independence, if $\widetilde{B} = \sum_{i,j} b_{ij} e_i^* \otimes e_j^*$ is the zero element in $V^{(0,2)}$, then $\widetilde{B}(x, y) = 0$ for all x, y, so

$$b_{k\ell} = \widetilde{B}(e_k, e_\ell) = 0 \quad \text{for } 1 \leq k, \ell \leq n. \qquad \square$$

A similar discussion shows that the space $V^{(0,r)}$ of rank-r tensors has dimension

$$\dim(V^{(0,r)}) = \dim(V^* \otimes \cdots \otimes V^*) = \dim(V)^r = n^r.$$

If $\mathfrak{X} = \{e_1, \dots, e_n\}$ is a basis for V and $\{e_i^*\}$ is the dual basis in V^*, the monomials

$$e_{i_1}^* \otimes \cdots \otimes e_{i_r}^* \quad 1 \leq i_1, \dots, i_r \leq n$$

are a basis for $V^{(0,r)}$.

THEOREM 3.6 (Change of Basis). *Given $B \in V^* \otimes V^*$ and a basis $\mathfrak{X}$ in V, we describe B by its matrix as in* (3.2). *If $\mathfrak{Y} = \{f_j\}$ is another basis and if*

$$f_j = \mathrm{id}(f_j) = \sum_k s_{kj} e_k \quad \textit{for } 1 \leq j \leq n\,,$$

then $S = [S_{ij}] = [\text{id}]_{\mathfrak{X}\mathfrak{Y}}$ *is the transition matrix for basis vectors and we have*

$$\begin{aligned}([B]_{\mathfrak{Y}})_{ij} &= B(f_i, f_j) = B(\sum_k S_{ki}e_k, \sum_\ell S_{\ell j}e_\ell) \\ &= \sum_{k,\ell} S_{ki}B_{k\ell}S_{\ell j} = \sum_{k,\ell}(S^\mathsf{T})_{ik}B_{k\ell}S_{\ell j} \\ &= (S^\mathsf{T}[B]_{\mathfrak{X}}S)_{ij}.\end{aligned}$$

REMARK 3.7. We can also write this as $[B]_{\mathfrak{Y}} = P[B]_{\mathfrak{X}}P^\mathsf{T}$, taking $P = S^\mathsf{T} = [\text{id}]^\mathsf{T}_{\mathfrak{X}\mathfrak{Y}}$ where $id : V \to V$ is the (linear) identity map on V. (Recall the notational conventions set forth in equation (2.8) of Section 2.4 in LA I.) Thus, change of basis is effected by "congruence" of matrices $A \to PAP^\mathsf{T}$ with $\det(P) \neq 0$. This differs considerably from the "similarity transforms" $A \to SAS^{-1}$ that describe the effect of change of basis on the matrix $[T]_{\mathfrak{Y}\mathfrak{X}}$ of a linear operator $T : V \to V$. Note that S^T is generally not equal to S^{-1}, so congruence and similarity are not the same thing. The difference between these concepts will emerge when we seek "normal forms" for various kinds of bilinear (or sesquilinear) forms. ○

DEFINITION 3.8. A bilinear form B is *nondegenerate* if

$$B(v, V) = 0 \Rightarrow v = 0 \quad \text{and} \quad B(V, v) = 0 \Rightarrow v = 0.$$

If B is either symmetric or antisymmetric, we only need the one-sided version. The *radical* of B is the subspace

$$\text{rad}(B) = \{v \in V : B(v, v') = 0 \text{ for all } v' \in V\},$$

which measures the degree of degeneracy of the form B. The *B-orthocomplement* of a subspace $W \subseteq V$ is defined to be

$$W^{\perp,B} = \{v \in V : B(v, W) = (0)\}.$$

REMARK 3.9. Obviously, $W^{\perp,B}$ is a subspace, and when B is symmetric or antisymmetric, the conditions $B(v, W) = 0$ and $B(W, v) = 0$ determine the same "complementary" subspace $W^{\perp,B}$. In particular, nondegeneracy means that $V^{\perp,B} = \{0\}$, and in general $V^{\perp,B}$ is equal to the radical of B. ○

The notion of "nondegeneracy" is a little ambiguous when the bilinear form B is neither symmetric nor antisymmetric: Is there a difference between "right-nondegenerate," in which $B(V, y) = 0 \Rightarrow y = 0$, and "left-nondegenerate," in which $B(x, V) = 0 \Rightarrow x = 0$? The answer is no. In fact, if we impose a basis $\mathfrak{X}$ on V and view vectors $x, y \in V$ as $n \times 1$ columns, we may write $B(x, y) = [x]^\mathsf{T}_{\mathfrak{X}}[B]_{\mathfrak{X}}[y]_{\mathfrak{X}}$, and if $[B]_{\mathfrak{X}}$ is singular, there would be some $y \neq 0$ such that $[B]_{\mathfrak{X}}[y]_{\mathfrak{X}} = 0$. Hence $B(V, y) = 0$ if B is right-nondegenerate. Thus, $[B]_{\mathfrak{X}}$ must be nonsingular if B is right-nondegenerate. But this works in both directions, making this condition both necessary and sufficient. The same argument applies to left-nondegeneracy, so the two nondegeneracy conditions are equivalent. We have proved the following lemma:

LEMMA 3.10. *If B is a bilinear form on V, it is right-nondegenerate if and only if $[B]_{\mathfrak{X}}$ is nonsingular, and likewise for left-nondegeneracy.*

Thus, we may drop the "left/right" conditions on nondegeneracy.

LEMMA 3.11. *If B is a nondegenerate bilinear form on a finite dimensional vector space V, B determines a natural linear bijection $J_B : V \to V^*$ given by*

$$\langle J_B(v), v'\rangle = B(v', v) \quad \text{for } v' \in V.$$

PROOF. If $|V| < \infty$, any nondegenerate bilinear form B mediates a natural bijection $J = J_B : V \to V^*$ that identifies each vector $v \in V$ with a functional $J(v)$ in V' such that

$$\langle J(v), v'\rangle = B(v', v) \qquad \text{for all } v, v' \in V.$$

This map is clearly $\mathbb{K}$-linear and $J(w) = 0 \Rightarrow B(V, w) = 0 \Rightarrow w = 0$ by nondegeneracy of B, so J is one-to-one. It is also a bijection, hence an isomorphism, because $\dim(V) = \dim(V^*)$. □

Hereafter, it will often be convenient to abbreviate $\dim(V) = |V|$ in discussing B-orthocomplements of subspaces in V.

THEOREM 3.12 (Dimension of B-Orthocomplements). *If B is a nondegenerate bilinear form on a finite dimensional space V, the dimension of the B-orthocomplement $M^{\perp,B} = \{w : B(V, w) = 0\}$ of a subspace is related to that of M by*

$$\text{(3.3)} \qquad \text{DIMENSION FORMULA:} \quad |M| + |M^{\perp,B}| = |V|,$$

even though we need not have $M \cap M^{\perp,B} = (0)$ or $V = M \oplus M^{\perp,B}$.

This is essentially a consequence of previous results about *annihilators* $M^\circ \subseteq V^*$ of subspaces $M \subseteq V$ (discussed in LA I, Exercises 3.25 - 3.26). For completeness, here is a self-contained proof of this relation.

In Section 2.3 of LA I we defined the "annihilator" M° of a subspace $M^\circ \subseteq V$ to be a subspace in V^* with

$$M^\circ = \{\ell \in V^* : \langle \ell, M\rangle = 0\}$$

and explained why

$$(M^\circ)^\circ = M \quad \text{and} \quad |V| = |M| + |M^\circ|$$

when $|V| < \infty$. The annihilator M° is analogous to the orthogonal complement $M^\perp$ in an inner product space, but it lives in the dual space V^* instead of V. It has the advantage that M° makes sense in *any* vector space V, whether or not it is equipped with an inner product or a nondegenerate bilinear form. (Also, orthogonal complements $M^\perp$ depend on the particular inner product on V, while the annihilator M° has an absolute meaning all moving observers can agree upon.)

EXERCISE 3.13. When V is equipped with a nondegenerate bilinear form B, we may invoke the natural isomorphism $V \cong V^*$ it induces (as in Lemma 3.11) to identify an annihilator M° in V^* with a uniquely defined subspace $J_B^{-1}(M^\circ)$ in V. From the definitions, verify that $M^\circ \subseteq V^*$ becomes the B-orthocomplement $M^{\perp,B} \subseteq V$ under these identifications.

With this in mind, the dimension formula (3.3) is an immediate consequence of the following exercise regarding annihilators.

EXERCISE 3.14. If V is finite dimensional and M is a subspace, let $\mathfrak{X} = \{v_1, \ldots, v_r, v_{r+1}, \ldots, v_n\}$ be a basis in V such that $\mathbb{K}$-span$\{x_1 \ldots x_r\} = M$ and let $\mathfrak{X}^* = \{v_1^*, \ldots, e_n^*\}$ be the basis in V^* dual to $\mathfrak{X}$ in V.

(a) Verify that the annihilator M° is equal to $\mathbb{K}$-span$\{v_{r+1}^*, \ldots, v_r^*\}$.

(b) Use this and the remarks of Exercise 3.13 to show that

$$|M^{\perp,B}| = |M^\circ| = n - (r+1) = |V| - |M|$$

as asserted in equation (3.3).

Dealing with Degeneracy of *B*. If B is degenerate, so the radical rad(B) is nonzero, the role of the radical can be eliminated for most practical purposes, allowing us to focus on nondegenerate forms.

EXERCISE 3.15. Let $M = \text{rad}(B)$ and form the quotient space $\widetilde{V} = V/M$. Show the following.

(a) B induces a well-defined bilinear form $\widetilde{B} : \widetilde{V} \times \widetilde{V} \to \mathbb{K}$ if we let

$$\widetilde{B}(x + M, y + M) = B(x, y) \quad \text{for all } x, y \in V.$$

(b) $\widetilde{B}$ is symmetric (or antisymmetric) $\Leftrightarrow$ B is.

(c) Prove that $\widetilde{B}$ is nondegenerate on the quotient space $\widetilde{V} = V/M$.

EXERCISE 3.16. Given $n \times n$ matrices A, B, show that

$$\mathbf{x}^{\mathsf{T}} B \mathbf{y} = \mathbf{x}^{\mathsf{T}} A \mathbf{y} \text{ for all } \mathbf{x}, \mathbf{y} \in \mathbb{K}^n \text{ if and only if } A = B.$$

3.2. Canonical Models for Bilinear Forms

In doing calculations, it is natural to work with the matrices $[B]_{\mathfrak{X}}$ that represent B with respect to various bases and seek bases yielding the simplest possible form. If a bilinear form B is represented by $A = [B]_{\mathfrak{X}}$ for some basis, we must examine the effect of a change of basis $\mathfrak{X} \to \mathfrak{Y}$ and describe $[B]_{\mathfrak{Y}}$ in terms of the transition matrix $S = [\text{id}]_{\mathfrak{X}\mathfrak{Y}}$ that tells us how to write vectors in the $\mathfrak{Y}$-basis in terms of vectors in $\mathfrak{X}$, as in (3.2). Writing $\mathfrak{X} = \{e_i\}$ and $\mathfrak{Y} = \{f_j\}$, let $S = [s_{ij}] = [\text{id}]_{\mathfrak{X}\mathfrak{Y}}$ be the matrix of coefficients such that

$$f_j = \text{id}(f_j) = \sum_k s_{kj} e_k \quad \text{for } 1 \le j \le n. \tag{3.4}$$

Obviously, $\det(S) \neq 0$ because this system of vector equations must be invertible. In Theorem 3.6, we worked out the effect of this basis change:

$$[B]_{\mathfrak{Y}} = S^{\mathsf{T}} [B]_{\mathfrak{X}} S \quad \text{where } S = [\text{id}]_{\mathfrak{X}\mathfrak{Y}}\,,$$

and if we let $P = S^{\mathsf{T}}$, this becomes

$$[B]_{\mathfrak{Y}} = P [B]_{\mathfrak{X}} P^{\mathsf{T}} \qquad \text{with } P = S^{\mathsf{T}}. \tag{3.5}$$

We now show that there is only a limited number of possible "canonical forms" for nondegenerate bilinear forms on a vector space of dim $V = n$, at least when

B is either symmetric or antisymmetric (the forms of greatest interest in applications). We might also ask whether these canonical forms are unique. (Answer: not very.)

The Automorphism Group of a Bilinear Form B. If a vector space is equipped with a nondegenerate bilinear form B, a natural (and important) symmetry group, the *automorphism group* $\mathrm{Aut}(B) \subseteq \mathrm{GL}_{\mathbb{K}}(V)$, comes along with it. It consists of the invertible linear maps $T : V \to V$ that "leave the form invariant," with

$$B(T(x), T(y)) = B(x, y) \quad \text{for all vectors } x, y \in V.$$

We have encountered such automorphism groups before, by various names. For example,

1. The REAL-ORTHOGONAL GROUP $\mathrm{O}(n)$ consists of the invertible linear maps T on $\mathbb{R}^n$ that preserve the usual inner product,

$$B(\mathbf{x}, \mathbf{y}) = \sum_{i=1}^{n} x_i y_i \quad \text{for } \mathbf{x}, \mathbf{y} \in \mathbb{R}^n.$$

   As explained in Theorem 6.68 of LA I, the automorphisms that preserve this symmetric bilinear form are precisely the linear *rigid motions* on Euclidean space, those that leave invariant lengths of vectors and distances between them, so that

$$\|T(\mathbf{x})\| = \|\mathbf{x}\| \quad \text{and} \quad \|T(\mathbf{x}) - T(\mathbf{y})\| = \|\mathbf{x} - \mathbf{y}\|$$

   for $\mathbf{x}, \mathbf{y} \in \mathbb{R}^n$, where

$$\|\mathbf{x}\| = \Big(\sum_{i=1}^{n} |x_i|^2\Big)^{1/2} \quad \text{(Pythagoras' formula).}$$

2. The UNITARY GROUP $\mathrm{U}(n)$ is the group of invertible linear operators on $V = \mathbb{C}^n$ that preserve the (Hermitian, sesquilinear) standard inner product

$$B(\mathbf{z}, \mathbf{w}) = \sum_{k=1}^{n} z_k \overline{w_k}$$

   on complex n-space. For these operators, the following conditions are equivalent (see Theorem 6.68 and Proposition 6.71 of LA I):

$$\begin{aligned} T \in \mathrm{U}(n) &\Leftrightarrow B\big(T(\mathbf{z}),\, T(\mathbf{w})\big) = B(\mathbf{z}, \mathbf{w}) \\ &\Leftrightarrow \|T(\mathbf{z})\| = \|\mathbf{z}\| \\ &\Leftrightarrow \|T(\mathbf{z}) - T(\mathbf{w})\| = \|\mathbf{z} - \mathbf{w}\| \end{aligned}$$

   for $\mathbf{z}, \mathbf{w} \in \mathbb{C}^n$, where

$$\|\mathbf{z}\| = B(\mathbf{z}, \mathbf{z})^{1/2} = \Big(\sum_{i=1}^{n} |z_i|^2\Big)^{1/2} \quad \text{(Pythagoras' formula for } \mathbb{C}^n\text{).}$$

EXERCISE 3.17. Explain why $\mathrm{U}(n)$ is a closed and bounded subset in matrix space $\mathrm{M}(n, \mathbb{C}) \cong \mathbb{C}^{n^2}$ — i.e.,

(a) There is a bound $M > 0$ such that all entries in a unitary matrix A satisfy $|a_{ij}| \leq M$, and

(b) If a sequence $\{A_n\}$ of unitary matrices converges to a limit in matrix space $\mathrm{M}(n, \mathbb{C})$, say with $A_n \to A$ as $n \to \infty$, the limit A is also a unitary matrix.

HINT. In (a) you must find a common upper bound M for the size $|a_{ij}|$ of the (complex) entries in any unitary $n \times n$ matrix. In (b) recall the discussion of limits of matrices and linear operators in LA I (Sections 5.3 - 5.4). For matrices, we have $A^{(n)} \to A$ as $n \to \infty \Leftrightarrow A_{ij}^{(n)} \to A_{ij}$ in $\mathbb{C}$ as $n \to \infty$ for all i, j.

3. The COMPLEX ORTHOGONAL GROUP $\mathrm{O}(n, \mathbb{C})$ is the automorphism group of the following bilinear form on complex n-space $\mathbb{C}^n$:

$$B(\mathbf{z}, \mathbf{w}) = \sum_{k=1}^{n} z_k w_k \quad (\mathbf{z}, \mathbf{w} \in \mathbb{C}^n).$$

This is bilinear over $\mathbb{K} = \mathbb{C}$, but is *not* an inner product; it is not conjugate-linear in the entry $\mathbf{w}$ because w_k appears in B instead of $\overline{w_k}$. Furthermore, not all vectors have $B(\mathbf{z}, \mathbf{z}) \geq 0$ (try $\mathbf{z} = (1, i)$ in $\mathbb{C}^2$).

In the present section, we will systematically examine the canonical forms and associated automorphism groups for nondegenerate symmetric or antisymmetric forms over $\mathbb{K} = \mathbb{R}$ or $\mathbb{C}$. The number of possibilities is surprisingly small.

DEFINITION 3.18. The *automorphism group* of a nondegenerate symmetric or antisymmetric form $B : V \times V \to \mathbb{K}$ is

$$\text{(3.6)} \quad \mathrm{Aut}(B) = \{T \in \mathrm{GL}_{\mathbb{K}}(V) : B\big(T(v), T(w)\big) = B(v, w) \quad \text{for all } v, w \in V\}.$$

Here, $\mathrm{GL}_{\mathbb{K}}(V) = \{T : \det(T) \neq 0\}$ is the *general linear group* on V, consisting of all invertible $\mathbb{K}$-linear operator $T : V \to V$. $\mathrm{Aut}(B)$ is a *group* because it contains the identity element $I = \mathrm{id}_V$, the composition product $S \circ T$ of any two elements, and the inverse T^{-1} of any element.

Given a basis $\mathfrak{X} = \{e_i\}$ for V, each linear operator $T \in \mathrm{Aut}(B)$ corresponds to an invertible matrix $[T]_{\mathfrak{X}\mathfrak{X}}$ such that $T(e_j) = \sum_k t_{kj} e_k$, and these matrices form a group

$$G_{B,\mathfrak{X}} = \big\{ [T]_{\mathfrak{X}\mathfrak{X}} : T \in \mathrm{Aut}(B) \big\}$$

under matrix multiplication $(\cdot)$. The group $(\mathrm{Aut}(B), \circ)$ and the matrix group $(G_{B,\mathfrak{X}}, \cdot\,)$ are isomorphic and are often identified.

Matrices in $G_{B,\mathfrak{X}}$ are characterized by the following algebraic properties:

$$\text{(3.7)} \qquad G_{B,\mathfrak{X}} = \big\{E \in \mathrm{GL}(n, \mathbb{K}) : E^{\mathsf{T}} [B]_{\mathfrak{X}} E = [B]_{\mathfrak{X}}\big\}.$$

This follows because $[T(x)]_{\mathfrak{X}} = [T]_{\mathfrak{X}\mathfrak{X}} \cdot [x]_{\mathfrak{X}}$ for $x \in V$. Hence,

$$\begin{aligned}
T \in \mathrm{Aut}(B) &\Leftrightarrow B(x,y) = B\big(T(x), T(y)\big) \quad \text{for all } x, y \in V \\
&\Leftrightarrow [x]_{\mathfrak{X}}^{\mathrm{T}}[B]_{\mathfrak{X}}[y]_{\mathfrak{X}} = [T(x)]_{\mathfrak{X}}^{\mathrm{T}}[B]_{\mathfrak{X}}\,[T(y)]_{\mathfrak{X}} \\
&\qquad = [x]_{\mathfrak{X}}^{\mathrm{T}}\left([T]_{\mathfrak{X}\mathfrak{X}}^{\mathrm{T}}\,[B]_{\mathfrak{X}}[T]_{\mathfrak{X}\mathfrak{X}}\right)[y]_{\mathfrak{X}} \quad \text{for } x, y \in V \\
&\Leftrightarrow [B]_{\mathfrak{X}} = [T]_{\mathfrak{X}\mathfrak{X}}^{\mathrm{T}}[B]_{\mathfrak{X}}\,[T]_{\mathfrak{X}\mathfrak{X}}\,,
\end{aligned}$$

so T is an automorphism of the bilinear form B if and only if the matrix $[T]_{\mathfrak{X}\mathfrak{X}}$ satisfies the identity

$$[B]_{\mathfrak{X}} = [T]_{\mathfrak{X}\mathfrak{X}}^{\mathrm{T}}[B]_{\mathfrak{X}}[T]_{\mathfrak{X}\mathfrak{X}}\,,$$

and this must be true for any basis $\mathfrak{X}$. Matrices in $G_{B,\mathfrak{X}}$ are precisely the matrix realizations (with respect to basis $\mathfrak{X}$) of the various automorphisms in Aut(B).

EXERCISE 3.19. If B is a nondegenerate bilinear form, show that $G_B = \mathrm{Aut}(B)$ is a subgroup in the general linear group $\mathrm{GL}_{\mathbb{K}}(V)$ — i.e., that (a) $I \in G_B$, (b) $T_1, T_2 \in G_B \Rightarrow T_1 \circ T_2 \in G_B$, and (c) $T \in G_B \Rightarrow T^{-1} \in G_B$.

We can also assess the effect of a change of basis $\mathfrak{X} \to \mathfrak{Y}$. The matrix group $G_{B,\mathfrak{Y}}$ is a conjugate

$$G_{B,\mathfrak{Y}} = A \cdot G_{B,\mathfrak{X}} \cdot A^{-1}$$

of $G_{B,\mathfrak{X}}$ for some $A \in \mathrm{GL}(n, \mathbb{K})$.

EXERCISE 3.20. If $\mathfrak{X}$, $\mathfrak{Y}$ are bases in V, define $G_{B,\mathfrak{X}}$ and $G_{B,\mathfrak{Y}}$ as in (3.7) and prove that

$$G_{B,\mathfrak{Y}} = SG_{B,\mathfrak{X}}S^{-1} \quad \text{where } S = [\mathrm{id}]_{\mathfrak{Y}\mathfrak{X}} \text{ and } S^{-1} = [\mathrm{id}]_{\mathfrak{X}\mathfrak{Y}}.$$

Recall that $S = [S]_{\mathfrak{X}\mathfrak{Y}}$ is the matrix such that $f_i = \mathrm{id}(f_i) = \sum_{k=1}^{n} s_{ki}e_k$ if $\mathfrak{X} = \{e_i\}$ and $\mathfrak{Y} = \{f_j\}$.

The *general linear group* $\mathrm{GL}_{\mathbb{K}}(V)$ in which all these automorphism groups live is defined by the condition $\det(T) \neq 0$, which makes no reference to a bilinear form. The *special linear group* on V, $\mathrm{SL}_{\mathbb{K}}(V) = \{T \in \mathrm{GL}_{\mathbb{K}}(V) : \det(T) = 1\}$, is another "classical group" that does not arise as the automorphism group of a bilinear form B. All the other classical groups of physics and geometry are automorphism groups of particular bilinear forms on V or their intersections with $\mathrm{SL}_{\mathbb{K}}(V)$. We determine the congruence classes of nondegenerate bilinear forms according to whether B is symmetric or antisymmetric and whether the ground field is $\mathbb{K} = \mathbb{R}$ or $\mathbb{C}$, always assuming B is nondegenerate. The analysis is the same for antisymmetric forms over $\mathbb{K} = \mathbb{R}$ or $\mathbb{C}$, so there are really only three cases to deal with.

Canonical Forms. Case 1: *B* Symmetric, $\mathbb{K} = \mathbb{R}$. If B is a nondegenerate symmetric bilinear form on a vector space over $\mathbb{R}$ with $\dim(V) = n$, there are just $n + 1$ possible canonical forms.

THEOREM 3.21 (B symmetric; $\mathbb{K} = \mathbb{R}$). *There is an $\mathbb{R}$-basis $\mathfrak{X} \subseteq V$ such that the matrix describing B has the form*

$$(3.8) \qquad [B]_{\mathfrak{X}} = \begin{pmatrix} \boxed{I_{p\times p}} & 0 \\ 0 & \boxed{-I_{q\times q}} \end{pmatrix} \quad \text{with } p+q=n=\dim(V).$$

In this case, we say B has SIGNATURE (p,q). *The possibilities $p = 0$ and $p = n$ are allowed.*

PROOF. First observe that we have a polarization identity for symmetric B which determines $B(v,w)$ from homogeneous expressions of the form $B(u,u)$, just as with inner products over $\mathbb{R}$:

$$(3.9) \qquad B(v,w) = \frac{1}{2}\left[\, B(v+w, v+w) - B(v,v) - B(w,w)\,\right]$$

for all $v, w \in V$.

DEFINITION 3.22. The map $Q(v) = B(v,v)$ from V to $\mathbb{R}$ is the *quadratic form* associated with a nondegenerate symmetric bilinear form $B : V\times V \to \mathbb{R}$. Note that $Q(\lambda v) = B(\lambda v, \lambda v) = \lambda^2 Q(v,v)$, and the quadratic form $Q : V \to \mathbb{R}$ determines the full bilinear form $B : V \times V \to \mathbb{K}$ via the polarization identity (3.9).

We proceed by arguing by induction on $n = |V|$. The case $n = 1$ is trival, so we assume Theorem 3.21 true for all spaces of dimension $\leq n$ and consider a form B on a space with $|V| = n+1$. Since $B \not\equiv 0$, it follows from (3.9) that there is some $v_1 \neq 0$ such that $B(v_1, v_1) \neq 0$, and after scaling v_1 by some $a \neq 0$, we can ensure that $B(v_1, v_1) = \pm 1$. But because $\mathbb{K} = \mathbb{R}$, we cannot control whether the outcome will be $+1$ or -1.

Let $M_1 = \mathbb{R} \cdot v_1$ and

$$M_1^{\perp,B} = \{v \in V : B(v, v_1) = 0\}.$$

We have $M_1 \cap M_1^{\perp,B} = \{0\}$ because any w in the intersection must have the form $w = c_1 v_1$, $c_1 \in \mathbb{R}$. But $w \in M_1^{\perp,B}$ too, so $0 = B(w,w) = c_1^2 B(v_1,v_1) = \pm c_1^2$, hence $c_1 = 0$ and $w = 0$. Therefore $M_1 \oplus M_1^{\perp,B} = V$ because $|W| + |W^{\perp,B}| = |V|$ for any $W \subseteq V$ (Exercise 3.14). For an alternative proof, recall the general dimension formula $|W_1 + W_2| = |W_1| + |W_2| - |W_1 \cap W_2|$ for subspaces W_1, W_2 in a finite-dimensional vector space V.

If B_1 is the restriction of B to $M_1^{\perp,B}$, we claim that $B_1 : M_1^{\perp,B} \times M_1^{\perp,B} \to \mathbb{R}$ is nondegenerate on this lower-dimensional subspace. Otherwise, there would be an $x \in M_1^{\perp,B}$ such that $B\big(x, M_1^{\perp,B}\big) = 0$. But since $x \in M_1^{\perp,B}$ too, we would also have $B(x, M_1) = 0$, and therefore by additivity of B in each entry,

$$B(x, V) = B\big(x\,,\, M_1^{\perp,B} + M_1\big) = 0.$$

Nondegenerancy of B on V then forces $x = 0$.

We may therefore continue by induction on $\dim(V)$. Choosing a suitable basis $\mathfrak{X}' = \{v_2, \dots, v_{n+1}\}$ in $M_1^{\perp,B}$ and $\mathfrak{X} = \{v_1, v_2, \dots, v_{n+1}\}$ in V, we get

$$[B]_{\mathfrak{X}} = \begin{pmatrix} \boxed{\pm 1} & \cdots & 0 \\ 0 & \boxed{I_{p\times p}} & \\ 0 & & \boxed{-I_{q\times q}} \end{pmatrix} \quad \text{with } p+q = n-1.$$

If the top left entry is -1, we may switch vectors $e_1 \leftrightarrow e_{p+1}$, which replaces $[B]_{\mathfrak{X}}$ with $[B]_{\mathfrak{Y}} = E^{\mathrm{T}}[B]_{\mathfrak{X}}E$ where E is the following permutation matrix (the zero on the diagonal is in column $p+1$):

$$E = \begin{pmatrix} 0 & 0 & \cdots & 0 & 1 & 0 & \cdots & 0 \\ 0 & +1 & & & 0 & & & 0 \\ \vdots & & \ddots & & \vdots & & & \\ 0 & & & +1 & & & & \\ 1 & & & & 0 & & & \\ 0 & & & & \vdots & -1 & & \\ \vdots & & & & 0 & & \ddots & 0 \\ 0 & & & & 0 & & & -1 \end{pmatrix}$$

Note that $E^{\mathrm{T}} = E$ for this permutation matrix so that $[B]_{\mathfrak{Y}}$ has the block-diagonal form (3.8). That concludes the proof of Theorem 3.21. □

Later on we will describe an algorithmic procedure for putting B into canonical form $\mathrm{diag}(+1, \dots, +1, -1, \dots, -1)$; these algorithms work the same way over $\mathbb{K} = \mathbb{R}$ or $\mathbb{C}$. We will also see that an antisymmetric B cannot be diagonalized by any congruence, but it nevertheless has a different (and equally useful) canonical form.

THE REAL ORTHOGONAL GROUPS $\mathrm{O}(p,q)$, $p+q=n$. The outcome in Theorem 3.21 breaks into $n+1$ possibilities (including $p=0$ and $q=0$). If $\mathfrak{X}$ is a basis such that $[B]_{\mathfrak{X}}$ has the standard form (3.8), then $A \in G_{B,\mathfrak{X}}$ if and only if

$$A^{\mathrm{T}} \left(\begin{array}{c|c} I_{p\times p} & 0 \\ \hline 0 & -I_{q\times q} \end{array} \right) A = \left(\begin{array}{c|c} I_{p\times p} & 0 \\ 0 & -I_{q\times q} \end{array} \right), \tag{3.10}$$

which can be written concisely as

$$A^{\mathrm{T}}JA = J \quad \text{where} \quad J = \left(\begin{array}{c|c} I_{p\times p} & 0 \\ \hline 0 & -I_{q\times q} \end{array} \right).$$

Members of this family of matrix groups over $\mathbb{R}$ are denoted by $\mathrm{O}(p,q)$, and each contains as a subgroup the SPECIAL ORTHOGONAL GROUP:

$$\mathrm{SO}(p,q) = \mathrm{O}(p,q) \cap \mathrm{SL}(n,\mathbb{R}).$$

Several of the groups $\mathrm{O}(p,q)$ and $\mathrm{SO}(p,q)$ are of particular interest.

THE REAL-ORTHOGONAL GROUPS $\mathrm{O}(n,0) = \mathrm{O}(n)$ AND $\mathrm{SO}(n)$. With respect to the standard basis in $\mathbb{R}^n$, we have $B_{\mathfrak{X}} = I_{n\times n}$, so $J = I_{n\times n}$ in (3.10) and

$$\mathrm{O}(n,0) = G_{B,\mathfrak{X}} = \{A : A^{\mathrm{T}}A = A^{\mathrm{T}}IA = I\}.$$

The set of multiplication operators $L_A(\mathbf{x}) = A \cdot \mathbf{x}$ for $A \in \mathrm{O}(n, 0)$ is just the familar group of orthogonal linear transformations on $\mathbb{R}^n$, traditionally denoted $\mathrm{O}(n)$. This group is a closed and bounded set in matrix space $\mathrm{M}(n, \mathbb{R}) \cong \mathbb{R}^{n^2}$.

THE LORENTZ GROUP $\mathrm{O}(n-1, 1)$. This is the group of space-time symmetries at the center of Einstein's theory of special relativity for $n-1$ space dimensions $x_1, \ldots, x_{n-1}$ and one time dimension x_n (which is generally labeled "t" by physicists). For a suitably chosen basis $\mathfrak{X}$ in $\mathbb{R}^n$, the matrix describing an arbitrary nondegenerate symmetric bilinear form B of signature $(n-1, 1)$ becomes

$$[B]_{\mathfrak{X}} = \begin{pmatrix} 1 & & & 0 \\ & \ddots & & \\ & & 1 & \\ 0 & & & -1 \end{pmatrix}, \tag{3.11}$$

and the associated quadratic form is

$$B(x, x) = [x]_{\mathfrak{X}}^{\mathrm{T}} [B]_{\mathfrak{X}} [x]_{\mathfrak{X}} = x_1^2 + \cdots + x_{n-1}^2 - x_n^2.$$

REMARK 3.23. The physicists' version of this is a little different:

$$B(x, x) = x_1^2 + \cdots + x_{n-1}^2 - c^2 t^2 ,$$

where c is the speed of light. But the numerical value of c depends on the physical units used to describe it — *feet per second*, etc. – and one can always choose the units of *length* and *time* to make the experimentally measured speed of light have numerical value $c = 1$. For instance we could take t = *seconds* and measure lengths in *light-seconds* = the distance a light ray travels in one second; or we could measure t in *years* and lengths in *light-years*. Either way, the numerical value of the speed of light is $c = 1$. ○

From (3.11), it is clear that A is in $\mathrm{O}(n-1, 1)$ if and only if

$$A^{\mathrm{T}} \left(\begin{array}{c|c} I_{n-1} & 0 \\ \hline 0 & -1 \end{array} \right) A = \left(\begin{array}{c|c} I_{n-1} & 0 \\ \hline 0 & -1 \end{array} \right). \tag{3.12}$$

$\mathrm{O}(n-1, 1)$ contains the subgroup $\mathrm{SO}(n-1, 1) = \mathrm{O}(n-1, 1) \cap \mathrm{SL}(n, \mathbb{R})$ which correspond to *proper Lorentz transformations*, those with $\det(A) = +1$. Within the group $\mathrm{SO}(n-1, 1)$, we find a copy $\widetilde{\mathrm{SO}}(n-1)$ of the usual orthogonal group $\mathrm{SO}(n-1)$ embedded in $\mathrm{M}(n, \mathbb{R})$ via the one-to-one homomorphism

$$A \in \widetilde{\mathrm{SO}}(n-1) \text{ in } \mathrm{M}(n-1, \mathbb{R}) \mapsto \left(\begin{array}{c|c} A & 0 \\ \hline 0 & 1 \end{array} \right) \in \mathrm{SO}(n-1, 1) \text{ in } \mathrm{M}(n, \mathbb{R}).$$

The subgroup $\widetilde{\mathrm{SO}}(n-1)$, called the *Galilean group*, acts only on the "space coordinates" $x_1, \ldots, x_{n-1}$ in $\mathbb{R}^n$, leaving the time coordinate $t = x_n$ fixed.

The following family of matrices in $\mathrm{O}(n-1, 1)$ is of particular interest in understanding the meaning of special relativity:

$$A = \begin{pmatrix} 1/\sqrt{1-v^2} & 0 & & 0 & -v/\sqrt{1-v^2} \\ 0 & 1 & \ddots & & 0 \\ \vdots & & \ddots & & \vdots \\ 0 & & & 1 & 0 \\ -v/\sqrt{1-v^2} & 0 & \cdots & 0 & 1/\sqrt{1-v^2} \end{pmatrix}. \tag{3.13}$$

When we employ units that make the speed of light $c = 1$, the parameter v must have values $|v| < 1$ to prevent the corner entries in this array from having physically meaningless imaginary values; as $v \to 1$, these entries blow up, so $\mathrm{SO}(n-1,1)$ is indeed an unbounded set in matrix space $\mathrm{M}(n,\mathbb{R})$.

In special relativity, an *event* is described by a point $(\mathbf{x}, t)$ in space-time $\mathbb{R}^{n-1}\times\mathbb{R}$ that specifies the location $\mathbf{x}$ and the time t at which the event occurred. Now suppose two observers are moving through space at constant velocity with respect to one another (*no acceleration* as time passes). Each will use his or her own frame of reference in observing an event to assign space-time coordinates to it. The matrix A in (3.13) tells us how to make the (relativistic) transition from the values $(\mathbf{x}, t)$ seen by Observer #1 to those recorded by Observer #2:[1]

$$\begin{pmatrix} \mathbf{x}' \\ t' \end{pmatrix} = A \cdot \begin{pmatrix} \mathbf{x} \\ t \end{pmatrix}$$

EXERCISE 3.24. Verify that the matrices in (3.14) all lie in $\mathrm{SO}(n-1,1)$. Be sure to check that $\det(A) = +1$.

HINT. Show that (3.14) $\Rightarrow$ $\det(A)^2 = 1$, so $\det(A) = \pm 1$, and then argue that $\det(I) = 1$ and $\det(A)$ is a continuous function of the real-valued parameter $-1 < v < +1$.

EXERCISE 3.25. Show that

$$B = \begin{pmatrix} \cosh(y) & 0 & 0 & \sinh(y) \\ 0 & 1 & 0 & 0 \\ 0 & 0 & 1 & 0 \\ \sinh(y) & 0 & 0 & \cosh(y) \end{pmatrix}$$

is in $\mathrm{SO}(3,1)$ for all $y \in \mathbb{R}$.

REMARK 3.26. If we work with physical units that do not make $c = 1$, as assumed in (3.13), we must replace "$\sqrt{1-v^2}$" everywhere it appears with

$$\sqrt{1 - \left(\frac{v}{c}\right)^2}$$

in which the speed of light c appears explicitly. ○

Sylvester's Theorem: Invariance of the Signature on $\mathcal{O}symbfit(p,q)$. One way to compute the signature would be to find a basis that puts $[B]_{\mathfrak{X}}$ into the block-diagonal form (3.10), but how do we know the signature does not depend on the basis used to compute it? That it does not is the subject of the next theorem.

Proving this amounts to showing that the signature is a *congruence invariant*: you cannot transform

$$\begin{pmatrix} \boxed{I_{p\times p}} & 0 \\ 0 & \boxed{-I_{q\times q}} \end{pmatrix} \longrightarrow \begin{pmatrix} \boxed{I_{p'\times p'}} & 0 \\ 0 & \boxed{-I_{q'\times q'}} \end{pmatrix}$$

[1] To keep things simple, the transition matrix (3.13) describes what happens when Observer #2 is moving with velocity v in the positive x_1-direction, as seen by Observer #1, so that $x_1' = x_1 - vt, x_2' = x_2, \ldots, x_{n-1}' = x_{n-1}$. The general formula is more complicated.

unless $p' = p$ and $q' = q$. This fact is often referred to as "Sylvesters's law of inertia."

THEOREM 3.27 (Sylvester). *If A is a nondegenerate real symmetric $n \times n$ matrix, there is some $P \in \mathrm{GL}(n, \mathbb{R})$ such that $P^{\mathrm{T}}AP = \mathrm{diag}(1, \dots, 1, -1, \dots, -1)$. The number p of $+1$ entries and the canonical form* (3.10) *are uniquely determined.*

PROOF. Existence of a diagonalization has already been proved in Theorem 3.21. If $B(\mathbf{x}, \mathbf{y}) = \sum_{i,j} x_i A_{ij} y_j = \mathbf{x}^{\mathrm{T}} A \mathbf{y}$ is a nondegenerate symmetric bilinear form on $\mathbb{R}^n$, so $[B] = [A_{ij}]$ with respect to the standard basis, then there is a basis $\mathfrak{X}$ such that $[B]_{\mathfrak{X}} = \mathrm{diag}(1, \dots, 1, -1, \dots, -1)$. Suppose $p = \#(\text{entries} = +1)$ for $\mathfrak{X}$ and that there is another diagonalizing basis $\mathfrak{Y}$ such that $p' = \#(\text{entries} = +1)$ is $\neq p$. We may assume $p < p'$. Writing $\mathfrak{X} = \{v_1, \dots, v_p, v_{p+1}, \dots, v_n\}$ and $\mathfrak{Y} = \{w_1, \dots, w_{p'}, w_{p'+1}, \dots, w_n\}$, define $L : V \to \mathbb{R}^{p-p'+n}$ via

$$L(\mathbf{x}) = \big(B(\mathbf{x}, v_1), \dots, B(\mathbf{x}, v_p), B(\mathbf{x}, w_{p'+1}), \dots, B(\mathbf{x}, w_n)\big).$$

Then $\mathrm{rank}(L)$ is at most $\dim(\mathbb{R}^{p-p'+n}) = p - p' + n < n$. Hence, $\dim(\ker(L)) = \dim(V) - \mathrm{rank}(L) > 0$, and there is some $v_0 \neq 0$ in V such that $L(v_0) = 0$. That means

$$B(v_0, v_i) = 0 \text{ for } 1 \leq i \leq p \quad \text{and} \quad B(v_0, w_i) = 0 \quad \text{for } p' + 1 \leq i \leq n.$$

Writing v_0 in terms of the two bases, we have $v_0 = \sum_{j=1}^n a_j v_j = \sum_{k=1}^n b_k w_k$.

For $i \leq p$, we get

$$\begin{aligned} 0 = B(v_0, v_i) &= B\Big(\sum_j a_j v_j\,,\, v_i\Big) = \sum_j a_j B(v_j, v_i) \\ &= \sum_j a_j \delta_{ij} = a_i = a_i B(v_i, v_i) \end{aligned}$$

because $[B]_{\mathfrak{X}} = \mathrm{diag}(1, \dots, 1, -1, \dots, -1)$. But $B(v_i, v_i) > 0$ for $i \leq p$ while $B(v_0, v_i) = 0$, so we conclude that $a_i = 0$ for $0 \leq i \leq p$. Similarly, $b_j = 0$ for $p' + 1 \leq j \leq n$. It follows that $a_i \neq 0$ for some $p' < i \leq n$, and hence

$$\begin{aligned} B(v_0, v_0) &= B\Big(\sum_{j=1}^n a_j v_j\,,\, \sum_{\ell=1}^n a_\ell v_\ell\Big) = \sum_{j=1}^n a_j^2 B(v_j, v_j) \\ &= \sum_{j=p+1}^n a_j^2 B(v_j, v_j) < 0. \end{aligned}$$

Furthermore,

$$\begin{aligned} B(v_0, v_0) = B\Big(\sum_{j=1}^n b_j w_j\,,\, \sum_{\ell=1}^n b_\ell w_\ell\Big) &= \sum_{j=1}^n b_j^2 B(w_j, w_j) \\ &= \sum_{j=1}^{p'} b_j^2 B(w_j, w_j) > 0. \end{aligned}$$

Thus $B(v_0, v_0) < 0$ and $B(v_0, v_0) > 0$, which is a contradiction. □

COROLLARY 3.28. *Two nonsingular symmetric matrices in $M(n,\mathbb{R})$ are congruent via $A \to P^{\mathsf{T}}AP$ for some $P \in \mathrm{GL}(n,\mathbb{R})$ if and only if they have the same signature (p,q).*

A Diagonalization Algorithm. Let A be a symmetric $n \times n$ matrix with entries from a field $\mathbb{K}$ not of characteristic two (so $1+1 \neq 0$ in $\mathbb{K}$). We know that there are matrices $Q, D \in \mathrm{M}(n,\mathbb{K})$ such that Q is invertible and $Q^{\mathsf{T}}AQ = D$ is diagonal. We now give a method for computing suitable Q and diagonal form D via elementary row and column operations; a short additional step then yields the signature (p,q) when $\mathbb{K} = \mathbb{R}$.

The effect of an elementary column operation on A is obtained by right multiplication $A \to AE$ by a suitable "elementary matrix" E (recall sections 1.1 and 4.2 of LA I). Furthermore, the same elementary operation on rows is effected by a left multiplication $A \to E^{\mathsf{T}}A$ using the same E. If we perform an elementary operation on rows followed by the same elementary operation on columns, the net result is the congruence transformation

$$A \to E^{\mathsf{T}}AE.$$

(Since matrix multiplication is associative, we have $E^{\mathsf{T}}(AE) = (E^{\mathsf{T}}A)E$, so no parentheses are needed in this expression.)

Now suppose that Q is an invertible matrix such that $Q^{\mathsf{T}}AQ = D$ is diagonal. Any invertible Q is a product of elementary matrices, say $Q = E_1E_2\cdots E_k$. Hence

$$D = Q^{\mathsf{T}}AQ = E_k^{\mathsf{T}}E_{k-1}^{\mathsf{T}}\cdots E_1^{\mathsf{T}}AE_1E_2\cdots E_k.$$

Putting these observations together we get the following theorem:

THEOREM 3.29. *A suitably chosen sequence of paired elementary row and column operations can transform any real symmetric matrix A into a diagonal matrix D. Furthermore, if $E_1,\dots,E_k$ are the appropriate elementary matrices that effect the necessary column operations (indexed in the order performed), then $Q^{\mathsf{T}}AQ = D$ if we take $Q = E_1E_2\cdots E_k$.*

The same idea works for symmetric bilinear forms over any field $\mathbb{K}$ with characteristic $\neq 2$.

COROLLARY 3.30. *If $A \in \mathrm{M}(n,\mathbb{K})$ is a symmetric matrix with entries in a field not of characteristic $= 2$, then A is congruent to a diagonal matrix.*

EXAMPLE 3.31. Taking A to be the symmetric matrix in $\mathrm{M}(3,\mathbb{R})$,

$$A = \begin{pmatrix} 1 & -1 & 3 \\ -1 & 2 & 1 \\ 3 & 1 & 1 \end{pmatrix},$$

we apply the procedure just described to find an invertible matrix Q such that $Q^{\mathsf{T}}AQ = D$ is diagonal.

DISCUSSION. We begin by eliminating all nonzero entries in the first row and first column except for the entry a_{11}. To this end, we start by performing the column operation $\mathrm{Col}_2 \to \mathrm{Col}_2 + \mathrm{Col}_1$; this yields a new matrix to which we apply the same operation on rows, $\mathrm{Row}_2 \to \mathrm{Row}_2 + \mathrm{Row}_1$. These first steps yield

$$A = \begin{pmatrix} 1 & -1 & 3 \\ -1 & 2 & 1 \\ 3 & 1 & 1 \end{pmatrix} \to \begin{pmatrix} 1 & 0 & 3 \\ -1 & 1 & 1 \\ 3 & 4 & 1 \end{pmatrix} \to \begin{pmatrix} 1 & 0 & 3 \\ 0 & 1 & 4 \\ 3 & 4 & 1 \end{pmatrix} = E_1^{\mathsf{T}} A E_1$$

where

$$E_1 = \begin{pmatrix} 1 & 0 & 0 \\ 1 & 1 & 0 \\ 0 & 0 & 1 \end{pmatrix}.$$

The second round of moves is $\mathrm{Col}_3 \to \mathrm{Col}_3 - 3 \cdot \mathrm{Col}_1$ followed by the same row operations

$$\begin{pmatrix} 1 & 0 & 3 \\ 0 & 1 & 4 \\ 3 & 4 & 1 \end{pmatrix} \to \begin{pmatrix} 1 & 0 & 0 \\ 0 & 1 & 4 \\ 3 & 4 & -8 \end{pmatrix} \to \begin{pmatrix} 1 & 0 & 0 \\ 0 & 1 & 4 \\ 0 & 4 & -8 \end{pmatrix} = E_2^{\mathsf{T}} E_1^{\mathsf{T}} A E_1 E_2$$

where

$$E_2 = \begin{pmatrix} 1 & 0 & 0 \\ 0 & 1 & 0 \\ -3 & 0 & 1 \end{pmatrix}.$$

Finally, we achieve a diagonal form by applying $\mathrm{Col}_3 \to \mathrm{Col}_3 - 4 \cdot \mathrm{Col}_2$ and then the corresponding operation on rows to get

$$E_3^{\mathsf{T}} E_2^{\mathsf{T}} E_1^{\mathsf{T}} A E_1 E_2 E_3 = \begin{pmatrix} 1 & 0 & 0 \\ 0 & 1 & 0 \\ 0 & 0 & -24 \end{pmatrix} \quad \text{where } E_3 = \begin{pmatrix} 1 & 0 & 0 \\ 0 & 1 & 0 \\ 0 & -4 & 1 \end{pmatrix}.$$

Since the outcome is a diagonal matrix, the process is complete. To summarize: by taking

$$Q = E_1 E_2 E_3 = \begin{pmatrix} 1 & 1 & -7 \\ 0 & 1 & -4 \\ 0 & 0 & 1 \end{pmatrix},$$

we get a diagonal form

$$D = Q^{\mathsf{T}} A Q = \begin{pmatrix} 1 & 0 & 0 \\ 0 & 1 & 0 \\ 0 & 0 & -24 \end{pmatrix}.$$

To obtain the canonical form (3.8), we need one more pair of operations

$$\mathrm{Row}_3 \to \frac{1}{\sqrt{24}} \cdot \mathrm{Row}_3 \quad \text{and} \quad \mathrm{Col}_3 \to \frac{1}{\sqrt{24}} \cdot \mathrm{Col}_3,$$

both of which correspond to the (diagonal) elementary matrix

$$E_4 = \begin{pmatrix} 1 & 0 & 0 \\ 0 & 1 & 0 \\ 0 & 0 & \frac{1}{\sqrt{24}} \end{pmatrix}.$$

Therefore, the canonical form is

$$\operatorname{diag}(1,1,1,-1) = \widetilde{Q}^{\mathrm{T}} A \widetilde{Q} \quad \text{where } \widetilde{Q} = E_4 \cdot \cdots \cdot E_1.$$

○

This example also shows that the diagonal form of a real symmetric matrix achieved through congruence transformations $A \to Q^{\mathrm{T}}AQ$ is not unique; both $\operatorname{diag}(1,1,1,-24)$ and $\operatorname{diag}(1,1,1,-1)$ are congruent to A. Only the sigature (3.1) is a true congruence invariant.

The Gauss-Seidel Algorithm. In Section 4.2 of LA I, we showed that the inverse A^{-1} of an invertible matrix can be obtained multiplying on the left by a sequence of elementary matrices (or equivalently, by executing the corresponding sequence of elementary row operations on the matrix). We also developed the *Gauss-Seidel Algorithm* which does this efficiently.

> GAUSS-SEIDEL ALGORITHM. *Starting with the $n \times 2n$ augmented matrix $[A : I_{n\times n}]$, apply row operations to bring the left-hand block into reduced echelon form, which must equal $I_{n\times n}$ since A is invertible. Applying the same moves to the entire $n \times 2n$ augmented matrix, we arrive at a matrix $[I_{n\times n} : A^{-1}]$ whose right-hand block is the desired inverse.*

An algorithm similar to Gauss-Seidel yields a matrix Q such that $Q^{\mathrm{T}}AQ = D$ is diagonal; the signature (r, s) can then be determined by inspection as in the last steps of Example 3.31. The reader should justify the method, illustrated below, for computing an appropriate Q without recording each elementary matrix separately. Starting with an augmented $n \times 2n$ matrix $[A : I_{n\times n}]$, we apply paired row and column operations to drive the left-hand block into diagonal form, but we apply them to the *entire augmented matrix*. When the left-hand block achieves diagonal form D, the right-hand block in $[D : Q^{\mathrm{T}}]$ is a matrix such that $Q^{\mathrm{T}}AQ = D$. The steps are worked out in detail below; we leave the reader to verify that $Q^{\mathrm{T}}AQ = D$.

Starting with $\mathrm{Col}_2 \to \mathrm{Col}_2 + \mathrm{Col}_1$ and then the corresponding operation on rows, we get

$$[A : I] = \left(\begin{array}{ccc|ccc} 1 & -1 & 3 & 1 & 0 & 0 \\ -1 & 2 & 1 & 0 & 1 & 0 \\ 3 & 1 & 1 & 0 & 0 & 1 \end{array}\right) \xrightarrow{\text{paired R/C opns.}} \left(\begin{array}{ccc|ccc} 1 & 0 & 3 & 1 & 0 & 0 \\ 0 & 1 & 4 & 1 & 1 & 0 \\ 3 & 4 & 1 & 0 & 0 & 1 \end{array}\right)$$

$$\xrightarrow{\text{paired R/C opns.}} \left(\begin{array}{ccc|ccc} 1 & 0 & 0 & 1 & 0 & 0 \\ 0 & 1 & 4 & 1 & 1 & 0 \\ 0 & 4 & -8 & -3 & 0 & 1 \end{array}\right)$$

$$\xrightarrow{\text{paired R/C opns.}} \left(\begin{array}{ccc|ccc} 1 & 0 & 0 & 1 & 0 & 0 \\ 0 & 1 & 0 & 1 & 1 & 0 \\ 0 & 0 & -24 & -7 & -4 & 1 \end{array}\right) \to [D : Q^{\mathrm{T}}].$$

Therefore,

$$Q^{\mathrm{T}} = \begin{pmatrix} 1 & 0 & 0 \\ 1 & 1 & 0 \\ -7 & -4 & 1 \end{pmatrix}, \qquad Q = \begin{pmatrix} 1 & 1 & -7 \\ 0 & 1 & -4 \\ 0 & 0 & 1 \end{pmatrix},$$

and a diagonalized form $Q^{\mathsf{T}}AQ$ is

$$D = \begin{pmatrix} 1 & 0 & 0 \\ 0 & 1 & 0 \\ 0 & 0 & -24 \end{pmatrix}.$$

This concludes our discussion of Example 3.31.

EXERCISE 3.32. On $V = M(2, \mathbb{R})$, consider the symmetric bilinear form $B(P, Q) = \mathrm{Tr}(PQ)$, the trace of the product matrix PQ.

(a) Explain why this form is symmetric.
(b) Is this form nondegenerate? Explain.
(c) Determine the matrix $[B]_{\mathfrak{X}}$ of this form with respect to the ordered basis $\mathfrak{X} = \{E_{11}, E_{12}, E_{21}, E_{22}\}$ of matrix units.
(d) Determine the signature of this bilinear form.

HINT. For calculations, use the standard basis of "matrix units" E_{ij}, $1 \leq i, j \leq 2$ in V whose entries are zero except for a "1" in the (i, j) spot. Recall that $E_{ij} \cdot E_{k\ell} = \delta_{jk} E_{i\ell}$ where δ_{ij} = Kronecker delta.

Canonical Forms. Case 2: *B* Symmetric, $\mathbb{K} = \mathbb{C}$. In this setting, there is just one canonical form for each dimension.

THEOREM 3.33 (B Symmetric; $\mathbb{K} = \mathbb{C}$). *If B is a nondegenerate, symmetric bilinear form over $\mathbb{K} = \mathbb{C}$, there is a basis $\mathfrak{X}$ such that $[B]_{\mathfrak{X}} = I_{n\times n}$. In coordinates, for this basis we have*

$$B(\mathbf{z}, \mathbf{w}) = \sum_{j=1}^{n} z_j w_j \quad \textit{(no conjugate, even though } \mathbb{K} = \mathbb{C}\textit{)}.$$

PROOF. By our discussion for $\mathbb{K} = \mathbb{R}$, we know we can put B in diagonal form $[B]_{\mathfrak{X}} = \mathrm{diag}(\lambda_1, \dots, \lambda_n)$ with each $\lambda_i \neq 0$ since B is nondegenerate. Now take square roots in $\mathbb{C}$ and let $P = \mathrm{diag}(1/\sqrt{\lambda_1}, \dots, 1/\sqrt{\lambda_n})$ to get $P^{\mathsf{T}}[B]_{\mathfrak{X}}P = I_{n\times n}$. □

There are various symmetric nondegenerate bilinear forms B on $\mathbb{C}^n$, but the associated automorphism groups $\mathrm{Aut}(B) \subseteq \mathrm{GL}(n, \mathbb{C})$ are essentially the same: we have seen that there is always a basis $\mathfrak{X}$ such that $[B]_{\mathfrak{X}} = I$, and then the matrix realization $G_{B,\mathfrak{X}}$ of $\mathrm{Aut}(B)$ becomes the *complex orthogonal group* encountered earlier:

$$\mathrm{O}(n, \mathbb{C}) = G_{B,\mathfrak{X}} = \{A \in \mathrm{M}(n, \mathbb{C}) : A^{\mathsf{T}} I_{n\times n} A = I_{n\times n}\},$$

in which $A^{\mathsf{T}} = A^{-1}$. (Notice the use of A^{T} here and not the adjoint $A^* = \overline{A^{\mathsf{T}}}$, even though $\mathbb{K} = \mathbb{C}$.) Among the subgroups of $\mathrm{SO}(n, \mathbb{C})$, we find the *special orthogonal group* over $\mathbb{C}$,

$$\mathrm{SO}(n, \mathbb{C}) = \mathrm{O}(n, \mathbb{C}) \cap \mathrm{SL}(n, \mathbb{C}).$$

Both $\mathrm{O}(n, \mathbb{C})$ and $\mathrm{SO}(n, \mathbb{C})$ are closed, unbounded subsets in $\mathrm{M}(n, \mathbb{C})$.

EXERCISE 3.34. Taking $n = 2$, show that $\mathrm{SO}(2, \mathbb{C})$ has the following properties:

(a) $\mathrm{SO}(2, \mathbb{C})$ is abelian and isomorphic to the direct product group $S^1 \times \mathbb{R}$ in which $S^1 = \{z \in \mathbb{C} : |z| = 1\}$ and the product operation on pairs (z, t) is

$$(z, t) * (z't') = (zz', t + t').$$

Find a bijection $\phi : S^1 \times \mathbb{R} \to \mathrm{SO}(2, \mathbb{C})$ that implements the desired group isomorphism so that

$$\phi((z, t) * (z', t')) = \phi(z, t) \cdot \phi(z', t') \quad \text{(matrix product)}$$

and $\phi(1, 0) = I_{2\times 2}$.

(b) Show that $A \in \mathrm{M}(2, \mathbb{C})$ is in $\mathrm{SO}(2, \mathbb{C})$ if and only if

$$A = \begin{pmatrix} a & b \\ -b & a \end{pmatrix}$$

with $a, b \in \mathbb{C}$ and $a^2 + b^2 = 1$.

(c) Show that $\mathrm{SO}(2, \mathbb{C})$ is an unbounded subset in $\mathrm{M}(2, \mathbb{C})$ and, hence, that $\mathrm{SO}(n, \mathbb{C})$ is unbounded in $\mathrm{M}(n, \mathbb{C})$ for all $n \geq 2$ because we may embed $\mathrm{SO}(2, \mathbb{C})$ in $\mathrm{SO}(n, \mathbb{C})$ via

$$A \in \mathrm{SO}(2, \mathbb{C}) \mapsto \left(\begin{array}{c|ccc} A & 0 & \cdots & 0 \\ \hline 0 & 1 & & 0 \\ \vdots & & \ddots & \\ 0 & 0 & & 1 \end{array}\right).$$

HINT. In (b) write $A = [a, b; c, d]$. The identities $A^{\mathrm{T}}A = I$ and $\det(A) = 1$ yield five equations in the complex unknowns a, b, c, d, but only four remain when duplicates are deleted. There is still some redundancy in the remaining system of equations, which can nevertheless be solved by algebraic elimination despite its nonlinearity. In (c) use the supnorm $\|A\| = \max_{i,j}\{|A_{ij}|\}$ on matrices to discuss bounded sets in matrix space.

REMARK 3.35. A similar problem was posed in Section 6.5 of LA I (see Euler's theorem 6.84) regarding the group of real matrices $\mathrm{SO}(3) \subseteq \mathrm{M}(3, \mathbb{R})$. That came down to understanding the action of the group of real orthogonal matrices $\mathrm{SO}(2)$ on the plane $\mathbb{R}^2$. The groups $\mathrm{O}(2, C)$ and $\mathrm{SO}(2, \mathbb{C})$ acting as $\mathbb{C}$-linear operators on $\mathbb{C}^2$ are the "complexified" analogs of the real orthogonal groups acting on $\mathbb{R}^2$. ○

EXERCISE 3.36. Which scalar matrices λI ($\lambda \in \mathbb{C}$) lie in $\mathrm{SO}(n, \mathbb{C})$ or in $\mathrm{O}(n, \mathbb{C})$? Which diagonal matrices lie in these groups?

EXERCISE 3.37. Prove that $\mathrm{SO}(n, \mathbb{C})$ is a *closed subset* in $\mathrm{M}(n, \mathbb{C})$, so that

> *If a sequence $\{A^{(n)}\}$ in $\mathrm{SO}(n, \mathbb{C})$ converges to a limit A in $\mathrm{M}(n, \mathbb{C})$, the limit A must also be in $\mathrm{SO}(n, \mathbb{C})$.*

(One cannot escape from $\mathrm{SO}(n, \mathbb{C})$ by taking a limit.)

HINT. Recall the discussion of matrix and operator limits in LA I, Sections 5.3 - 5.4. Convergence $A^{(n)} \to A$ as $n \to \infty$ means $A_{ij}^{(n)} \to A_{ij}$ in $\mathbb{C}$ for each matrix entry.

Canonical Forms: Case 3: *B* Antisymmetric, $\mathbb{K} = \mathbb{R}$ or $\mathbb{C}$. In the antisymmetric case, the same argument applies whether $\mathbb{K} = \mathbb{R}$ or $\mathbb{C}$. By antisymmetry we have $B(v, v) = 0$ for all v, and if $W \subseteq V$, the B-orthocomplement $W^{\perp,B} = \{v : B(v, W) = 0\}$ need not be a subspace complementary to W. We might even have $W^{\perp,B} \supseteq W$, although the identity $\dim(W) + \dim(W^{\perp,B}) = \dim(V)$ remains valid.

THEOREM 3.38 (B antisymmetric; $\mathbb{K} = \mathbb{R}$ or $\mathbb{C}$). *If B is a nondegenerate antisymmetric form over $\mathbb{K} = \mathbb{R}$ or $\mathbb{C}$, there is a basis $\mathfrak{X}$ such that*

$$[B]_{\mathfrak{X}} = J = \begin{pmatrix} 0 & I_{m\times m} \\ -I_{m\times m} & 0 \end{pmatrix}.$$

In particular, $\dim_{\mathbb{K}}(V)$ must be even if V carries a nondegenerate antisymmetric bilinear form.

PROOF. Recall that $\dim(W)+\dim\left(W^{\perp,B}\right) = \dim(V)$ for any nondegenerate bilinear form B on V and any subspace $W \subseteq V$. Fix $v_1 \neq 0$. Then $M_1 = (\mathbb{K}v_1)^{\perp,B}$ has dimension $n-1$ if $\dim(V) = n$, but it includes $\mathbb{K}v_1 \subsetneq (\mathbb{K}v_1)^{\perp,B}$ because $B(v, v) \equiv 0$. Now take any $v_2 \notin M_1$ (so $v_2 \neq 0$); scale it to get $B(v_1, v_2) = -1$ and let $M_2 = (\mathbb{K}v_2)^{\perp,B}$. Again, nondegeneracy implies $\dim(M_2) = n-1 = \dim(M_1)$. But $M_2 \neq M_1$ since $v_2 \in M_2$ and $v_2 \notin M_1$. The basic dimension formula

$$|W_1 + W_2| = |W_1| + |W_2| - |W_1 \cap W_2| \quad \text{for subspaces } W_1, W_2 \subseteq V$$

then implies that $\dim(M_1 \cap M_2) = n - 2$. The space $M = M_1 \cap M_2$ is B-orthogonal to $\mathbb{K}$-span$\{v_1, v_2\}$ by definition of these vectors. Furthermore, $B|_M$ is antisymmetric and nondegenerate. In fact, we already know that $B(M, v_1) = B(M, v_2) = 0$ and $V = \mathbb{K}v_1 \oplus \mathbb{K}v_2 \oplus M$, so if $B(w, M) = 0$ for some $w \in M$, then $B(w, V) = B(w, \mathbb{R}v_1 + \mathbb{R}v_2 + M) = 0$ and $w = 0$ by nondegeneracy. Finally, if $N = \mathbb{K}$-span$\{v_1, v_2\}$, we have $V = N \oplus M$. (Why?)

We can now argue by induction on dimension: $n = \dim(V)$ must be even, and there is a basis $\mathfrak{X}_0 = \{v_3, \dots, v_n\}$ in M such that

$$[B|_M]_{\mathfrak{X}_0} = \begin{pmatrix} \boxed{R} & & 0 \\ & \ddots & \\ 0 & & \boxed{R} \end{pmatrix}$$

with

$$R = \begin{pmatrix} 0 & 1 \\ -1 & 0 \end{pmatrix}.$$

Hence, $\mathfrak{X} = \{v_1, v_2\} \cup \mathfrak{X}_0$ is a basis for V such that

$$[B]_{\mathfrak{X}} = \left(\begin{array}{c|c} R & 0 \\ \hline 0 & [B|_M]_{\mathfrak{X}_0} \end{array}\right) = \begin{pmatrix} \boxed{R} & & & 0 \\ & \boxed{R} & & \\ & & \ddots & \\ 0 & & & \boxed{R} \end{pmatrix}.$$

A simple relabeling of basis vectors (corresponding to some permutation matrix E such that $E^{\mathrm{T}} = E^{-1}$) gives the standard form

$$E^{\mathrm{T}}[B]_{\mathfrak{X}}E = [B]_{\mathfrak{Y}} = \left(\begin{array}{c|c} 0 & I_{m\times m} \\ \hline -I_{m\times m} & 0 \end{array}\right)$$

where $m = \frac{1}{2}\dim(V)$. □

A skew-symmetric nondegenerate form B imposes a *symplectic structure* on V. The dimension $\dim_{\mathbb{K}}(V)$ must be even, and as we saw earlier, there is just one such nondegenerate structure up to congruence of the representative matrix.

DEFINITION 3.39. The automorphism group Aut(B) of a nondegenerate skew-symmetric form on V is called a *symplectic group.*

If $\mathfrak{X}$ is a basis that puts B into standard form, we have

$$B(x,y) = [x]_{\mathfrak{X}}^{\mathrm{T}}[B]_{\mathfrak{X}}[y]_{\mathfrak{X}} = [x]_{\mathfrak{X}}^{\mathrm{T}}\,J\,[y]_{\mathfrak{X}} \quad \text{where} \quad J = \begin{pmatrix} 0 & I_{m\times m} \\ -I_{m\times m} & 0 \end{pmatrix}.$$

By (3.7), T is an operator in Aut(B) if and only $B(x,y) = B(Tx,Ty)$ for x, y in V. If $A = [T]_{\mathfrak{X}\mathfrak{X}}$, we claim that

$$A \text{ is in } G_{\mathfrak{X},B} \Leftrightarrow A^{\mathrm{T}}JA = J.$$

In fact, $B(x,y) = [x]_{\mathfrak{X}}^{\mathrm{T}}\,J\,[y]_{\mathfrak{X}}$ by definition of Aut(B) since $B(x,y) = B(Tx,Ty)$ for $x, y \in V$. This happens if and only if

$$[x]_{\mathfrak{X}}^{\mathrm{T}}\,J\,[y]_{\mathfrak{X}} = [Tx]_{\mathfrak{X}}^{\mathrm{T}}\,J\,[Ty]_{\mathfrak{X}} = [x]_{\mathfrak{X}}^{\mathrm{T}}[T]_{\mathfrak{X}\mathfrak{X}}^{\mathrm{T}}\,J\,[T]_{\mathfrak{X}\mathfrak{X}}[y]_{\mathfrak{X}}.$$

Since this is true for all x, y, we have

$$T \in \mathrm{Aut}(B) \Leftrightarrow A = [T]_{\mathfrak{X}\mathfrak{X}} \in G_{B,\mathfrak{X}} \Leftrightarrow A^{\mathrm{T}}JA = J\,,$$

as claimed. The corresponding matrix group is the classical SYMPLECTIC GROUP OF DEGREE $m = \frac{1}{2}\dim(V)$:

$$\mathrm{Sp}(n,\mathbb{K}) = G_{B,\mathfrak{X}} = \{A \in \mathrm{M}(n,\mathbb{K}) : A^{\mathrm{T}}JA = J\}.$$

The related matrix

$$J' = \begin{pmatrix} \boxed{R} & & 0 \\ & \ddots & \\ 0 & & \boxed{R} \end{pmatrix} \quad \text{with} \quad R = \begin{pmatrix} 0 & 1 \\ -1 & 0 \end{pmatrix}$$

is a GL-conjugate of J with $J' = CJC^{-1}$ for some $C \in \mathrm{GL}(2m,\mathbb{R})$, and the algebraic condition

$$A^{\mathrm{T}}J'A = J'$$

determines a subgroup $G' = \{A : A^{\mathrm{T}}\,J'\,A = J'\}$ that is a conjugate of (hence isomorphic to) the matrix group $G_{B,\mathfrak{X}} = \mathrm{Sp}(n,\mathbb{K})$ above. The commutation relations involving J and J' have both been used in the literature to describe the matrices associated with elements of $\text{Aut}(B)$. The version involving J is often favored by physicists in discussions of Hamiltonian systems in classical mechanics and in quantum mechanics.

NOTE. We have $\det(A) \neq 0$ automatically because

$$\det(J) = \det(A^{\mathrm{T}}JA) = \det(A)^2 \cdot \det J \Rightarrow (\det(A))^2 = 1$$

for $A \in \text{Sp}(n, \mathbb{K})$, so $\det(A) = \pm 1$ whether the underlying field is $\mathbb{R}$ or $\mathbb{C}$. The only scalar matrices λI in $\text{Sp}(n, \mathbb{K})$ are those such that $\lambda^2 = 1$. Incidentally, $\det(J) = 1$ because m row transpositions send $J \to \text{diag}(-I_{m\times m}, I_{m\times m})$ which has $\det = (-1)^m$; but transposing those m rows creates another factor of $(-1)^m$, making $\det(J) = +1$. ○

3.3. Sesquilinear Forms ($\mathbb{K} = \mathbb{C}$)

Finally, we take up sesquilinear forms $B : V \times V \to \mathbb{C}$ (which are only defined on complex vector spaces). These are linear functions of the first entry in $B(v, w)$ but conjugate-linear in the second so that $B(x, \lambda y) = \overline{\lambda} B(x, y)$ and $B(\lambda x, y) = \lambda B(x, y)$. There is a limited number of possibilities.

LEMMA 3.40. *A sesquilinear form on V cannot be symmetric or antisymmetric unless it is zero.*

PROOF. If B were symmetric, we would have

$$\lambda B(x, y) = B(\lambda x, y) = B(y, \lambda x) = \overline{\lambda} B(y, x) = \overline{\lambda} B(x, y)$$

for all $\lambda \in \mathbb{C}$, which is impossible unless $B(x, y) \equiv 0$. A similar argument works when a sesquilinear B is antisymmetric. □

Thus, the only natural symmetries of sesquilinear forms over $\mathbb{C}$ are

1. HERMITIAN SYMMETRY. $B(x, y) = \overline{B(y, x)}$.
2. SKEW-HERMITIAN SYMMETRY. $B(x, y) = -\overline{B(y, x)}$.

However, if B is Hermitian, then iB (where $i = \sqrt{-1}$) is skew-Hermitian and vice-versa, so once we analyze Hermitian sesquilinear forms, there is nothing new to say about skew-Hermitian forms.

The sesquilinear forms on V are a vector space over $\mathbb{C}$. Every such form is uniquely a sum $B = B_H + B_S$ of a Hermitian and skew-Hermitian form:

$$B(v, w) = \frac{B(v, w) + \overline{B(w, v)}}{2} + \frac{B(v, w) - \overline{B(w, v)}}{2} \quad \text{for all } v, w \in V$$

As usual, a sesquilinear form B is determined by its matrix representation relative to a basis $\mathfrak{X} = \{e_1, \dots, e_n\}$ in V given by

$$[B]_{\mathfrak{X}} = [B_{ij}] \quad \text{where } B_{ij} = B(e_i, e_j).$$

Given any basis $\mathfrak{X}$, the form B is

1. Nondegenerate if and only if $[B]_{\mathfrak{X}}$ is nonsingular (nonzero determinant).

2. Hermitian symmetric if and only if $[B]_{\mathfrak{X}}$ is self-adjoint ($= [B]^*_{\mathfrak{X}}$).
3. The correspondence $B \to [B]_{\mathfrak{X}}$ is a $\mathbb{C}$-linear isomorphism between the vector space of sesquilinear forms on V and matrix space $\mathrm{M}(n, \mathbb{C})$.

Sesquilinear Change of Basis Formula. The change of basis formula is a bit different from that for bilinear forms. If $\mathfrak{Y} = \{f_j\}$ is another basis related to $\mathfrak{X} = \{e_i\}$ via

$$f_i = \sum_{j=1}^{n} s_{ji} e_j \quad \text{where} \quad S = [\mathrm{id}]_{\mathfrak{X}\mathfrak{Y}} ,$$

we have

$$\begin{aligned}([B]_{\mathfrak{Y}})_{ij} &= B(f_i, f_j) = B\Big(\sum_k s_{ki} e_k \,,\, \sum_\ell s_{\ell j} e_\ell\Big) \\ &= \sum_{k,l} s_{ki} \overline{s_{\ell j}}\, ([B]_{\mathfrak{X}})_{k\ell} = (S^{\mathrm{T}} [B]_{\mathfrak{X}} \overline{S})_{ij}\end{aligned}$$

where $\overline{S}$ is the conjugate matrix: $\overline{S}_{ij} = \overline{s_{ij}}$. Letting $P = \overline{S}$, we may rewrite the outcome as

$$[B]_{\mathfrak{Y}} = P^* [B]_{\mathfrak{X}} P \tag{3.14}$$

where $\det(P) \neq 0$ and $P^* = (\overline{P})^{\mathrm{T}}$. In terms of the transition matrix S between bases, we have $P = \overline{S} = \overline{[\mathrm{id}]_{\mathfrak{X}\mathfrak{Y}}}$.

Here P^* need not be equal to P^{-1}, so P need not be a unitary matrix in $\mathrm{M}(n, \mathbb{C})$. Formula (3.14) differs from that for orthogonal matrices in that P^{T} has been replaced by P^*.

EXERCISE 3.41. Prove the basic properties 1–3 of sesquilinear forms listed above.

EXERCISE 3.42. If B is sesquilinear, $\mathfrak{X}$ is a basis in V, and $x = \sum_i x_i e_i$, $y = \sum_j y_j e_j$ in V, show that

$$B(x, y) = [x]^{\mathrm{T}}_{\mathfrak{X}} [B]_{\mathfrak{X}} \overline{[y]_{\mathfrak{X}}} \quad \text{so that} \quad B(x, y) = \sum_{ij} x_i B_{ij} \overline{y_j}.$$

DEFINITION 3.43. An *inner product* is a sesquilinear form that is

1. HERMITIAN. $B(x, y) = \overline{B(y, x)}$.
2. POSITIVE DEFINITE. $B(x, x) \geq 0$ for all x.
3. NONDEGENERATE. $B(x, V) = (0) \Leftrightarrow x = 0$.

If B is Hermitian as in (1), the associated quadratic form $Q(x) = B(x, x)$ has real values for all $x \in V$. In general, conditions (2) and (3) amount to saying $B(x, x) \geq 0$ and $B(x, x) = 0 \Rightarrow x = 0$ — i.e., the form is *strictly* positive definite. This equivalence follows from the polarization identity for sesquilinear forms.

LEMMA 3.44 (Polarization Identity). *If B is any sesquilinear form on a complex vector space, then*

$$B(v,w) = \frac{1}{4}\Big[\sum_{k=0}^{3} i^k \cdot B(v+i^k w, v+i^k w)\Big] \quad \text{where } i=\sqrt{-1}.$$

In particular, $B(x,y)$ is nondegenerate if $B(x,x) \neq 0$ for some $x \in V$.

PROOF. Expand the sum. □

PROPOSITION 3.45. *Every nondegenerate Hermitian sesquilinear form B on a complex vector space can be put into one of the following canonical forms by a suitable choice of basis in V.*

$$[B]_{\mathfrak{X}} = \begin{pmatrix} I_{p\times p} & 0 \\ 0 & -I_{q\times q} \end{pmatrix} \quad (\textit{forms of type } (p,q)) \tag{3.15}$$

where $p+q=n=\dim V$ and $p,q \geq 0$. If $\mathbf{z} = \sum_i z_i e_i$, $\mathbf{w} = \sum_j w_j e_j$ with respect to a basis $\mathfrak{X}$ such that $[B]_{\mathfrak{X}}$ has canonical form, we get the standard realization of a sesquilinear form of type (p,q).

$$B_{p,q}(\mathbf{z},\mathbf{w}) = \sum_{i=1}^{p} z_i\overline{w_i} - \sum_{i=p+1}^{n} z_i\overline{w_i} \quad (\textit{with } p+q=n) \tag{3.16}$$

in which $V = \mathbb{C}^n$.

When $p=n$, we obtain the standard inner product $(\mathbf{z},\mathbf{w})$ on $\mathbb{C}^n$

$$B_{n,0}(\mathbf{z},\mathbf{w}) = \sum_{j=1}^{n} z_j\overline{w_j} = (\mathbf{z},\mathbf{w}) \quad \text{for } \mathbf{z},\mathbf{w} \in \mathbb{C}^n,$$

and if $q=n$, we get

$$B_{0,n}(\mathbf{z},\mathbf{w}) = -\sum_{i=1}^{n} z_i\overline{w_i} = -(\mathbf{z},\mathbf{w}).$$

PROOF. If B is nondegenerate Hermitian and $v \neq 0$, there must be some $w \in V$ such that $B(v,w) \neq 0$, but by the polarization identity, nondegeneracy of B implies there is some v such that $B(v,v) \neq 0$. Hermitian symmetry implies $B(v,v)$ is real, and if $c \in \mathbb{C}$, we have $B(cv,cv) = |c|^2 B(v,v)$; thus, by rescaling, we can make $B(v,v) = \pm 1$, depending on whether $B(v,v)$ is positive or negative.

We now argue by induction on $\dim(V)$, assuming $\dim(V) = n$ and that $[B]_{\mathfrak{X}}$ can be put into canonical form for any nondegenerate Hermitian form B_0 on any complex vector space of dimension $< n$. As in the preceding paragraph, we can find a vector v_1 such that $B(v_1,v_1) = \pm 1$; let $M_1 = \mathbb{R}v_1$ and relabel

$$M_1^{\perp,B} = M_1^{\perp} = M_0.$$

By nondegeneracy of B, we have $\dim(M_0) = n-1$, and then $M_1 \cap M_0 = (0)$ because $w \in M_1 \cap M_1^{\perp} \Rightarrow w = cv_1$ and also $0 = (w,v_1) = c(v_1,v_1)$, which implies $c = 0$. Thus, $V = M_1 \oplus M_0$.

The restricted form $B_0 = B|_{M_0}$ is also nondegenerate because if $B(w, M_0)$ is zero for some nonzero $w \in M_0$, then $B(w, V) = B\big(w, M_1 + M_0\big) = (0)$ too, contrary to nondegeneracy of B on V. We conclude that there is a basis

$$\mathfrak{X} = \{e_1 = v_1, e_2, \dots, e_n\} \text{ in } V = M_1 \oplus M_0$$

such that

$$[B]_{\mathfrak{X}} = \begin{pmatrix} \mu_1 & 0 & \\ 0 & \boxed{I_{p'\times p'}} & 0 \\ 0 & 0 & \boxed{-I_{q'\times q'}} \end{pmatrix}$$

with $p' + q' = n - 1$.

Finally, apply a permutation matrix E (relabel basis vectors) to get a new basis $\mathfrak{Y}$ such that

$$[B]_{\mathfrak{Y}} = E^*[B]_{\mathfrak{X}}E = \begin{pmatrix} I_{p\times p} & 0 \\ 0 & -I_{q\times q} \end{pmatrix}$$

with $p + q = n = \dim(V)$ and $p, q \geq 0$. This reduces to $[B]_{\mathfrak{X}} = I_{n\times n}$ (or $-I_{n\times n}$) when $p = n$ (or $p = 0$). □

Proposition 3.45 implies that there are just $n+1$ congruence classes of nondegenerate Hermitian sesquilinear forms on a complex vector space of dimension n, which are distinguished by their signatures (p, q). Convenient models of the possible automorphism groups

$$\text{Aut}(B) = \{T : \det(T) \neq 0 \quad \text{and} \quad B(T(v), T(w)) = B(v, w) \text{ for all } v, w\}$$

are provided by the matrix groups $G_{B,\mathfrak{X}}$ determined by a basis $\mathfrak{X}$ that puts B into canonical form. This yields the UNITARY GROUPS OF TYPE (p, q), for which $\text{Aut}(B)$ is isomorphic to the matrix group

$$\text{(3.17)} \qquad \mathrm{U}(p, q) = \{A : A^* J_{p,q} A = J_{p,q}\} \quad \text{with } J_{p,q} = \begin{pmatrix} \boxed{I_{p\times p}} & 0 \\ 0 & \boxed{-I_{q\times q}} \end{pmatrix}.$$

There is a slight twist in the correspondence between operators $T \in \text{Aut}(B)$ and matrices $A = [T]_{\mathfrak{X}\mathfrak{X}} \in G_{B,\mathfrak{X}}$ in $\mathrm{U}(p, q)$.

EXERCISE 3.46. Let B be nondegenerate Hermitian sesquilinear and $\mathfrak{X} = \{e_i\}$ be a basis such that $[B]_{\mathfrak{X}}$ is in canonical form (3.15). If $A = [T]_{\mathfrak{X}\mathfrak{X}}$ is the matrix associated with $T \in \text{Aut}(B)$ and the basis $\mathfrak{X}$ in V, verify that the complex conjugate $\overline{A}$ satisfies the identity in (3.17), and conversely if $A \in \mathrm{U}(p, q)$, then $\overline{A} = \big([T]_{\mathfrak{X}\mathfrak{X}}\big)$ for some $T \in \text{Aut}(B)$.

Thus, the correspondence $\Phi : T \mapsto A = \big([T]_{\mathfrak{X}\mathfrak{X}}\big)^{-}$ (rather than $T \mapsto A = [T]_{\mathfrak{X}\mathfrak{X}}$) is a bijection between $\text{Aut}(B)$ and the matrix group $\mathrm{U}(p, q) \subseteq \mathrm{M}(n, \mathbb{C})$ such that

$$\Phi(T_1 \circ T_2) = \Phi(T_1) \cdot \Phi(T_2) \quad \text{(matrix product)}.$$

Hence Φ is a group isomorphism between $\text{Aut}(B)$ and $\mathrm{U}(p, q)$.

When $p = n$ and we identify $V \cong \mathbb{C}^n$ via a basis such that $[B]_{\mathfrak{X}} = I_{n\times n}$, Aut$(B)$ becomes the classical group of unitary operators on $\mathbb{C}^n$, and the group of matrices $G_{B,\mathfrak{X}}$ is the familiar group of unitary matrices

$$\mathrm{U}(n) = \mathrm{U}(n,0) = \{A \in \mathrm{M}(n,\mathbb{C}) : A^*A = A^*I_{n\times n}A = I_{n\times n}\}$$

which includes the *special unitary group*

$$\mathrm{SU}(n) = \mathrm{U}(n) \cap \mathrm{SL}(n,\mathbb{C}) \subseteq \mathrm{U}(n)$$

as a closed subgroup.

Similarly $\mathrm{U}(p,q)$ contains the SPECIAL UNITARY GROUP OF TYPE (p,q),

$$\mathrm{SU}(p,q) = \mathrm{U}(p,q) \cap \mathrm{SL}(n,\mathbb{C}).$$

For $A \in \mathrm{U}(p,q)$, the identity $A^*J_{p,q}A = J_{p,q}$ implies

$$|\det(A)|^2 \det(J_{p,q}) = \det(J_{p.q}),$$

so $|\det(A)| \equiv 1$ and $\det(A)$ lies on the unit circle $S^1 = \{z : |z| = 1\}$ in $\mathbb{C}$.

We know that matrices in $\mathrm{U}(n)$ are orthogonally diagonalizable over $\mathbb{C}$ since they are normal operators ($A^*A = AA^*$, so $A^*A = I \Leftrightarrow AA^* = I$). And because $\|Ax\|^2 = \|x\|^2$ for all x, all eigenvalues λ_i have absolute value 1, so for unitary matrices (or operators) the spectrum $\mathrm{sp}_{\mathbb{C}}(A)$ is a subset of the unit circle $S^1 = \{z \in \mathbb{C} : |z| = 1\}$. Furthermore, $\mathrm{U}(n)$ contains a copy $S^1 \cong \{\lambda I_{n\times n} : |\lambda| = 1\}$ of the unit circle, which is a group under the usual multiplication of complex numbers because $|zw| = |z| \cdot |w|$ and $|z| = 1 \Rightarrow |1/z| = 1$. In $\mathrm{SU}(n)$, however, the only scalar matrices are those of the form λI where λ is an nth root of unity, $\lambda = e^{2\pi ik/n}$ with $0 \leq k \leq n-1$.

Notice the parallel between the orthogonal groups over $\mathbb{K} = \mathbb{R}$ and the unitary groups over $\mathbb{K} = \mathbb{C}$.

1. $\mathrm{SO}(p,q)$ and $\mathrm{O}(p,q)$ over $\mathbb{R}$ are the "real points" in the groups $\mathrm{SU}(p,q)$ and $\mathrm{U}(p,q)$. In fact, we have

$$\mathrm{O}(p,q) = \mathrm{U}(p,q) \cap \big(\mathrm{M}(n,\mathbb{R}) + i0\big)$$

   when we identify $\mathrm{M}(n,\mathbb{C}) = \mathrm{M}(n,\mathbb{R}) + \sqrt{-1}\,\mathrm{M}(n,\mathbb{R})$ by splitting $A = [z_{ij}]$ as $[x_{ij}] + \sqrt{-1}\,[y_{ij}]$ when $z_{ij} = x_{ij} + \sqrt{-1}\,y_{ij}$.
2. We also recognize $\mathrm{SO}(n)$ and $\mathrm{O}(n)$ as the real points in the complex matrix groups $\mathrm{SO}(n,\mathbb{C})$ and $\mathrm{O}(n,\mathbb{C})$. At the same time, these subgroups are also the real points in the unitary groups $\mathrm{SU}(n)$ and $\mathrm{U}(n)$.

EXERCISE 3.47. Prove that $\mathrm{U}(n)$ is a closed, bounded subset when we identify $\mathrm{M}(n,\mathbb{C}) \approx \mathbb{C}^{n^2}$. Hence, it is a compact matrix group.

EXERCISE 3.48. If $1 < p < n$, prove that $\mathrm{U}(p,q)$ and $\mathrm{SU}(p,q)$ are closed but unbounded subsets in matrix space $\mathrm{M}(n,\mathbb{C})$.

EXERCISE 3.49. Verify that matrix group $\mathrm{U}(p,q)$ is in fact the matrix model $G_{B,\mathfrak{X}}$ of the automorphism group $\mathrm{Aut}(B_{p,q})$ with respect to any basis that puts $B = B_{p,q}$ into canonical form (3.17).

Additional Exercises

Section 3.1. Bilinear, Quadratic, and Multilinear Forms

1. TRUE/FALSE QUESTIONS ("True" if the statement is *always* true.)

(a) If a bilinear form B on a finite dimensional vector space is symmetric, then its matrix $[B]_{\mathfrak{X}}$ with respect to any basis $\mathfrak{X}$ in V is symmetric so that $[B]_{\mathfrak{X}}^{\mathrm{T}} = [B]_{\mathfrak{X}}$.

(b) For an arbitrary bilinear form on a vector space over $\mathbb{R}$ or $\mathbb{C}$, the left- and right-handed nondegeneracy conditions

$$B(x, V) = \{0\} \Rightarrow x = 0$$
$$B(V, x) = \{0\} \Rightarrow x = 0$$

are equivalent.

(c) If B is a bilinear form on a vector space such that $\dim(V) > 1$ and $x \neq 0$ in V, there is some $y \in V$ such that $y \neq 0$ but $B(x, y) = 0$.

(d) If $B : V \times \cdots \times V \to \mathbb{K}$ is a tensor of rank-k on a vector space V, then $B(v_1, \dots, v_k) = 0$ if one of the inputs $v_1, \dots, v_k$ is zero.

CONGRUENCE OF MATRICES. Recall that matrices $A, A' \in \mathrm{M}(n, \mathbb{K})$ are *congruent*, indicated by writing $A' \cong A$, if there is a nonsingular matrix P such that $A' = P^{\mathrm{T}}AP$. This happens $\Leftrightarrow$ they are matrix realizations of same bilinear form B on an n-dimensional vector space with respect to different bases, so

$$A = [B]_{\mathfrak{X}} \text{ and } A' = [B]_{\mathfrak{Y}} \quad \text{for bases } \mathfrak{X}, \mathfrak{Y} \text{ in } V.$$

(e) If two matrices in $\mathrm{M}(n, \mathbb{K})$ are congruent, so $B = Q^{\mathrm{T}}AQ$, they must have the same eigenvalues in $\mathbb{K}$.

(f) Any symmetric matrix $A \in \mathrm{M}(n, \mathbb{K})$ is congruent to a diagonal matrix.

(g) Every quadratic form is a bilinear form.

2. Which of the following functions with inputs $\mathbf{x}, \mathbf{y}, \mathbf{z}$ from $V = \mathbb{R}^n$ have the right properties to be rank-3 tensors in $V^{(0,2)} = V^* \otimes V^* \otimes V^*$?

(a) $B(\mathbf{x}, \mathbf{y}, \mathbf{z}) = 3x_1x_2x_3 - x_3z_1y_4, \quad V = \mathbb{R}^4$

(b) $B(\mathbf{x}, \mathbf{y}, \mathbf{z}) = 3x_1^2x_2y_4 + 2x_3z_3, \quad V = \mathbb{R}^4$

(c) $B(\mathbf{x}, \mathbf{y}, \mathbf{z}) = \det\begin{pmatrix} x_1 & x_2 & x_3 \\ y_1 & y_2 & y_3 \\ z_1 & z_2 & z_3 \end{pmatrix}, \quad V = \mathbb{R}^3.$

3. The map $B : \mathbb{C}^4 \times \mathbb{C}^4 \to \mathbb{C}$ given by

$$B(\mathbf{v}, \mathbf{w}) = v_1w_1 + v_2w_2 - v_3w_3 - v_4w_4$$

is easily seen the be bilinear (a rank-2 tensor). Find the 4×4 matrix $B_{\mathfrak{X}}$ that describes it with respect thc standard basis $\mathbf{e}_1, \dots, \mathbf{e}_4$ in $\mathbb{C}^4$. Is this matrix symmetric? Antisymmetic? Is B a sesquilinear form on $\mathbb{C}^4$?

4. Let $B : V \times V \to \mathbb{R}$ be a bilinear form on a vector space V over $\mathbb{R}$.

(a) Is

$$\widetilde{B}(x + iy, x' + iy') = \big(B(x, x') - B(y, y')\big) + i\,\big(B(x, y') + B(x', y)\big)$$

a complex-bilinear form on the complexification $V_{\mathbb{C}} = V + iV$?

(b) Is $\widetilde{B}$ symmetric if B is symmetric?

(c) We say that $\widetilde{B}$ is nondegenerate if $\widetilde{B}(x, V_{\mathbb{C}}) = (0) \Rightarrow x = 0$ in $V_{\mathbb{C}}$. Is $\widetilde{B}$ nondegenerate on $V_{\mathbb{C}}$ if B is nondegenerate on V?

5. Regarding the columns C_1, C_2 of a matrix $A \in \mathrm{M}(2, \mathbb{K})$ as vectors in $V = \mathbb{K}^2$, is the map $B(C_1, C_2) = \det(A)$ a bilinear form on V? If so, is B symmetric? Antisymmetric?

6. If a matrix A is congruent to a diagonal matrix D with $D = Q^{\mathrm{T}}AQ$, is A necessarily a symmetric matrix?

7. If B is a symmetric bilinear form on vector space over a field $\mathbb{K}$, the associated *quadratic form* $Q : V \to \mathbb{K}$ is given by $Q(x) = B(x, x)$ for all $x \in V$. Prove the *polarization identity*

$$B(x, y) = \frac{1}{2}[Q(x + y, x + y) - Q(x, x) - Q(y, y)] \quad \text{for } x, y \in V\,,$$

which allows us to reconstruct $B(x, y)$ from Q.

Section 3.2. Canonical Models for Bilinear Forms

1. TRUE/FALSE QUESTIONS ("True" if the statement is *always* true.)

(a) When $\mathbb{K} = \mathbb{C}$, every symmetric matrix can be diagonalized by a similarity transformation $A \to SAS^{-1}$ (S invertible).

(b) If $\mathbb{K} = \mathbb{R}$ or $\mathbb{C}$, every symmetric matrix in $\mathrm{M}(n, \mathbb{K})$ is congruent to a diagonal matrix.

2. Which pairs from the following list of matrices are *congruent*?

$$(a)\ A = \begin{pmatrix} 1 & 1 & -2 \\ -1 & 2 & 1 \\ 3 & 1 & 1 \end{pmatrix} \quad (b)\ B = \begin{pmatrix} 1 & 2 & 1 \\ 2 & 3 & 2 \\ 3 & 1 & 1 \end{pmatrix} \quad (c)\ C = \begin{pmatrix} 1 & 0 & 1 \\ 0 & 1 & 2 \\ 1 & 2 & 1 \end{pmatrix}$$

Give your reasoning.

HINT. You could use Sylvester's theorem to compare symmetric matrices. You might also consider the ranks of the matrices.

3. Determine which pairs of matrices in Additional Exercise 2 are in the same *similarity class*.

4. Let V be an inner product space over $\mathbb{R}$ and $B(x, y)$ a symmetric bilinear form on V.

(a) Prove that there is an *orthonormal basis* $\mathfrak{X}$ in V such that $[B]_{\mathfrak{X}}$ is a diagonal matrix.

(b) What can you say about these diagonal entries? (In particular, are they necessarily related to the eigenvalues of $[B]_{\mathfrak{X}}$?

5. Let $A = \begin{pmatrix} a & b \\ c & d \end{pmatrix}$ be a real-orthogonal matrix in O(2). Explain why the following hold:

(a) $\det(A) = +1$ or -1.

(b) $|\lambda| = 1$ for any complex eigenvalues of A.

(c) $\tilde{L}_A : \mathbb{C}^2 \to \mathbb{C}^2$ is diagonalizable.

6. If $\det(A) = +1$ and $A \neq I$, we have shown (Euler's theorem, LA I) that L_A acts on $\mathbb{R}^2$ as a rotation about the origin. Now suppose $\det(A) = ad - bc = -1$.

(b) Explain why

$$a^2 + b^2 = 1 \quad c^2 + d^2 = 1 \quad ac + bd = 0.$$

Then show that $A = \begin{pmatrix} \cos\theta & \sin\theta \\ \sin\theta & -\cos\theta \end{pmatrix}$ for some real θ.

(c) Show that L_A is reflection across some line L through the origin, with θ related to the angle between L and the $+x$-axis.

HINT. Recall the discussion of Euler's theorem 6.48 in LA I.

7. A matrix A is in the complex orthogonal group $O(n, \mathbb{C})$ if and only if $\omega(A\mathbf{z}, A\mathbf{w}) = \omega(\mathbf{z}, \mathbf{w})$ for the canonical symmetric bilinear form on $\mathbb{C}^n$,

$$\omega(\mathbf{z}, \mathbf{w}) = \sum_{j=1}^{n} z_j w_j \quad (\mathbf{z}.\mathbf{w} \in \mathbb{C}^n),$$

whose signature is $(n, 0)$. Now consider the related symmetric bilinear form

$$\widetilde{\omega}(\mathbf{z}, \mathbf{w}) = \omega(i\mathbf{z}, i\mathbf{w})$$

where $i = \sqrt{-1}$. Explain why the following hold:

(a) The matrix $[\widetilde{\omega}]_{\mathfrak{X}}$ of this form with respect to the standard basis in $\mathbb{C}^n$ is $-I_{n\times n}$.

(b) What does this mean for the product AA^{T}?

(c) What is the signature of $\widetilde{\omega}$?

8. In discussing symplectic groups (the automorphism groups of nondegenerate skew-symmetric bilinear forms B over $\mathbb{R}$ or $\mathbb{C}$), we mentioned two canonical forms in common use, namely

$$[B]_{\mathfrak{X}} = J = \begin{pmatrix} 0 & I_{m\times m} \\ -I_{m\times m} & 0 \end{pmatrix}$$

and (for a different basis)

$$[B]_{\mathfrak{Y}} = \begin{pmatrix} \boxed{R} & & 0 \\ & \ddots & \\ 0 & & \boxed{R} \end{pmatrix} \quad \text{where} \quad R = \begin{pmatrix} 0 & 1 \\ -1 & 0 \end{pmatrix}.$$

Restricting attention to the simpler case of 4×4 matrices, find a permutation matrix P such that $P^{\mathrm{T}}[B]_{\mathfrak{X}}P = [B]_{\mathfrak{Y}}$.

HINT. In this case the desired change of basis can be achieved by a simple relabeling of the basis vectors in $\mathfrak{X}$ without taking any linear combinations (which is why P is a permutation matrix).

Section 3.3. Sesquilinear Forms Over $\mathbb{C}$

1. TRUE/FALSE QUESTIONS ("True" if the statement is *always* true.)
   (a) For bilinear forms on a vector space over $\mathbb{R}$, there is no difference between "symmetry" and "Hermitian-symmetry."
   (b) Let A be a skew-adjoint complex matrix (so $A^* = -A$) and define $B : \mathbb{C}^n \times \mathbb{C}^n \to \mathbb{C}$ letting
   $$B(\mathbf{z}, \mathbf{w}) = \mathbf{z}^{\mathrm{T}} A \mathbf{w}\,,$$
   regarding 1×1 matrices as scalars and vectors in $\mathbb{C}^n$ as $n \times 1$ column vectors. Then B is a sesquilinear form on the complex vector space $V = \mathbb{C}^n$.
2. If $(\mathbf{z}, \mathbf{w})$ is the standard inner product on $\mathbb{C}^n$ and A is a complex matrix, what properties must A have if
   $$B(\mathbf{z}, \mathbf{w}) = \big(A\mathbf{z}, \mathbf{w}\big) \quad (\mathbf{z}, \mathbf{w} \in V = \mathbb{C}^n)$$
   is to be the following:
   (a) A sesquilinear form on V.
   (b) A nondegenerate sesquilinear form on V (so $B(V, \mathbf{w}) = 0$ if and only if $\mathbf{w} = 0$).
   (c) A Hermitian symmetric sesquilinear form on V.
   (d) A skew-Hermitian sesquilinear form on V.
   (e) An inner product on V.

CHAPTER 4

Tensor Fields, Manifolds, and Vector Calculus

... or, what they never told you in Calculus I-III

An Overview of This Chapter. This chapter explores some aspects of linear algebra that lie at the heart of modern differential geometry, and it ends with a reinterpretation of many of the main results of multivariate calculus. This is a vast subject, so the presentations of this chapter will not be as fully developed as in preceding chapters.

Euclidean n-dimensional space E_n is a featureless space — a line, a plane, etc. — perhaps equipped with a *metric* — a distance function $d(x, y)$ between points. By marking an origin and imposing coordinates, we can model E_n as n-tuples of real numbers $\mathbf{x} = (x_1, \dots, x_n)$ in $\mathbb{R}^n$. In addition to allowing us to describe locations of points p as coordinate n-tuples, $\mathbb{R}^n$ comes equipped with certain algebraic operations — *scaling* and *vector addition* — which make Euclidean space into a vector space over $\mathbb{R}$. This extra structure, unknown to Euclid, was inspired by the "parallelogram law" for vector addition that arose in physics as the correct law for adding forces. In the late 1800's, additional operations on coordinate space $\mathbb{R}^n$ were introduced, such as the inner product $(\mathbf{x}, \mathbf{y}) = \sum_{i=1}^{n} x_i y_i$ (for arbitrary dimensions) and the cross product $\mathbf{x}\times\mathbf{y}$ (which makes sense only in $n = 3$ dimensions).

The tangent space TE_p to E_n at a base point p was thought of as a copy of the vector space $\mathbb{R}^n$ attached to E_n at p, whose elements in TE_p are described as pairs $(p, \mathbf{x})$ where p is a base point in E_n and $\mathbf{x}$ a vector in $\mathbb{R}^n$. There is a separate tangent space attached to each base point in E. In calculus, tangent vectors based at p were thought of as "arrows" attached to the base point, which can be scaled and added to other tangent vectors attached to the same base point via the rules

$$\lambda \cdot (p, \mathbf{x}) = (p, \lambda\mathbf{x}) \quad \text{and} \quad (p, \mathbf{x}_1) + (p, \mathbf{x}_2) = (p, \mathbf{x}_1 + \mathbf{x}_2).$$

There is, however, no meaningful way to add tangent vectors $(p, \mathbf{x}) \in \mathrm{TE}_p$ *and* $(q, \mathbf{y}) \in \mathrm{TE}_q$ *attached to different base points* $p \neq q$. When elements of TE_p are viewed as arrows, those arrows might represent force vectors acting on objects located at p or the velocity of a moving particle as it passes through p, for example. They could also represent more general fields: an electric, magnetic, or gravitational field, or a field of stress tensors that pervade some solid medium, etc.

Many mathematical models are concerned with *fields of vectors* $\mathbf{F}(\mathbf{x})$ defined on E_n (or some open subset). These assign a tangent vector $\mathbf{F}(p) \in \mathrm{TE}_p$ to each base point, and the coordinates imposed on E_n determine basis vectors

$\{\mathbf{e}_1, \dots, \mathbf{e}_n\}$ in the tangent space TE_p that allow us to describe any tangent vector at p as a linear combination $\mathbf{a} = a_1(p)\mathbf{e}_1 + \cdots + a_n(p)\mathbf{e}_n$ ($a_i \in \mathbb{R}$). Similarly, we can describe a field of vectors on E_n as

$$\mathbf{F}(p) = F_1(p)\,\mathbf{e}_1 + \cdots + F_n(p)\,\mathbf{e}_n$$

whose scalar-valued coefficients $F_k(p)$ vary with the base point. If $\mathbf{a}, \mathbf{b}$ are in TE_p and $n = 3$, their scalar multiples and vector sums are given by

$$\lambda\mathbf{a} = \sum_{i=1}^{n} \lambda a_i\,\mathbf{e}_i \quad \text{and} \quad \mathbf{a}+\mathbf{b} = \sum_{i=1}^{3}(a_i + b_i)\,\mathbf{e}_i\,,$$

and their cross product is

$$\begin{aligned}\mathbf{a}\times\mathbf{b} &= \det\begin{pmatrix} \mathbf{e}_1 & \mathbf{e}_2 & \mathbf{e}_3 \\ a_1 & a_2 & a_3 \\ b_1 & b_2 & b_3 \end{pmatrix} \\ &= (a_2b_3 - a_3b_2)\,\mathbf{e}_1 - (a_1b_3 - a_3b_1)\,\mathbf{e}_2 + (a_1b_2 - a_2b_1)\,\mathbf{e}_3.\end{aligned}$$

By definition, *scalar fields* on E_n are just scalar-valued functions $F : E_n \to \mathbb{R}$.

This coordinate description of vector fields on $\mathbb{R}^3$ or $\mathbb{R}^n$ is used in calculus to define the standard "vector operations" on vector fields $\mathbf{y} = \mathbf{F}(\mathbf{x}) = \sum_{i=1}^n F_i(\mathbf{x})\,\mathbf{e}_i$ that have differentiable coefficients. Writing $D_{x_i}\phi(p)$ for the ith partial derivative $\partial\phi/\partial x_i(p)$ of a scalar-valued function $\phi : \mathbb{R}^m \to \mathbb{R}$, we define the following operations:

1. GRADIENT:

$$\mathbf{grad}\,F(p) = \nabla F(p) = \sum_{i=1}^{n} D_{x_i}F(p)\cdot\mathbf{e}_i$$

for scalar fields $F : \mathbb{R}^n \to \mathbb{R}$.

2. CURL:

$$\begin{aligned}\mathbf{curl}\,\mathbf{F}(p) = \nabla\times\mathbf{F}(p) &= \det\begin{pmatrix} \mathbf{e}_1 & \mathbf{e}_2 & \mathbf{e}_3 \\ \partial/\partial x_1 & \partial/\partial x_2 & \partial/\partial x_3 \\ F_1 & F_2 & F_3 \end{pmatrix} \\ &= \left(\frac{\partial F_3}{\partial x_2} - \frac{\partial F_2}{\partial x_3}\right)\mathbf{e}_1 - \left(\frac{\partial F_3}{\partial x_1} - \frac{\partial F_1}{\partial x_3}\right)\mathbf{e}_2 + \left(\frac{\partial F_2}{\partial x_1} - \frac{\partial F_1}{\partial x_2}\right)\mathbf{e}_3\end{aligned}$$

for smooth vector fields $\mathbf{F} = F_1\,\mathbf{e}_1 + F_2\,\mathbf{e}_2 + F_3\,\mathbf{e}_3$. This definition only works for fields on 3-dimensional space $\mathbb{R}^3$. Although, as we will see in Section 4.4, it can be adapted to deal with vector fields on two-dimensional space.

3. DIVERGENCE:

$$\begin{aligned}\mathbf{div}\,\mathbf{F}(p) = \nabla\circ\mathbf{F}(p) &= \frac{\partial F_1}{\partial x_1}(p)\,\mathbf{e}_1 + \frac{\partial F_2}{\partial x_1}(p)\,\mathbf{e}_2 + \frac{\partial F_3}{\partial x_3}(p)\,\mathbf{e}_3 \\ &= \mathrm{Trace}\left[\frac{\partial F_i}{\partial x_j}(p)\right]\end{aligned}$$

for smooth vector fields $\mathbf{F} : \mathbb{R}^3 \to \mathbb{R}^3$. The result is a scalar field. This too works only for fields on 3-dimensional Euclidean space, but it can be adapted to work for fields on $\mathbb{R}^2$.

Various questions lurk in the background of these definitions.

QUESTIONS. *All these operators are described in terms of Cartesian coordinates imposed on the blank slate of Euclidean space. But how do we know that the numerical results of our calculations are anything more than artifacts of the coordinates we imposed to do the calculation?*

For instance:

1. What happens in polar coordinates or other coordinate systems?
2. How are bases $\{\mathbf{e}_1, \dots, \mathbf{e}_m\}$ in the tangent spaces TE_p induced by different coordinates systems on E_n?
3. How should the vector operations of calculus
$$\mathbf{grad} = \nabla F \quad \mathbf{curl} = \nabla \times \mathbf{F} \quad \mathbf{div} = \nabla \circ \mathbf{F}$$
be described in non-Cartesian coordinates?
4. Finally, do vector fields and the operations on them have physical or geometric meaning independent of any particular system of coordinates?

Physicists and mathematicians also wondered why these particular combinations of the n^2 second order partial derivatives keep showing up in so many applications. Or, what analogs of the vector operators *div*, *grad*, *curl* that pervade the physical sciences might exist in dimensions $n \geq 4$, where the classical definitions of **curl** and **div** no longer make sense?

Featureless Euclidean space is a very limited setting for developing mathematical physics and for understanding the possible meaning of "tangent vectors." It is simply too special to reveal the subtleties involved. For instance, suppose you wanted to model phenomena occuring on the spherical surface of the Earth.

- What does it mean to speak of a vector field on the two dimensional sphere
$$S^2 = \{x \in \mathbb{R}^3 \,:\, x_1^2 + x_2^2 + x_3^2 = 1\},$$
a smooth two-dimensional hypersurface in $\mathbb{R}^3$? How about calculations involving vector fields on the 3-dimensional sphere
$$S^3 = \{\mathbf{x} \in \mathbb{R}^4 \,:\, ||\mathbf{x}||^2 = 1\},$$
a curved 3-dimensional hypersurface that lives in 4-dimensional Euclidean space?
- What becomes of the classical vector operations on vector fields in higher dimensional settings?

The same questions apply to vector fields defined on any curved lower dimensional hypersurface embedded in $\mathbb{R}^n$.

There are deeper and more fundamental issues to deal with in a consistent re-imagining of calculus. Suppose we regard S^2 or S^3 as the entire universe of discourse. For instance, in general relativity the universe can be viewed as a 3-dimensional sphere S^3 with a metric structure that evolves with time. There is no outside space in which this object lives, and any reference to it is pure metaphysics, divorced from phenomena observable from within S^3. "Tangent vectors" or "tangent spaces" viewed as arrows or hyperplanes attached to base points $p \in S^3$ and extending into the "surrounding space" simply have no meaning. There is no "outside" to the universe. So let's see how calculus might be developed using concepts intrinsic to S^2 or S^3, or more general spaces one might use to model phenomena.

Underlying everything is this issue: How can you deal with such questions in a coordinate-independent way? A satisfactory answer emerged only in the first quarter of the 20th century, based on the concept of a "differentiable manifold." Differential geometry, the area of mathematics that deals with these issues, is a hybrid of hard analysis and advanced topics in linear algebra (tensors, differential forms) which will also be explored in this chapter.

4.1. Tangent Vectors, Cotangent Vectors, and Tensors

The definition of a manifold begins with an m-dimensional *locally Euclidean space* M; a set of points that can be covered with *coordinate charts* (U_α, x_α), each consisting of an open set $U_\alpha \subseteq M$, the *chart domain*, and a continuous *chart map* $x_\alpha : U_\alpha \to \mathbb{R}^m$ that assigns Euclidean coordinates $\mathbf{x} = x_\alpha(u) = (x_1, \dots, x_m)$ in $\mathbb{R}^m$ to each point $u \in U_\alpha$. To make M into a differentiable manifold, we require the following:

(4.1)

1. The chart domains U_α cover all of M.
2. The images $V_\alpha = x_\alpha(U_\alpha)$ are open sets in $\mathbb{R}^m$ for each index α.
3. The chart map x_α is a bicontinuous bijection (a *homeomorphism*) from $U_\alpha \subseteq M$ to the open set $V_\alpha = x_\alpha(U_\alpha)$ in coordinate space $\mathbb{R}^m$.

The chart map $x_\alpha : U_\alpha \to V_\alpha \subseteq \mathbb{R}^m$ and its inverse $x_\alpha^{-1} : V_\alpha \to U_\alpha \subseteq M$ are both continuous maps. Thus a locally Euclidean space M looks like $\mathbb{R}^m$ near any base point $p \in M$; in particular, Cartesian coordinates can be imposed on M near p. The idea is obviously inspired by the way the spherical surface of the Earth is mapped in a geographic atlas, each page being a "chart" that describes the features in a particular region. There may of course be overlap between the regions covered by certain pages, but the way those regions fit together can be read by comparing the appropriate pages.

But more is needed to do calculus on M — in order to discuss derivatives of functions defined on M, the charts (U_α, x_α) should be "differentiably related" whenever two chart domains U_α and U_β overlap.

DEFINITION 4.1. For $1 \leq k \leq \infty$, a *$\mathcal{C}^{(k)}$-structure* on an m-dimensional locally Euclidean space M is a family $\mathcal{U}$ of charts (U_α, x_α), labeled by indices α in some index set I, such that the chart domains U_α cover M and the coordinates on them are *differentiably related* whenever two chart domains overlap, so the

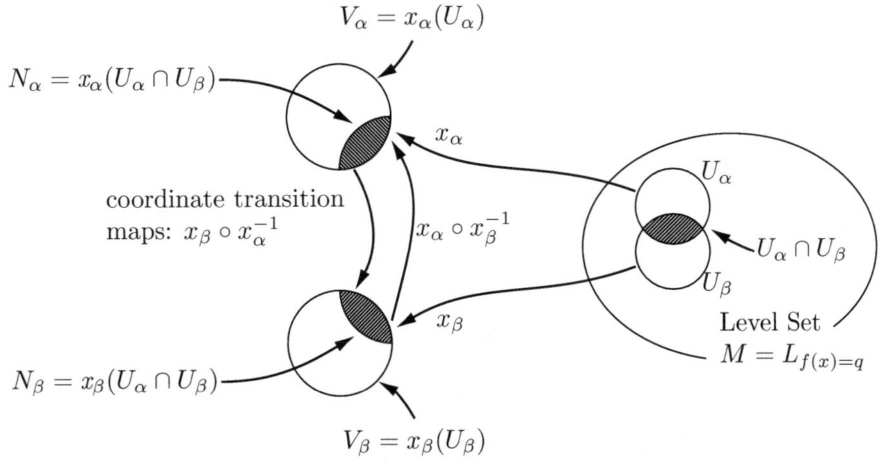

FIGURE 4.1. Coordinate transition maps $x_\alpha \circ y_\beta^{-1}$ and $y_\beta \circ x_\alpha^{-1}$ between charts (U_α, x_α) and (U_β, y_β) and their (shaded) domains of definition N_α, N_β in $\mathbb{R}^m$ are shown. The domains N_α, N_β live in different copies of coordinate space $\mathbb{R}^m$; both correspond to the intersection $U_\alpha \cap U_\beta$ of the chart domains, which is an open set in the locally Euclidean space M.

coordinate transition maps between (U_α, x_α) and (U_β, y_β),

$$x_\alpha \circ y_\beta^{-1} : \mathbb{R}^m \to \mathbb{R}^m \quad \text{and} \quad y_\beta \circ x_\alpha^{-1} : \mathbb{R}^m \to \mathbb{R}^m ,$$

are of class $\mathcal{C}^{(k)}$ on the open sets in coordinate space $\mathbb{R}^m$ where they are defined. Such a family of charts $\mathcal{U}$ is a "founding family" for a differentiable structure on M that will make it a *differentiable manifold*. The meaning of the coordinate transition maps is illustrated in Figure 4.1.

Here is a basic example of how this works.

EXAMPLE 4.2. Consider the two-dimensional sphere $M = S^2$ in $\mathbb{R}^3$:

$$M = \{\mathbf{x} \in \mathbb{R}^3 : x_1^2 + x_2^2 + x_3^2 = 1\}.$$

One family of charts that cover M is given by the pairs $(H_i^\pm, P_i)$ where $H_i^\pm$ are the six open hemispheres

$$H_i^+ = \{\mathbf{x} \in S^2 : x_i > 0\} \quad \text{and} \quad H_i^- = \{\mathbf{x} \in S^2 : x_i < 0\} \quad (1 \le i \le 3)$$

and P_i = restriction to the sphere S^2 of the projection maps $P_i : \mathbb{R}^3 \to \mathbb{R}^2$

$$P_1(\mathbf{x}) = (x_2, x_3), \ \dots \ , P_3(\mathbf{x}) = (x_1, x_2).$$

Each P_i simply erases one of the coordinates in $\mathbf{x} = (x_1, x_2, x_3)$, as indicated in Figure 4.2.

In each case, range(P_i) is the open unit disc $D = \{(s_1, s_2) \in \mathbb{R}^2 : s_1^2 + s_2^2 < 1\}$. Obviously the restriction $P_i|_{H_i^\pm} : H_i^\pm \to D$ is one-to-one and continuous onto D

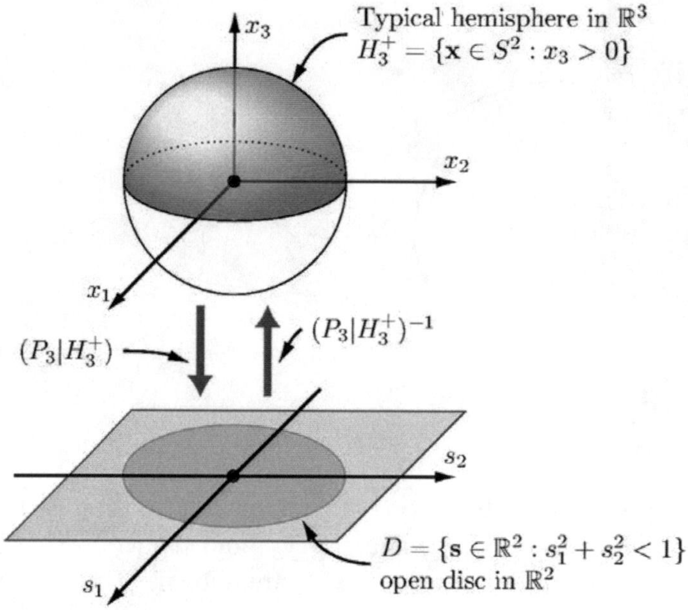

FIGURE 4.2. One of the charts (x_α, U_α) that determine the standard manifold structure for the smooth hypersurface $M = S^2$, a two-dimensional sphere embedded in $\mathbb{R}^3$. The chart domain is the hemisphere

$$U_\alpha = H_3^+ = \{\mathbf{x} : \|\mathbf{x}\| = 1 \text{ and } x_3 > 0\}.$$

The chart map $x_\alpha = P_3|_{H_3^+}$ sending $\mathbf{x} \to (x_1, x_2)$ projects the hemisphere U_α onto an open disc D of radius $r = 1$ in the x_1, x_2-plane.

and is a $\mathcal{C}^\infty$ map $P_i : \mathbb{R}^3 \to \mathbb{R}^2$ whose inverse can be calculated explicitly. For instance when $i = 3$, we have

$$\left(P_3|_{H_3^+}\right)^{-1}(s_1, s_2) = \left(s_1, s_2, \sqrt{1 - (s_1^2 + s_2^2)}\right) \quad \text{for all } \mathbf{s} \in D,$$

which is clearly a $\mathcal{C}^\infty$ map from $D \subseteq \mathbb{R}^2$ onto $H_3^+ \subseteq \mathbb{R}^3$. The coordinate transition maps between two such charts, say $x_\alpha = P_3|_{H_3^+}$ and $y_\beta = P_2|_{H_2^-}$, are then

$$x_\alpha \circ y_\beta^{-1}(\mathbf{s}) = P_3\left(s_1, -\sqrt{1 - (s_1^2 + s_2^2)}, s_2\right) = \left(s_1, -\sqrt{1 - (s_1^2 + s_2^2)}\right)$$

and

$$y_\beta \circ x_\alpha^{-1}(\mathbf{s}) = P_2\left(s_1, s_2, +\sqrt{1 - (s_1^2 + s_2^2)}\right) = \left(s_1, +\sqrt{1 - (s_1^2 + s_2^2)}\right)$$

for $\mathbf{s} = (s_1, s_2) \in D$. Both are $\mathcal{C}^\infty$ maps from $\mathbb{R}^2 \to \mathbb{R}^2$.

A map $\mathbf{y} = f(\mathbf{x}) = \big(f_1(\mathbf{x}), \dots, f_n(\mathbf{x})\big)$ from $\mathbb{R}^m \to \mathbb{R}^n$ is "of class $\mathcal{C}^{(k)}$" on an open set $U \subseteq \mathbb{R}^m$ if all partial derivatives

$$D_x^\alpha f_i(\mathbf{x}) = D_{x_1}^{\alpha_1} \cdots D_{x_m}^{\alpha_m} f_i(\mathbf{x}) \quad \alpha = (\alpha_1, \dots, \alpha_m) \in \mathbb{Z}_+^m$$

of scalar components $f_i : U \to \mathbb{R}$ of the map $\mathbf{y} = f(\mathbf{x})$ exist and are continuous for $1 \le i \le m$ and for all m-tuples $\alpha = (\alpha_1, \dots, \alpha_m)$ with $|\alpha| = \alpha_1 + \cdots + \alpha_m \le k$. The "exponents" $\alpha = (\alpha_1, \dots, \alpha_m)$ are called *multi-indices*, and D_x^α is a partial derivative of *total degree* $|\alpha| = \alpha_1 + \cdots + \alpha_m$. By convention, $D_{x_i}^{\alpha_i} = I$ (the identity operator) if $\alpha_i = 0$, and in particular the multi-index $\alpha = (0, \dots, 0)$ yields $D_x^\alpha = I$ (the identity operator).

Once we have a "founding set" of $\mathcal{C}^\infty$-related charts $\mathcal{U} = \{(U_\alpha, x_\alpha) : \alpha \in I\}$ whose domains cover M, it determines a *maximal atlas* of consistent differentiably related charts on M.

> *The* MAXIMAL ATLAS $\overline{\mathcal{U}}$ *determined by* $\mathcal{U}$ *consists of all charts* (ϕ, W) *with* $\phi : W \to \mathbb{R}^m$, *as in (4.1), such that* (ϕ, W) *is differentiably related to all the founding charts* (U_α, x_α) *in* $\mathcal{U}$.

The following observation shows that this makes sense.

LEMMA 4.3. *All charts* (ϕ, W), (ψ, W') *in the maximal atlas* $\overline{\mathcal{U}}$ *are* $\mathcal{C}^{(k)}$-*related to each other, so coordinate transition maps* $\phi \circ \psi^{-1}$ *and* $\psi \circ \phi^{-1}$ *from* $\mathbb{R}^m \to \mathbb{R}^m$ *are* $C^{(k)}$ *wherever the chart domains* W, W' *overlap.*

PROOF. This follows from the classical chain rule: if $p \in W \cap W'$ and (U_α, x_α) is a founding chart in $\overline{\mathcal{U}}$ that contains p, we can split

$$\phi \circ \psi^{-1} = (\phi \circ x_\alpha^{-1}) \circ (x_\alpha \circ \psi^{-1})$$

at points near p. The factors $(\phi \circ x_\alpha^{-1})$, $(x_\alpha \circ \psi^{-1})$ are both $\mathcal{C}^{(k)}$ maps between copies of $\mathbb{R}^m$ because ϕ, ψ are $\mathcal{C}^\infty$-related to the founding chart (U_α, x_α). □

The maximal atlas $\overline{\mathcal{U}}$ in Example 4.2 was determined by the six founding charts, but $\overline{\mathcal{U}}$ includes many other charts. For instance, determinations of spherical coordinates can be defined on various open sets $U \subseteq S^2$ that avoid the north and south poles $N = (0, 0, 1)$ and $S = (0, 0, -1)$ to obtain admissible coordinate charts in the maximal atlas. For example, spherical coordinates (θ, ϕ) can be defined on the open hemisphere H_1^+ as shown in Figure 4.3: If $x_1 > 0$ for $\mathbf{x} = (x_1, x_2, x_3) \in S^2$, we can define a chart (z_γ, U_γ) on $U_\gamma = H_1^+$ by taking

$$(\theta, \phi) = z_\gamma(\mathbf{x}) = \left(\arctan\left(\frac{x_2}{x_1}\right), \arcsin(x_3) \right),$$

so the transition map from $(U_\alpha, x_\alpha) = (P_3, H_3^+)$ to (z_γ, U_γ) with $U_\gamma = H_1^+$ is

$$\begin{aligned}(\theta, \phi) = z_\gamma \circ x_\alpha^{-1}(s_1, s_2) &= z_\gamma\left(s_1, s_2, \sqrt{1 - (s_1^2 + s_2^2)}\right) \\ &= \left(\arctan\left(\frac{s_2}{s_1}\right), \arcsin\sqrt{1 - (s_1^2 + s_2^2)} \right).\end{aligned}$$

The inverse map can be handled similarly; both are $\mathcal{C}^\infty$ from $\mathbb{R}^2 \to \mathbb{R}^2$.

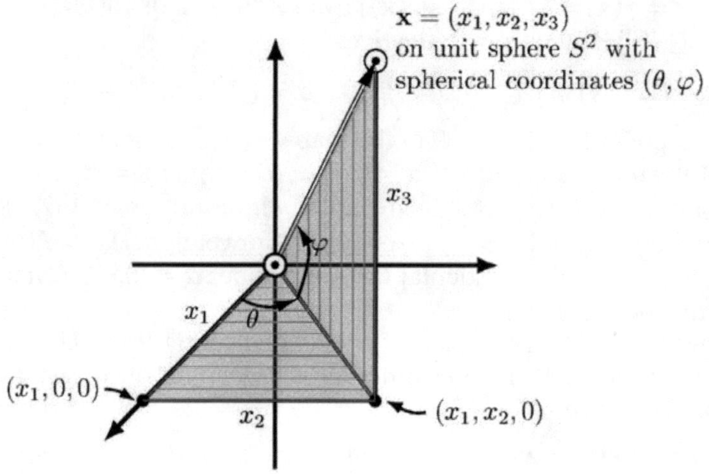

FIGURE 4.3. Spherical coordinates (θ, ϕ) on the two-dimensional sphere $S^2 = \{\mathbf{x} \in \mathbb{R}^3 : x_1^2 + x_2^2 + x_3^2 = 1\}$ are assigned as shown. We have chosen a somewhat non-standard definition of ϕ, one that agrees with the way latitude is assigned on the surface of the Earth. Both angles are ambiguous up to a multiple of 2π radians ($= 360°$); to get definite values, our convention restricts $-\pi < \theta < \pi$ and $-\pi < \phi < \pi$. The angles θ and ϕ cannot be defined at the north pole $N = (0, 0, 1)$ or south pole $S = (0, 0, -1)$ on the sphere.

EXAMPLE 4.4 (Stereographic Projections on S^2). Actually, the differentiable structure of the two-dimensional sphere S^2, and even that of the n-dimensional sphere

$$S^n = \{\mathbf{x} \in \mathbb{R}^{n+1} : x_1^2 + \cdots + x_{n+1}^2 = 1\} \quad \text{in } \mathbb{R}^n ,$$

can be determined by just two founding charts that impose Euclidean coordinates on the "punctured spheres" $U_+ = S^n \sim \{S\}$ and $U_- = S^n \sim \{N\}$ obtained by deleting the south pole $S = (0, \ldots, -1)$ or the north pole $N = (0, \ldots, 1)$ and mapping these open subsets of S^n by stereographic projection onto the n-dimensional tangent planes passing through the opposite pole N or S. Stereographic projection will be discussed at some length in Exercise 5.21 of Chapter 5, so we won't say more about it here.

Smooth Functions and Mappings on Manifold *M*. Once a manifold M is equipped with a differentiable structure (a maximal atlas of differentiably related covering charts), we can define the following:

1. *Smooth Functions* $f : M \to \mathbb{R}$.
2. *Smooth Parametrized Curves* $\gamma : \mathbb{R} \to M$.
3. *Smooth Mappings* $\phi : M \to N$ between differentiable manifolds M and N.

DEFINITION 4.5. A function $f : M \to \mathbb{R}$ is of *class* $\mathcal{C}^\infty$, indicated by writing $f \in \mathcal{C}^\infty(M)$, if $f \circ x_\alpha^{-1} : V_\alpha \to \mathbb{R}$ is $\mathcal{C}^\infty$ for every chart — i.e., f is smooth (of class $\mathcal{C}^\infty$ on M) if it is smooth in the classical sense when described in local chart coordinates. The same definition, applied to any open set $U \subseteq M$, determines the space $\mathcal{C}^\infty(U)$ of *smooth scalar valued functions* on U. These are not only infinite-dimensional vector spaces, but they are also associative algebras since they are closed under forming pointwise products $(f \cdot h)(u) = f(u)h(u)$, as well as scalar multiples $\lambda \cdot f$, sums $f_1 + f_2$, and linear combinations $\sum_{i=1}^r c_i f_i$. By the classical chain rule, smoothness of $f : M \to N$ does not depend on any particular choice of local coordinates.

We now call attention to certain special families of functions associated with a base point $p \in M$.

DEFINITION 4.6 (The Local Algebra $\mathcal{C}^\infty(p)$). The *local algebra* $\mathcal{C}^\infty(p)$ at a base point $p \in M$ is the associative algebra of $\mathcal{C}^\infty$ functions defined at and near p. A typical element in $\mathcal{C}^\infty(p)$ is therefore a pair (f, U) involving an open neighborhood U of p and a $\mathcal{C}^\infty$ function $f : U \to \mathbb{R}$. We then define $(+)$, pointwise product $(\cdot)$, and scaling operations

$$\begin{aligned}(f_1, U_1) + (f_2, U_2) &= (f_1 + f_2, U_1 \cap U_2)\\ (f_1, U_1) \cdot (f_2, U_2) &= (f_1 \cdot f_2, U_1 \cap U_2) \quad \text{(pointwise product)}\\ \lambda \cdot (f, U) &= (\lambda \cdot f, U) \quad \text{for all } \lambda \in \mathbb{R}\end{aligned}$$

on $\mathcal{C}^\infty(p)$. The zero element in this vector space is the pair $\mathbf{0} = (0, M)$, and $\mathbf{1} = (1, M)$ is the multiplicative identity element. Note that $\mathcal{C}^\infty(p)$ includes the algebra $\mathcal{C}^\infty(M)$ of globally defined $\mathcal{C}^\infty$ functions on M, as well as $\mathcal{C}^\infty(U)$ for any open set U that contains p.

DEFINITION 4.7. A map $\phi : M \to N$ between differentiable manifolds M, N is *smooth near* p if it becomes a $\mathcal{C}^\infty$ map from $\mathbb{R}^m \to \mathbb{R}^n$ when described in local coordinates, so that

$$y_\beta \circ \phi \circ x_\alpha^{-1} : \mathbb{R}^m \to \mathbb{R}^n \quad \text{is } \mathcal{C}^\infty \text{ in the classical sense}$$

for all charts (U_α, x_α) about p and (U_β, y_β) about $q = \phi(p)$. The map is $\mathcal{C}^\infty$ on an open subset (or on all of M) if it is $\mathcal{C}^\infty$ near each base point in U.

DEFINITION 4.8. A *parametrized curve* $\gamma : \mathbb{R} \to M$ is of class $\mathcal{C}^{(k)}$ if, for all charts (U_α, x_α) on M, $x_\alpha \circ \gamma(t)$ is a $\mathcal{C}^{(k)}$-map from $\mathbb{R} \to \mathbb{R}^m$ on some open interval $(a, b) \subseteq \mathbb{R}$. It is of class $\mathcal{C}^{(k)}$ on a closed interval $[a, b]$ if there is a slightly larger interval $(a - \epsilon, b + \epsilon)$ on which γ is defined and of class $\mathcal{C}^{(k)}$. Thus, in local coordinates, γ becomes a classical $\mathcal{C}^{(k)}$ curve

$$\tilde{\gamma}(t) = x_\alpha \circ \gamma(t) = (x_1(t), \dots, x_m(t))$$

from $\mathbb{R} \to \mathbb{R}^m$ whose scalar components $x_k(t)$ are differentiable of class $\mathcal{C}^{(k)}$.

The classical derivative $d\gamma/dt$ of a smooth curve $\gamma : [a,b] \to \mathbb{R}^m$ is a vector in $\mathbb{R}^n$:

$$\frac{d\gamma}{dt}(t_0) = \lim_{\Delta t \to 0} \frac{\Delta\gamma}{\Delta t} = \lim_{\Delta t \to 0} \frac{\gamma(t_0 + \Delta t) - \gamma(t_0)}{\Delta t} \tag{4.2}$$
$$= \left(\frac{dx_1}{dt}(t_0), \dots, \frac{dx_m}{dt}(t_0)\right).$$

But this makes no sense for smooth curves $\gamma : [a,b] \to M$ with values in a manifold; M is not a vector space, and differences of $\Delta\gamma$ of points in M are undefined. (How would you interpret the difference quotients $\Delta\gamma/\Delta t$ in (4.2) for a curve that lives within some smooth curved hypersurface M in $\mathbb{R}^n$ without making any reference to the surrounding space $\mathbb{R}^n$?) We can, however, make sense of $\gamma'(t_0)$ as an *operator* $\gamma'(t_0) : \mathcal{C}^\infty(p) \to \mathbb{R}$ that acts as a directional derivative on the local algebra of $\mathcal{C}^\infty$ functions defined near $p = \gamma(t_0)$, letting

$$\langle \gamma'(t_0), f \rangle = \frac{d}{dt}\{f \circ \gamma(t)\}\big|_{t=t_0}. \tag{4.3}$$

The operator $\gamma'(t_0)$ sends f to the time-derivative of the values $f(\gamma(t))$ seen as t increases past t_0; it is a linear operator from $\mathcal{C}^\infty(p) \to \mathbb{R}$.

Given a chart (U_α, x_α) about p, the action of $\gamma'(t_0)$ on functions can be computed by writing $f \circ \gamma(t)$ as a composite $(f \circ x_\alpha^{-1}) \circ (x_\alpha \circ \gamma) : \mathbb{R} \to \mathbb{R}^m \to \mathbb{R}$ and applying the classical chain rule. If $x_\alpha \circ \gamma(t) = (x_1(t), \dots, x_m(t))$, we get

$$\begin{aligned}
\langle \gamma'(t_0), f \rangle &= \frac{d}{dt}\{f \circ \gamma(t)\}\big|_{t=t_0} \\
&= \frac{d}{dt}\{(f \circ x_\alpha^{-1}) \circ (x_\alpha \circ \gamma)\}\big|_{t=t_0} \\
&= \sum_{i=1}^m D_{x_i}(f \circ x_\alpha^{-1})(x_\alpha \circ \gamma(t_0)) \cdot \frac{dx_i}{dt}(t_0) \\
&= \sum_{i=1}^m \left(D_{x_i}(f \circ x_\alpha^{-1})\right)(x_\alpha(p)) \cdot \frac{dx_i}{dt}(t_0) \\
&= \sum_{i=1}^m \frac{\partial f}{\partial x_i}(p) \cdot \frac{dx_i}{dt}(t_0).
\end{aligned} \tag{4.4}$$

NOTATION. Here, as in earlier chapters, we use "bracket notation" $\langle \ell, v \rangle$ to indicate the result $\ell(v)$ when we combine a vector $v \in V$ with a dual vector $\ell \in V^*$. This "bracket notation" turns out to be a good idea. ○

DEFINITION 4.9. Any linear map $\ell : \mathcal{C}^\infty(p) \to \mathbb{R}$ with the property

$$\langle \ell, f \cdot h \rangle = \langle \ell, f \rangle \cdot h(p) + f(p) \cdot \langle \ell, h \rangle$$

is referred to as a *derivation* on the local algebra.

If $\gamma : (a,b) \to M$ is a smooth curve passing through $p \in M$ when $t = t_0$, the directional derivative along γ, $\gamma'(t_0) : \mathcal{C}^\infty(p) \to \mathbb{R}$, is an example of a derivation on $\mathcal{C}^\infty(p)$. With some effort, it can be shown that *all* derivations on the local algebra $\mathcal{C}^\infty(p)$ arise in this manner, but we omit these technical

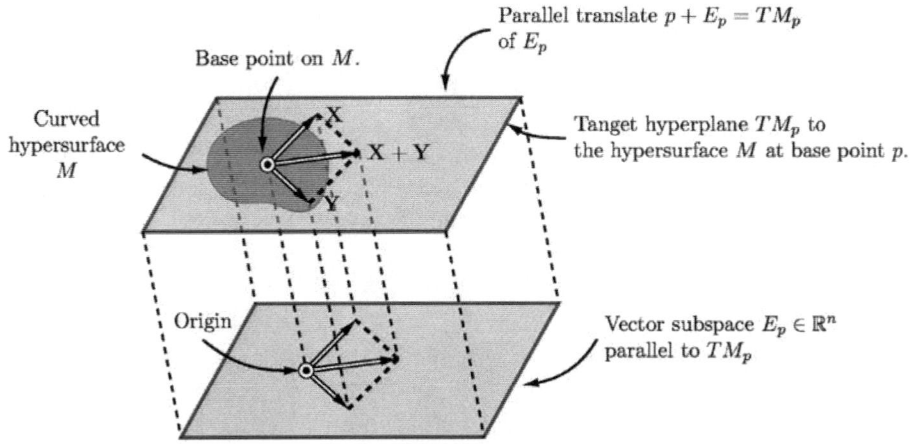

FIGURE 4.4. The tangent space TM_p to a smooth m-dimensional hypersurface $M \subseteq \mathbb{R}^n$ is a translate $p + E$ of some m-dimensional vector subspace $E \subseteq \mathbb{R}^n$. Elements $\mathbf{v} = (p, \mathbf{a})$, $\mathbf{w} = (p, \mathbf{b})$ in TM_p are indicated by arrows attached to the base point p. Scaling and vector addition in TM_p are accomplished by parallel transport transferring arrows back to the corresponding vectors in E, which can be added via the usual parallelogram law for vector addition. This makes TM_p an m-dimensional vector space.

details. Given this, we arrive at the modern definition of tangent vector and tangent space for a differentiable manifold M.

Tangent Vectors and the Tangent Spaces $\mathbf{TM}_p$. Classically, manifolds were taken to be smooth hypersurfaces M of some dimension m embedded in a larger Euclidean space $\mathbb{R}^n$. Elements of the tangent space TM_p at a base point $p \in M$ were "tangent vectors," thought of as pairs $(p, \mathbf{a})$ consisting of a base point $p \in M$ and an arrow $\mathbf{a} \in \mathbb{R}^m$ attached to p. Vectors in TM_p make up an m-dimensional hyperplane attached to M at p, which is a translate $\text{TM}_p = p + E$ centered at p of some m-dimensional vector subspace $E \subseteq \mathbb{R}^n$. Algebraic structure — addition and scaling operations on tangent vectors in TM_p — was introduced by translating arrows $p + \mathbf{a}$, $p + \mathbf{b}$ attached to p back to vectors $\mathbf{a}, \mathbf{b}$ in the vector subspace E, where they can be added or scaled. Their sum $\mathbf{a} + \mathbf{b}$ was then translated back to the pair $(p, \mathbf{a} + \mathbf{b})$ at p. In effect, sums and scalar multiples of tangent vectors at p were defined to be

$$(p, \mathbf{a}) + (p, \mathbf{b}) = (p, \mathbf{a} + \mathbf{b})$$
$$\lambda \cdot (p, \mathbf{a}) = (p, \lambda \cdot \mathbf{a}).$$

These operations are *not* defined for tangent vectors attached to different base points in M.

Some Objections: Finding the Right Definition.

1. None of this makes sense if we regard M as our entire universe of discourse and are forbidden to refer to objects lying in some mythical all-encompassing Euclidean space.
2. A more intrinsic approach (still regarding M as an embedded hypersurface in $\mathbb{R}^n$) is to recall that the velocity vector of a moving point in $\mathbb{R}^n$ whose position is $\gamma(t)$ at time t is given by
$$\frac{d\gamma}{dt}(t_0) = \lim_{\Delta t\to 0} \frac{\Delta\gamma}{\Delta t} = \lim_{\Delta t\to 0} \frac{\gamma(t_0+\Delta t)-\gamma(t_0)}{\Delta t}$$
(derivative taken in $\mathbb{R}^n$). When this velocity vector is viewed as an arrow attached to $p = \gamma(t_0)$ in M, it always lies in the classical tangent space to M at p because $\gamma(t)$ remains within M at all times. If we interpret tangent vectors at p as velocity vectors for curves passing through p, this defines tangent vectors in terms of objects intrinsic to M — smooth parametrized curves in M. But the derivative $d\gamma/dt$ is a limit of difference quotients $\Delta\gamma/\Delta t$, which can't be computed without referring to the surrounding space $\mathbb{R}^n$.
3. The tangent space to M at p is supposed to be a vector space. If smooth curves γ_1, γ_2 in M pass through p when $t = t_0$, it is not immediately obvious how to find a curve $\eta(t)$ such that $\eta(t_0) = p$ and $\eta'(t_0)$ is the sum $\gamma_1'(t_0) + \gamma_2'(t_0)$ of the tangent vectors. There is also an ambiguous relationship between curves through p and tangent vectors at p since different curves through p can determine the same classical tangent vector $d\gamma/dt$.

Another approach is already implicit in these comments. We shall view tangent vectors $\gamma'(t_0)$ as "directional derivatives" of functions $f : M \to \mathbb{R}$ defined near $p = \gamma(t_0)$, obtained by computing the time derivative of the values of f seen by the moving point $\gamma(t)$. From this point of view, $\gamma'(t_0)$ becomes a *linear operator*
$$\gamma'(t_0) : f \to \langle\gamma'(t_0), f\rangle = \frac{d}{dt}\{f(\gamma(t))\}\big|_{t=t_0}$$
acting on elements f in the local algebra $\mathcal{C}^\infty(p)$ defined in (4.3). The objects $f, \gamma(t)$, and $\gamma'(t_0)$ are now defined on M without assuming that M lives within some "universal Euclidean space."

DEFINITION 4.10. If M is a smooth manifold and $p \in M$ is a base point, *tangent vectors* at p are regarded as the derivations on the local algebra $\mathcal{C}^\infty(p)$, and the *tangent space* TM_p at p is the set of all such derivations. It is a vector space because linear combinations of derivations on $\mathcal{C}^\infty(p)$ are again derivations.

Derivations on $\mathcal{C}^\infty(p)$ determined by directional derivatives along curves in M have the following properties:

1. DERIVATION PROPERTY. Directional derivatives $\gamma'(t_0)$ along smooth curves through p act as linear maps from $\mathcal{C}^\infty(p)$ to $\mathbb{R}$ — i.e., they are *linear functionals* in the dual space V^* of the ∞-dimensional vector

space $V = \mathcal{C}^\infty(p)$. The classical product formula for derivatives in one variable,

$$\frac{d}{dt}\{(f \cdot h)\}(t) = \frac{df}{dt} \cdot h(t) + f(t) \cdot \frac{dh}{dt},$$

shows that directional derivatives along a curve are also *derivations on* $\mathcal{C}^\infty(p)$ because we have

$$(f \cdot h)(\gamma(t)) = \big(f \circ \gamma(t)\big) \cdot \big(h \circ \gamma(t)\big)$$

for all t. Then

$$\begin{aligned}\langle \gamma'(t_0), f \cdot h \rangle &= \frac{d}{dt}\{(f \circ \gamma(t)) \cdot (h \circ \gamma(t))\}\big|_{t=t_0} \\ &= \frac{d}{dt}\{f \circ \gamma(t)\}\big|_{t=t_0} \cdot h(p) + f(p) \cdot \frac{d}{dt}\{h \circ \gamma(t)\}\big|_{t=t_0} \qquad (4.5) \\ &= \langle \gamma'(t_0), f \rangle \cdot h(p) + f(p) \cdot \langle \gamma'(t_0), h \rangle\end{aligned}$$

for all $f, h \in \mathcal{C}^\infty(p)$.

2. Vector Space Structure. Sums $\ell_1 + \ell_2$ and scalar multiples $\lambda \cdot \ell$ of derivations are again derivations on $\mathcal{C}^\infty(p)$, so the tangent space TM_p is a vector space over $\mathbb{R}$.
3. Local Operators Property. Directional derivatives $\gamma'(t_0)$ are *local operators* on $\mathcal{C}^\infty(p)$: the outcome $\langle \gamma'(t_0), f \rangle$ depends only on the behavior of f in the immediate vicinity of $p = \gamma(t_0)$. In particular, if $f \equiv h$ on any open neighborhood of p, then $\langle \gamma'(t_0), f \rangle = \langle \gamma'(t_0), h \rangle$, and if $f \equiv 0$ near p, we get $\langle \gamma'(t_0), f \rangle = 0$.

Which curves passing through p when $t = 0$ determine the same operation on $\mathcal{C}^\infty(p)$? If we describe such curves $\gamma(t), \eta(t)$ in local coordinates, say with

$$\begin{aligned}\mathbf{x}(t) &= x_\alpha \circ \gamma(t) = \big(x_1(t), \dots, x_m(t)\big) \\ \mathbf{y}(t) &= x_\alpha \circ \eta(t) = \big(y_1(t), \dots, y_m(t)\big),\end{aligned}$$

then $\gamma'(t_0) = \eta'(t_0)$ are equal as operations on $\mathcal{C}^\infty(p)$ if and only if the "first-order terms" agree, so that

$$\begin{aligned}&\mathbf{x}(t_0) = \mathbf{y}(t_0) \\ &\frac{dx_1}{dt}(t_0) = \frac{dy_1}{dt}(t_0), \dots, \frac{dx_m}{dt}(t_0) = \frac{dy_m}{dt}(t_0).\end{aligned}$$

This is clear from equation (4.4), from which it also follows that the higher order derivatives of $\mathbf{x}(t)$ and $\mathbf{y}(t)$ are irrelevant.

These definitions are framed in a way that does not favor any particular system of local coordinates near p in the maximal atlas. In this sense, all definitions so far are coordinate-independent.

Differential Operators $\partial/\partial x_i$ on a Manifold. Given a chart (U_α, x_α) on M, certain derivations on $\mathcal{C}^\infty(p)$ correspond to the familiar partial derivatives of calculus. If p in M corresponds under the chart map to $\mathbf{p} = x_\alpha(p) = (p_1, \dots, p_m)$

in coordinate space $\mathbb{R}^m$, we can define parametrized straight lines $\tilde{\gamma}_i(t)$ in $\mathbb{R}^m$ such that $\tilde{\gamma}_i(0) = \mathbf{p}$ by

$$\tilde{\gamma}_i(t) = \mathbf{p} + t\mathbf{e}_i = (p_1, \dots, p_i + t, \dots, p_m) = \big(x_1(t), \dots, x_m(t)\big)$$

for $1 \le i \le m$, where $\mathbf{e}_1, \dots, \mathbf{e}_m$ are the standard basis vectors in $\mathbb{R}^m$. These curves in coordinate space can be transferred to $\mathcal{C}^\infty$ curves $\gamma_i(t) = x_\alpha^{-1} \circ \tilde{\gamma}_i(t)$ in M that pass through p when $t = 0$. Directional derivatives along these curves are then derivations $\gamma_i'(0)$ on $\mathcal{C}^\infty(p)$, which will hereafter be denoted by the symbol

$$\left(\frac{\partial}{\partial x_i}\bigg|_p\right) : \mathcal{C}^\infty(p) \to \mathbb{R} \quad \text{for } 1 \le i \le m.$$

These operators act on functions that live on the manifold M, while the classical partial derivatives D_{x_i} act on functions that live on coordinate space $\mathbb{R}^m$.

The effect of $\big(\partial/\partial x_i|_p\big)$ on an element $f \in \mathcal{C}^\infty(p)$ will often be indicated using "bracket notation," writing

$$\frac{\partial f}{\partial x_i}(p) = \left\langle \frac{\partial}{\partial x_i}\bigg|_p , f \right\rangle \quad \text{for } f \in \mathcal{C}^\infty(U_\alpha) \text{ and } 1 \le i \le m. \tag{4.6}$$

Note also that a chart (U_α, x_α) simultaneously determines correlated tangent vectors $X_p = \big(\partial/\partial x_i|_p\big)$ for $1 \le i \le m$ at *every* base point $p \in U_\alpha$. The result is a "field of tangent vectors" on the chart domain U_α for each i.

If we let the base point p vary within M, we obtain scalar-valued functions $\partial f/\partial x_i(u) = \langle \gamma_i'(0), f\rangle$ defined for all u in the chart domain U_α. A simplified version of (4.4) tells us how to compute these partial derivatives $\partial f/\partial x_i$ on the manifold.

LEMMA 4.11. *If (U_α, x_α) is a chart on M and $f \in \mathcal{C}^\infty(U_\alpha)$, the partial derivatives $\partial f/\partial x_i$ on the manifold are given by*

$$\frac{\partial f}{\partial x_i}(p) = (D_{x_i}(f \circ x_\alpha^{-1})) \circ x_\alpha(p) \quad 1 \le i \le n$$

for all $p \in U_\alpha$. Thus we get $\partial f/\partial x_i$ on a chart in M in three steps:

- *Transfer $f(u)$ in $\mathcal{C}^\infty(M)$ to a $\mathcal{C}^\infty$ function $(f \circ x_\alpha^{-1})(\mathbf{x})$ on $\mathbb{R}^m$.*
- *Take the classical partial derivative $D_{x_i}(f \circ x_\alpha^{-1})(\mathbf{x})$ on $\mathbb{R}^m$.*
- *Bring the result back to M to get*

$$\frac{\partial f}{\partial x_i}(u) = (D_{x_i}(f \circ x_\alpha^{-1}))(x_\alpha(u))$$

for $u \in U_\alpha$.

PROOF. Write $\mathbf{p} = x_\alpha(p)$ and $\mathbf{x}(t) = x_\alpha \circ \gamma_i(t) = \big(x_1(t), \dots, x_m(t)\big)$. Then factor $f \circ \gamma_i = (f \circ x_\alpha^{-1}) \circ (x_\alpha \circ \gamma_i)$ to get maps $\mathbb{R} \to \mathbb{R}^m \to \mathbb{R}$, to which we may apply the classical chain rule to get

$$\begin{aligned}
\frac{\partial f}{\partial x_i}(p) &= \left\langle \frac{\partial}{\partial x_i}\bigg|_p , f \right\rangle = \langle \gamma_i'(0), f\rangle \\
&= \frac{d}{dt}\{f \circ \gamma_i(t)\}\Big|_{t=0} = \frac{d}{dt}\{(f \circ x_\alpha^{-1}) \circ (x_\alpha(\gamma_i(t)))\}\big|_{t=0} \\
&= \frac{d}{dt}\{(f \circ x_\alpha^{-1})(p_1, \dots, p_i + t, \dots, p_m)\}\big|_{t=0} \\
&= \sum_{j=1}^{m} \left[(D_{x_j}(f \circ x_\alpha^{-1}))(p_1, \dots, p_i + t, \dots, p_m) \cdot \frac{dx_j}{dt}(t) \right]\bigg|_{t=0} \\
&= 0 + \cdots + 1 \cdot D_{x_i}(f \circ x_\alpha^{-1})(\mathbf{p}) + \cdots + 0 = D_{x_i}(f \circ x_\alpha^{-1}) \circ x_\alpha(p)
\end{aligned}$$

because $x_\alpha \circ \gamma_i(t) = (p_1, \dots, p_i + t, \dots, p_m)$ if $\mathbf{p} = x_\alpha(p) = x_\alpha \circ \gamma(0)$. □

We will often invoke the following example.

EXAMPLE 4.12. If (U_α, x_α) is a chart on a differentiable manifold, the chart map x_α can be written as

$$\mathbf{x} = x_\alpha(u) = \big(X_1(u), \dots, X_m(u)\big),$$

whose components $X_k(u)$ are scalar-valued $\mathcal{C}^\infty$ functions $X_k : U_\alpha \to \mathbb{R}$ on the chart domain U_α. It is amusing to compute $\partial X_k/\partial x_j$ on U_α as an exercise in understanding how the notation works. We claim that

$$\frac{\partial X_k}{\partial x_j}(u) \equiv \delta_{kj} \quad \text{for all } u \in U_\alpha \text{ and all } j, k \tag{4.7}$$

where δ_{kj} is the *Kronecker delta symbol*, equal to 1 if $k = j$ and $= 0$ when $j \neq k$.

DISCUSSION. Since

$$\mathbf{x} = (x_1, \dots, x_m) = x_\alpha(x_\alpha^{-1}(\mathbf{x})) = \big(X_1(x_\alpha^{-1}(\mathbf{x})), \dots, X_m(x_\alpha^{-1}(\mathbf{x}))\big),$$

we have $X_k \circ x_\alpha^{-1}(\mathbf{x}) \equiv x_k$ for all $\mathbf{x}$ in the open set $V_\alpha = x_\alpha(U_\alpha) \subseteq \mathbb{R}^m$. Applying Lemma 4.11 and standard facts from calculus, we get

$$\begin{aligned}
\frac{\partial X_k}{\partial x_j}(u) &= \left\langle \frac{\partial}{\partial x_j}\bigg|_u , X_k \right\rangle = \big(D_{x_j}(X_k \circ x_\alpha^{-1})\big)(x_\alpha(u)) \\
&= D_{x_j}(x_k)\big|_{\mathbf{x}=x_\alpha(u)} = \delta_{kj}
\end{aligned}$$

for all $u \in U_\alpha$. ○

If (U_α, x_α) is a chart on M and $p \in M$, we can evaluate the directional derivative of a function $f \in \mathcal{C}^\infty(p)$ along a smooth curve $\gamma(t)$ through p by passing the problem over to a coordinate space where the outcome can be determined using the familiar tools of multivariate calculus. The final result is an easily applied formula involving the functions $\partial f/\partial x_i$ on chart domains in M.

COROLLARY 4.13. *Let (U_α, x_α) be a chart on M, p a base point in U_α, and $f \in \mathcal{C}^\infty(p)$. If $\gamma(t)$ is a $\mathcal{C}^\infty$ curve such that $\gamma(0) = p$, whose description in local coordinates has the form*

$$x_\alpha(\gamma(t)) = \big(x_i(t), \dots, x_m(t)\big) \quad \textit{with } \mathcal{C}^\infty \textit{ coefficients } x_i(t),$$

then the tangent vector $X_p = \gamma'(0)$ *determined by differentiating along* $\gamma(t)$ *is given by*

$$\begin{aligned}\langle X_p, f\rangle = \langle \gamma'(0)\,,\, f\rangle &= \sum_{j=1}^{m} \frac{dx_j}{dt}(0)\cdot \frac{\partial f}{\partial x_j}(p)\\ &= \Big\langle \sum_{j=1}^{m} \frac{dx_j}{dt}(0)\cdot\Big(\frac{\partial}{\partial x_j}\Big|_p\Big), f\Big\rangle\end{aligned}$$

for all $f \in \mathcal{C}^\infty(p)$*. Hence*

$$X_p = \sum_{j=1}^{m} \frac{dx_j}{dt}(0)\cdot\Big(\frac{\partial}{\partial x_j}\Big|_p\Big) \quad \textit{in } \mathrm{TM}_p. \tag{4.8}$$

PROOF. After factoring $f \circ \gamma = (f \circ x_\alpha^{-1}) \circ (x_\alpha \circ \gamma)$, apply the chain rule to get

$$\begin{aligned}\langle \gamma'(t), f\rangle = \frac{d}{dt}\{f\circ\gamma(t)\} &= \frac{d}{dt}\{f\circ x_\alpha^{-1}(x_1(t),\dots,x_m(t))\}\\ &= \sum_{j=1}^{m} \frac{dx_j}{dt}(t)\cdot\frac{\partial f}{\partial x_j}(\gamma(t)).\end{aligned}$$

Then set $t = 0$. □

COROLLARY 4.14. *If* (U_α, x_α) *is a chart on M, the vectors*

$$\mathfrak{X} = \Big\{\Big(\frac{\partial}{\partial x_1}\Big|_p\Big), \dots, \Big(\frac{\partial}{\partial x_m}\Big|_p\Big)\Big\}$$

are a basis for the tangent space TM_p *for every* $p \in U_\alpha$*. In particular,* $\dim(\mathrm{TM}_p) = m$*, so the space of derivations on* $\mathcal{C}^\infty(p)$ *is finite dimensional even though the local algebra* $\mathcal{C}^\infty(p)$ *itself is infinite dimensional.*

PROOF. With some effort, one can prove that every local derivation $D : \mathcal{C}^\infty(p) \to \mathbb{R}$ is a directional derivative $\gamma'(0)$ along some (not necessarily unique) smooth curve such that $\gamma(0) = p$. The identity (4.8) shows that the vectors in $\mathfrak{X}$ span TM_p. They are also independent: if there are c_i such that $0 = \sum_{j=1}^{m} c_i(\partial/\partial x_i|_u)$ for all $u \in U_\alpha$, we may apply this functional on $\mathcal{C}^\infty(U_\alpha)$ to each of the scalar components $X_k(u)$ of the chart map $x_\alpha(u) = (X_1(u), \dots, X_m(u))$ to get

$$0 = \sum_{i=1}^{m} c_i\Big\langle\frac{\partial}{\partial x_i}\Big|_u, X_k\Big\rangle = \sum_{i=1}^{m} c_i\cdot\delta_{ik} = c_k \quad \text{for } k = 1, 2, \dots, m.$$

Thus, $\mathfrak{X}$ is a basis for TM_p at every base point $p \in U_\alpha$. □

Change of Coordinates. At each point u in the chart domain U_α, a chart (U_α, x_α) determines basis vectors $(\partial/\partial x_i|_u)$ and partial derivatives $\partial f/\partial x_i(u)$. A different chart (U_β, y_β) about p will assign other basis vectors $(\partial/\partial y_i|_u)$ and partial derivatives $\partial/\partial y_i(u)$ at points where the domains overlap. We often need to pass descriptions of these constructs from one coordinate system to another.

Given overlapping charts (U_α, x_α) and (U_β, y_β) on M, the scalar components in $x_\alpha = (X_1, \dots, X_m)$ and $y_\beta = (Y_1, \dots, Y_m)$ are $\mathcal{C}^\infty$ functions on the chart domains. For future reference, we determine the *coordinate transition maps* $x_\alpha \circ y_\beta^{-1}$ and $y_\beta \circ x_\alpha^{-1}$ shown in Figure 1.1. They are defined on the open sets in $\mathbb{R}^m$ that correspond to $U = U_\alpha \cap U_\beta$ in M under the chart maps x_α and y_β. A point $\mathbf{x} \in x_\alpha(U)$ maps to

$$\mathbf{y} = (y_1, \dots, y_n) = y_\beta \circ x_\alpha^{-1}(\mathbf{x}) = \big(Y_1 \circ x_\alpha^{-1}(\mathbf{x}), \dots, Y_m \circ x_\alpha^{-1}(\mathbf{x})\big)$$

in $y_\beta(U)$, so $y_j = Y_j \circ x_\alpha(\mathbf{x})$. The vector-valued map $\mathbf{y} = F(\mathbf{x}) = y_\beta \circ x_\alpha^{-1}(\mathbf{x})$ from $\mathbb{R}^m$ to $\mathbb{R}^m$ is $\mathcal{C}^\infty$, and its classical Jacobian matrix is

$$(4.9)\quad [DF(\mathbf{x})] = \big[D_{x_j}(y_i)\big] = \big[D_{x_j}(Y_i \circ x_\alpha^{-1})(\mathbf{x})\big] = \frac{\partial\big(Y_1 \circ x_\alpha^{-1}, \dots, Y_m \circ x_\alpha^{-1}\big)}{\partial(x_1, \dots, x_n)}$$

for $\mathbf{x} \in x_\alpha(U)$. In the reverse direction, we have

$$\mathbf{x} = (x_1, \dots, x_m) = G(\mathbf{y}) = x_\alpha \circ y_\beta^{-1}(\mathbf{y}) = \big(X_1 \circ y_\beta^{-1}, \dots, X_m \circ y_\beta^{-1}\big),$$

and this inverse map has Jacobian matrix

$$(4.10)\qquad [DG(\mathbf{y})] = \big[D_{y_j}(x_i)\big] = \big[D_{y_j}(X_i \circ y_\beta^{-1})(\mathbf{y})\big] \quad \text{for } \mathbf{y} \in y_\beta(U).$$

The transformation law (4.11) below will let us write derivatives $\partial f/\partial x_i$ computed with respect to one chart on M in terms of the partial derivatives $\partial f/\partial y_j$ for another chart. We will make heavy use of these laws. Note that the resulting formulas are cast entirely in terms of functions that live *on the manifold M*.

THEOREM 4.15 (Change of Variable for Partial Derivatives). *Let (U_α, x_α) and (U_β, y_β) be overlapping charts on a manifold M. For any $\mathcal{C}^\infty$ function f on M, we have*

$$(4.11)\qquad \frac{\partial f}{\partial x_i}(u) = \sum_{j=1}^m \frac{\partial f}{\partial y_j}(u) \cdot \frac{\partial Y_j}{\partial x_i}(u) \quad \textit{for all } u \in U_\alpha \cap U_\beta$$

where $Y_j \in \mathcal{C}^\infty(U_\beta)$ are the scalar components of the chart map $y_\beta = (Y_1, \dots, Y_m)$.

PROOF. The proof is a mildly strenuous exercise in applying the classical chain rule (and our definition of the operators $\partial/\partial x_i$ on the manifold). For $u \in U_\alpha \cap U_\beta$, we have

$$\frac{\partial f}{\partial x_i}(u) = \big(D_{x_i}(f \circ x_\alpha^{-1})\big) \circ x_\alpha(u) = \big[D_{x_i}\big((f \circ y_\beta^{-1}) \circ (y_\beta \circ x_\alpha^{-1})\big)\big] \circ x_\alpha(u).$$

Insertion of $y_\beta^{-1} \circ y_\beta$ is a crucial step that makes the map $f \circ x_\alpha^{-1} : \mathbb{R}^m \to \mathbb{R}$ a composite of smooth maps $\mathbb{R}^m \xrightarrow{\phi} \mathbb{R}^m \xrightarrow{F} \mathbb{R}$ between Euclidean spaces, with $F = f \circ y_\beta^{-1}$ and $\phi = y_\beta \circ x_\alpha^{-1}$. The classical chain rule can then be applied to obtain

$$D_{x_i}(F \circ \phi)(\mathbf{x}) = \sum_{j=1}^{m} D_{y_j}F(\phi(\mathbf{x})) \cdot D_{x_i}y_j(\mathbf{x})$$

where $\mathbf{y} = \phi(\mathbf{x}) = \big(y_1(\mathbf{x}), \dots, y_m(\mathbf{x})\big)$ maps $\mathbb{R}^m \to \mathbb{R}^m$. Now set $F = f \circ y_\beta^{-1}$ and $\mathbf{y} = \phi(\mathbf{x}) = y_\beta \circ x_\alpha^{-1}(\mathbf{x}) = \big(y_1(\mathbf{x}), \dots, y_m(\mathbf{x})\big)$ mapping $\mathbb{R}^m \to \mathbb{R}^m$ to get

$$\begin{aligned}
\frac{\partial f}{\partial x_i}(u) &= \sum_{j=1}^{m} \big[(D_{y_j}(f \circ y_\beta^{-1})) \circ (y_\beta \circ x_\alpha^{-1})\big](x_\alpha(u)) \cdot \big[D_{x_i}((y_\beta \circ x_\alpha^{-1})_j)\big](x_\alpha(u)) \\
&= \sum_{j=1}^{m} \big[((D_{y_j}(f \circ y_\beta^{-1})) \circ y_\beta)(u)\big] \cdot \big[D_{x_i}(Y_j \circ x_\alpha^{-1})\big](x_\alpha(u)) \\
&= \sum_{j=1}^{m} \frac{\partial f}{\partial y_j}(u) \cdot \frac{\partial Y_j}{\partial x_i}(u) \quad \text{(definition of } \partial f/\partial y_j \text{ and } \partial Y_j/\partial x_i\text{)}
\end{aligned}$$

on N. □

COROLLARY 4.16 (Change of Variable Formula for Operators). *The operators $\partial/\partial x_i$ and $\partial/\partial y_j$ determined by overlapping charts (U_α, x_α) and (U_β, y_β) on a manifold M transform in the following way:*

$$\left.\frac{\partial}{\partial x_i}\right|_u = \sum_{j=1}^{m} \frac{\partial Y_j}{\partial x_i}(u) \cdot \left.\frac{\partial}{\partial y_j}\right|_u \quad \textit{for } 1 \le i, j \le m \textit{ and } u \in U_\alpha \cap U_\beta. \tag{4.12}$$

These "change of variable" formulas are easy to remember: the Y_j and y_j should "cancel" when the formula is written correctly, leaving only terms that involve $\partial/\partial x_i$.

Given overlapping charts $(U_\alpha, x_\alpha), (U_\beta, y_\beta)$ on M, entries in the Jacobian matrices (4.9) and (4.10) are $\mathcal{C}^\infty$ scalar-valued functions on open subsets of coordinate space $\mathbb{R}^m$:

$$\left[\frac{\partial y_i}{\partial x_j}(\mathbf{x})\right] = \big[D_{x_j}(Y_i \circ x_\alpha^{-1})\big] \quad \text{at points } \mathbf{x} \in x_\alpha(U_\alpha) \subseteq \mathbb{R}^m$$

$$\left[\frac{\partial x_i}{\partial y_j}(\mathbf{y})\right] = \big[D_{y_j}(X_i \circ y_\beta^{-1})\big] \quad \text{at points } \mathbf{y} \in y_\beta(U_\beta) \subseteq \mathbb{R}^m.$$

These matrices are inverses of each other when evaluated at base points $\mathbf{x}, \mathbf{y}$ in $\mathbb{R}^m$ that correspond under the transition maps. This relation between Jacobian matrices becomes clearer if we move these matrix-valued functions from coordinate space $\mathbb{R}^m$ to the manifold M itself. The resulting matrix-valued functions *defined on M*

$$\left[\frac{\partial Y_i}{\partial x_j}\right] = \left[\frac{\partial y_i}{\partial x_j} \circ x_\alpha\right] = \left[\frac{\partial (Y_i \circ x_\alpha^{-1})}{\partial x_j} \circ x_\alpha\right] \quad \text{(by Lemma 4.11)}$$

$$\left[\frac{\partial X_i}{\partial y_j}\right] = \left[\frac{\partial x_i}{\partial y_j} \circ y_\beta\right] = \left[\frac{\partial (X_i \circ y_\beta^{-1})}{\partial y_j} \circ y_\beta\right] \quad \text{(Lemma 4.11 again)}$$

are then matrix-inverses of each other at every base point in $U_\alpha \cap U_\beta$.

EXERCISE 4.17. Prove that the transferred Jacobian matrices on M are inverses of each other at every base point in $U_\alpha \cap U_\beta$, so

$$\left[\frac{\partial Y_i}{\partial x_j}(u)\right] \cdot \left[\frac{\partial X_i}{\partial y_j}(u)\right] = I_{m\times m} \quad \text{for all } u \in U_\alpha \cap U_\beta\,. \tag{4.13}$$

HINT. By Example 4.12, $\partial Y_i/\partial y_j = \delta_{ij}$; rewrite $\partial/\partial y_j$ in terms of the $\partial/\partial x_k$ as in (4.12).

Vector Fields as Differential Operators on M. On M, or any open subset thereof, a *vector field* $\widetilde{X}$ is a map that assigns a tangent vector $X_p \in \mathrm{TM}_p$ at each $p \in M$. By Corollary 4.13, on any chart domain U_α, there are uniquely determined coefficients $c_i(u)$ such that

$$\widetilde{X}_u = \sum_{i=1}^{m} c_i(u) \cdot \left(\frac{\partial}{\partial x_i}\Big|_u\right) \quad \text{for } u \in U_\alpha.$$

These coefficients will change if we pass to another coordinate chart, but smoothness of the coefficients is always preserved.

LEMMA 4.18. *If $X_u = \sum c_i(u) \cdot (\frac{\partial}{\partial x_i}|_u)$ for u near base point $p \in U_\alpha$ when described in local coordinates (U_α, x_α), then in any other chart (U_β, y_β) containing p, we have*

$$X_u = \sum_{j=1}^{m} d_j(u) \cdot \left(\frac{\partial}{\partial y_j}\Big|_u\right)$$

with coefficients

$$d_j(u) = \sum_{k=1}^{m} c_k(u) \cdot \frac{\partial Y_j}{\partial x_k}(u)\,,$$

where the Y_j are the scalar components of $y_\beta = \big(Y_1(u), \dots, Y_m(u)\big)$.

PROOF. This is immediate from equation (4.11). □

A *smooth vector field* $\widetilde{X}$ is a field $p \mapsto X_p$ of tangent vectors on M whose description in local coordinates has $\mathcal{C}^\infty$ coefficients for every chart. It follows from Lemma 4.18 that smoothness of $\widetilde{X}$ on M has a coordinate-independent meaning because the scalar components $Y_j(u)$ of y_β and their derivatives $\partial Y_j/\partial x_k$ are smooth functions on M.

COROLLARY 4.19. *If $\widetilde{X}$ is a vector field on M whose description in local coordinates is $X_u = \sum c_i(u) \cdot (\frac{\partial}{\partial x_i}|_u)$ with coefficients that are $\mathcal{C}^\infty$ near p, then the same will be true on $U_\alpha \cap U_\beta$ for any other chart (U_β, y_β).*

The set of smooth vector fields on M is denoted $\mathcal{D}^{(1,0)}(M)$, which becomes a vector space over $\mathbb{R}$ if we define

$$(\widetilde{X} + \widetilde{Y})_p = X_p + Y_p \quad \text{and} \quad (\lambda \widetilde{X})_p = \lambda \cdot X_p \quad \text{for } p \in M.$$

This ∞-dimensional space is also a "$\mathcal{C}^\infty(M)$-module" because there is a natural action $\mathcal{C}^\infty(M) \times \mathcal{D}^{(1,0)}(M) \to \mathcal{D}^{(1,0)}(M)$ defined by pointwise multiplication,

$$(f \cdot \widetilde{X})_p = f(p) \cdot X_p \quad \text{for all } p.$$

Similarly, an action $\mathcal{D}^{(1,0)} \times \mathcal{C}^\infty(M) \to \mathcal{C}^\infty(M)$ is obtained by letting a vector field $\widetilde{X}$ act on functions in the following way:

$$\widetilde{X}f(u) = \langle X_u, f \rangle \quad \text{for all } u \in M. \tag{4.14}$$

When $\widetilde{X}f$ is described in local chart coordinates, $\widetilde{X}$ becomes an operator that acts on smooth functions on coordinate space $\mathbb{R}^m$ as a first-order partial differential operator with variable coefficients and no scalar term $a_0(u)I$. In fact, if (U_α, x_α) is a chart on M and we describe $\widetilde{X}$ in chart coordinates, we get

$$X_u = \sum_i c_i(u) \cdot \left(\frac{\partial}{\partial x_i}\bigg|_u \right) \quad \text{with } c_i(u) \in \mathcal{C}^\infty(U_\alpha),$$

and then

$$\begin{aligned} \widetilde{X}f(u) = \langle X_u, f \rangle &= \sum_{i=1}^m c_i(u) \cdot \left\langle \frac{\partial}{\partial x_i}\bigg|_u, f \right\rangle \\ &= \sum_{i=1}^m c_i(u) \cdot \frac{\partial f}{\partial x_i}(u) \end{aligned}$$

is in $\mathcal{C}^\infty(U_\alpha)$. When smooth vector fields are regarded as differential operators on M, they have their own "global derivation" property. In addition to being linear operators $\widetilde{X} : \mathcal{C}^\infty(M) \to \mathcal{C}^\infty(M)$, they are linear derivations on $\mathcal{C}^\infty(M)$ with

$$\widetilde{X}(f \cdot h)(u) = \widetilde{X}f(u) \cdot h(u) + f(u) \cdot \widetilde{X}h(u)$$

for $u \in M$ and $f, h \in \mathcal{C}^\infty(M)$. This follows directly from Definition 4.5. Another important property involves the *support sets* of functions f in $\mathcal{C}^\infty(M)$,

$$\begin{aligned} \operatorname{supp}(f) &= \text{Closure in } M \text{ of the set } \{u \in M : f(u) \neq 0\} \\ &= \text{Complement in } M \text{ of the open set } \{u \in M : f \equiv 0 \text{ near } u\}. \end{aligned}$$

PROPOSITION 4.20 (Reduction of Supports). *The action of a smooth vector field $\widetilde{X}$ can only decrease the support set* supp(f) *of a function $f \in \mathcal{C}^\infty(M)$ so that*

$$f \equiv 0 \text{ on an open set } U \subseteq M \Rightarrow \widetilde{X}(f) \equiv 0 \text{ on } U.$$

PROOF. If $f \equiv 0$ near p, write $\widetilde{X}$ in local chart coordinates to see $\widetilde{X}f \equiv 0$ near p. □

By looking at differences $f - h$, we see that

$$f \equiv h \text{ on an open set } U \subseteq M \text{ implies } \widetilde{X} f \equiv \widetilde{X} h$$

on U for all $f, h \in \mathcal{C}^\infty(M)$. Thus, smooth vector fields $\widetilde{X}$ are *local operators* on $\mathcal{C}^\infty(M)$: the value of $\widetilde{X} f$ at a point p is determined only by the behavior of f in the immediate vicinity of p.

The Differential of a Map Between Manifolds. If $\phi : M \to N$ is a $\mathcal{C}^\infty$ map, $p \in M$, and $q = \phi(p)$ in N, then under the action of ϕ, we have the following:

- Points $p \in M$ get moved forward to points $q = \phi(p)$ in N.
- Functions $f \in \mathcal{C}^\infty(N)$ get "pulled back" to functions $\phi^{\mathsf{T}} f = f \circ \phi$ in $\mathcal{C}^\infty(M)$.
- Tangent vectors X_p in TM_p get "pushed forward" to vectors $(d\phi)_p(X_p)$ in TN_q by the *differential of* ϕ at p, the linear map $(d\phi)_p$ from TM_p to TM_q defined as follows.

DEFINITION 4.21 (The Differential $d\phi_p : \mathrm{TM}_p \to \mathrm{TN}_{\phi(p)}$). For $X_p \in \mathrm{TM}_p$ and $f \in \mathcal{C}^\infty_N(\phi(p))$, we define $(d\phi)_p X_p$ in $\mathrm{TN}_{\phi(p)}$ to be the unique tangent vector Y_q at $q = \phi(p)$ such that

$$\langle Y_q, f \rangle = \langle\, (d\phi)_p X_p \,,\, f \,\rangle = \langle X_p, f \circ \phi \rangle$$

for all $\mathcal{C}^\infty$ functions f defined on N near $\phi(p)$.

It is obvious that $Y_q = (d\phi)_p X_p$ is a dervation on $\mathcal{C}^\infty_N(\phi(p))$, so it is in TN_q; it is also obvious that $(d\phi)_p$ is a linear map from $\mathrm{TM}_p \to \mathrm{TN}_{\phi(p)}$, which we call the "differential of ϕ at p." None of these definitions refer to coordinates on M or N.

To calculate the effect of $(d\phi)_p$, we impose local charts (U_α, x_α) about p and (U_β, y_β) about $q = \phi(p)$ which determine bases $\mathfrak{X} = \{(\partial/\partial x_i|_p)\}$ and $\mathfrak{Y} = \{(\partial/\partial y_j|_q)\}$ in TM_p and TN_q. Since $(d\phi)_p : \mathrm{TM}_p \to \mathrm{TN}_{\phi(p)}$ is linear, its action is completely determined once we know what it does to these basis vectors. The end result is a law similar to the change of coordinates rule in Corollary 4.16 to Theorem 4.15.

THEOREM 4.22 (Transformation Law for Tangent Vectors). *If $\phi : M \to N$ is a $\mathcal{C}^\infty$ map, $p \in M$, and $q = \phi(p)$, let (U_α, x_α) and (U_β, y_β) be charts about p and q respectively. If the scalar components of the chart map y_β on N are given by $y_\beta(v) = \big(Y_1(v), \dots, Y_n(v)\big)$ for $v \in U_\beta$, then*

$$(d\phi)_p\Big(\frac{\partial}{\partial x_i}\Big|_p\Big) = \sum_{j=1}^{n} \frac{\partial(Y_j \circ \phi)}{\partial x_i}(p) \cdot \Big(\frac{\partial}{\partial y_j}\Big|_{\phi(p)}\Big) \quad \textit{for } 1 \le i \le m. \tag{4.15}$$

PROOF. On any chart (U_β, y_β) in N, the scalar components of y_β have the property

$$\partial Y_i / \partial y_j \equiv \delta_{ij} \quad \text{(Kronecker delta)}$$

throughout the chart domain U_β. The image in $\mathrm{TN}_{\phi(p)}$ of any basis vector $(\partial/\partial x_i|_p)$ in TM_p can be written

$$(d\phi)_p\Big(\frac{\partial}{\partial x_i}\Big|_p\Big) = \sum_{j=1}^n c_j(q)\cdot\Big(\frac{\partial}{\partial y_j}\Big|_q\Big) \quad (q = \phi(p)).$$

By (4.7) the coefficients $c_j(q)$ can be recovered by bracketing with the scalar components Y_j of the chart map y_β, so

$$\begin{aligned} c_j(q) &= \Big\langle (d\phi)_p\Big(\frac{\partial}{\partial x_i}\Big|_p\Big),\, Y_j\Big\rangle \\ &= \Big\langle \Big(\frac{\partial}{\partial x_i}\Big|_p\Big),\, Y_j\circ\phi\Big\rangle \quad \text{(Definition 4.21 of } (d\phi)_p) \\ &= \frac{\partial(Y_i\circ\phi)}{\partial x_i}(p). \end{aligned}$$

That proves the identity (4.15). □

Note the formal cancellation of "y_j" and "Y_j" when the formula is written correctly. Note also that we recover the change of variable formula (Corollary 4.16) when $M = N$ and $\phi = \mathrm{id}_M$.

EXERCISE 4.23. Let $M \overset{\psi}{\to} N \overset{\phi}{\to} Q$ be $\mathcal{C}^\infty$ maps between manifolds. If $p \in M$ and $q = \psi(p)$ in N, explain why the composite $\phi\circ\psi : M \to Q$ is a $\mathcal{C}^\infty$ map and prove that

$$d(\phi\circ\psi)_p = (d\phi)_{\psi(p)}\circ(d\psi)_p$$

as maps $\mathrm{TM}_p \to \mathrm{TN}_{\psi(p)} \to \mathrm{TQ}_{\phi(\psi(p))}$.

Next, we show how to calculate the differential $(d\phi)_p : \mathrm{TM}_p \to \mathrm{TN}_{\phi(p)}$ of a smooth mapping between manifolds.

EXAMPLE 4.24. The complex variable "squaring map" $w = \phi(z) = z^2$ takes the form

$$u + iv = \phi(x,y) = (x+iy)^2 = (x^2-y^2) + i(2xy) = (x^2-y^2, 2xy)$$

in Cartesian coordinates when we write $z = x+iy$ and $w = u+iv$. This is a map $\phi : M \to N$ between two copies of the complex plane $\mathbb{C}$, each of which can be regarded as a $\mathcal{C}^\infty$ manifold whose structure is determined by a single global chart:

- For M, take (U_α, x_α) with $U_\alpha = M$ and $(x,y) = x_\alpha(z) = (X(z), Y(z))$ in $\mathbb{R}^2$, where $X(z) = x$ and $Y(z) = y$ if $z = x+iy$.
- For N, take (U_β, y_β) with $U_\beta = N$ and $(u,v) = y_\beta(w) = (U(w), V(w))$ in $\mathbb{R}^2$, where $U(w) = u$ and $V(w) = v$ if $w = u+iv$.

At a typical point $z = x+iy$ in M, the identity $(u,v) = y_\beta\circ\phi\circ x_\alpha^{-1}(x,y)$ forces the scalar components of the chart maps $x_\alpha = (X,Y)$ and $y_\beta = (U,V)$ to satisfy the relations

$$u = X^2(z) - Y^2(z) = x^2 - y^2$$
$$v = 2X(z)Y(z) = 2xy$$

where $w = u + iv = z^2$. The players involved are shown in the following diagram:

$$\begin{array}{ccccc} x+iy=z & \mathbb{C} & \xrightarrow{\ \ \phi\ \ } & \mathbb{C} & w=u+iv=z^2 \\ \downarrow & x_\alpha \downarrow & & \downarrow y_\beta & \downarrow \\ (X(z),Y(z))=(x,y) & \mathbb{R}^2 & \xrightarrow{\Phi = y_\beta\circ\phi\circ x_\alpha^{-1}} & \mathbb{R}^2 & (u,v)=(U(w),V(w)) \end{array}$$

where $U(w) = x^2 - y^2$ and $V(w) = 2xy$.

If $p \in M$ and $q = \phi(p) \in N$, the charts x_α, y_β determine basis vectors $(\partial/\partial x|_p), (\partial/\partial y|_p)$ and $(\partial/\partial u|_q), (\partial/\partial v|_q)$ in the tangent spaces $\mathrm{TM}_p, \mathrm{TN}_q$, which are two-dimensional over $\mathbb{R}$. To compute $(d\phi)_p$, we describe ϕ in these local coordinates. The $\mathcal{C}^\infty$ map $\Phi = y_\beta \circ \phi \circ x_\alpha^{-1} : \mathbb{R}^2 \to \mathbb{R}^2$ between coordinate spaces is

$$(u, v) = \Phi(x, y) = y_\beta \circ \phi \circ x_\alpha^{-1}(x, y) = y_\beta(\phi(x + iy)) = (x^2 - y^2\,, 2xy),$$

so its Jacobian matrix is

$$[D\Phi(x, y)] = \frac{\partial(u, v)}{\partial(x, y)} = \begin{bmatrix} 2x & -2y \\ 2y & 2x \end{bmatrix}.$$

If $p = x + iy$ in M, so $x_\alpha(p) = (X(p), Y(p)) = (x, y)$, basis vectors in TM_p are transformed to vectors in $\mathrm{TN}_{\phi(p)}$ as in (4.15):

$$(d\phi)_p\Big(\frac{\partial}{\partial x}\Big|_p\Big) = \frac{\partial(U\circ\phi)}{\partial x}(p)\cdot\Big(\frac{\partial}{\partial u}\Big|_{\phi(p)}\Big) + \frac{\partial(V\circ\phi)}{\partial x}(p)\cdot\Big(\frac{\partial}{\partial v}\Big|_{\phi(p)}\Big).$$

But on $U_\alpha \subseteq M$, we have $\partial X/\partial x \equiv 1, \partial X/\partial y \equiv 0$, etc. (courtesy of Example 4.12). Hence

$$\frac{\partial(U\circ\phi)}{\partial x} = \frac{\partial}{\partial x}(X^2 - Y^2) = 2X\cdot\frac{\partial X}{\partial x} - 2Y\cdot\frac{\partial Y}{\partial x} = 2X$$
$$\frac{\partial(V\circ\phi)}{\partial x} = \frac{\partial}{\partial x}(2XY) = 2Y\cdot\frac{\partial X}{\partial x} + 2X\cdot\frac{\partial Y}{\partial x} = 2Y\,,$$

so at $p = x + iy$ in M, we have

$$(4.16)\qquad \begin{aligned} (d\phi)_p\Big(\frac{\partial}{\partial x}\Big|_p\Big) &= 2X(p)\cdot\Big(\frac{\partial}{\partial u}\Big|_{\phi(p)}\Big) + 2Y(p)\cdot\Big(\frac{\partial}{\partial v}\Big|_{\phi(p)}\Big) \\ &= 2x\cdot\Big(\frac{\partial}{\partial u}\Big|_{\phi(p)}\Big) + 2y\cdot\Big(\frac{\partial}{\partial v}\Big|_{\phi(p)}\Big) \end{aligned}$$

where $\phi(p) = p^2$. As a numerical example, suppose $p = 2 + i$ in M, so $\phi(p) = (2 + i)^2 = 3 + 4i$, and (4.16) becomes

$$(d\phi)_p\Big(\frac{\partial}{\partial x}\Big|_p\Big) = 4\cdot\Big(\frac{\partial}{\partial u}\Big|_{\phi(p)}\Big) + 2\cdot\Big(\frac{\partial}{\partial v}\Big|_{\phi(p)}\Big).$$

A similar calculation yields

$$\begin{aligned}(d\phi)_p\Big(\frac{\partial}{\partial y}\Big|_p\Big) &= -2Y(p)\cdot\Big(\frac{\partial}{\partial u}\Big|_{\phi(p)}\Big) + 2X(p)\cdot\Big(\frac{\partial}{\partial v}\Big|_{\phi(p)}\Big)\\ &= -2\cdot\Big(\frac{\partial}{\partial u}\Big|_{\phi(p)}\Big) + 4\cdot\Big(\frac{\partial}{\partial v}\Big|_{\phi(p)}\Big).\end{aligned} \tag{4.17}$$

Given this action on basis vectors, the action of $(d\phi)_p : \mathrm{TM}_p \to \mathrm{TN}_{\phi(p)}$ on arbitrary tangent vectors is determined by linearity.

4.2. Cotangent Vectors and Differential Forms

As in Chapter 3 of LA I, the *dual space* V^* of a vector space V over $\mathbb{R}$ is the space of linear functionals on V — i.e., the linear maps $\ell : V \to \mathbb{R}$. If $\dim(V) < \infty$, then $\dim(V^*) = \dim(V)$, and any basis $\mathfrak{X} = \{e_1, \dots, e_n\}$ in V induces a *dual basis* $\mathfrak{X}^* = \{e_1^*, \dots, e_n^*\}$ in V^* that is determined by the property

$$\langle e_i^*, e_j\rangle = \delta_{ij} \quad \text{(Kronecker delta symbol).}$$

If $v = \sum_{i=1}^n c_i e_i$ in V, the dual functional e_k^* reads the kth coefficient c_k of v, so $\langle e_k^*, v\rangle = c_k$, and if $\ell = \sum_{j=1}^n d_j e_j^*$ in V^*, its coefficients can be found by bracketing ℓ with the basis vectors e_j to get

$$d_j = \Big\langle \sum_{k=1}^m d_k e_k^*, e_j\Big\rangle = \langle \ell, e_j\rangle \quad \text{for } 1 \le j \le n.$$

For differentiable manifolds, the tangent spaces TM_p and their dual spaces TM_p^* both play crucial roles in differential geometry.

DEFINITION 4.25. The *cotangent space* at $p \in M$ is the dual space TM_p^* to the tangent space TM_p. For reasons that will gradually emerge, this space is often referred to as the space of *1-forms*, or *rank-1 differential forms*, on TM_p and is often denoted by $\bigwedge^1(\mathrm{TM}_p^*)$. By definition, the field of scalars $\mathbb{R}$ is regarded as the space of *rank-0 differential forms* on TM_p, and we denote this 1-dimensional space by $\bigwedge^0(\mathrm{TM}_p^*)$.

NOTATION. Given a chart (U_α, x_α) on M and a point $p \in M$, we get a basis

$$\mathfrak{X}_p = \Big\{\Big(\frac{\partial}{\partial x_i}\Big|_p\Big)\Big\} \text{ in } \mathrm{TM}_p \quad \text{and a dual basis} \quad \mathfrak{X}_p^* = \Big\{\Big(\frac{\partial}{\partial x_i}\Big|_p\Big)^*\Big\} \text{ in } \mathrm{TM}_p^*.$$

For various reasons, these dual vectors have come to be denoted by other symbols that may seem peculiar at first, but the notation will grow on you as its advantages become apparent. Hereafter, we shall write

$$(dx_i)_p \equiv \Big(\frac{\partial}{\partial x_i}\Big|_p\Big)^* \tag{4.18}$$

so the dual basis determined by a chart can be written succinctly as

$$(dx_1)_p, \dots, (dx_m)_p \quad \text{instead of} \quad \left(\frac{\partial}{\partial x_1}\bigg|_p\right)^*, \dots, \left(\frac{\partial}{\partial x_m}\bigg|_p\right)^*.$$

By definition of "dual basis," we then have

$$\left\langle (dx_i)_p, \left(\frac{\partial}{\partial x_j}\bigg|_p\right)\right\rangle = \delta_{ij} \quad \text{(Kronecker delta)} \tag{4.19}$$

for $1 \leq i, j \leq m$. ○

The Canonical *d*-Operator on Functions. Each $f \in \mathcal{C}^\infty(p)$ determines an element $(df)_p \in \mathrm{TM}_p^*$, the *exterior derivative of f at p*. This is defined via a construction that makes no mention of local coordinates.

DEFINITION 4.26 (Exterior Derivative of a Function). If $f : M \to \mathbb{R}$ is a smooth function defined near base point p in a manifold M, its *exterior derivative* at p is the linear map $(df)_p : \mathrm{TM}_p \to \mathbb{R}$, a cotangent vector in TM_p^* determined by the identity

$$\langle (df)_p, X_p\rangle = \langle X_p, f\rangle \tag{4.20}$$

for all $f \in \mathcal{C}^\infty(p)$ and $X_p \in \mathrm{TM}_p$.

On any chart domain U_α, the outcome is described in local coordinates by writing $(df)_p$ as a linear combination of the dual basis vectors $(dx_1)_p, \dots, (dx_m)_p$ determined by the chart (U_α, x_α).

PROPOSITION 4.27. *Given a chart (U_α, x_α) and a function $f \in \mathcal{C}^\infty(U_\alpha)$, the rank-0 exterior derivative $(df)_p \in \mathrm{TM}_p^*$ has the following description in local chart cooordinates at every base point $p \in U_\alpha$:*

$$(df)_p = \sum_{i=1}^{m} \frac{\partial f}{\partial x_i}(p) \cdot (dx_i)_p. \tag{4.21}$$

PROOF. At each $p \in U_\alpha$, there exist unique coefficients such that $(df)_p = \sum_{i=1}^m c_i(u)\,(dx_i)_p$ in TM_p^*. To determine $c_i(p)$, simply apply $(df)_p$ to the basis vectors $(\partial/\partial x_i|_p)$ in TM_p; by (4.18), (4.20), and (4.21), we get

$$c_i(p) = \left\langle (df)_p, \frac{\partial}{\partial x_i}\bigg|_p\right\rangle = \left\langle \frac{\partial}{\partial x_i}\bigg|_p, f\right\rangle = \frac{\partial f}{\partial x_i}(p)$$

for $1 \leq i \leq m$. □

It is important to observe that $(df)_p$ has the same general form in all coordinate systems. For example, given Cartesian coordinates $\mathbf{x} = x_\alpha(u) = (x, y)$ and polar coordinates $\mathbf{y} = y_\beta(u) = (r, \theta)$ on an open subset $U \subseteq M = \mathbb{R}^2$, we have

$$(df)_u = \frac{\partial f}{\partial x}(u) \cdot (dx)_u + \frac{\partial f}{\partial y}(u) \cdot (dy)_u = \frac{\partial f}{\partial r}(u) \cdot (dr)_u + \frac{\partial f}{\partial \theta}(u) \cdot (d\theta)_u$$

for all u where the charts overlap. Notice the resemblance between the description of $(df)_p$ in (4.21) and the classical gradient $\nabla f(p) = \sum_{i=1}^n D_{x_i} f(p)\,\mathbf{e}_i$ of a smooth function $f : \mathbb{R}^n \to \mathbb{R}$. This is not an accident. In the theory of manifolds, the classical gradient operation ∇f is obtained by applying the exterior derivative $d : \mathcal{C}^\infty(M) \to \bigwedge^1(\mathrm{TM}_p^*)$; then ∇f becomes the 1-form df of (4.21).

Every cotangent vector $\omega_p \in \mathrm{TM}_p^*$ is equal to $(df)_p$ for some $f \in \mathcal{C}^\infty(p)$, but the f is not unique.

LEMMA 4.28. *The map $d : \mathcal{C}^\infty(p) \to \mathrm{TM}_p^*$ that sends $f \mapsto (df)_p$ is linear and surjective but not one-to-one. The nontrival kernel is*

$$\ker(d) = \left\{ f \in \mathcal{C}^\infty(p) : \frac{\partial f}{\partial x_1}(p) = \cdots = \frac{\partial f}{\partial x_m}(p) = 0 \right\}.$$

Thus $(df)_p = 0$ in $\bigwedge^1(\mathrm{TM}_p^*)$ if and only if p is a critical point for $f : \mathbb{R}^m \to \mathbb{R}$ when f is described in local coordinates.

PROOF. If we have $\partial f/\partial x_i(p) = 0$ for $1 \leq i \leq m$ for one chart (U_α, x_α) about p, then by (4.21) this must be true for every other chart (U_β, y_β) about p. Thus, the property "*p is a critical point for f*" is independent of any choice of local coordinates.

Surjectivity follows for an interesting reason. The scalar component functions $X_i(u)$ of the chart map $x_\alpha(u) = (X_1(u), \dots, X_m(u))$ are in $\mathcal{C}^\infty(U_\alpha)$. But by (4.7) we get

$$(dX_i)_u = (dx_i)_u \quad \text{for } 1 \leq i \leq m. \tag{4.22}$$

In fact, when we apply $(dX_i)_p \in \mathrm{TM}_p^*$ to a typical basis vector in TM_p, equation (4.7) yields

$$(dX_i)_u = \sum_{j=1}^{m} \frac{\partial X_i}{\partial x_j}(u) \cdot (dx_j)_u = \sum_{j=1}^{m} \delta_{ij} \cdot (dx_j)_u = (dx_i)_u.$$

Thus $(dX_i)_u$ and the dual vector $(dx_i)_u = \left(\partial/\partial x_i|_u\right)^*$ determined by the chart (U_α, x_α) are the same element in TM_u^* for all $u \in U_\alpha$, despite their very different origins. It is now clear that the range of $d : \mathcal{C}^\infty(p) \to \mathrm{TM}_p^*$ contains a basis for TM_p^*, and therefore d is surjective. □

Let $\omega : p \to \omega_p \in \mathrm{TM}_p^*$ be a field of 1-forms on M. Given a chart (U_α, x_α), there are unique coefficients $c_j(p)$ such that

$$\omega_p = \sum_{j=1}^{m} c_j(p) \cdot (dx_j)_p \quad \text{in } \mathrm{TM}_p^* \text{ for } p \in U_\alpha. \tag{4.23}$$

By (4.19) the coefficients can be recovered by bracketing ω_p with the basis vectors $\{(\partial/\partial x_i|_p)\}$ in TM_p determined by the chart.

DEFINITION 4.29. We say ω is a *smooth field of 1-forms*, or a *rank-1 differential form* on M, if for every chart (U_α, x_α) the coefficient functions $c_k(u)$ are in $\mathcal{C}^\infty(U_\alpha)$. As we will show in Theorem 4.30 below, this definition is coordinate-independent. The set of such fields on M is denoted by $\mathcal{D}^{(0,1)}(M)$, or sometimes by $\bigwedge^1 M$ depending on the context. This becomes an infinite-dimensional vector space if sums and scalar multiples in $\mathcal{D}^{(0,1)}(M)$ are given by

$$(\lambda\omega)_p = \lambda \cdot \omega_p \quad \text{and} \quad (\omega + \mu)_p = \omega_p + \mu_p \quad \text{in } \mathrm{TM}_p^*$$

for all $p \in M$ and fields $\omega, \mu \in \mathcal{D}^{(0,1)}(M)$. This space of 1-forms is also a $\mathcal{C}^\infty(M)$-module under the action $\mathcal{C}^\infty(M) \times \mathcal{D}^{(0,1)}(M) \to \mathcal{D}^{(0,1)}(M)$ given by

$$(f \cdot \omega)_p = f(p) \cdot \omega_p \quad \text{for } p \in M, f \in \mathcal{C}^\infty(M), \omega \in \mathcal{D}^{(0,1)}(M).$$

The following change of variable law for 1-forms implies that smoothness is a coordinate-independent property of a field of cotangent vectors. The proof follows easily from the previously established transformation laws for tangent vectors, Theorem 4.15 or Corollary 4.16.

THEOREM 4.30 (Change of Variable for Cotangent Vectors.). *Cotangent vectors $(dx_i)_u$ and $(dy_j)_u$ determined by charts (U_α, x_α) and (U_β, y_β) on a manifold M transform in the following way:*

$$(dy_i)_u = \sum_{j=1}^{m} \frac{\partial Y_i}{\partial x_j}(u) \cdot (dx_j)_u \quad \text{in } \mathrm{TM}_p^*, 1 \le i, j \le m$$

for all base points $u \in U_\alpha \cap U_\beta$. Here the Y_i are the scalar components of the chart map $y_\beta = (Y_1, \dots, Y_m)$.

PROOF. There are unique coefficients such that $(dy_i)_u = \sum_{j=1}^m c_j(u)\cdot(dx_j)_u$. If the scalar components of y_β are $y_\beta(u) = \big(Y_1(u), \dots, Y_m(u)\big)$, then by (4.19) and (4.22), bracketing with the dual basis vectors $\big(\partial/\partial x_k|_u\big)$ yields

$$\begin{aligned} c_k(u) &= \left\langle (dy_i)_u, \frac{\partial}{\partial x_k}\Big|_u \right\rangle = \left\langle (dY_i)_u, \frac{\partial}{\partial x_k}\Big|_u \right\rangle \\ &= \left\langle \frac{\partial}{\partial x_k}\Big|_u, Y_i \right\rangle = \frac{\partial Y_i}{\partial x_k}(u) \quad \text{for } 1 \le i \le m, \end{aligned}$$

proving the formula. □

A law similar to Corollary 4.16 for tangent vectors describes the action by which a $\mathcal{C}^\infty$ map $\phi : M \to N$ "pulls back" cotangent vectors from base points in N to base points in M. We won't pursue this here since it won't be needed in later discussion, but it is a fundamental result in differential geometry.

Exterior Derivatives (an Overview). Equation (4.21) defines a linear operator $d = d_0 : \bigwedge^0 M \to \bigwedge^1 M$ called the *rank-0 exterior derivative*, where $\bigwedge^0 M = \mathcal{C}^\infty(M)$ is the space of rank-0 differential forms. By Definition 4.26, $(df)_p$ acts on elements $X_p \in \mathrm{TM}_p$ via

$$\langle (df)_p, X_p \rangle = \langle X_p, f \rangle \quad \text{for all } f \in \mathcal{C}^\infty(p), p \in M, X_p \in \mathrm{TM}_p. \tag{4.24}$$

This d-operator has its own derivation property:

$$\big(d(fh)\big)_u = (df)_u \cdot h(u) + f(u) \cdot (dh)_u \quad \text{for all } u \in M \text{ and } f, h \in \mathcal{C}^\infty(M).$$

EXERCISE 4.31. If $f(u), h(u)$ are $\mathcal{C}^\infty$ functions defined on a chart x_α in a manifold M, verify that

$$(d(f\cdot h))_u = (df)_u h(u) + f(u)(dh)_u$$

for all $u \in U_\alpha$.

We will eventually define a hierarchy of spaces of smooth tensor fields

$$\bigwedge^k M = \text{(the space of smooth rank } k\text{-differential forms on } M)$$

together with associated rank-k exterior derivatives $d = d_k : \bigwedge^k M \to \bigwedge^{k+1} M$ such that

$$\bigwedge^0 M \xrightarrow{d=d_0} \bigwedge^1 M \xrightarrow{d=d_1} \bigwedge^2 M \xrightarrow{d=d_2} \cdots \xrightarrow{d=d_{m-1}} \bigwedge^m M, \tag{4.25}$$

where $m = \dim(M)$. In all dimensions, these "d-operators" d_k play the roles occupied by the vector operators *div, grad, curl* of calculus when $m = 3$, but a few more ideas must be developed before that connection can be explained.

Primitives of 1-Forms. If ω is a smooth 1-form on an open set $U \subseteq M$, a *primitive* of ω is an $f \in \mathcal{C}^\infty(U)$ such that $df = \omega$ on U. If f is such a primitive on U, consider a chart (U_α, x_α) about p; replacing $U_\alpha \to U \cap U_\alpha$, we may assume $U_\alpha \subseteq U$. On U_α, we can describe ω in local coordinates:

$$\omega_u = \sum_{i=1}^m w_i(u)\cdot (dx_i)_u \quad \text{with } w_i \in \mathcal{C}^\infty(U_\alpha).$$

By (4.21), $df = \omega$ on U_α means that

$$(df)_u = \sum_{i=1}^m \frac{\partial f}{\partial x_i}(u)\,(dx_i)_u \quad \text{is equal to} \quad \omega_u = \sum_{i=1}^m w_i(u)\,(dx_i)_u$$

for all $u \in U_\alpha$. The coefficients must agree, so if there is an f such that $df = \omega$, the coefficient functions $w_i(u)$ must be related to the partial derivatives of f via the following system of partial differential equations:

$$w_i(u) = \frac{\partial f}{\partial x_i}(u) \quad \text{for all } u \in U_\alpha \text{ and } 1 \le i \le m.$$

This system does not always admit solutions, even locally. In fact, the coefficients $w_i(u)$ of ω in local coordinates must satisfy "consistency condition" if the equation $df = w$ is to have any local solutions; within the chart domain we must have

$$\frac{\partial w_i}{\partial x_j} - \frac{\partial w_j}{\partial x_i} \equiv 0 \quad \text{on } U_\alpha \text{ for } 1 \le i < j \le m. \tag{4.26}$$

These conditions are *necessary* because if f is of class at least $C^{(2)}$, its mixed 2nd order partial derivatives must agree, so the $w_i(u)$ satisfy a system of $\frac{1}{2}(m^2 - m)$ equations on $\mathbb{R}^m$:

$$\frac{\partial w_i}{\partial x_j} = \frac{\partial}{\partial x_j}\left(\frac{\partial f}{\partial x_i}\right) = \frac{\partial}{\partial x_i}\left(\frac{\partial f}{\partial x_j}\right) = \frac{\partial w_j}{\partial x_i} \quad \text{for } 1 \le i, j \le m.$$

In order to defer a long digression, we list the basic facts about local and global primitives of 1-forms $\omega \in \bigwedge^1(M)$ below.

1. The consistency conditions for ω take the same form (4.26) in every system of local coordinates. (This is fairly obvious, as noted in the remarks following Proposition 4.27).
2. Using line integrals $\int_\gamma w$ of 1-forms along $C^{(1)}$ curves, one can prove the existence of local solutions near any base point $p \in U_\alpha$ if the conditions (4.26) are satisfied.
3. Conditions (4.26) do not by themselves imply the existence of a "global solution" f in $\mathcal{C}^\infty(U)$ to the identity $df = \omega$. The geometry of U can present obstructions to existence of global solutions for certain forms ω even when (4.26) holds (see Example 4.34 below).
4. If ω is a smooth 1-form on a manifold M and the equation $df = \omega$ has a $\mathcal{C}^\infty$ solution on an open subset $U \subseteq M$ that is *connected*, all other solutions on U are obtained by adding an arbitrary constant to f.

EXERCISE 4.32. If f is a $\mathcal{C}^\infty$ function near $p \in M$, prove that the following properties are equivalent:

(a) $\dfrac{\partial f}{\partial x_i} \equiv 0$ near p for $1 \leq i \leq m$.

(b) $f \equiv$ (constant) on some open set containing p.

HINT. When transferred to coordinate space via a chart about p, this local result can be proved by calculus methods.

REMARK 4.33. It follows from the definition of "connectedness" that $f \equiv$ (constant) on any connected open set $U \subseteq M$ such that $df \equiv 0$ on U. ○

EXAMPLE 4.34. Let M be the "punctured plane" $\mathbb{R}^2 \sim \{\mathbf{0}\}$. The *angle-variable function*

$$\theta(x, y) = \arcsin\left(\frac{y}{\sqrt{x^2 + y^2}}\right)$$

is multiple-valued, but on any open half-plane $H \subseteq M$ bounded by a line through the origin, there are smooth single-valued $\mathcal{C}^\infty$ determinations of θ. When angles are measured in radians, these functions $\theta(x, y)$ can only differ on H by an added constant of the the form $c = 2\pi n$, $n \in \mathbb{Z}$, because it is connected. Since $d(f + c\mathfrak{1}) = (df)$, all smooth determinations of $\theta(x, y)$ on H must have the same single-valued exterior derivative $\omega = d\theta$ in $\bigwedge^1 M$, namely

$$\omega_{\mathbf{x}} = \left(\frac{-y}{x^2 + y^2}\right) dx + \left(\frac{x}{x^2 + y^2}\right) dy \quad \text{for } \mathbf{x} = (x, y) \in H \tag{4.27}$$

when ω is described in Euclidean coordinates on H. In particular, ω has local primitives near every $p \neq 0$ in M, but the multiple-valued nature of the primitive $\theta(x, y)$ on the punctured plane M prevents us from piecing together these local solutions to get a globally defined $\mathcal{C}^\infty$ function $f : M \to \mathbb{R}$ such that $df = \omega$ throughout M.

EXERCISE 4.35. The punctured plane $M = \mathbb{R}^2 \sim \{\mathbf{0}\}$ is covered by the four open half-planes:

$$H_1 = \{(x,y) : y > 0\} \qquad H_2 = \{(x,y) : x > 0\}$$
$$H_3 = \{(x,y) : y < 0\} \qquad H_4 = \{(x,y) : x < 0\}.$$

For each half-plane, give an explicit single-valued $\mathcal{C}^\infty$ determination of the angle variable $\theta(x,y)$. In each case show that the exterior derivative $\omega = d\theta$ is in fact equal to (4.27).

HINT. You may have to express your answer in terms of the functions *arcsin, arctan,* or *arccos* from calculus, depending on which H_i you examine.

EXERCISE 4.36. The following 1-forms on $M_2 = \mathbb{R}^2 \sim \{\mathbf{0}\}$ are described with respect to the standard Euclidean coordinates on this space. Identify those that can have well-defined local primitives:

(a) $\omega = \left(\dfrac{y}{x^2+y^2}\right)\cdot(dx) + \left(\dfrac{x}{x^2+y^2}\right)\cdot(dy)$

(b) $\omega = \left(\dfrac{x}{x^2+y^2}\right)\cdot(dx) + \left(\dfrac{y}{x^2+y^2}\right)\cdot(dy)$

(c) $\omega = \dfrac{x}{(x^2+y^2)^{3/2}}\cdot(dx) + \dfrac{y}{(x^2+y^2)^{3/2}}\cdot(dy)$

(d) $\omega = \ln(x^2+y^2)\cdot(dx) + \ln(x^2+y^2)\cdot(dy)$

(e) $\omega = (x^2-y^2)\cdot(dx) + 2xy\cdot(dy)$

(f) $\omega = (2xy)\cdot(dx) - (x^2-y^2)\cdot(dy)$

EXERCISE 4.37. Verify that the following 1-form on $M = \mathbb{R}^3 \sim \{\mathbf{0}\}$

$$\omega = \frac{1}{(x^2+y^2+z^2)^{3/2}}\cdot(x(dx)+y(dy)+z(dz))$$

has as a primitive the famed "$1/r$ potential"

$$\phi(\mathbf{x}) = \frac{1}{r} \quad \text{in which} \quad r = \|\mathbf{x}\| = \sqrt{x^2+y^2+z^2}.$$

Interpretations of 1-Forms and Their Primitives. In calculus, "vector fields" on $\mathbb{R}^n$ (or open subsets thereof) are represented as $\mathbf{F}(\mathbf{x}) = \sum_{i=1}^n F_i(\mathbf{x})\,\mathbf{e}_i$ where $\mathbf{e}_1, \dots, \mathbf{e}_n$ are the standard basis vectors in $\mathbb{R}^n$. But the nature of these basis vectors is seldom mentioned. This is a problem because fields appearing in applications are not all of the same type. Some examples:

1. VELOCITY FIELDS $\widetilde{X}$ on $M = \mathbb{R}^n$ must be regarded as fields of *tangent vectors*. In fact, if $\gamma(t)$ is a curve of class $\mathcal{C}^{(1)}$, its vector derivative $\gamma'(t)$ is the instantaneous velocity of the moving point $p = \gamma(t)$, and this is a tangent vector in TM_p. Therefore, at any $p \in M$, the basis vectors $\mathbf{e}_i$ in the identity
$$\gamma'(t) = v_1\,\mathbf{e}_1 + \cdots + v_n\,\mathbf{e}_m \quad (v_i \in \mathbb{R})$$
should also be interpreted as vectors in TM_p. The instantaneous velocities of particles in a fluid flow at a particular moment in time provide

a physical example of a smooth vector field of tangent vectors on an open subset $M \subseteq \mathbb{R}^n$.

2. GRADIENT FIELDS ∇f determined by a function $f : M \to \mathbb{R}$ are classically presented in the form
$$\nabla f(\mathbf{x}) = D_{x_1} f(\mathbf{x})\,\mathbf{e}_1 + \cdots + D_{x_m} f(\mathbf{x})\,\mathbf{e}_m.$$
But it should be apparent by now that there are many reasons to interpret every gradient field ∇f as a smoothly varying field of *cotangent vectors*, not tangent vectors, and then the basis vectors $\mathbf{e}_i$ must also be regarded as cotangent vectors. Formula (4.21) strongly suggests the following interpretation: taking global coordinates $x_\alpha(\mathbf{x}) = (x_1, \dots, x_m)$ on Euclidean space M (and $U_\alpha = \mathbb{R}^m$), we have basis vectors $\mathfrak{X}_p^* = \{(dx_1)_p, \dots, (dx_m)_p\}$ in TM_p^* for every base point p. In the standard chart coordinates on $\mathbb{R}^m$, the exterior derivative d mapping $\bigwedge^0 M \to \bigwedge^1 M$ takes the form
$$(df)_p = \sum_{i=1}^{m} \frac{\partial f}{\partial x_i}(p) \cdot (dx_i)_p \quad \text{for all } p \in \mathbb{R}^m.$$
If $f \in \mathcal{C}^\infty(\mathbb{R}^m)$ and we identify the standard basis vectors $\mathbf{e}_1, \dots, \mathbf{e}_m$ attached to $p \in \mathbb{R}^m$ with the dual vectors $\mathbf{e}_1^* = (dx_1)_p, \dots, \mathbf{e}_m^* = (dx_m)_p$ in the cotangent space TM_p^*, then the exterior derivative $(df)_p$ becomes the classical gradient $\nabla f(p)$ everywhere on $\mathbb{R}^m$.

3. ELECTRIC FIELDS $\mathbf{E}(\mathbf{x}) = E_1(\mathbf{x})\,\mathbf{e}_1 + \cdots + E_m(\mathbf{x})\mathbf{e}_m$ are also fields of cotangent vectors. It has long been realized by physicists that, at least locally, $\mathbf{E}$ fields are gradients $\mathbf{E} = \nabla\phi$ of scalar "potential functions" $\phi : M \to \mathbb{R}$. If $M = \mathbb{R}^m$ and we identify the $\mathbf{e}_i$ at a base point p with the 1-forms $(dx_i)_p$ determined by the standard coordinate chart on $\mathbb{R}^m$, then the classical statement $\mathbf{E} = \nabla\phi$ becomes a statement about the exterior derivative of ϕ,
$$\mathbf{E}(\mathbf{x}) = (d\phi)_\mathbf{x} = D_{x_1}\phi(\mathbf{x})\,(dx_1)_\mathbf{x} + \cdots + D_{x_m}\phi(\mathbf{x}) \cdot (dx_m)_\mathbf{x}$$
for all $\mathbf{x} \in \mathbb{R}^m$.

4. MAGNETIC FIELDS are described in introductory courses as "fields of vectors," but the "vectors" involved are something quite different, neither tangent vectors nor cotangent vectors. Magnetic fields are in fact represented by fields of rank-2 antisymmetric tensors, which assign at each base point p a nondegenerate antisymmetric bilinear form B_p on each tangent space TM_p. We will see how these arise in the scheme (4.25), but understanding this last statement requires an excursion into multilinear algebra which will be elaborated in the next section.

4.3. Differential Forms on *M* and Their Exterior Derivatives

In Chapter 3, we discussed bilinear forms $\omega : V \times V \to \mathbb{K}$ for $\mathbb{K} = \mathbb{R}$ or $\mathbb{C}$, but to deal with the issues that arise in differential geometry and various areas of applied mathematics, it becomes necessary to get a deeper understanding of multilinear algebra, particularly multilinear forms on vector spaces (aka *tensors*) and fields of tensors on differentiable manifolds.

It may be helpful to regard a rank-k tensor ω on V as a "black box" that accepts as an input an ordered list $(v_1, \dots, v_k)$ of vectors from V and produces a scalar $\omega(v_1, \dots, v_k) \in \mathbb{K}$ as its output; the *rank* $\mathrm{rk}(\omega)$ of the tensor is the number of inputs to the box. The workings of the box are subject to just one rule:

> *For each $1 \le i \le k$, the output $\omega(v_1, \dots, v_k)$ must be a linear function of $v_i \in V$ when the other inputs v_j $(j \neq i)$ are held fixed.*

The bilinear forms examined in Chapter 3 are all rank-2 tensors, and in particular any inner product on a vector space V over $\mathbb{R}$ is a rank-2 tensor on V. The rank-1 tensors on V are just the linear functionals $\ell : V \to \mathbb{K}$ in the dual space V^*, and we define the rank-0 tensors on V to be the scalars in the ground field $\mathbb{K}$.

The *rank-k tensors* on V are the multilinear forms $\omega : V \times \cdots \times V \to \mathbb{R}$ with k inputs, which are often referred to as the *k-linear forms*. We make these into a vector space by imposing the following addition and scaling operations:

$$\begin{aligned}(\omega_1 + \omega_2)(v_1, \dots, v_k) &= \omega_1(v_1, \dots, v_k) + \omega_2(v_1, \dots, v_k)\\ (\lambda \cdot \omega)(v_1, \dots v_k) &= \lambda \cdot \omega(v_1, \dots, v_k).\end{aligned}$$

The resulting vector space of "tensors of type $(0, k)$" is denoted by $V^{(0,k)}$ with $V^{(0,1)} = V^*$ and $V^{(0,0)} = \mathbb{K}$, so

$$\dim(V^{(0,1)}) = \dim(V^*) = \dim(V) \quad \text{and} \quad \dim(V^{(0,0)}) = 1.$$

Tensors of arbitrary mixed type $V^{(r,k)}$ on V can be defined, but we won't need them in this account, with one exception: in this notation, the space V itself is denoted by $V = V^{(1,0)}$, and its elements are the *tensors of type* (1,0).

If $\mathfrak{X} = \{e_i\}$ is a basis for V and $\mathfrak{X}^* = \{e_i^*\}$ is the dual basis in V^*, the space $V^{(0,k)}$ of k-linear forms $\omega : V \times \cdots \times V \to \mathbb{R}$ is a vector space of dimension $\dim(V^{(0,k)}) = m^k$ if $\dim(V) = m$. We will prove this by showing that it is spanned by "monomials," which are "tensor products" of vectors in the dual basis $\mathfrak{X}^*$. If $I = (i_1, \dots, i_k)$ is a multi-index with each $i_j \in [1, m]$, the corresponding *monomial* $e_I^* \in V^{(0,k)}$ is the k-linear map

$$e_I^* = e_{i_1}^* \otimes \cdots \otimes e_{i_k}^* \quad \textit{that sends} \quad (v_1, \dots, v_k) \mapsto \prod_{j=1}^{k} \langle e_{i_j}^*, v_j \rangle \tag{4.28}$$

where the $v_j \in V$. These operators are easily seen to be rank-k multilinear forms and are a basis for $V^{(0,k)}$, so every rank-k multilinear form can be written uniquely as $\omega = \sum_{I \in [1,m]^k} c_I e_I^*$ with $c_I \in \mathbb{R}$. Except for the number of players on the field, the proof is the same as that given in Chapter 3 to compute the dimensions of spaces of bilinear forms.

Our main interest here will be in tensors that act on the tangent spaces to a differentiable manifold M, so at $p \in M$ we consider tensors that act on $V = \mathrm{TM}_p$. However, many results discussed below are purely algebraic in nature and are true for tensors on arbitrary vector spaces that might be attached to base points in M. A *tensor field of rank-k* on M assigns a k-linear form $\omega_u \in \mathrm{TM}_u^{(0,k)}$ at each base point $u \in M$. Given a chart (U_α, x_α) on M and base point $u \in U_\alpha$, we have bases $\{\partial/\partial x_i|_u\}$ in TM_u and dual bases $\{(dx_i)_u\}$ in TM_u^*, and then there are uniquely determined coefficients $c_I(u)$ on U_α for $I = (i_1, \dots, i_k) \in [1, m]^k$ such that

$$\omega_u = \sum_I c_I(u)\,(dx_I)_u = \sum_I c_I(u) \cdot (dx_{i_1})_u \otimes \cdots \otimes (dx_{i_k})_u \quad \text{for } u \in U_\alpha.$$

A tensor field ω on M is *smooth* if the coefficients $c_I(u)$ are in $\mathcal{C}^\infty(U_\alpha)$ for all charts on M. A routine (but messy) calculation involving the change of basis formula shows that smoothness of ω described in one chart implies smoothness with respect to any other chart in the maximal atlas.

The space $\mathcal{D}^{(0,k)}(M)$ of smooth rank-k tensor fields is an infinite-dimensional vector space; it is also a $\mathcal{C}^\infty(M)$-module under the action

$$\mathcal{C}^\infty(G) \times \mathcal{D}^{(0,k)}(M) \to \mathcal{D}^{(0,k)}(M) \quad \text{given by} \quad (f\omega)_u = f(u) \cdot \omega_u$$

for $u \in M$ and $f \in \mathcal{C}^\infty(M)$. When $n = 1$, the space $\mathcal{D}^{(0,1)}(M)$ is precisely the space $\bigwedge^1 M$ of smooth 1-forms on M, and $\mathcal{D}^{(0,0)}(M)$ is equal to $\bigwedge^0 M = \mathcal{C}^\infty(M)$ by definition. The space $\mathcal{D}^{(0,2)}(M)$ of smooth rank-2 tensor fields consists of smoothly varying fields of bilinear forms on the tangent spaces TM_p.

EXAMPLE 4.38 (Riemannian Structure on M). On any vector space V over $\mathbb{R}$, an *inner product* $g(v_1, v_2)$ is a particular type of rank-2 tensor in $V^{(0,2)}$, but there are many possible inner products on V. A *Riemannian structure* on a manifold M is a smooth field of inner products $p \mapsto g_p$ with values $g_p \in \mathrm{TM}_p^{(0,2)}$. When M is equipped with this extra structure, we can define the following:

- LENGTH $\|X_p\| = \sqrt{g_p(X_p, X_p)}$ of any vector $X_p \in \mathrm{TM}_p$. This determines a vector space norm on each tangent space that allows us to speak of the "length" of a tangent vector.
- ORTHOGONALITY of vectors in $X_p, Y_p \in \mathrm{TM}_p$, which we interpret to mean $g_p(X_p, Y_p) = 0$. The *angle* between two nonzero tangent vectors at p is then

$$\cos\big(\theta(X_p, Y_p)\big) = \frac{g_p(X_p, Y_p)}{\|X_p\|\,\|Y_p\|}.$$

Lengths, angles, and orthogonality of tangent vectors cannot be defined in the absence of a Riemannian structure on M.

DISCUSSION. In particular, it now becomes meaningful to speak of *orthonormal* bases in each tangent space TM_p, as well as more exotic constructs such as a "field of orthogonal frames": a family of smooth vector fields $\{\widetilde{X}_1, \dots, \widetilde{X}_m\}$ on M such that

$$\mathfrak{X}_p = \{(\widetilde{X}_1)_p, \dots, (\widetilde{X}_m)_p\} \text{ is an orthonormal basis in } \mathrm{TM}_p$$

with respect to the inner product $g_p : \mathrm{TM}_p \times \mathrm{TM}_p \to \mathbb{R}$ at each base point p.

Furthermore, if $\gamma : [a,b] \to \mathbb{R}$ is a $\mathcal{C}^\infty$ curve, or merely one of class $\mathcal{C}^{(1)}$, its length has a natural definition as the Riemann integral:

$$\text{ARC LENGTH:} \quad L(\gamma) = \int_a^b \|\gamma'(t)\|\, dt = \int_a^b \sqrt{g_{\gamma(t)}(\gamma'(t), \gamma'(t))}\, dt$$

where $\gamma'(t)$ is the tangent vector to the curve at $p = \gamma(t)$. The change of variable formula for Riemann integrals shows that the value of $L(\gamma)$ is unchanged, with $L(\eta) = L(\gamma)$, if η is any orientation-preserving reparametrization of γ such that

$$\eta = \gamma \circ \phi \text{ where } \phi : [c,d] \to [a,b] \text{ is in } \mathcal{C}^{(1)} \text{ with } \frac{d\phi}{ds} > 0 \text{ for all } s\,.$$

For orientation-reversing reparametrizations, we get $L(\eta) = (-1) \cdot L(\gamma)$.

With considerably more effort, one can show that if $p \in M$, there is an open neighborhood U such that any $q \in U$ can be connected to p by a *geodesic* — a $\mathcal{C}^{(1)}$ curve $\gamma_0 : [a,b] \to U$ such that $\gamma_0(a) = p, \gamma_0(b) = q, \gamma_0(t) \in U$ for all t and

$$L(\gamma_0) \le L(\gamma) \quad \text{for } \textit{any } \mathcal{C}^{(1)} \text{ curve } \eta \text{ in } U \text{ that connects } p \text{ to } q.$$

These "minimal length" curves in M are the analogs of straight line segments when Euclidean space $\mathbb{R}^n$ is equipped with the Euclidean Riemannian structure, obtained by taking the standard global chart $x_\alpha(\mathbf{x}) = (x_1, \dots, x_n)$ on $U_\alpha = \mathbb{R}^n$ and defining $g_p = \sum_{i=1}^n (dx_i)_p \otimes (dx_i)_p$ so that

$$g_p\left(\frac{\partial}{\partial x_i}\bigg|_p, \frac{\partial}{\partial x_j}\bigg|_p\right) = \delta_{ij} \quad \text{(Kronecker).}$$

The bases $\mathfrak{X}_p = \{(\partial/\partial x_i|_p)\}$ induced in TM_p by the chart coordinates are then orthonormal bases with respect to the inner product g_p in each tangent space to $\mathbb{R}^n$.

This notion of arc length also induces a natural metric on any Riemannian manifold:

$$d_M(p,q) = \inf\{L_g(\gamma) : \gamma \text{ any } \mathcal{C}^\infty \text{ curve such that } \gamma(a) = p, \gamma(b) = q\}.$$

(However, verifying that d_M satisfies the triangle inequality takes considerable effort.) While inf{...} is actually achieved by a unique $\mathcal{C}^\infty$ curve (a geodesic) for all q sufficiently close to p, this need not be true for points far from p in the manifold. Even if the minimal length $d_M(p,q)$ is achieved, the geodesic connecting p to q might not be unique if p and q are widely separated. Think of the north and south poles on the unit sphere $S^2 \subseteq \mathbb{R}^3$. The sphere inherits a natural Riemannian structure from the surrounding Euclidean space (because the restriction of an inner product on V to a vector subspace W is an inner product on W), and the length-minimizing geodesics in the sphere S^2 are segments of great circles (intersections of the sphere with planes through the origin). Every great circle from N to S is a geodesic with the same length $L(\gamma) = \pi$. ○

We won't have time to explore the geometry of Riemannian manifolds in these *Notes*, but we emphasize that there are many examples. An excellent account of this subject is given in the book *Riemannian Geometry* **[1]** by Manfredo do Carmo. All $\mathcal{C}^\infty$ manifolds embedded in a Euclidean space $\mathbb{R}^n$, such as the spheres S^n of various dimensions or smooth level hypersurfaces determined via the implicit function theorem, inherit a natural Riemannian structure induced by the standard Riemannian structure on $\mathbb{R}^n$ described above. This Riemannian structure will, however, depend on how the manifold is embedded in $\mathbb{R}^n$.

Action of Permutation Groups S_n on Tensors. Permutations $\sigma \in S_k$ act as linear operators on the space of rank-k tensors $t \in V^{(0,k)}$ if we let

$$\sigma \cdot t(v_1, \dots, v_k) = t\big(v_{\sigma(1)}, \dots, v_{\sigma(k)}\big) \quad \text{for } v_j \in V.$$

Notice that in evaluating the action of $\sigma \cdot t$, the inputs $v_1, \dots, v_k$ are permuted by $\sigma \in S_k$ before being fed into the tensor t. Then $\sigma : V^{(0,k)} \to V^{(0,k)}$ is a linear map and we obtain a "left action of S_k on tensors," which means that

$$(\sigma\tau) \cdot t = \sigma \cdot (\tau \cdot t) \quad \text{for all } \sigma, \tau \in S_k \text{ and } t \in V^{(0,k)}. \tag{4.29}$$

REMARK 4.39. Many find the proof of this "covariance" property confusing. For a straightforward argument that does the job, observe that

$$\begin{aligned}
\sigma \cdot (\tau \cdot t)(v_1, \dots, v_k) &= (\tau \cdot t)\big(v_{\sigma(1)}, \dots, v_{\sigma(k)}\big) \\
&= (\tau \cdot t)(w_1, \dots, w_k)\big|_{w_1 = v_{\sigma(1)}, \dots, w_k = v_{\sigma(k)}} \\
&= t\big(w_{\tau(1)}, \dots, w_{\tau(k)}\big)\big|_{w_i = v_{\sigma(i)}} \\
&= t\big(v_{\sigma(\tau(1))}, \dots, v_{\sigma(\tau(k))}\big) \quad (\text{since } w_{\tau(i)} = v_{\sigma(\tau(i))}) \\
&= t\big(v_{(\sigma\tau)(1)}, \dots, v_{(\sigma\tau)(k)}\big) \\
&= (\sigma\tau) \cdot t(v_1, \dots, v_k)
\end{aligned}$$

for all $v_i \in V$. ○

EXERCISE 4.40. If $\ell_1, \dots, \ell_k \in V^*$, we have defined their tensor product to be the k-linear form $\ell_1 \otimes \cdots \otimes \ell_k$ as in (4.28). Verify that the action of S_k on such monomials is given by

$$\sigma \cdot (\ell_1 \otimes \cdots \otimes \ell_k) = \ell_{\sigma^{-1}(1)} \otimes \cdots \otimes \ell_{\sigma^{-1}(k)}.$$

(Compare with the initial definition, and note the role σ^{-1} plays here.)

All of this applies to manifolds M if we take $V = \mathrm{TM}_u$ at base points $u \in M$. If ω is a tensor field in $\mathcal{D}^{(0,k)}(M)$, so $\omega_u \in \mathrm{TM}_u^{(0,k)}$ for all base points in M, we let σ act independently on each tangent space (but not on the base points to which they are attached). Then $(\sigma \cdot \omega)_u = \sigma \cdot (\omega_u)$ in $\mathrm{TM}_u^{(0,k)}$ for each $u \in M$, and we get a left-action of the group S_k by linear operators

$$\sigma : \mathcal{D}^{(0,k)}(M) \to \mathcal{D}^{(0,k)}(M)$$

on the space of smooth rank-k tensor fields.

Antisymmetric tensors and tensor fields on manifolds are objects of special interest in calculus on manifolds.

DEFINITION 4.41 (Symmetric and Antisymmetric Tensor Fields). If V is a finite-dimensional vector space, a tensor ω in $V^{(0,k)}$ is *symmetric* or *antisymmetric* if

$$\sigma \cdot \omega = \omega \text{ for all } \sigma \in S_k \quad \text{or} \quad \sigma \cdot \omega = \text{sgn}(\sigma) \cdot \omega \tag{4.30}$$

where $\text{sgn}(\sigma)$ is the signature of the permutation $\sigma \in S_k$. The symmetric tensors of rank-k are a vector subspace $S^k(V) \subseteq V^{(0,k)}$; the antisymmetric tensors are also a vector subspace, denoted by $\bigwedge^k V$. When $k \geq 2$, $S^k(V) \cap \bigwedge^k(V) = \{0\}$. When $k = 0, 1$ the definition (4.30) is ambiguous, so tensors in $V^{(0,1)} = V^*$ and $V^{(0,0)} = \mathbb{K}$ are regarded as being both symmetric and antisymmetric; in particular, $S^1 V = \bigwedge^1 V = V^*$ and $S^0 V = \bigwedge^0 V = \mathbb{K}$.

For a tensor field $\omega \in \mathcal{D}^{(0,k)}(M)$ on a manifold M, (anti-)symmetry means that ω_p is an (anti-)symmetric tensor at each base point $p \in M$. Smooth fields of antisymmetric rank-k tensors are of special interest. They are called *k-forms*, or *differential forms of rank-k*, and the space of all such tensor fields is denoted $\bigwedge^k M$. This is a vector subspace of $\mathcal{D}^{(0,k)}(M)$, and both become $\mathcal{C}^\infty(M)$-modules if we let $\mathcal{C}^\infty(M)$ act in the following way on rank-k tensors:

$$(f \cdot \omega)_u(v_1, \dots, v_k) = f(u) \cdot \omega_u(v_1, \dots, v_k) \quad \text{for } v_1, \dots, v_k \in \text{TM}_u, u \in M.$$

The space of *symmetric tensor fields* of rank-k is denoted $S^k M$, and the symmetric rank-k tensors on TM_p are denoted by $S^k(\text{TM}_p)$.

Many tensors on a vector space V are neither symmetric nor antisymmetric. Antisymmetric tensors can be created by "antisymmetrizing" arbitrary rank-k tensors in $V^{(0,k)}$ via a surjective linear projection map

$$\text{Alt} : V^{(0,k)} \to \bigwedge^k V \subseteq V^{(0,k)}$$

defined as follows:

$$\text{Alt}(\omega) = \frac{1}{k!} \sum_{\sigma \in S_k} \text{sgn}(\sigma)\,(\sigma \cdot \omega) \quad \text{for all } \omega \in V^{(0,k)}. \tag{4.31}$$

The map *Alt* is linear; the fudge factor "$1/k!$" is needed to make *Alt* a projection operator, with $(Alt)^2 = Alt$. We obtain a similar "symmetrization operator" $S : V^{(0,k)} \to S^k(V)$ by dropping the factor "sgn(σ)" in equation (4.31), but symmetrization will not play a role in the present narrative.

On a manifold M, doing this to $V = \text{TM}_u^{(0,k)}$ at each base point converts any smooth tensor field $\omega \in \mathcal{D}^{(0,k)}(M)$ into a smooth field of rank-k antisymmetric tensors; the resulting tensor field $\text{Alt}(\omega)$ is then given by

$$\big(\text{Alt}(\omega)\big)_u = \text{Alt}(\omega_u) \quad \text{for all } u \in M, \omega \in \mathcal{D}^{(0,k)}(M). \tag{4.32}$$

Below, we list without proof the basic facts about antisymmetrization.

THEOREM 4.42. *If V is a vector space, the map* Alt $: V^{(0,k)} \to V^{(0,k)}$ *in* (4.31) *is linear, and the following hold:*

1. $(\mathrm{Alt})^2 = \mathrm{Alt} \circ \mathrm{Alt}$ *is equal to* Alt, *so* Alt *is a projection onto its range in* $V^{(0,k)}$.
2. $\mathrm{Alt}(V^{(0,k)}) \subseteq \bigwedge^k(V^{(0,k)})$, *and if ω was an antisymmetric tensor on V to begin with, then* $\mathrm{Alt}(\omega) = \omega$. *Thus,* Alt $=$ id *on* $\bigwedge^k V$, *and range*(Alt) *is all of* $\bigwedge^k V$.

As an operator on tensor fields, Alt *maps* $\mathcal{D}^{(0,k)}(M) \to \bigwedge^k M$ *if we take* $V = \mathrm{TM}_u^{(0,k)}$ *in* (4.31) *at every base point.*

Products of Tensors and Tensor Fields. The *tensor product* $\omega \otimes \mu$ of $\omega \in V^{(0,k)}$ and $\mu \in V^{(0,\ell)}$ is a rank-$(k+\ell)$ tensor on V such that

$$\omega \otimes \mu(v_1, \dots, v_k, v_{k+1}, \dots, v_{k+\ell}) = \omega(v_1, \dots, v_k) \cdot \mu(v_{k+1}, \dots, v_{k+\ell})$$

The $(\otimes)$ operation sends pairs of tensors (ω, μ) in the Cartesian product space $V^{(0,k)} \times V^{(0,\ell)}$ to tensors $\mu \otimes \nu$ of rank $k+\ell$ in $V^{(0,k+\ell)}$. It is easily seen to be a linear function of each input to $\omega \otimes \mu$ when the other is held fixed and is an associative operation in the sense that

$$\omega \otimes (\mu \otimes \tau) = (\omega \otimes \mu) \otimes \tau \quad \text{in } V^{k+\ell+m}.$$

It is not a commutative operation because because $\omega \otimes \mu$ and $\mu \otimes \omega$ are generally different elments of $V^{(0,k+\ell)}$.

Wedge Product of Antisymmetric Tensors. The *wedge product* $\omega \wedge \mu$ is a bilinear map of antisymmetric tensors

$$\bigwedge^k V \times \bigwedge^\ell V \to \bigwedge^{k+\ell} V$$

that takes $\omega \in \bigwedge^k V$ and $\mu \in \bigwedge^\ell V$ to an antisymmetric tensor of rank $k+\ell$:

$$\omega \wedge \mu = \frac{(k+\ell)!}{k!\,\ell!} \mathrm{Alt}(\omega \otimes \mu). \tag{4.33}$$

Note that the ordinary tensor product $\omega \otimes \mu$ will have rank $k+\ell$, but it need not be antisymmetric *even if both factors were antisymmetric*, so $\omega \otimes \mu$ has to be antisymmetrized to end up in $\bigwedge^{k+\ell} V$. The algebraic properties of the wedge product are listed below. We start with a simple observation that the transpose operation $A \mapsto A^{\mathrm{T}}$ on linear operators induces a natural "transpose operation" on tensors in $V^{(0,k)}$.

LEMMA 4.43. *If V is a finite-dimensional vector space and $A : V \to V$ is a linear operator, its transpose acts on the dual space V^* with*

$$\langle A^{\mathrm{T}}\ell, v\rangle = \langle \ell, Av\rangle \quad \text{for } v \in V, \ell \in V^*.$$

This construction can be extended to a transpose operation $A^{\mathrm{T}} : V^{(0,k)} \to V^{(0,k)}$ *on arbitrary rank-k tensors by letting A act independently on each input vector v_i in $(v_1, \dots, v_k)$ before ω is applied:*

$$A^{\mathrm{T}}\omega(v_1, \dots, v_k) = \omega(A(v_1), \dots, A(v_k)). \tag{4.34}$$

The operator A^{T} leaves the subspace $\bigwedge^k V$ invariant, so A^{T} induces a linear operator on the space of antisymmetric tensors $A^{\mathrm{T}} : \bigwedge^k V \to \bigwedge^k V$ as well as on the full space $V^{(0,k)}$ of rank-k tensors.

Much of what follows is based on the interaction between the wedge product and transpose A^{T} operations on (anitsymmetric) tensors.

THEOREM 4.44 (Algebraic Properties of the Wedge Product). *Let V be a finite-dimensional vector space with basis $\mathfrak{X} = \{e_i\}$ and dual basis $\mathfrak{X}^* = \{e_i^*\}$ in V^*. The wedge product has the following algebraic properties:*

1. ANTISYMMETRY OF WEDGE. *If ϕ_1, ϕ_2 are in V^** (rank-1 *tensors on V), their wedge product is an antisymmetric tensor in $V^{(0,2)}$,*

$$\phi_1 \wedge \phi_2 = \frac{1}{2}(\phi_1 \otimes \phi_2 - \phi_2 \otimes \phi_1) = (-1) \cdot (\phi_2 \wedge \phi_1).$$

In particular, $\phi \wedge \phi = 0$ in $V^{(0,2)}$ for every rank-1 tensor $\phi \in V^*$. If $\omega \in \bigwedge^k V$ and $\mu \in \bigwedge^\ell V$, we have the more general commutation relations

$$\omega \wedge \mu = (-1)^{kl} \mu \wedge \omega. \tag{4.35}$$

2. ASSOCIATIVITY OF WEDGE. *If $\omega \in \bigwedge^k V$, $\mu \in \bigwedge^\ell V$, $\tau \in \bigwedge^m V$, then $\omega \wedge (\mu \wedge \tau) = (\omega \wedge \mu) \wedge \tau$. Associativity implies that when we form the wedge product $\omega_1 \wedge \cdots \wedge \omega_r$ of several antisymmetric tensors, we don't have to worry about where to put the parentheses. The proof of associativity also yields a fact that greatly simplifies many calculations involving iterated wedge products:*

$$\omega \wedge (\mu \wedge \tau) = (\omega \wedge \mu) \wedge \tau = \frac{(k + \ell + m)!}{k!\ \ell!\ m!} \operatorname{Alt}(\omega \otimes \mu \otimes \tau). \tag{4.36}$$

Notice that evaluating our original definition of $\omega \wedge (\mu \wedge \tau)$,

$$\omega \wedge (\mu \wedge \tau) = \frac{(k + \ell + m)!}{k!\ \ell!\ m!} \operatorname{Alt}(\omega \otimes (\operatorname{Alt}(\mu \otimes \tau)),$$

involves two applications of "Alt," but the last expression in (4.36) requires only one.

3. EVALUATING MONOMIAL WEDGE PRODUCTS. *If $\mathfrak{X} = \{e_1, \dots, e_m\}$ is a basis in V and $\mathfrak{X}^* = \{e_1^*, \dots, e_m^*\} \subseteq V^*$ is the dual basis, we can define a rank-k antisymmetric tensor*

$$e_J^* = e_{j_1}^* \wedge \cdots \wedge e_{j_k}^* \quad \text{for any multi-index } J = (j_1, \dots, j_k)$$

without any need to interpolate parentheses among the factors since the wedge product is associative. One important computational fact is

$$e_1^* \wedge \cdots \wedge e_m^*(e_1, \dots, e_m) = 1\,.$$

More generally, we can evaluate

$$e_{j_1}^* \wedge \cdots \wedge e_{j_k}^*(e_{i_1}, \dots, e_{i_k}) = e_J^*(e_{i_1}, \dots, e_{i_k})$$

for arbitrary multi-indices I, J in $[1, m]^k$ (ordered or not). The outcome is zero if there are any repeated entries in the k-tuple $J = (j_1, \dots, j_k)$; otherwise, there is a unique permutation $\sigma \in S_k$ such that $I = \sigma \cdot J = (j_{\sigma(1)}, \dots, j_{\sigma(k)})$ is an ordered multi-index with $j_{\sigma(1)} < \cdots < i_{\sigma(k)}$. The corresponding "ordered monomials" σ_I turn out to be a basis for $\bigwedge^k V$.

THEOREM 4.45 (The Basis Theorem). *Given a basis $\mathfrak{X} = \{e_i\}$ in V and dual basis $\mathfrak{X}^* = \{e_i^*\}$ in V^*, we define the set $\mathcal{E}_k$ of ordered k-tuples*

$$I = (i_1, \dots, i_k) \quad \text{with} \quad 1 \le i_1 < \cdots < i_k \le m = \dim(V).$$

The corresponding "ordered monomials" in $\bigwedge^k V$

$$e_I^* = e_{i_1}^* \wedge \cdots \wedge e_{i_k}^*, \quad (I = (i_1 < \cdots < i_k) \in \mathcal{E}_k)$$

are a basis for the space of k-forms $\bigwedge^k V$. In particular,

$$\dim\left(\bigwedge^k V\right) = \binom{m}{k} = \#(\text{multi-indices in } \mathcal{E}_k).$$

When $k = 0$, we have $\dim\left(\bigwedge^0 V\right) = \dim_{\mathbb{R}}(\mathbb{R}) = 1$. It is important to notice that the space $\bigwedge^m V$ of antisymmetric tensors of maximal rank $m = \dim(V)$ is also one-dimensional, and $\bigwedge^k V = \{0\}$ for all $k > m = \dim(V)$.

EXERCISE 4.46. If $\phi_1, \phi_2 \in V^*$, verify by example that $\phi_1 \otimes \phi_2$ need not equal $\phi_2 \otimes \phi_1$ as elements of $V^{(0,2)}$, so "$\otimes$" is not commutative. Remember, equality of rank-2 tensors means they have the same action on all inputs (v_1, v_2) and that the inputs are *ordered lists*, so $(v_1, v_2) \neq (v_2, v_1)$.

EXERCISE 4.47. Let V be a finite-dimensional vector space, $\mathfrak{X} = \{e_i\}$ a basis, and $\mathfrak{X}^* = \{e_i^*\}$ the dual basis. In Chapter 3, we explained how every bilinear form $B : V \times V \to \mathbb{R}$ is described by an $m \times m$ matrix ($m = \dim V$)

$$[B]_{\mathfrak{X}} \quad \text{with entries} \quad B_{ij} = B(e_i, e_j).$$

Given two vectors $\phi_1 = \sum_{i=1}^m a_i e_i^*$, $\phi_2 = \sum_{i=1}^m b_i e_i^*$ in V^*, we get the bilincar forms $\phi_1 \otimes \phi_2$ and $\phi_2 \otimes \phi_1$ on V, using the rules of Theorem 4.44.

(a) Rewrite $\phi_1 \wedge \phi_2$ as a linear combination $\sum_{I \in \mathcal{E}_k} c_I e_I^*$ of standard basis vectors in $\bigwedge^2 V$. Compute the coefficients c_I in terms of $\{a_i\}$ and $\{b_i\}$.

(b) Compute the associated matrices $[\phi_1 \wedge \phi_2]_{\mathfrak{X}}$ and $[\phi_2 \wedge \phi_1]_{\mathfrak{X}}$. Are they equal?

Computations with antisymmetric tensors often reduce to working with wedge products $\phi_1 \wedge \cdots \wedge \phi_k$ of rank-1 tensors $\phi \in V^*$. The following simple formula, easily derived by induction from the associative law in Theorem 4.44, provides a way to directly evaluate such products:

$$\phi_1 \wedge \cdots \wedge \phi_k = \frac{(1 + \cdots + 1)!}{1! \cdots 1!} \cdot \mathrm{Alt}(\phi_1 \otimes \cdots \otimes \phi_k) = k!\ \mathrm{Alt}(\phi_1 \otimes \cdots \otimes \phi_k)$$

Note that the factor out front is now $k!$, not $\frac{1}{k!}$ as in (4.31).

If $\{e_i\}$ is a basis for V and $\{e_i^*\}$ the dual basis in V^*, one can evaluate a wedge product $e_J^* = e_{j_1}^* \wedge \cdots \wedge e_{j_k}^*$ for an arbitrary multi-index $J = (j_1, \dots, j_k)$ of length k with $j_s \in [1, m]$, whose entries need not be increasing or even distinct.

EXERCISE 4.48. If $\{\phi_1^*, \dots, \phi_r^*\}$ are vectors in V^*, use the algebraic rules governing the "$\wedge$" operation to explain why $\phi_1^* \wedge \cdots \wedge \phi_r^* = 0$ in the following circumstances:

(a) There is a repeated vector, say with $\phi_i^* = \phi_j^*$ for indices $i \neq j$.
(b) One of the vectors ϕ_i^* is the zero vector in V^*.
(c) The vectors $\{\phi_i^*\}$ are not linearly independent in V^*.
(d) $r > \dim(V^*)$.

In particular, $\bigwedge^r V = (0)$ for $r > \dim(V)$.

EXERCISE 4.49. Let V ve a vector space with basis $\{e_1, \dots, e_m\}$ and dual basis $\{e_1^*, \dots, e_m^*\}$ in V^*, and let $e_J^* = e_{j_1}^* \wedge \cdots \wedge e_{j_k}^*$ be an arbitrary monomial in the dual basis vectors with $j_1, \dots, j_k \in [1, m]$ but not necessarily with $j_1 < \cdots < j_k$.

(a) If we interchange adjacent factors $e_{j_s}^* \leftrightarrow e_{j_{s+1}}^*$ in the wedge product, explain why e_J^* becomes $-e_J^*$.
(b) If the indices $(j_1, \dots, j_k)$ are *distinct*, there is a unique permutation σ in S_k such that $I = \sigma \cdot J = \big(j_{\sigma(1)}, \dots, j_{\sigma(k)}\big)$ has the same indices listed in increasing order $j_{\sigma(1)} < \cdots < j_{\sigma(r)}$, so e_I^* is a standard basis vector in $\bigwedge^k V$. Explain why

$$e_J^* = e_{\sigma(J)} = \operatorname{sgn}(\sigma) \cdot e_I^*.$$

HINT. In any case, e_J^* is either zero or is $\pm e_I^*$ for a unique standard basis vector $e_I^* \in \bigwedge^k V$. If there are identical factors in e_J^*, repeated swapping of adjacent factors will bring the identical factors together, where they annihilate each other because $\phi \wedge \phi = 0$. If the factors in e_J^* are distinct, some permutation of the entries yields a multi-index I in $\mathcal{E}_k$. Then $J = \sigma \cdot I$ for some $\sigma \in S_k$ and $e_J^* = \pm e_I^*$. The problem in (b) is to show that the $\pm$ sign is given by the signature $\operatorname{sgn}(\sigma)$.

Calculating Wedge Products on a Chart. If $J = \{j_1, \dots, j_k\}$ is an arbitrary multi-index of length k (entries not necessarily increasing or distinct), we can form the wedge product $(dx)_J = (dx_{j_1}) \wedge \cdots \wedge (dx_{j_k})$ of 1-forms (dx_i) determined by the chart (U_α, x_α). But in view of the anti-commutation relations for wedge products, we get $e_J^* \equiv 0$ on U_α unless the entries in J are distinct, and if they are distinct, then $e_J^* = \pm e_I^*$ for a unique standard basis vector e_I^*, $I \in \mathcal{E}_k$. One last thing should be noted. In discussing wedge products $\omega \wedge \mu$ of differential forms on M, we often encounter "weighted" expressions like $f\,(dx_I) \wedge h\,(dx_J)$ involving $\mathcal{C}^\infty$ functions f, h on the chart domain U_α. By definition, $(\omega \wedge \mu)_u = \omega_u \wedge \mu_u$ at all base points. Hence,

$$(f\omega) \wedge (h\mu) = (fh) \cdot (\omega \wedge \mu) \tag{4.37}$$

as differential forms on M.

EXAMPLE 4.50. Let U_α be an open set in $M = \mathbb{R}^2$ on which Cartesian coordinates $x_\alpha(\mathbf{x}) = (x, y)$ and polar coordinates $y_\beta(\mathbf{x}) = (r, \theta)$ are both defined. The smooth 2-forms $dx \wedge dy$ and $dr \wedge d\theta$ each provide a basis for the 1-dimensional space $\bigwedge^2(\mathrm{TM}_p)$ at each $p \in U_\alpha$. We shall rewrite the smooth 2-form $dx \wedge dy$ as $F(r, \theta) \cdot (dr \wedge d\theta)$ using the algebraic properties of Theorem 4.44 and the change of variable formula for cotangent vectors (Theorem 4.30).

DISCUSSION. The chart maps have scalar components $x_\alpha(\mathbf{x}) = (x, y) = \big(X(\mathbf{x}), Y(\mathbf{x})\big)$ and $y_\beta(\mathbf{x}) = (r, \theta) = \big(R(\mathbf{x}), \Theta(\mathbf{x})\big)$ for $\mathbf{x} \in M$, and the coordinate transition maps in either direction are given by

$$\begin{aligned} &X = R\cos(\Theta),\ Y = R\sin(\Theta) && \text{for } x_\alpha \circ y_\beta^{-1} \\ &R = (X^2 + Y^2)^{1/2},\ \Theta = \arctan(Y/X) && \text{for } y_\beta \circ x_\alpha^{-1}. \end{aligned} \tag{4.38}$$

Apply the exterior derivative to the scalar component functions X, Y in (4.38) to get

$$\begin{aligned} dx = dX &= \frac{\partial X}{\partial r} \cdot dr + \frac{\partial X}{\partial \theta} \cdot d\theta \\ &= \frac{\partial}{\partial r}\{R\cos\Theta\} \cdot dr + \frac{\partial}{\partial \theta}\{R\sin\Theta\} \cdot d\theta \\ &= \left[\frac{\partial R}{\partial r}\cos(\Theta) - R\sin(\Theta) \cdot \frac{\partial \Theta}{\partial r}\right] \cdot dr \\ &\quad + \left[\frac{\partial R}{\partial \theta}\sin(\Theta) + R\cos(\Theta) \cdot \frac{\partial \Theta}{\partial \theta}\right] \cdot d\theta \end{aligned}$$

where $dX = (dx)$ is the exterior derivative of the scalar component $X : U_\alpha \to \mathbb{R}$ in the chart map x_α. By (4.22), (dx) and dX are the same smooth 1-form on U_α, and by (4.7) we get

$$\frac{\partial R}{\partial r} \equiv 1,\ \frac{\partial \Theta}{\partial \theta} \equiv 1,\ \frac{\partial R}{\partial \theta} \equiv \frac{\partial \Theta}{\partial r} \equiv 0$$

on U_α, so on the chart domain we have

$$dx = dX = \cos(\Theta) \cdot dr - R\sin(\Theta) \cdot d\theta.$$

The same sort of calculation yields

$$dy = dY = \sin(\Theta) \cdot dr + R\cos(\Theta) \cdot d\theta.$$

Adopting the time-honored abuse of notation that ignores the distinction between a function $r = R(\mathbf{x})$ and its values r, and likewise for the variable θ, we get

$$\begin{aligned} dx \wedge dy &= [\cos(\theta) \cdot dr - r\sin(\theta) \cdot d\theta] \wedge [\sin(\theta) \cdot dr + r\cos(\theta) \cdot d\theta\,] \\ &= \sin(\theta)\cos(\theta) \cdot (dr \wedge dr) - r\sin^2(\theta) \cdot (d\theta \wedge dr) \\ &\quad + r\cos^2(\theta) \cdot (dr \wedge d\theta) - r^2\sin(\theta)\cos(\theta) \cdot (d\theta \wedge d\theta) \\ &= -r\sin^2(\theta) \cdot (d\theta \wedge dr) + r\cos^2(\theta) \cdot (dr \wedge d\theta) \\ &= r(dr \wedge d\theta) \end{aligned}$$

because $dr \wedge dr = d\theta \wedge d\theta \equiv 0$ and $d\theta \wedge dr = -dr \wedge d\theta$ by antisymmetry. ○

EXERCISE 4.51. On $\mathbb{R}^2$, compute $dr \wedge d\theta$ in terms of $dx \wedge dy$. On $\mathbb{R}^3$, compute $dx \wedge dy \wedge dz$ in terms of $d\rho \wedge d\phi \wedge d\theta$ (spherical coordinates).

REMARK 4.52. Recall that spherical coordinates (ρ, θ, ϕ) and Cartesian coordinates (x, y, z) on $\mathbb{R}^3$ are related via

$$z = \rho \sin(\phi) \quad y = \rho \sin(\theta) \cos(\phi) \quad x = \rho \cos(\theta) \cos(\phi)$$

(see Figure 4.3). ○

Exterior Derivative of Higher Rank k-Forms. Smooth k-forms (smooth rank-k differential forms) ω are tensor fields whose values ω_p lie in $\bigwedge^k(\mathrm{TM}_p)$ for each $p \in M$. Given any chart (U_α, x_α), such a field is uniquely described on U_α as a sum

$$\omega_u = \sum_{I \in \mathcal{E}_k} c_I(u) \cdot (dx_{i_1})_u \wedge \cdots \wedge (dx_{i_k})_u = \sum_{I \in \mathcal{E}_k} c_I(u) \cdot (dx_I)_u \quad \text{for } u \in U_\alpha\,,$$

and smoothness means the c_I are in $\mathcal{C}^\infty(U_\alpha)$ for all charts. The general exterior derivatives d_k on a manifold with $\dim(M) = m$

$$\bigwedge^0 M \xrightarrow{d=d_0} \bigwedge^1 M \xrightarrow{d=d_1} \bigwedge^2 M \xrightarrow{d=d_2} \cdots \xrightarrow{d=d_{m-1}} \bigwedge^m M \xrightarrow{d=d_m} 0$$

are operations on rank-k differential forms for $k = 0, 1, 2, \ldots$ The rank-0 derivative d_0 has been defined in a coordinate-free way on $\bigwedge^0 M = \mathcal{C}^\infty(M)$ in (4.20) for $f \in \mathcal{C}^\infty(M)$; the result $(df) \in \bigwedge^1 M$ is the 1-form such that

$$\langle (df)_p, X_p \rangle = \langle X_p, f \rangle \quad \text{for all } X_p \in \mathrm{TM}_p,\, f \in \mathcal{C}^\infty(M). \tag{4.39}$$

As in (4.21), in local chart coordinates, df takes the form

$$(df)_p = \sum_{i=1}^{n} \frac{\partial f}{\partial x_i}(p) \cdot (dx_i)_p \quad \text{for all } p \in U_\alpha.$$

Coordinate-free formulas exist describing the action of exterior derivatives d_k of all ranks $k = 0, 1, 2, \ldots$ In rank-0, the formula (4.39) for d_0 is simple, but such descriptions become more complicated and unintuitive for $k \geq 1$. There is, however, a fairly straightforward way to describe $d\omega$ in local chart coordinates for ω of any rank k, and this is the approach we shall pursue. The only problem lies in showing that we get the same element $d\omega_p$ in $\bigwedge^{k+1}(\mathrm{TM}_p)$ at all base points in M, no matter which local chart (U_α, x_α) we use to compute it. (The proof that the definition is chart-independent is fairly arduous, and we won't have time to prove it in this brief survey.) □

Here is the definition of $d\omega$ in local coordinates.

DEFINITION 4.53 (General Exterior Derivative d_k). If (U_α, x_α) is any chart and $\omega \in \bigwedge^k M$, there are unique $\mathcal{C}^\infty$ coefficients $c_I(u)$ such that

$$\omega_u = \sum_{I \in \mathcal{E}_k} c_I(u) \cdot (dx_I)_u = \sum_{I \in \mathcal{E}_k} c_I(u) \cdot (dx_{i_1} \wedge \cdots \wedge dx_{i_k})$$

for all $u \in U_\alpha$. Then the *exterior derivative* $d\omega \in \bigwedge^{k+1} M$ is given on U_α by

$$
\begin{aligned}
d\omega &= \sum_{I\in\mathcal{E}_k} (dc_I)\wedge(dx_I) = \sum_{I\in\mathcal{E}_k}\Big(\sum_{j=1}^m \frac{\partial c_I}{\partial x_j}\,dx_j\Big)\wedge(dx_I) \\
&= \sum_{I\in\mathcal{E}_k}\sum_{j=1}^m \frac{\partial c_I}{\partial x_j}\cdot(dx_j)\wedge(dx_{i_1}\wedge\cdots\wedge dx_{i_k})
\end{aligned}
\tag{4.40}
$$

where dc_I is the usual exterior derivative (4.39) of the scalar function $c_I(u)$.

The chart determines a basis $\mathfrak{X}_p^* = \{(dx_1)_p, \dots, (dx_m)_p\}$ in TM_p^* for each $p \in U_\alpha$ and "standard" basis vectors in $\bigwedge^k(\mathrm{TM}_p^*)$,

$$e_I^* = dx_I = (dx_{i_1}\wedge\cdots\wedge dx_{i_k}) \quad \text{for } I = (i_1 < \cdots < i_k) \text{ in } \mathcal{E}_k.$$

A monomial $dx_j \wedge (dx_{i_1}\wedge\cdots\wedge dx_{i_k})$ in (4.40) will be zero in $\bigwedge^{k+1}(\mathrm{TM}_p^*)$ if j is one of the entries in I; otherwise it will be nonzero but not a standard basis vector in $\bigwedge^{k+1}(\mathrm{TM}_u^*)$ unless $j < i_1 < \cdots < i_k$. However, it is easy to rewrite any monomial $dx_j\wedge(dx_{i_1}\wedge\cdots\wedge dx_{i_k})$ appearing in (4.40) as a sum involving ordered monomials using the rules provided in Theorem 4.44 and Exercise 4.47.

REMARK 4.54. An annoying technical issue arises in applying (4.40). This definition presumes that ω has been presented as a sum of monomials $c_I\, e_I^*$ involving the *standard basis vectors* with $I \in \mathcal{E}_k$. If we wish to compute the exterior derivative of a k-form $\omega = \sum_{J\in[1,m]^k} c_J\cdot e_{j_1}^*\wedge\cdots\wedge e_{j_k}^*$ involving *unordered* monomials e_J^*, it seems we would have to rewrite each summand in terms of the standard (ordered) basis monomials e_I^* with $I \in \mathcal{E}_k$ before applying (4.40). This could be done using the commutation rule $dx_i \wedge dx_j = -dx_j \wedge dx_i$, repeatedly transposing adjacent factors to get $dx_J = \pm dx_I$ for some $I \in \mathcal{E}_k$. That would be a terrible nuisance, but the following observation saves the day. ○

LEMMA 4.55 (Extended Exterior Derivative). *Formula* (4.40) *remains valid for k-forms on a chart domain U_α,*

$$\omega = \sum_{J\in[1,m]^k} c_J\cdot e_J^* \quad (c_J \in \mathcal{C}^\infty(U_\alpha)),$$

even if they involve monomials e_J^* that are unordered or have repeated indices (in which case the term is zero).

PROOF. If an arbitrary multi-index $J \in [1,m]^k$ has a repeated entry, say $j_r = j_s$, then $d(f\,dx_J) = (df)\wedge dx_J = 0$ and does not contribute to the sum (4.40). Otherwise there is a unique permutation $\sigma \in S_k$ such that $J = \sigma\cdot I$ for an ordered index $I \in \mathcal{E}_k$, and then for any $f \in \mathcal{C}^\infty(U_\alpha)$, we have

$$
\begin{aligned}
d(f\cdot dx_J) &= d(f\cdot \mathrm{sgn}(\sigma)\,dx_I) \\
&= \mathrm{sgn}(\sigma)\cdot(df\wedge dx_I) \quad \text{(definition of } d\text{-operator)} \\
&= df\wedge \mathrm{sgn}(\sigma)\cdot dx_I = df\wedge dx_J.
\end{aligned}
$$

□

We will take advantage of this very useful fact below.

We now list the basic properties of the operators $d_k : \bigwedge^k M \to \bigwedge^{k+1} M$. The proofs are complicated, so we will not go into them here in order to press on toward reinterpreting multivariate calculus in terms of differential forms.

THEOREM 4.56 (Basic Properties of Exterior Derivatives). *Although $d_k\omega$ is defined on M using its description on a typical coordinate chart (U_α, x_α), formula* (4.40) *determines the same differential form $(d_k\omega)_p$ in $\bigwedge^{k+1} M$ for every chart (U_α, x_α) containing p. Thus,* (4.40) *yields a well-defined differential form $d_k\omega$ in $\bigwedge^{k+1} M$ for every $\omega \in \bigwedge^k M$. Furthermore, we have the following:*

1. THE RANK-0 CASE. *The rank-0 derivative $d_0 : \bigwedge^0 M \to \bigwedge^1 M$ is just the d-operator discussed earlier in* (4.24) *such that $\langle d_0 f, X_p\rangle = \langle X_p, f\rangle$ for $X_p \in \mathrm{TM}_p, f \in \mathcal{C}^\infty(p)$.*
2. LINEARITY. *Each d_k is a linear operator from $\bigwedge^k M \to \bigwedge^{k+1} M$.*
3. GENERAL DERIVATION PROPERTY. *If $\omega_1 \in \bigwedge^k M$ and $\omega_2 \in \bigwedge^\ell M$, then*
$$d_{k+\ell}(\omega_1 \wedge \omega_2) = d_k(\omega_1) \wedge \omega_2 + (-1)^k \omega_1 \wedge d_\ell(\omega_2)$$
in $\bigwedge^{k+\ell+1} M$. This property governs the interaction between exterior derivatives and wedge products of differential forms on M.
4. *By far the most interesting property is $d^2 = 0$, which is a shorthand version of the statement*

$$d_{k+1} \circ d_k(\omega) = 0 \text{ in } \textstyle\bigwedge^{k+2} M \quad \text{for all } \omega \in \bigwedge^k M\,. \tag{4.41}$$

This follows from equality of mixed second-order partial derivatives of class $\mathcal{C}^{(2)}$ functions on $\mathbb{R}^m$.

Parts 1 and 2 are trivial; here is a proof that $d^2 = 0$.

LEMMA 4.57. *For each $k = 0, 1, 2, \ldots$ we have $0 = d^2 = d_{k+1} \circ d_k$.*

PROOF. It suffices to show $d^2\omega = 0$ on a coordinate chart (U_α, x_α). In local coordinates, $\omega \in \bigwedge^k M$ is a sum over $I \in \mathcal{E}_k$ of terms like $f \cdot dx_I$. Since the exterior derivative d_k is linear, $d_k\omega$ is a sum of terms
$$d(f dx_I) = \sum_{i=1}^m \frac{\partial f}{\partial x_i} \cdot (dx_i \wedge dx_I),$$
and then $d^2(\omega) = d_{k+1}(d_k\omega)$ consists of terms
$$d(d\omega) = \sum_{j=1}^m \sum_{i=1}^m \frac{\partial^2 f}{\partial x_j \partial x_i} \cdot dx_j \wedge (dx_i \wedge dx_I) \qquad \text{with } I \in \mathcal{E}_k.$$
The monomials $(dx_i \wedge dx_I)$ appearing in $d(f dx_i)$ might not be in standard form, but formula (4.40) nevertheless determines the same differential form $(d\omega)_p$ in $\bigwedge^{k+1}(\mathrm{TM}_p^*)$ for every chart U_α containing p.

Since $\dfrac{\partial^2 f}{\partial x_j \partial x_i} = \dfrac{\partial^2 f}{\partial x_i \partial x_j}$, anticommutativity of wedge products makes the terms

$$\frac{\partial^2 f}{\partial x_j \partial x_i}(dx_j \wedge dx_i \wedge dx_I)$$

and

$$\frac{\partial^2 f}{\partial x_i \partial x_j}(dx_i \wedge dx_j \wedge dx_I) = (-1)\frac{\partial^2 f}{\partial x_i \partial x_j}(dx_j \wedge dx_i \wedge dx_I)$$

in the double sum cancel in pairs for $i \neq j$, and when $j = i$, we have $dx_j \wedge dx_i = dx_i \wedge dx_i \equiv 0$ on U_α. Thus, $d(d\omega) = 0$ on any chart, as claimed. □

The following observation is also useful in calculations involving exterior derivatives.

LEMMA 4.58. *If $f \in \mathcal{C}^\infty(M)$ and $\omega \in \bigwedge^k M$ with $\omega = \sum_{I \in \mathcal{E}_k} c_I(u) \cdot (dx_I)$ in local coordinates, then the exterior derivative of the combined k-form $f\omega$ is described on the chart domain U_α by*

$$d(f\omega) = \sum_{I \in \mathcal{E}_k} c_I(u) \cdot (df) \wedge (dx_I) + f(u) \cdot (dc_I) \wedge (dx_I) \tag{4.42}$$

in $\bigwedge^{k+1}(U)$.

PROOF. The exterior derivative takes the form

$$\begin{aligned} d(f\omega) &= d\Big(\sum_{I \in \mathcal{E}_k} (f \cdot c_I)(u) \cdot (dx_I)\Big) = \sum_{I \in \mathcal{E}_k} d((fc_I) \cdot (dx_I)) \\ &= \sum_{I \in \mathcal{E}_k} d(fc_I) \wedge (dx_I) \quad \text{(by (4.40))} \end{aligned}$$

in local coordinates. Since d is a derivation on $\mathcal{C}^\infty(U_\alpha)$, we get (4.42). □

Primitives of k-Forms. We say that $\mu \in \bigwedge^k M$ is a *primitive* for ω in $\bigwedge^{k+1} M$ if $d_k\mu = \omega$ on M. A primitive for ω might not exist, but the condition $d^2 \equiv 0$ imposes a necessary condition for ω to have locally defined primitives, namely

$$d\omega = d_{k+1}\omega \equiv 0 \quad \text{on } M.$$

In fact, if $\omega = d\mu$ on some open set $U \subseteq M$, then $d\omega = d^2\mu = 0$ on U. For rank-1 forms, this identity is equivalent to the set of "consistency conditions" (4.26) mentioned earlier. It suffices to verify this on a typical chart, so suppose $d_1\omega = 0$ in $\bigwedge^2(U_\alpha)$ for $\omega = \sum_{i=1}^m c_i \cdot dx_i$ in $\bigwedge^1(U_\alpha)$. Then $dx_i \wedge dx_j = 0$ if $i = j$, so

$$\begin{aligned} d\omega &= \sum_{i=1}^m dc_i \wedge dx_i = \sum_i \sum_j \left(\frac{\partial c_i}{\partial x_j}\right) \cdot (dx_j \wedge dx_i) \\ &= \sum_{1 \le i < j \le m} \left(\frac{\partial c_j}{\partial x_i} - \frac{\partial c_i}{\partial x_j}\right) \cdot (dx_i \wedge dx_j). \end{aligned}$$

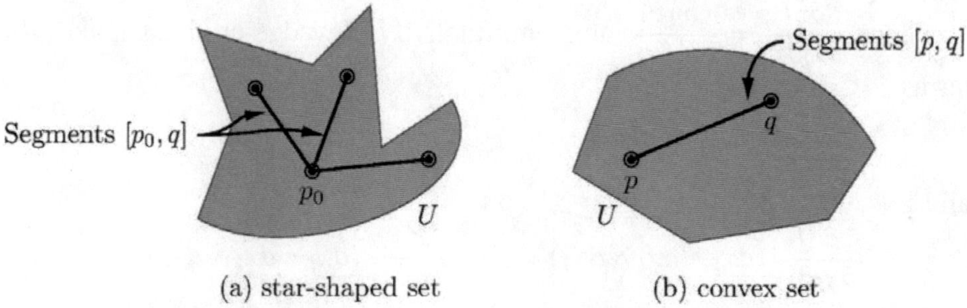

FIGURE 4.5. Definition of *convex* and *star-shaped* open sets in $\mathbb{R}^n$. Here $[\,p, q\,]$ is the straight line segment in $\mathbb{R}^n$ connecting p and q.

This sum over basis vectors in $\bigwedge^2(\mathrm{TM}_p)$ is zero at each $p \in U_\alpha$ if and only if (4.26) holds.

The higher-rank analogues of these consistency conditions assert that if $\omega = d_{k-1}\mu$ for some $\mu \in \bigwedge^{k-1} M$, then on any chart the coefficients $c_I \in \mathcal{C}^\infty(U_\alpha)$ in $\omega = \sum_{I \in \mathcal{E}_k} c_I(u) \cdot (dx_I)$ must satisfy a system of partial differential equations embodied in the concise statement

$$d_k\omega \equiv 0 \quad \text{in } \bigwedge^{k+1}(U_\alpha).$$

In local coordinates, this system of differential equations becomes increasingly complicated for $k = 2, 3, \ldots$

EXERCISE 4.59. Write out the system of partial differential equations satisfied on a chart (U_α, x_α) by the coefficients of a 2-form in $\bigwedge^2(U_\alpha)$:

$$\omega = \sum_{1 \le i < j \le m} c_{ij}(u) \cdot (dx_i \wedge dx_j)$$

if $d_2\omega \equiv 0$ in $\bigwedge^3(U_\alpha)$.

Poincaré's Lemma. The necessary condition $d\omega = 0$ actually ensures that solutions of the equation $\omega = d\mu$ exist locally near any base point in U_α. There are two fundamental results about solutions of $d\mu = \omega$. The first asserts that *local* solutions exist if $d\omega \equiv 0$; it is a consequence of Poincaré's lemma, proved below, which also provides information about existence of *global* solutions.

THEOREM 4.60. *Local primitives of $\omega \in \bigwedge^k M$ exist near a point $p \in M$ if and only if the necessary condition $d_k\omega \equiv 0$ in $\bigwedge^{k+1} M$ is satisfied.*

Once we know that $d\omega \equiv 0$ throughout M, the geometry of the manifold must be taken into account in seeking *global* solutions of $d\mu \equiv \omega$ on all of M. There is also the problem of *constructing* the desired local and global solutions μ when they exist; this is actually addressed within the proof of Poincare's lemma.

The proof of Theorem 4.60 requires some preparation. A set $E \subseteq \mathbb{R}^n$ is *convex* if for any pair of points $\mathbf{p}, \mathbf{q} \in E$, the "line segment"

$$[\mathbf{p}, \mathbf{q}] = \{r\mathbf{p} + s\mathbf{q} : 0 \leq r, s \leq 1 \text{ and } r + s = 1\}$$

connecting $\mathbf{p}$ to $\mathbf{q}$ lies entirely within E; more generally, E is *star-shaped* if there is some $\mathbf{p} \in E$ from which every point $\mathbf{x} \in E$ is "visible," with $[\mathbf{p}, \mathbf{x}] \subseteq E$ — see Figure 4.5.

LEMMA 4.61 (Poincaré Lemma). *If E is a convex or star-shaped open set in $\mathbb{R}^n$ and $\omega \in \bigwedge^k E$ satisfies the condition $d_k\omega \equiv 0$ throughout E, then there is a global solution $\mu \in \bigwedge^{k-1} E$ to the equation $d_{k-1}\mu \equiv \omega$, so ω has a global primitive on E.*

REMARK 4.62. A $\mathcal{C}^\infty$ map $\phi : M \to N$ between manifolds induces a natural linear map $\delta\phi : \bigwedge^k N \to \bigwedge^k M$ that "pulls back" smooth k-forms on N to smooth k-forms on M; furthermore, $\delta\phi$ is a bijective linear isomorphism if ϕ is a diffeomorphism between M and N (invertible, with $\mathcal{C}^\infty$ inverse). Finally, $\delta\phi$ "intertwines" the exterior derivatives d^M and d^N on M and N so that $\delta\phi(d^N(\omega)) = d^M(\delta\phi(\omega))$ — i.e., the following diagram is commutative:

$$\begin{array}{ccc} \bigwedge^k N & \xrightarrow{(\delta\phi)} & \bigwedge^k M \\ d^N \downarrow & & \downarrow d^M \\ \bigwedge^{k+1} N & \xrightarrow{(\delta\phi)} & \bigwedge^{k+1} M. \end{array} \tag{4.43}$$

The significance of the diagram (4.43) deserves further discussion, but for the moment, a few general comments about pullbacks of forms will suffice. The map $\delta\phi$ is a generalization to higher rank differential forms of the natural linear map $(\delta\phi)_{\phi(p)} : \mathrm{TN}^*_{\phi(p)} \to \mathrm{TM}^*_p$ between cotangent spaces that transfers 1-forms in TN^*_q to 1-forms in TM^*_p. If $\phi(p) = q$ and $k = 1$, $(\delta\phi)_q$ is given by

$$\langle (\delta\phi)_q\omega_q \, , X_p \rangle = \langle (\delta\phi)_{\phi(p)}\omega_{\phi(p)} \, , X_p \rangle = \langle \omega_{\phi(p)} \, , (d\phi)_p X_p \rangle$$

for all $\omega_q \in \mathrm{TM}^*_q$ and $X_p \in \mathrm{TM}_p$. Thus, $(\delta\phi)_{\phi(p)}$ is just the transpose

$$(\delta\phi)_{\phi(p)} = (d\phi)^{\mathrm{T}}_p : \mathrm{TN}^*_{\phi(p)} \to \mathrm{TM}^*_p$$

of the differential $(d\phi)_p : \mathrm{TM}_p \to \mathrm{TN}_{\phi(p)}$ introduced in Definition 4.21. Both $(\delta\phi)_{\phi(p)}$ and $(d\phi)_p$ are "natural maps," defined without reference to any particular system of local coordinates. ○

Transferring Calculations Between M and $\mathbb{R}^m$. If (U_α, x_α) is a chart on a manifold M, the chart map $x_\alpha : U_\alpha \to V_\alpha = x_\alpha(U_\alpha)$ is a diffeomorphism between open sets in M and $N = \mathbb{R}^m$ (which is also a manifold). We may therefore use the chart map $\phi = x_\alpha$ to transfer questions about smooth differential forms on $U_\alpha \subseteq M$ to the corresponding problems on $V_\alpha \subseteq \mathbb{R}^m$. By commutativity of the diagram in (4.43), solutions of $d^M\mu = \omega$ on U_α correspond to solutions of $d^N\mu' = \omega'$ for the pulled-back forms $\mu' = \delta\phi(\mu)$ and $\omega' = \delta\phi(\omega)$. Likewise, the necessary conditions $d^M\omega = 0$ and $d^N\omega = 0$ for existence of primitives match up under this correspondence.

Now observe that if $p \in M$, there are always (small) "star-shaped charts" (U_α, x_α) about p such that $V_\alpha = x_\alpha(U_\alpha)$ is a star-shaped neighborhood of $\mathbf{p} = x_\alpha(p)$ in $\mathbb{R}^m$ — in fact, any chart about p contains an open neighborhood U such that $p \in U \subseteq U_\alpha$ and $x_\alpha(U)$ is a (convex) open ball in $\mathbb{R}^n$. If we are lucky, we can sometimes find quite large star-shaped charts on M. For instance, if $M = \mathbb{R}^2$, the "cut plane"

$$E = \mathbb{R}^2 \sim \{(x, y) \in \mathbb{R}^2 : -\infty < x \le 0,\ y = 0\}$$

is star-shaped with respect to any point p on the positive x-axis. Poincaré's lemma guarantees existence of global solutions for $d\mu = \omega$ throughout any star-shaped chart domain on which $d\omega \equiv 0$.

Proof of Poincaré's Lemma.[1] We now provide a long (but elementary) proof of Poincaré's lemma.

PROOF. We may obviously assume that the "central point" $p \in E$ is the origin in $\mathbb{R}^n$. If $\omega = \sum_{I \in \mathcal{E}_k} w_I(u) \cdot (dx_I)_u$ on E, we will define "rank-lowering" linear maps $I_k : \bigwedge^k E \to \bigwedge^{k-1}(E)$ for $k = 1, 2, \ldots$ such that

$$d_{k-1} I_k(\omega) + I_{k+1} d_k(\omega) = \omega \quad \text{on } E \text{ for } \omega \in \bigwedge^k E\,, \tag{4.44}$$

as in the diagram below.

$$\begin{array}{ccc} \bigwedge^k E & \xrightarrow{I_k} & \bigwedge^{k-1} E \\ d_k \downarrow & & \downarrow d_{k-1} \\ \bigwedge^{j+1} E & \xrightarrow{I_{k+1}} & \bigwedge^k E \end{array} \tag{4.45}$$

If $d_k(\omega) = 0$, then $I_{k+1} d_k(\omega) = 0$ and $\mu = I_k(\omega) \in \bigwedge^{k-1} E$ is the desired solution of $d_{k-1}\mu = \omega$. Since I_k and d_k are linear operators, we may restrict attention to weighted monomial k-forms on E,

$$\omega = w_I(u) \cdot dx_I = w_I(u) \cdot (dx_{i_1} \wedge \cdots \wedge dx_{i_k}) \quad \text{for } I \in \mathcal{E}_k,\ w_I \in \mathcal{C}^\infty(E).$$

Primitives of ω resemble antiderivatives of scalar-valued functions, so it is not surprising to find that $I_k(\omega)$ can be constructed by taking integrals that involve the coefficients $w_I(u)$ and applying a basic result from advanced calculus.

THEOREM. (DIFFERENTIATING UNDER THE INTEGRAL). *Let $B \subseteq \mathbb{R}^m$ be a closed rectangular set and $f : \mathbb{R}^m \times \mathbb{R} \to \mathbb{R}$ a function defined on some open set $U \supseteq B \times [a, b\,]$. If the partial derivatives $\partial f / \partial x_i(\mathbf{x}, y)$ exist and are continuous on U for $1 \le i \le m$, then $H(\mathbf{x}) = \int_a^b f(\mathbf{x}, t)\, dt$ has continuous partial derivatives*

$$\frac{\partial H}{\partial x_i}(\mathbf{x}) = \int_a^b \frac{\partial f}{\partial x_i}(\mathbf{x}, t)\, dt$$

on an open neighborhood of B for $1 \le i \le m$.

[1] Adapted (with corrections) from M. Spivak, *Calculus on Manifolds*, W. A. Benjamin, 1965.

If the partial derivatives $D_{\mathbf{x}}^{\alpha} f(\mathbf{x}, t)$ in the $\mathbf{x}$-variables exist and are continuous on U for multi-indices of degree $|\alpha| = \alpha_1 + \cdots + \alpha_m \leq k$, then repeated differentiation under the integral shows that $H(\mathbf{x})$ is of class $\mathcal{C}^{(k)}$ on an open neighborhood of B with

$$D_{\mathbf{x}}^{\alpha} H(\mathbf{x}) = \int_a^b D_{\mathbf{x}}^{\alpha} f(\mathbf{x}, t)\, dt$$

for $|\alpha| \leq k$.

For any weighted monomial $\omega = w_I(\mathbf{x}) \cdot dx_I$ with $I = (i_1 < \cdots < i_r) \in \mathcal{E}_r$, $r \geq 1$, we define the rank-lowering operator I_r, taking $I_r\omega(\mathbf{x}) \in \bigwedge^{r-1}(E)$ to be the following sum of integrals:

(4.46)

$$I_r\omega(\mathbf{x}) = \sum_{\ell=1}^{r} (-1)^{\ell-1}\Big(x_{i_\ell} \cdot \int_0^1 w_I(t\mathbf{x})\, t^{r-1}\, dt\Big) \cdot (dx_{i_1} \wedge \cdots \wedge \widehat{dx_{i_\ell}} \wedge \cdots \wedge dx_{i_r})$$

where the "hat" over dx_{i_ℓ} indicates an omitted factor. Hereafter, we shall simplify notation by relabeling the variables as $x_1, \ldots, x_m$ so $dx_I = dx_{i_1} \wedge \cdots \wedge dx_{i_r}$ becomes $dx_1 \wedge \cdots \wedge dx_r$.

Taking $r = k$, differentiation under the integral shows that $I_k\omega(\mathbf{x})$ is a smooth $(k-1)$-form on E. To compute the second term $d_{k-1}I_k\omega$ in (4.44), we first find $D_{x_s}\{I_k\omega(\mathbf{x})\}$ for $1 \leq s \leq m$ by differentiating under the integral. By the product formula for derivatives, we get

$$\begin{aligned} D_{x_s}\Big(x_\ell \cdot \int_0^1 w_I(t\mathbf{x})\, t^{k-1}\, dt\Big) &= x_\ell \cdot \int_0^1 D_{x_s}\{w_I(t\mathbf{x})\}\, t^{k-1}\, dt + \delta_{s,\ell} \cdot \Big(\int_0^1 w_I(t\mathbf{x})\, t^{k-1}\, dt\Big) \\ &= x_\ell \cdot \int_0^1 D_{x_s} w_I(t\mathbf{x})\, t^{k}\, dt + \delta_{s,\ell} \cdot \Big(\int_0^1 w_I(t\mathbf{x})\, t^{k-1}\, dt\Big) \end{aligned} \tag{4.47}$$

because $D_{x_s}\{w_I(t\mathbf{x})\} = t \cdot D_{x_s} w_I(t\mathbf{x})$. Here $\delta_{s,\ell}$ is the Kronecker delta symbol ($= 1$ if $s = \ell$ and zero otherwise). Noting that $dx_s \wedge dx_1 \wedge \cdots \wedge dx_k = 0$ if $s \in [1, k]$, Definition 4.53 of exterior derivative d_{k-1} yields

$$\begin{aligned} d_{k-1}I_k\omega(\mathbf{x}) &= \sum_{\ell=1}^{k}\sum_{s=1}^{m} D_{x_s}(\cdots) \cdot (-1)^{\ell-1} dx_s \wedge (dx_1 \wedge \cdots \wedge \widehat{dx_\ell} \wedge \cdots \wedge dx_k) \\ &= \sum_{\ell=1}^{k}\sum_{s=1}^{m} D_{x_s}(\cdots) \cdot (-1)^{\ell+k}\, (dx_1 \wedge \cdots \wedge \widehat{dx_\ell} \wedge \cdots \wedge dx_k) \wedge dx_s \end{aligned} \tag{4.48}$$

(because $(-1)^{-2} = 1$ and shifting dx_s across $k-1$ vectors produces a factor $(-1)^{k-1}$). When (4.47) is substituted in (4.48), the first term in (4.47) yields the following double sum:

$$\sum_{\ell=1}^{k}\sum_{s=1}^{m} x_\ell \cdot \left(\int_0^1 D_{x_s} w_I(t\mathbf{x})\, t^k\, dt\right) \cdot (-1)^{\ell+k}(dx_1 \wedge \cdots \wedge \widehat{dx_\ell} \wedge \cdots \wedge dx_k \wedge dx_s)$$

in which the nonzero terms are those with $s > k$ or $s = \ell$. Owing to the Kronecker delta in the second term of (4.47), substitution of this into (4.48) yields only k terms (those with $s = \ell$), and the second term becomes

$$\sum_{\ell=1}^{k}\left(\int_0^1 w_I(t\mathbf{x})\, t^{k-1}\, dt\right) \cdot (-1)^{\ell-1} dx_\ell \wedge (dx_1 \wedge \cdots \wedge \widehat{dx_\ell} \wedge \cdots \wedge dx_k)$$

$$= \sum_{\ell=1}^{k}\left(\int_0^1 w_I(t\mathbf{x})\, t^{k-1}\, dt\right) \cdot dx_I \;=\; k \cdot \left(\int_0^1 w_I(t\mathbf{x})\, t^{k-1}\, dt\right) dx_I$$

since $(-1)^{\ell-1} \cdot (-1)^{\ell-1} = 1$.

As for the second term in (4.44), by our labeling convention, we may write

$$d_k\omega = d_k(w_I(\mathbf{x}) \cdot dx_I) = \sum_{s=1}^{m} D_{x_s} w_I(\mathbf{x}) \cdot dx_s \wedge dx_I = \sum_{s=1}^{m} D_{x_s} w_I(\mathbf{x}) \cdot (-1)^k dx_I \wedge dx_s.$$

Notation must be interpreted carefully in applying I_r: in particular, when $r = k+1$, the monomial "$dx_{i_1} \wedge \cdots \wedge dx_{i_{k+1}}$" appearing in (4.46) gets interpreted as "$dx_I \wedge dx_s$." Then by definition of I_{k+1}, we get

$$I_{k+1} d_k(w_I \cdot dx_I) = I_{k+1}\left(\sum_{s=1}^{m} D_{x_s} w_I(\mathbf{x}) \cdot (-1)^k (dx_1 \wedge \cdots \wedge dx_k) \wedge dx_s\right)$$

$$= \sum_{\ell=1}^{k+1}\sum_{s=1}^{m}\left(x_\ell \cdot \int_0^1 D_{x_s} w_I(t\mathbf{x})\, t^{(k+1)-1}\, dt\right) \cdot (-1)^k(-1)^{\ell-1}$$

$$\cdot (dx_1 \wedge \cdots \wedge \widehat{dx_\ell} \wedge \cdots \wedge dx_k \wedge dx_s)$$

$$= \sum_{\ell=1}^{k}\sum_{s=1}^{m} (\cdots) + (\text{terms with } \ell = k+1)\,.$$

In the sum of terms with $\ell = k+1$, we have $(-1)^k(-1)^{(k+1)-1} = (-1)^k(-1)^k = 1$ since $dx_{k+1} = dx_s$. The monomial

$$(dx_1 \wedge \cdots \wedge \widehat{dx_\ell} \wedge \cdots \wedge dx_k \wedge dx_s) = (dx_1 \wedge \cdots \wedge dx_k \wedge \widehat{dx_{k+1}})$$

common to all these terms becomes $dx_1 \wedge \cdots \wedge dx_k = dx_I$, so the $\ell = k+1$ term becomes

$$\sum_{s=1}^{m}\left(x_s \cdot \int_0^1 D_{x_s} w_I(t\mathbf{x})\, t^k\, dt\right) \cdot (-1)^k(-1)^k(dx_1 \wedge \cdots \wedge dx_k)$$

$$= \sum_{s=1}^{m} (\cdots) \cdot (dx_1 \wedge \cdots \wedge dx_k) = \sum_{s=1}^{m} (\cdots) \cdot dx_I.$$

On the other hand, by the chain rule, we also have

$$D_{x_s}\{w_I(t\mathbf{x})\} = x_s \cdot w_I(t\mathbf{x})\,.$$

Hence,

$$\begin{aligned} w_I(\mathbf{x}) &= \left[w_I(t\mathbf{x})\, t^k \Big|_{t=0}^{1} \right] = \int_0^1 \frac{d}{dt}\{w_I(t\mathbf{x})\, t^k\}\, dt \\ &= \left(x_s \cdot \int_0^1 \sum_{s=1}^{m} D_{x_s} w_I(t\mathbf{x})\, t^k\, dt \right) + k \cdot \left(\int_0^1 w_I(t\mathbf{x})\, t^{k-1}\, dt \right). \end{aligned}$$

Substituting this into the $\ell = k+1$ term, we may rewrite

$$I_{k+1} d_k \omega = \sum_{\ell=1}^{k} \sum_{s=1}^{m} (\cdots) + w_I(\mathbf{x})\, dx_I - k \cdot \left(\int_0^1 w_I(\mathbf{x})\, t^{k-1}\, dt \right).$$

When this is combined with $d_{k-1} I_k \omega$, the double sums cancel and so do the intergrals involving w_I, leaving only

$$I_{k+1} d_k \omega + d_{k-1} I_k \omega = \omega.$$

That completes the proof of Poincaré's lemma. □

The identity (4.44) has a long history, which we can't spell out here. You might wonder how anyone deduced that an identity like (4.44) might hold and how the definition (4.45) of the integral operators $I_k : \bigwedge^k E \to \bigwedge^{k-1} E$ was discovered. Creative mathematics is all about doing things like that.

4.4. Div, Grad, Curl, and All That

We now consider what all this means on the Euclidean spaces $M = \mathbb{R}^2$ and $\mathbb{R}^3$ and its connection with the traditional vector operators $\nabla\circ =$ **div**, $\nabla\cdot =$ **grad**, and $\nabla\times =$ **curl** of calculus. When $M = \mathbb{R}^3$, the sequence of exterior derivatives

$$\bigwedge^0 M \xrightarrow{d_0} \bigwedge^1 M \xrightarrow{d_1} \bigwedge^2 M \xrightarrow{d_2} \bigwedge^3 M \xrightarrow{d_3} (0)$$

terminates because $\dim \bigwedge^k(\mathrm{TM}_p) = \binom{3}{k} = 0$ for $k > 3$.

Let $\{\mathbf{e}_i\}$ be the standard basis vectors in $V = \mathbb{R}^3$ and $\{\mathbf{e}_i^*\}$ the dual vectors in $V^* = \bigwedge^1 V$. Fix a base point p in $M = \mathbb{R}^3$. If we interpret the $\mathbf{e}_i$ as tangent vectors at p, so $\mathbf{e}_i =$ the directional derivative $\big(\partial/\partial x_i|_p\big)$ in TM_p, then the dual basis vectors are $\mathbf{e}_i^* = (dx_i)_p$ in TM_p^*. We have already seen that

$$\text{Classical Gradient: } \nabla f = \sum_{i=1}^{3} D_{x_i} f(\mathbf{x}) \cdot \mathbf{e}_i$$

should be interpreted as a rank-1 differential form, the exterior derivative of $f \in \bigwedge^0 M = \mathcal{C}^\infty(M)$, which involves the dual basis vectors $\mathbf{e}_i^* = (dx_i)_p$,

$$\nabla f(p) = \sum_{i=1}^{3} \frac{\partial f}{\partial x_i}(p)\, \mathbf{e}_i^* = \sum_{i=1}^{3} \frac{\partial f}{\partial x_i}(p)\, (dx_i)_p = (df)_p \quad \text{in } \mathrm{TM}_p^*$$

rather than as a tangent vector

$$\sum_{i=1}^{3} \frac{\partial f}{\partial x_i}(p)\,\mathbf{e}_i = \sum_{i=1}^{3} \frac{\partial f}{\partial x_i}(p)\left(\frac{\partial}{\partial x_i}\Big|_p\right) \quad \text{in } \mathrm{TM}_p\,.$$

As noted earlier, the exterior derivative $df = \nabla f$ takes the same general form in all coordinate systems. For instance, in spherical coordinates $y_\beta(\mathbf{x}) = (\rho, \theta, \phi)$ and Cartesian coordinates $x_\alpha(\mathbf{x}) = (x, y, z)$, we have

$$df = \frac{\partial f}{\partial x}\,dx + \cdots + \frac{\partial f}{\partial z}\,dz = \frac{\partial f}{\partial \rho}\,d\rho + \cdots + \frac{\partial f}{\partial \phi}\,d\phi\,.$$

Now consider a classical "vector field" on $\mathbb{R}^3$ of the sort encountered in calculus, which is usually written as

$$\mathbf{F}(\mathbf{x}) = F_1(\mathbf{x})\,\mathbf{i} + F_2(\mathbf{x})\,\mathbf{j} + F_3(\mathbf{x})\,\mathbf{k}$$

with smooth coefficients $F_i : \mathbb{R}^3 \to \mathbb{R}$. It is not clear how the basis vectors $\{\mathbf{i}, \mathbf{j}, \mathbf{k}\}$ in these classical narratives are to be interpreted — as tangent vectors attached to each base point p? Cotangent vectors? Something else? The answer depends on the physical nature of the "field" being modeled. For instance we have seen that an electric field $\mathbf{E}(p)$, being $\nabla\phi$ for some scalar potential function, should be regarded as a field of cotangent vectors and not as a field of tangent vectors.

Our interpretation of ∇f as a field of cotangent vectors suggests that, in this case at least, we might interpret the traditional basis vectors $\{\mathbf{i}, \mathbf{j}, \mathbf{k}\}$ as

$$\mathbf{i} = (dx)_p \quad \mathbf{j} = (dy)_p \quad \mathbf{k} = (dz)_p.$$

On the other hand, consider the operator **curl** $= \nabla\times$ which sends a classical vector field $\mathbf{F}$ to a new vector field $\nabla \times \mathbf{F}$ given by

$$\nabla \times \mathbf{F} = \mathbf{curl}(\mathbf{F}) = \det \begin{bmatrix} \mathbf{i} & \mathbf{j} & \mathbf{k} \\ \partial/\partial x & \partial/\partial y & \partial/\partial z \\ F_1 & F_2 & F_3 \end{bmatrix}$$

$$(4.49) \qquad = \left(\frac{\partial F_3}{\partial y} - \frac{\partial F_2}{\partial z}\right)\mathbf{i} + \left(\frac{\partial F_1}{\partial z} - \frac{\partial F_3}{\partial x}\right)\mathbf{j} + \left(\frac{\partial F_2}{\partial x} - \frac{\partial F_1}{\partial y}\right)\mathbf{k}\,.$$

Should we still interpret the symbols $\mathbf{i}, \mathbf{j}, \mathbf{k}$ the same way — as dual vectors in TM_p^*? The proper interpretation will depend on what type of tensor field $\mathbf{F}$ represents — a vector field in $\mathcal{D}^{(1,0)}(M)$, a field of cotangent vectors in $\mathcal{D}^{(0,1)}(M) = \bigwedge^1 M$, or whatever. If $\mathbf{F}$ represents a smooth 1-form (say an electric field in space), it turns out that $\nabla \times \mathbf{F}$ should be interpreted as the exterior derivative $d\mathbf{F}$, making $\nabla \times \mathbf{F}$ a smooth field of antisymmetric rank-2 tensors, a 2-form in $\bigwedge^2 M$.

To see that this is the correct interpretation, write $\mathbf{F} = F_1\,dx_1 + F_2\,dx_2 + F_3\,dx_3$, regarding $\mathbf{F}$ as a 1-form in $\bigwedge^1 M$. Its exterior derivative would then be

$$\begin{aligned}
d\mathbf{F} &= \sum_{i=1}^{3}(dF_i)\wedge dx_i = \sum_{i=1}^{3}\Big(\sum_{j=1}^{3}\frac{\partial F_i}{\partial x_j}\,dx_j\Big)\wedge dx_i \\
&= \sum_{i\neq j}\frac{\partial F_i}{\partial x_j}\cdot dx_j\wedge dx_i \quad (\text{since } dx_i\wedge dx_i = 0) \\
&= \sum_{1\le i<j\le 3}\left(\frac{\partial F_j}{\partial x_i}-\frac{\partial F_i}{\partial x_j}\right)\cdot dx_i\wedge dx_j \quad (\text{since } dx_j\wedge dx_i = -dx_i\wedge dx_j) \\
&= \left(\frac{\partial F_2}{\partial x_1}-\frac{\partial F_1}{\partial x_2}\right)\cdot dx_1\wedge dx_2 + \left(\frac{\partial F_3}{\partial x_2}-\frac{\partial F_2}{\partial x_3}\right)\cdot dx_2\wedge dx_3 \\
&\quad + \left(\frac{\partial F_3}{\partial x_1}-\frac{\partial F_1}{\partial x_3}\right)\cdot dx_1\wedge dx_3\,.
\end{aligned}$$

We get a match with the classical curl $\nabla\times\mathbf{F}$ if we interpret the "unit vectors" $\mathbf{i}, \mathbf{j}, \mathbf{k}$ as

$$\mathbf{i} = dx_2\wedge dx_3 \quad \mathbf{j} = dx_3\wedge dx_1 = -dx_1\wedge dx_3 \quad \mathbf{k} = dx_1\wedge dx_2.$$

Note the cyclic order $dx_1 \to dx_2 \to dx_3 \to dx_1$ employed here.

Up to a $\pm$ sign, these are the ordered basis monomials in $\bigwedge^2 M$ when M is equipped with Cartesian coordinates. However, these "unit vectors" are no longer "vectors" of the same type as the unit vectors $\mathbf{i}, \mathbf{j}, \mathbf{k}$ in the original field $F_1\,\mathbf{i} + F_2\,\mathbf{j} + F_3\,\mathbf{k} \in \bigwedge^1(\mathrm{TM}_p)$ because they are rank-2 antisymmetric tensors in $\bigwedge^2(\mathrm{TM}_p)$. It is an accident that the tensor $\nabla\times\mathbf{F}(p)$ can be mistaken for the same kind of object as $\mathbf{F}(p)$. This happens because

$$\dim \bigwedge^1(\mathrm{TM}_p) = \binom{3}{1} = \binom{3}{2} = \dim \bigwedge^2(\mathrm{TM}_p) = 3$$

on any three-dimensional manifold.

Since vector spaces of the same dimension are "isomorphic," it is tempting to regard them as being "the same," but they are not, and there is no natural way to identify them. While it is often useful to introduce bases and identify both spaces $\bigwedge^1(\mathrm{TM}_p)$ and $\bigwedge^2(\mathrm{TM}_p)$ with coordinate space $\mathbb{R}^3$ in order to perform calculations, this obscures the very different nature of these tensors, which have different ranks and do not transform the same way when described in other systems of local coordinates. Moreover, this dodge cannot work in higher dimensions, for if physical space were four-dimensional, we would have

$$\dim \bigwedge^1(\mathrm{TM}_p) = \binom{4}{1} = 4 \quad \text{while} \quad \dim \bigwedge^2(\mathrm{TM}_p) = \binom{4}{2} = 6.$$

Furthermore, in vector calculus on $\mathbb{R}^4$, a new "vector operator" would appear — the exterior derivative d_2 in

$$\bigwedge^0\mathbb{R}^4 \xrightarrow{d_0} \bigwedge^1\mathbb{R}^4 \xrightarrow{d_1} \bigwedge^2\mathbb{R}^4 \xrightarrow{d_2} \bigwedge^3\mathbb{R}^4 \xrightarrow{d_3} \bigwedge^4\mathbb{R}^4 \xrightarrow{d_4} (0)$$

which is not encompassed in the three-dimensional theory of classical Calculus. In contrast, the theory of differential forms makes perfect sense for tensors of arbitrary ranks in all dimensions, and it even makes sense on differentiable manifolds such as spheres $S^n \subseteq \mathbb{R}^{n+1}$ as well as other manifolds that have no obvious realizations as smooth hypersurfaces embedded in Euclidean spaces.

Finally, consider the divergence operator $\mathbf{div} = \nabla\circ$. In calculus, you were told that **div** acts on "vector fields" $\mathbf{F} = F_1\,\mathbf{i} + F_2\,\mathbf{j} + F_3\,\mathbf{k}$ to produce "scalar fields" via the formula

$$\nabla\circ\mathbf{F} = \operatorname{div}\mathbf{F} = \operatorname{Trace}\begin{bmatrix} \partial F_1/\partial x_1 & \partial F_1/\partial x_2 & \partial F_1/\partial x_3 \\ \partial F_2/\partial x_1 & \partial F_2/\partial x_2 & \partial F_2/\partial x_3 \\ \partial F_3/\partial x_1 & \partial F_3/\partial x_2 & \partial F_3/\partial x_3 \end{bmatrix} \tag{4.50}$$

$$= \frac{\partial F_1}{\partial x_1} + \frac{\partial F_2}{\partial x_2} + \frac{\partial F_3}{\partial x_3} \quad \text{(sum of diagonal entries).}$$

We shall make sense of this in terms of exterior derivatives by interpreting $\mathbf{F}$ as an antisymmetric rank-2 tensor field using labels $\mathbf{i},\mathbf{j},\mathbf{k}$ for the basis vectors in the three-dimensional space $\bigwedge^2(\mathrm{TM}_p)$,

$$\mathbf{i} = dx_2\wedge dx_3 \quad \mathbf{j} = dx_3\wedge dx_1 = -dx_1\wedge dx_3 \quad \mathbf{k} = dx_1\wedge dx_2.$$

Applying the exterior derivative $d_2 : \bigwedge^2(\mathbb{R}^3) \to \bigwedge^3(\mathbb{R}^3)$ to $\mathbf{F} = \sum_{I\in\mathcal{E}_2} F_I\,dx_I$, we get

$$\begin{aligned} d\mathbf{F} &= \sum_{I\in\mathcal{E}_2} (dF_I)\wedge dx_I \\ &= \Big(\sum_{i=1}^{3} \frac{\partial F_1}{\partial x_i}\,dx_i\Big)\wedge(dx_2\wedge dx_3) - \Big(\sum_{i=1}^{3} \frac{\partial F_2}{\partial x_i}\,dx_i\Big)\wedge(dx_1\wedge dx_3) \\ &\quad + \Big(\sum_{i=1}^{3} \frac{\partial F_3}{\partial x_i}\,dx_i\Big)\wedge(dx_1\wedge dx_2) \\ &= \Big(\frac{\partial F_1}{\partial x_1} + \frac{\partial F_2}{\partial x_2} + \frac{\partial F_3}{\partial x_3}\Big)\cdot dx_1\wedge dx_2\wedge dx_3 \quad \text{in } \textstyle\bigwedge^3(\mathbb{R}^3) \end{aligned}$$

because $dx_i\wedge dx_j\wedge dx_k = 0$ if any two of the indices i,j,k are equal. As an example, magnetic fields $\mathbf{B}(\mathbf{x})$ are represented by rank-2 antisymmetric tensor fields. An important physical law asserts that $d\mathbf{B} = \nabla\circ\mathbf{B}$ is identically zero in any region of space (which in effect says that "magnetic monopoles do not exist").

Since $\dim\bigwedge^3(\mathrm{TM}_p) = 1$ on $M = \mathbb{R}^3$, all elements of $\bigwedge^3(\mathrm{TM}_p)$ are scalar multiples of the single basis vector $(dx_1\wedge dx_2\wedge dx_3)_p$ and therefore might easily be confused with actual scalars (which are elements of the one-dimensional space $\bigwedge^0(\mathrm{TM}_p) = \mathbb{R}$). But they are *not* scalars. In fact, if we take a different Euclidean coordinate system on $M = \mathbb{R}^3$ with reversed orientation, say $(y_1,y_2,y_3) = y_\beta(\mathbf{x}) = (x_2,x_1,x_3)$, we find that

$$dy_I = dy_1\wedge dy_2\wedge dy_3 = dx_2\wedge dx_1\wedge dx_3 = -\,dx_1\wedge dx_2\wedge dx_3 = -dx_I$$

for all $I \in \mathcal{E}_3$, reversing the sign on the basis vector. That never happens for a true scalar field $f \in \bigwedge^0(\mathbb{R}^3) = \mathcal{C}^\infty(\mathbb{R}^3)$.

Summarizing these remarks, we have found the following:

- GRADIENT: **grad** $= \nabla\cdot$ acting on scalar fields should be interpreted as the exterior derivative $d_0 : \bigwedge^0(\mathbb{R}^3) \to \bigwedge^1(\mathbb{R}^3)$. The result is a smooth 1-form.
- CURL: **curl** $= \nabla\times$ acts on smooth 1-forms and should be interpreted as the exterior derivative $d_1 : \bigwedge^1(\mathbb{R}^3) \to \bigwedge^2(\mathbb{R}^3)$. The result is a smooth 2-form.
- DIVERGENCE: **div** $= \nabla\circ$ acts on smooth 2-forms and should be interpreted as the exterior derivative $d_2 : \bigwedge^2(\mathbb{R}^3) \to \bigwedge^3(\mathbb{R}^3)$. The resulting smooth 3-form can be written as $\phi(\mathbf{x})\cdot(dx_1 \wedge dx_2 \wedge dx_3)$ for some $\mathcal{C}^\infty$ scalar function $\phi : \mathbb{R}^3 \to \mathbb{R}$.

At the next level we have $d_3 : \bigwedge^3(\mathbb{R}^3) \to \bigwedge^4(\mathbb{R}^3) = (0)$. Interpreting the classical vector operators in this manner, we immediately obtain two classic vector identities for fields on $\mathbb{R}^3$:

$$\nabla\times(\nabla f) = d_1 \circ d_0(f) \equiv 0 \quad \text{and} \quad \nabla\circ(\nabla\times\mathbf{F}) = d_2 \circ d_1(\mathbf{F}) \equiv 0 \tag{4.51}$$

from the fundamental property $d^2 = d_{k+1} \circ d_k = 0$ of exterior derivatives.

Vector Operations in Two Dimensions. There are similar, but simpler, results for $M = \mathbb{R}^2$.

EXAMPLE 4.63 (The Two-Dimensional Case). In $M = \mathbb{R}^2$, there are only two nontrivial vector operators

$$\textstyle\bigwedge^0(\mathbb{R}^2) \xrightarrow{d_0} \bigwedge^1(\mathbb{R}^2) \xrightarrow{d_1} \bigwedge^2(\mathbb{R}^2) \xrightarrow{d_2} (0)$$

because $\bigwedge^k(\mathbb{R}^2) = (0)$ for $k \geq 3$. By definition of exterior derivative, we get

- **grad**, which maps $f \in \bigwedge^0(\mathbb{R}^2) = \mathcal{C}^\infty(\mathbb{R}^2)$ to

$$\nabla f = df = \frac{\partial f}{\partial x_1}\,dx_1 + \frac{\partial f}{\partial x_2}\,dx_2 \quad \text{in } \textstyle\bigwedge^1(\mathbb{R}^2)$$

- **curl**, which maps $\omega = F_1\,dx_1 + F_2\,dx_2$ in $\bigwedge^1(\mathbb{R}^2)$ to

$$\begin{aligned} d\omega &= \left(\frac{\partial F_1}{\partial x_1}\,dx_1 + \frac{\partial F_1}{\partial x_2}\,dx_2\right) \wedge dx_1 + \left(\frac{\partial F_2}{\partial x_1}\,dx_1 + \frac{\partial F_2}{\partial x_2}\,dx_2\right) \wedge dx_2 \\ &= \left(\frac{\partial F_2}{\partial x_1} - \frac{\partial F_1}{\partial x_2}\right)\cdot dx_1 \wedge dx_2 \quad \text{in } \textstyle\bigwedge^2(\mathbb{R}^2)\,. \end{aligned}$$

There is no two-dimensional analog of the divergence operator **div** since $d\omega = 0$ for every 2-form on $\mathbb{R}^2$. (Likewise for any two-dimensional differentiable manifold.)

It is now clear that the d-operators on the manifold $M = \mathbb{R}^m$

$$\bigwedge^0 M \xrightarrow{d_0} \bigwedge^1 M \xrightarrow{d_1} \bigwedge^2 M \xrightarrow{d_2} \dots \xrightarrow{d_{m-1}} \bigwedge^m M \xrightarrow{d_m} \bigwedge^{m+1} M = (0) \ ,$$

with $d \circ d = 0$ at every stage, are the natural higher-dimensional analogs of **div**, **grad**, **curl** in three dimensions.

In dealing with three dimensional space, physicists and mathematicians have adopted notation devised (circa 1890) by J. Willard Gibbs, who exploited the coincidence

$$\dim \bigwedge^0(\mathbb{R}^3) = \dim \bigwedge^3(\mathbb{R}^3) = 1 \quad \text{and} \quad \dim \bigwedge^1(\mathbb{R}^3) = \dim \bigwedge^2(\mathbb{R}^3) = 3$$

to identify these spaces with the usual coordinate spaces $\mathbb{R}^1$ and $\mathbb{R}^3$. He also exploited the fact that the manifold structure of $M = \mathbb{R}^3$ is determined by a *single chart* $x_\alpha(\mathbf{x}) = (x_1, x_2, x_3)$ which provides us with globally defined smooth k-forms:

For $k = 1$: $dx_1, \quad dx_2, \quad dx_3$
For $k = 2$: $dx_1 \wedge dx_2, \quad dx_2 \wedge dx_3, \quad dx_3 \wedge dx_1 = -\, dx_1 \wedge dx_3$
For $k = 3$: $dx_1 \wedge dx_2 \wedge dx_3$

that determine correlated basis vectors at all base points. This allows us to compare the spaces $\bigwedge^k(\mathrm{TM}_p)$ and $\bigwedge^k(\mathrm{TM}_q)$ at different base points. Thus we can identify spaces having the same dimension in a uniform way:

$$\overset{0}{\bigwedge}(\mathrm{TM}_p) \cong \overset{3}{\bigwedge}(\mathrm{TM}_p) \cong \mathbb{R} \quad \text{and} \quad \overset{1}{\bigwedge}(\mathrm{TM}_p) \cong \overset{2}{\bigwedge}(\mathrm{TM}_p) \cong \mathbb{R}^3$$

for all base points. This is really convenient in doing calculations, but it obscures the distinction between various kinds of "vectors" in $\mathbb{R}^2$ and "scalars" in $\mathbb{R}$. Most physicists and mathematicians have voted for convenience, although most were well aware that these identifications are only a shorthand description of what is actually going on.

Differential Forms and the Cross Product. Gibbs also introduced the *cross product* $\mathbf{v} \times \mathbf{w}$ of vectors in $\mathbb{R}^3$ as a simplified proxy for the wedge product of cotangent vectors. If we write the basis vectors in $\bigwedge^1(\mathrm{TM}_p)$ as

$$\mathbf{i} = (dx_1)_p \quad \mathbf{j} = (dx_2)_p \quad \mathbf{k} = (dx_3)_p$$

and those in $\bigwedge^2(\mathrm{TM}_p)$ as

$$\begin{aligned}
\mathbf{i}' &= \mathbf{j} \wedge \mathbf{k} = (dx_2 \wedge dx_3)_p \\
\mathbf{j}' &= \mathbf{k} \wedge \mathbf{i} = (dx_3 \wedge dx_1)_p = -\,(dx_1 \wedge dx_3)_p \\
\mathbf{k}' &= \mathbf{i} \wedge \mathbf{j} = (dx_1 \wedge dx_2)_p \ ,
\end{aligned}$$

and if we now identify basis vectors $\mathbf{i}' \sim \mathbf{i}$, $\mathbf{j}' \sim \mathbf{j}$, $\mathbf{k}' \sim \mathbf{k}$ in these three-dimensional spaces, these identities take the form

$$\mathbf{i} \wedge \mathbf{j} = \mathbf{k}' \sim \mathbf{k} \qquad \mathbf{j} \wedge \mathbf{k} = \mathbf{i}' \sim \mathbf{i} \qquad \mathbf{k} \wedge \mathbf{i} = \mathbf{j}' \sim \mathbf{j}. \tag{4.52}$$

These should look familiar. In calculus, the cross product of $\mathbf{v} = v_1\mathbf{i} + v_2\mathbf{j} + v_3\mathbf{k}$ and $\mathbf{w} = w_1\mathbf{i} + w_2\mathbf{j} + w_3\mathbf{k}$ in $\mathbb{R}^3$ is defined to be

$$\begin{aligned} \mathbf{v} \times \mathbf{w} &= \det \begin{bmatrix} \mathbf{i} & \mathbf{j} & \mathbf{k} \\ v_1 & v_2 & v_3 \\ w_1 & w_2 & w_3 \end{bmatrix} \\ &= (v_2w_3 - v_3w_2)\mathbf{i} + (v_3w_1 - v_1w_3)\mathbf{j} + (v_1w_2 - v_2w_1)\mathbf{k}. \end{aligned} \tag{4.53}$$

In particular,

$$\mathbf{i} \times \mathbf{j} = \mathbf{k} \qquad \mathbf{j} \times \mathbf{k} = \mathbf{i} \qquad \mathbf{k} \times \mathbf{i} = \mathbf{j}. \tag{4.54}$$

Under these identifications, the classical cross product *is* the wedge product. If we regard $\mathbf{v}, \mathbf{w}$ as cotangent vectors in $\bigwedge^1(\mathrm{TM}_p)$, then $\mathbf{v}\times\mathbf{w}$ is the 2-form $\mathbf{v} \wedge \mathbf{w}$ in $\bigwedge^2(\mathrm{TM}_p)$, which we then identify with a vector back in $\bigwedge^1(\mathrm{TM}_p) \cong \bigwedge^2(\mathrm{TM}_p)$ to get the classical cross product $\mathbf{v} \times \mathbf{w}$ in (4.53). Antisymmetry of the wedge product yields the familiar calculus identities

$$\mathbf{w} \times \mathbf{v} = -\mathbf{v} \times \mathbf{w} \qquad \mathbf{v} \times \mathbf{v} = \mathbf{0} \text{ for all } \mathbf{v}$$

which also follow from the calculus-style definition (4.54). But beware: the identifications that underlie this interpretation of $\mathbf{v} \times \mathbf{w}$ are only possible in three dimensional space, while the wedge product on $\bigwedge^k(\mathbb{R}^n)$ makes sense in *all* dimensions.

EXERCISE 4.64. For $k = 2, 3, 4$, compute components of the wedge product $\mathbf{v} \wedge \mathbf{w}$ of $\mathbf{v} = \sum_{i=1}^{l} v_i\mathbf{e}_i$ and $\mathbf{w} = \sum_{j=1}^{k} w_j\mathbf{e}_j$, interpreting $\mathbf{e}_i = (dx_i)$ in $\bigwedge^1(\mathbb{R}^k)$.

A geometric interpretation of the cross product $\mathbf{a} \times \mathbf{b}$ in $\mathbb{R}^3$ is shown in Figure 4.6. It is always perpendicular to the plane spanned by the two vectors, and its length $\|\mathbf{a}\times\mathbf{b}\|$ is the area of the parallelogram R determined by $\mathbf{a}$ and $\mathbf{b}$. The meaning of the *vector triple product* $(\mathbf{a}\times\mathbf{b}) \bullet \mathbf{c}$ of vectors in $\mathbb{R}^3$ is discussed in the following exercises. Its numerical value (with sign) is the *oriented 3-dimensional volume* Vol(P) of the parallelopiped

$$P = \{x_1\mathbf{a} + x_2\mathbf{b} + x_3\mathbf{c} \,:\, 0 \le x_i \le 1\}$$

whose edges are the vectors $\mathbf{a}, \mathbf{b}, \mathbf{c}$. The signed volume of P is $(+)$ if the vectors form a "right-handed system" — i.e., if you wrap your fingers around the axis perpendicular to $\mathbf{a}$ and $\mathbf{b}$ with your index finger pointing in the same direction as $\mathbf{a}$ and middle finger aligned with $\mathbf{b}$, your thumb should be pointing in the same direction as $\mathbf{c}$; otherwise the signed volume is $(-)$.

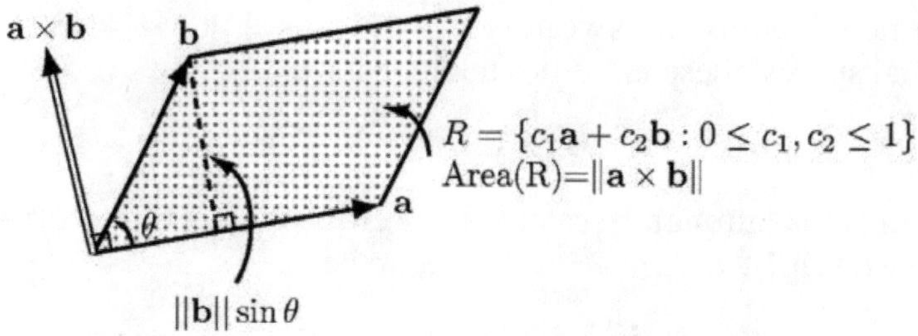

FIGURE 4.6. Geometric interpretation of the cross product $\mathbf{a}\times\mathbf{b}$ of two vectors in $\mathbb{R}^3$. The cross product is always perpendicular to the plane in $\mathbb{R}^3$ determined by the two vectors, and the length of the cross product is the area of the parallelogram R determined by the two vectors, $||\mathbf{a}\times\mathbf{b}|| = \text{Area}(R)$.

EXERCISE 4.65. Prove the following

(a) If $\mathbf{a}$ and $\mathbf{b}$ are nonzero, then $\mathbf{a}\times\mathbf{b} = \mathbf{0} \Leftrightarrow$ the vectors are collinear. Furthermore,

$$(\mathbf{a}\times\mathbf{b}) \bullet \mathbf{a} = 0 \quad \text{and} \quad (\mathbf{a}\times\mathbf{b}) \bullet \mathbf{b} = 0\,,$$

so $\mathbf{a}\times\mathbf{b}$ is orthogonal to every vector in the plane $\mathbb{R}\mathbf{a} + \mathbb{R}\mathbf{b}$.

(b) $\mathbf{a} \bullet (\mathbf{b}\times\mathbf{c}) = (\mathbf{a}\times\mathbf{b}) \bullet \mathbf{c}$.

(c) $||\mathbf{a}\times\mathbf{b}||^2 = ||\mathbf{a}||^2||\mathbf{b}||^2 \cdot (1-\cos^2\theta) = ||\mathbf{a}||^2||\mathbf{b}||^2 \cdot \sin^2(\theta)$ for the angle θ shown in Figure 4.6.

HINT. Show that the identity in (c) is equivalent to $||\mathbf{a}\times\mathbf{b}||^2 + |\mathbf{a}\bullet\mathbf{b}|^2$, which can be verified directly.

EXERCISE 4.66. If $\mathbf{a}, \mathbf{b}, \mathbf{c}$ are in $\mathbb{R}^3$, verify that the *triple product* $(\mathbf{a}\times\mathbf{b}) \bullet \mathbf{c}$, in which $(\bullet)$ is the usual inner product on $\mathbb{R}^3$, is the scalar

$$(\mathbf{a}\times\mathbf{b}) \bullet \mathbf{c} = \det \begin{pmatrix} a_1 & a_2 & a_3 \\ b_1 & b_2 & b_3 \\ c_1 & c_2 & c_3 \end{pmatrix}.$$

Explain why this is precisely the (oriented) volume of the parallelopiped P determined by the three vectors. This will be zero (P is degenerate) $\Leftrightarrow$ the vectors are linearly dependent.

EXERCISE 4.67. The triple product $\lambda = (\mathbf{a}\times\mathbf{b}) \bullet \mathbf{c}$ of vectors in $\mathbb{R}^3$ is a scalar. The space $\bigwedge^3(\mathbb{R}^3)$ is one-dimensional. If we interpret vectors as 1-forms in $\bigwedge^1(\mathbb{R}^3)$, prove that

$$\mathbf{a}\wedge\mathbf{b}\wedge\mathbf{c} = \lambda \cdot (dx_1)\wedge(dx_2)\wedge(dx_3)$$

in $\bigwedge^3(\mathbb{R}^3)$.

Additional Exercises

This chapter has been concerned with the behavior of partial derivatives $\partial f/\partial x_j$ of functions $f : M \to \mathbb{R}$ on a differentiable manifold M and the way calculations with them are managed by transferring them to Euclidean space $\mathbb{R}^m$ via the cordinate charts (U_α, x_α) that define the differentiable structure on M. That reduces these issues to questions about functions $f : \mathbb{R}^m \to \mathbb{R}$ and mappings $f : \mathbb{R}^m \to \mathbb{R}^n$ that should be familiar from multivariate calculus. As a preamble to the Additional Exercises for Section 4.1, we provide a very brief review of the issues involved in defining "derivatives" of functions of several variables.

You are familiar with the partial derivatives $D_{x_j}f(p) = \partial f/\partial x_j(p)$ and the more general directional derivatives of a function $f : \mathbb{R}^m \to \mathbb{R}$

$$D_u f(p) = \lim_{\Delta t \to 0} \frac{f(p + \Delta t \cdot u) - f(p)}{\Delta t}$$

for various base points and directions $u \in \mathbb{R}^m$. But mere existence of these at points in $\mathbb{R}^m$ is too weak a condition to be useful when $m \geq 2$. What we need is existence of the *total derivative* $Df(p)$, which is a linear map $Df(p) : \mathbb{R}^m \to \mathbb{R}$ that "closely approximates" the behavior of the nonlinear map f near p in the sense that

$$f(p + \Delta x) = f(p) + Df(p) \cdot \Delta x + \text{Error}(\Delta x)$$

where the error term becomes small compared to the displacement Δx from p as $||\Delta x|| \to 0$. That is,

$$\frac{||\text{Error}(\Delta x)||}{||\Delta x||} \to 0 \text{ as } ||\Delta x|| \to 0.$$

A similar definition applies to vector-valued maps $f : \mathbb{R}^m \to \mathbb{R}^n$.

Existence of the total derivative $Df(p)$ near a base point p implies that all partial derivatives and directional derivatives exist and are well behaved near p. Df is closely connected with the usual partial derivatives of f in the following ways:

- $Df(p)$ exists at a base point if the partial derivatives $D_{x_1}f, \ldots, D_{x_m}f$ exist and are continuous near p.
- When the scalar components of $f = \big(f_1(x), \ldots, f_n(x)\big)$ have continuous partial derivatives $D_{x_i}f_j$ on an open set $U \subseteq \mathbb{R}^m$, the total derivative Df exists at every base point and is represented by the Jacobian matrix

$$[Df(p)] = \left[\frac{\partial f_i}{\partial x_j}(p)\right] = \begin{pmatrix} \frac{\partial f_1}{\partial x_1} & \cdots & \frac{\partial f_1}{\partial x_m} \\ \vdots & & \vdots \\ \frac{\partial f_n}{\partial x_1} & \cdots & \frac{\partial f_n}{\partial x_m} \end{pmatrix}_{n \times m}$$

whose entries vary continuously with p.

Section 4.1. Tangent Vectors, Cotangent Vectors, and Tensors

1. TRUE/FALSE QUESTIONS ("True" if the statement is *always* true.)

(a) If $f : \mathbb{R}^2 \to \mathbb{R}$ has well-defined partial derivatives $\partial f/\partial x_1$ and $\partial f/\partial x_2$ at a particular base point $p \in \mathbb{R}^2$, then $f(x)$ must be continuous at p.

(b) If $f : \mathbb{R}^2 \to \mathbb{R}$ has well-defined partial derivatives $\partial f/\partial x_1$ and $\partial f/\partial x_2$ at a particular base point $p \in \mathbb{R}^2$, all directional derivatives $D_u f(p)$ exist.

(c) If $f : \mathbb{R}^2 \to \mathbb{R}$ has well-defined partial derivatives $\partial f/\partial x_1$ and $\partial f/\partial x_2$ at *all* base points $\mathbf{p} \in \mathbb{R}^2$, then $f(x)$ must be continuous on $\mathbb{R}^2$.

(d) If $f : \mathbb{R}^2 \to \mathbb{R}$ has well-defined partial derivatives $\partial f/\partial x_1$ and $\partial f/\partial x_2$ that are *continuous* throughout $\mathbb{R}^2$, then $f(x, y)$ must be continuous everywhere on $\mathbb{R}^2$.

HINT. You might want to look at Additional Exercise 4 below, which suggests that the search for a counterexamples is not entirely trivial.

(e) If the mixed partial derivatives $\partial^2 f/\partial x_1 \partial x_2$ and $\partial^2 f/\partial x_2 \partial x_1$ exist and are continuous on an open set $U \subseteq \mathbb{R}^2$, then

$$\frac{\partial^2 f}{\partial x_1 \partial x_2} = \frac{\partial^2 f}{\partial x_2 \partial x_1} \quad \text{throughout } U.$$

REMARK 4.68. After some obvious rewording, all of these questions apply to functions $f : \mathbb{R}^n \to \mathbb{R}$ (with the same answers) for all $n \geq 2$. The n-dimensional version of Item (e) is really important in differential geometry; the one-dimensional case is exceptional. In this chapter, we will circumvent these technical difficulties by confining attention to maps of "class $C^{(k)}$" (or class $\mathcal{C}^\infty$), for which all partial derivatives of degree $\leq k$ exist and are continuous on some open set $U \subseteq \mathbb{R}^m$. ○

2. If a map $f : \mathbb{R}^m \to \mathbb{R}$ has a well-defined total derivative $Df(p)$ at every base point in an open set U, prove from the definition of $Df(p)$ that f must be continuous at every base point.

3. If a map $f : \mathbb{R}^m \to \mathbb{R}$ has a well-defined total derivative $Df(p)$ at every base point in an open set U, prove from the definition of $Df(p)$ that directional derivatives

$$D_u f(p) = \lim_{\Delta t \to 0} \frac{f(p + \Delta t \cdot u) - f(p)}{\Delta t}$$

exist for all directions $u \in \mathbb{R}^m$ and all base points $p \in U$.

4. If $f : \mathbb{R}^2 \to \mathbb{R}$ is defined near the origin, with $f(0,0) = 0$ and

$$f(x,y) = \frac{x^2 y}{x^4 + y^2} \quad \text{for } (x,y) \neq (0,0)\,,$$

prove the following:

(a) The directional derivatives

$$D_u f(p) = \lim_{\Delta t \to 0} \frac{f(p + \Delta t \cdot u) - f(p)}{\Delta t}$$

exist at the origin $p = (0,0)$ in all directions u.

(b) Compute the directional derivative $\gamma'(0) \in \mathbb{R}^2$ along lines $\gamma(t) = (th, tk)$ through the origin in terms of the parameters h, k.

(c) The "total derivative" $Df(0,0)$ fails to exist because $f(x,y)$ isn't even continuous at the origin.

HINT. Show that directional derivatives along the lines through the origin $\gamma(t) = (th, tk)$ with $h, k \in \mathbb{R}$ exist and equal h^2/k when $k \neq 0$. (What if $k = 0$?) Then show that the limit of the f-values seen as we move toward the origin along the parabolic curve $(x,y) = (t, ct^2)$ is equal to $c^2/(1+c)$. These limits are not equal, as they would be if f were continuous.

NOTE. Even existence of all directional derivatives at the origin is not enough to ensure that the total derivative $Df(0,0)$ exists. In fact, $f(x,y)$ is not even continuous at the origin, which cannot happen when Df exists. All this reinforces the idea that the proper definition of differentiability of $f(x,y)$ at a base point is existence of the total derivative $Df(p)$, not existence of the individual partial derivatives $D_{x_j}(p)$. ○

5. In polar coordinates, the *logarithmic spiral* in $M = \mathbb{R}^2$ takes the form

$$(r, \theta) = \gamma(t) = (e^t, 2\pi t) \quad \text{for } t \in \mathbb{R}.$$

(a) Describe the tangent vector $\gamma'(t)$ in TM_p at $p = \gamma(t)$ in terms of the basis vectors in TM_p

$$\left(\frac{\partial}{\partial r}\Big|_p\right), \left(\frac{\partial}{\partial \theta}\Big|_p\right)$$

associated with polar coordinates $x_\beta = (r, \theta)$.

(b) Use the change of variable formula, Corollary 4.16, to rewrite $\gamma'(t)$ in terms of the basis vectors

$$\left(\frac{\partial}{\partial x}\Big|_p\right), \left(\frac{\partial}{\partial y}\Big|_p\right)$$

determined by the usual Cartesian coordinates $x_\alpha(u) = (x, y)$.

6. Suppose a function $f : M \to \mathbb{R}$ is defined near a base point p that lies in two overlapping coordinate charts (U_α, x_α) and (U_β, y_β). If all the partial derivatives $\partial f/\partial x_i(p)$ are zero for $1 \leq i \leq m$, explain why all partial derivatives $\partial f/\partial y_j(p)$ with respect to the y_β-coordinates must also be zero at p.

7. The complex exponential function

$$e^z = (e^x \cos y) + i(e^x \sin y) \quad \text{for } z = x + iy \in \mathbb{C}$$

can be regarded as a map

$$\phi(x, y) = \big(U(x, y), V(x, y)\big) = \big(e^x \cos(y)\,,\, e^x \sin(y)\big)$$

from $\mathbb{R}^2 \to \mathbb{R}^2$.

(a) Compute the Jacobian matrix

$$[d\phi(p)] = \begin{pmatrix} \dfrac{\partial U}{\partial x} & \dfrac{\partial U}{\partial y} \\ \dfrac{\partial V}{\partial x} & \dfrac{\partial V}{\partial y} \end{pmatrix}$$

that determines the total derivative at p.

(b) Compute the image vectors

$$(d\phi)_p\Big(\frac{\partial}{\partial x}\Big|_p\Big), \quad (d\phi)_p\Big(\frac{\partial}{\partial y}\Big|_p\Big)$$

in terms of the basis $\mathfrak{Y}$ in $\mathrm{TM}_{\phi(p)}$.

NOTE. Here $[(d\phi)_p]_{\mathfrak{X}\mathfrak{Y}}$ is the matrix of the "differential"

$$(d\phi)_p : \mathrm{TM}_p \to \mathrm{TM}_{\phi(p)}$$

of ϕ at p with respect to the bases

$$\mathfrak{X} = \Big\{\Big(\frac{\partial}{\partial x}\Big|_p\Big), \Big(\frac{\partial}{\partial y}\Big|_p\Big)\Big\} \qquad \text{in } \mathrm{TM}_p$$

$$\mathfrak{Y} = \Big\{\Big(\frac{\partial}{\partial x}\Big|_{\phi(p)}\Big), \Big(\frac{\partial}{\partial y}\Big|_{\phi(p)}\Big)\Big\} \quad \text{in } \mathrm{TM}_{\phi(p)}$$

determined by the globally defined coordinate chart $x_\alpha(u) = (x, y)$. ○

8. In the next chapter, in order to apply the implicit function theorem, we will want to find the rank r of the $n{\times}m$ Jacobian matrix $A = [Df(p)] = [\partial f_i/\partial x_j(p)]$ of a smooth map $f : \mathbb{R}^m \to \mathbb{R}^n$ at a typical base point p. It will also be necessary to explicitly identify r columns $C_{i_1}, \dots, C_{i_r}$ that are a basis for column space of A, which is the range $R(L_A)$ of $L_A : \mathbb{R}^m \to \mathbb{R}^n$. If

$$A = \begin{pmatrix} 3 & 8 & 3 & 2 & 5 \\ -1 & -1 & 1 & -1 & 0 \\ 4 & 10 & 5 & 1 & 6 \\ 2 & 3 & 2 & -1 & 1 \end{pmatrix}$$

(a) Determine $r = \operatorname{rank}(A)$ by any method.

(b) Find an explicit set of columns $C_{j_1}, \dots, C_{j_r}$ that are a basis for $R(L_A)$. (There may be several choices of such columns.)

(c) Find an explicit choice of rows $R_{i_1}, \dots, R_{i_r}$ such that the $r \times r$ submatrix A_{IJ} is nonsingular. (Entries in A_{IJ} lie in the spots where these rows R_i and columns C_j meet.)

Section 4.2. Cotangent Vectors and Differential Forms

1. TRUE/FALSE QUESTIONS ("True" if the statement is *always* true.)

(a) If $f : M \to \mathbb{R}$ is a $\mathcal{C}^\infty$ function defined near a base point in a manifold M, the exterior derivative $(df)_p$ is a tangent vector in TM_p for each $p \in M$.

(b) If p is a base point in a manifold M, every cotangent vector ω_p in TM_p^* is the exterior derivative $(df)_p$ for some $\mathcal{C}^\infty$ scalar-valued function f defined near p.

(c) If M is a differentiable manifold, the tangent space at any point $p \in M$ has the same dimension $\dim_{\mathbb{R}}(\text{TM}_p)$ as the coordinate charts (U_α, x_α) that determine the differentiable structure of M.

(d) Let p be a base point in a manifold M, (U_α, x_α) any coordinate chart about p, and $\mathfrak{X} = (\partial/\partial x_1|_p), \dots, (\partial/\partial x_m|_p)$ the standard basis vectors in TM_p determined by the chart. Then $\mathfrak{X}$ is a basis for the tangent space TM_p.

(e) Let p be a base point in a manifold M. Given any coordinate chart (U_α, x_α) about p, the scalar components of the chart map $(U_\alpha, x_\alpha) = (X_1, \dots, X_m)$ are $\mathcal{C}^\infty$ functions $X_k : U_\alpha \to \mathbb{R}$ whose exterior derivatives $(dX_k)_p$ are a basis for the cotangent space TM_p^*.

2. On $M = \mathbb{R}^3$, we have Cartesian coordinates $x_\alpha(u) = (x, y, z)$ and spherical coordinates $y_\beta(u) = (\rho, \theta, \phi)$ — see Figure 4.3. Let $\rho = R(u)$ be the Euclidean distance from $u \in \mathbb{R}^3$ to the origin.

(a) Write $R(u)$ as a function of the spherical coordinates of u, and describe the exterior derivatives $dR_u \in \text{TM}_u^*$ of $R(u)$ in terms of the 1-forms $(d\rho)_u, (d\theta)_u, (d\phi)_u$ determined by spherical coordinates.

(b) Rewrite this field of cotangent vectors in terms of the cotangent vectors $(dx)_u, (dy)_u, (dz)_u$ in TM_u^* determined by the Euclidean coordinate chart.

3. Same general question as in Additional Exercise 2 above, but this time compute the exterior derivative $(dX)_u$ of the x-variable $x = X(u)$ in terms of the spherical coordinate variables ρ, θ, ϕ. (Recall Figure 4.3.)

4. If we impose spherical coodinates (θ, ϕ) on the upper hemisphere $H_3^+ = \{\mathbf{x} \in S^2 : x_3 > 0\}$, the radial variable ρ is irrelevant because $\rho = \|\mathbf{x}\| = 1$ for all points on the sphere $\mathbb{S}^2$. The standard coordinate chart on the upper hemisphere H_3^+ is

$$x_\alpha(u) = P(x_1, x_2, x_3) = (x_1, x_2) \quad \text{in the unit disc } D \subseteq \mathbb{R}^2.$$

Rewrite the following rank-2 differential forms on S^2 in new coordinates:

(a) $(d\theta) \wedge (d\phi)$ in terms of $(dx_1) \wedge (dx_2)$.
(b) $(dx_1) \wedge (dx_2)$ in terms of $(d\theta) \wedge (d\phi)$.

Section 4.3. Differential Forms on *M* and Exterior Derivatives

1. TRUE/FALSE QUESTIONS ("True" if the statement is *always* true.)

(a) Every rank-k tensor ω on a finite-dimensional vector space V is either *symmetric* with

$$\omega\big(v_{\sigma(1)}, \dots, v_{\sigma(k)}\big) = \omega(v_1, \dots, v_k)$$

or *antisymmetric* with

$$\omega\big(v_{\sigma(1)}, \dots, v_{\sigma(k)}\big) = \operatorname{sgn}(\sigma) \cdot \omega(v_1, \dots, v_k),$$

for all permutations $\sigma \in S_k$.

(b) Every rank-k tensor $\omega \in V^{(0,k)}$ is a sum $\omega = \omega_a + \omega_s$ of an antisymmetric tensor and a symmetric tensor.

(c) Every rank-2 tensor $\omega(v_1, v_2)$ in $V^{(0,2)}$ has the symmetry property that allows you to move a scalar λ from one input to the other:

$$\omega(\lambda v_1, v_2) = \omega(v_1, \lambda v_2).$$

(d) A rank-r tensor $\omega(v_1, \dots, v_r)$ is zero if any of its inputs $v_i \in V$ is the zero vector in V.

(e) If $\dim(V) = n$, $\omega \in V^{(0,r)}$ is a tensor of rank $r \geq 2$ on V, and $1 \leq i < j \leq r$, then for all inputs $v_1, \dots, v_r$, we have
$\omega(v_1, \dots, v_j, \dots, v_i, \dots, v_r) = (-1) \cdot \omega(v_1, \dots, v_i, \dots, v_j, \dots, v_r)$,
i.e., transposing *any two inputs* to ω switches the sign ($\pm$) of the output.

(f) The output of a rank-k tensor an a finite-dimensional vector space is "translation invariant": for fixed $\mathbf{a} = (a_1, \dots, a_k)$ in V, we have

$$\omega(v_1 + a_1, \dots, v_k + a_k) = \omega(v_1, \dots, v_k)$$

for all inputs $v_1, \dots, v_k \in V$.

2. Explain why

$$\omega(\lambda_1 v_1, \dots, \lambda_k v_k) = \left(\prod_{i=1}^{k} \lambda_i\right) \cdot \omega(v_1, \dots, v_k) \quad \text{for } v_i \in V, \lambda_i \in \mathbb{K}$$

for any rank-k tensor ω on a vector space V.

3. If $\ell_1, \dots, \ell_k$ are in V^*, verify that the "tensor product operator"

$$\ell_1 \otimes \cdots \otimes \ell_k : (v_1, \dots, v_k) \to \prod_{j=1}^{k} \langle \ell_j, v_j \rangle$$

from $V \times \cdots \times V \to \mathbb{R}$ is a k-linear map and hence a tensor in $V^{(0,k)}$.

4. Given a chart (U_α, x_α) with $x_\alpha(u) = (x_1, \dots, x_m)$ on a manifold M, we may define rank-2 tensor fields $(dx_i \otimes dx_j)$ on U_α.

(a) If $i \neq j$, is $(dx_i \otimes dx_j)_p \neq 0$ for all $p \in U_\alpha$?

(b) If $i \neq j$, is $(dx_i \otimes dx_j)_p \neq (dx_j \otimes dx_i)_p$ for all $p \in U_\alpha$?

(c) Is $(dx_i \otimes dx_i)_p$ nonzero for all $p \in U_\alpha$?

Explain your answers.

5. Explain why a wedge product of vectors in V^*

$$\phi_1^* \wedge \cdots \wedge \phi_r^* \quad \text{with } 1 \leq r \leq n = \dim(V)$$

is zero unless the vectors ϕ_i are linearly independent.

6. If $\omega \in \bigwedge^1 M$, prove that $\omega \wedge \omega = 0$ in $\bigwedge^2 M$. If $\omega \in \bigwedge^2 M$, is $\omega \wedge \omega = 0$ in $\bigwedge^4 M$?

HINT. You can restrict attention to a chart domain, where we have $(dx_i \wedge dx_j) = (-1) \cdot (dx_j \wedge dx_i)$. For the second part, try a sum of two suitably chosen monomials $e_i^* \wedge e_j^*$.

7. We have noted the very useful fact that every rank-2 tensor ω (a bilinear form) on V can always be split as a sum $\omega_s + \omega_a$ of a symmetric tensor $\omega_s \in S^2(V)$ and an antisymmetric tensor $\omega_a \in \bigwedge^2(V)$. This is not always true for tensors of rank $r \geq 3$, and for large r, the dimension of the linear span $S^r(V) + \bigwedge^r(V)$ lags far behind $\dim\ V^{(0,r)} = (\dim\ V)^r$.

Prove that it is not always possible to decompose tensors of rank $= r > n = \dim(V)$ in $V^{(0,r)}$ as $\omega = (\text{antisymmetric}) + (\text{symmetric})$.

HINT. Antisymmetric tensors are zero when $r > n$, so existence of such a decomposition would force all tensors of rank $r > \dim V$ to be *symmetric*. Do you really believe that is so? In seeking a counterexample, try $V = \mathbb{R}^2$ and see if you can construct a rank-3 tensor on V that is not symmetric. In fact, if $\mathfrak{X} = \{\mathbf{e}_i\}$ is a basis in V with dual basis $\mathfrak{X}^* = \{\mathbf{e}_i^*\}$, try a few of the $2^3 = 8$ possible monomials $\mathbf{e}_{i_1}^* \otimes \mathbf{e}_{i_2}^* \otimes \mathbf{e}_{i_3}^*$ that are basis vectors for $V^{(0,3)}$. Are they all symmetric tensors?

8. DETERMINANTS AS TENSORS ON $\mathbb{K}^n$. If $A \in \mathrm{M}(n, \mathbb{K})$ and we regard its rows $R_i = (a_{i1}, \dots, a_{in}) \in \mathbb{K}^n$ as inputs to the map

$$\omega(R_1, \dots, R_n) = \det(A)\,,$$

explain why the following hold:

(a) $\omega : V \times \cdots \times V \to \mathbb{K}$ is a rank-n multilinear map on $V = \mathbb{K}^n$.

(b) The output $\omega(R_1, \dots, R_n)$ changes sign if two inputs are switched before applying ω.

(c) The output $\omega(R_1, \dots, R_n)$ gets multiplied by $\mathrm{sgn}(\sigma)$ if we apply an arbitrary permutation $\sigma \in S_n$ to the inputs, sending $R_i \to R_{\sigma(i)}$ for $1 \leq i \leq n$.

Thus ω is an antisymmetric tensor of rank $= \dim(V)$.

NOTE. $\bigwedge^n V$ is one-dimensional when $n = \dim(V)$; furthermore, if $\mathfrak{X} = \{e_j\}$ is a basis for V and $\mathfrak{X}^* = \{e_j^*\}$ is the dual basis in V^*, then $\omega = e_1^* \wedge \cdots \wedge e_n^*$ is a basis vector for $\bigwedge^n V$. Thus, the determinant of A is the unique scalar $d(A)$ such that $\det(A) = d(A) \cdot e_1^* \wedge \cdots \wedge e_n^*$ in $\bigwedge^n V$. This should be compared with the official definition of $\det(A)$ in equation (4.3) of LA I. ○

HINT. Review the subsection *Row Operations, Determinants, and Matrix Inverses* in Section 4.3 of LA I.

9. On $M = \mathbb{R}^2$, we have Cartesian coordinates $x_\alpha(u) = (x, y)$ and polar coordinates $y_\beta(u) = (r, \theta)$. Using the change of variable formula (Corollary 4.16) for tangent vectors,

(a) Rewrite the vector fields $(\partial/\partial x)$ and $(\partial/\partial y)$ in terms of $(\partial/\partial r)$ and $(\partial/\partial\theta)$.

(b) Regarding smooth vector fields $\widetilde{X}$ on M as operators that act on C^∞ functions $C^\infty(M)$ via

$$\widetilde{X}(f)(u) = \langle X_u, f\rangle \quad \text{for } f \in \mathcal{C}^\infty(M), u \in M\,,$$

express the following composite operators on functions

$$\frac{\partial^2 f}{\partial x^2} = \frac{\partial}{\partial x}\left(\frac{\partial f}{\partial x}\right) \quad \text{and} \quad \frac{\partial^2 f}{\partial y^2} = \frac{\partial}{\partial y}\left(\frac{\partial f}{\partial y}\right)$$

in terms of the operators $(\partial/\partial r)$ and $(\partial\,/\partial\theta)$.

(c) What form does the classical *Laplace operator*

$$\Delta^2 f = \frac{\partial^2 f}{\partial x^2} + \frac{\partial^2 f}{\partial y^2}$$

take when rewritten in terms of polar coordinates?

CALCULATIONS WITH VECTOR AND TENSOR FIELDS ON S^2. The next Additional Exercises show how to handle calculations involving vector and tensor fields on a manifold, the two-dimensional sphere $S^2 = \{\mathbf{x} : x_1^2 + x_2^2 + x_3^2 = 1\}$ in $\mathbb{R}^3$. One of the founding charts (U_α, x_α) that determine the manifold structure of S^2 has as its domain the upper hemisphere $M = H_3^+ = \{\mathbf{x} \in \mathbb{R}^3 : x_3 > 0\}$ with chart map x_α the projection $P_3(x_1, x_2, x_3) = (x_1, x_2) \in \mathbb{R}^2$ of M onto the unit disc $D = \{(x_1, x_2) : x_1^2 + x_2^2 < 1\}$ in the x_1, x_2-plane. (Recall Example 4.2 and the accompanying figure.) The calculations will also illustrate the use of the the change of variable formulas for tangent vectors (Theorem 4.15), for cotangent vectors (Theorem 4.30), and for a particularly important rank-2 tensor, the Euclidean metric tensor on $\mathbb{R}^2$ discussed in Example 4.38.

10. Define coordinate charts on the hemisphere $M \subseteq S^2$ as follows:

(i) (U_α, x_α) with $U_\alpha = M$ and chart map $x_\alpha(\mathbf{x}) = P(\mathbf{x}) = (x, y)$ for points $\mathbf{x} = (x, y, z) \in S^2$. (This is of course one of the "founding charts" by which differentiable manifold structure was imposed on S^2.)

(ii) Define (U_β, y_β) letting $U_\beta = M$ and $x_\beta(u) = (r, \theta)$ the map that sends $u = (x, y, z) \in S^2$ to the polar coordinates (r, θ) of the projected point $P(u) = (x, y)$ in the x, y-plane.

Both charts determine smooth fields of tangent vectors on $M \subseteq S^2$:

$$X = \frac{\partial}{\partial x}\bigg|_u, Y = \frac{\partial}{\partial y}\bigg|_u \quad \text{and} \quad V = \frac{\partial}{\partial r_u}\bigg|_u, W = \frac{\partial}{\partial \theta}\bigg|_u$$

for $u \in M$. Use the change of variable formula for tangent vectors, Corollary 4.16, to rewrite the vector field $\partial/\partial\theta$ in terms of $\partial/\partial x$ and $\partial/\partial y$ on M:

$$\frac{\partial}{\partial \theta} = A(u)\frac{\partial}{\partial x}\bigg|_u + B(u)\frac{\partial}{\partial y}\bigg|_u .$$

Write the coefficients $A(u)$ and $B(u)$ in terms of the scalar components of the chart map $y_\beta(u) = (R(u), \Theta(u))$ and their derivatives.

11. In the setting of Additional Exercise 10, take (U_α, x_α) to be the same, but let y_β be the hemisphere $U_\beta = M$ equipped with *spherical coordinates* $y_\beta(u) = (\theta, \phi)$. Use the change of variable formula for tangent vectors on S^2 to rewrite

$$\frac{\partial}{\partial \phi}\bigg|_u \text{ in terms of } \frac{\partial}{\partial x}\bigg|_u \text{ and } \frac{\partial}{\partial y}\bigg|_u .$$

As in the previous exercise, write the coefficients $A(u), B(u)$ in terms of the components of the chart map $y_\beta(u) = (\theta, \phi) = (\Theta(u), \Phi(u))$.

NOTE. For $(x, y, z) \in \mathbb{R}^3$ we have

$$x = \cos(\theta)\cos(\phi) \quad y = \sin(\theta)\cos(\phi) \quad x = \cos(\phi)$$

since $\|\mathbf{x}\| = 1$ on S^2 (see Figure 4.3). Keep in mind our convention $-\pi/2 < \phi < \pi/2$ regarding the "azimuth angle" ϕ. ○

CALCULATIONS INVOLVING DIFFERENTIAL FORMS ON S^2. The coordinates $x_\alpha(u) = (x, y)$ and $y_\beta = (\theta, \phi)$ imposed on S^2 in Additional Exercise 11 also determine differential forms (smooth fields of cotangent vectors) on the sphere, which are denoted by

$$(dx)_u, (dy)_u \quad \text{and} \quad (d\theta)_u, (d\phi)_u ,$$

respectively. Below we illustrate use of the change of variable formula for cotangent vectors, Theorem 4.30. Recall that these fields of cotangent vectors are the exterior derivatives of the scalar components of chart maps $x_\alpha = (X(u), Y(u))$ and $y_\beta = (R(u), \Theta(u))$ with

$$dx_u = (dX)_u, dy_u = (dY)_u \quad \text{and} \quad d\theta_u = (d\Theta)_u, d\phi_u = (d\Phi)_u.$$

(Recall the discussion surrounding equation (4.22) of differential forms $(dx_i)_p$ associated with a chart *vs* the exterior derivatives $(dX)_p$ of the scalar components of the chart map $x_\alpha = (X_1(u), \dots, X_m(u))$.)

12. For points $u \in M = \mathbb{R}^3$, let $\mathbf{x} = x_\alpha(u) = (x_1, x_2, x_3)$ be the standard Euclidean coordinates and define new coordinates, letting $\mathbf{y} = y_\beta(u) = (y_1, y_2, y_3)$ where $\mathbf{y} = L_A(\mathbf{x}) = A \cdot \mathbf{x}$ for the nonsingular matrix

$$A = \begin{pmatrix} 0 & 2 & -1 \\ 1 & 2 & 2 \\ 3 & 0 & -1 \end{pmatrix}.$$

(a) Determine the entries in the Jacobian matrix $[\partial y_i/\partial x_j(u)]$ at a typical point $u \in M$.

(b) Find the scalar-valued functions $A(u)$ and $B(u)$ on M such that

$$(dy_1 \wedge dy_2 \wedge dy_3)_u = A(u) \cdot (dx_1 \wedge dx_2 \wedge dx_3)_u$$
$$(dx_1 \wedge dx_2 \wedge dx_3)_u = B(u) \cdot (dy_1 \wedge dy_2 \wedge dy_3)_u$$

in $\bigwedge^3(\mathbb{R}^3)$ by rewriting (dy_i) in terms of (dx_i) (or vice-versa) using the change of variable formula for cotangent vectors (Theorem 4.30). Use the commutation relations to write products $dx_I = dx_{i_1} \wedge dx_{i_2} \wedge dx_{i_3}$ as an ordered monomial $\pm(dx_1 \wedge dx_2 \wedge dx_3)$, and similarly rewrite $dy_J = dy_{j_1} \wedge dy_{j_2} \wedge dy_{j_3} = \pm dy_1 \wedge dy_2 \wedge dy_3$.

13. Rewrite the smooth differential forms $d\theta$ and $d\phi$ on $M = S^2$ in terms of the forms dx and dy, and vice-versa.

14. As shown in Figure 4.3, spherical coordinates $x_\beta(p) = (\rho, \theta, \phi)$ on $M = \mathbb{R}^3$ are related to Cartesian coordinates via

$$z = \rho \sin(\phi) \qquad y = \rho \sin(\theta)\cos(\phi) \qquad x = \rho\cos(\theta)\cos(\phi).$$

Since $\bigwedge^3 M$ is one-dimensional, the maximal rank forms $dx \wedge dy \wedge dz$ and $d\rho \wedge d\theta \wedge d\phi$ are scalar multiples of each other at each base point $p \in M$, but the scale factor $c(p)$ may vary with p. Find the weight function $A(\rho, \theta, \phi) = c(y_\beta(p))$ such that

$$(dx \wedge dy \wedge dz) = A(\rho, \theta, \phi) \cdot (d\rho \wedge d\theta \wedge d\phi).$$

HINT. When you pass from Cartesian to spherical coordinates, each of the differential forms dx, dy, dz becomes a sum with three terms — $3^3 = 27$ terms in all. But all rank-3 wedgeups of $d\rho, d\theta, d\phi$ containing repeated factors are zero, so the only ones that survive are $3! = 6$ permutations of the symbol string $(d\rho \wedge d\theta \wedge d\phi)$ (and if you are willing to ignore a $(\pm)$ sign, there is just one survivor: $d\rho \wedge d\theta \wedge d\phi$).

CHANGE OF VARIABLE FOR TENSORS. If two charts (U_α, x_α), (U_β, y_β) overlap on a manifold, a field ω of rank-2 tensors will have different descriptions

$$\omega = \sum_{j_1, j_2 = 1}^{m} c^{(\alpha)}_{j_1, j_2}(u) \cdot (dx_{j_1} \otimes dx_{j_2})_u$$
$$= \sum_{i_1, i_2 = 1}^{m} c^{(\beta)}_{i_1, i_2}(u) \cdot (dy_{i_1} \otimes dy_{i_2})_u$$

(m^2 terms in each sum). On the open set $U_\alpha \cap U_\beta$ in M where chart domains overlap, the change of variable formula (Theorem 4.30) allows us to pass from a description in y_β-coordinates to one in x_α-coordinates. Simply replace (dy_i) wherever it appears by the weighted sum involving the forms dx_j,

$$dy_i = \sum_{k=1}^{m} \frac{\partial Y_i}{\partial x_j}(u)\,(dx_j)_u\,,$$

in which the $Y_i(u)$ are the scalar components of the chart map $y_\beta = (Y_1, \ldots, Y_m)$.

Each factor (dy_i) appearing in ω becomes a sum of forms (dx_j) associated with the x_α-chart. Using multilinearity of ω, we can then expand the resulting product of sums as a sum of simple monomials $(dx_i) \otimes (dx_j)$ because

$$\Big(\sum_{i=1}^{m} a_i\,(dx_i)\Big) \otimes \Big(\sum_{j=1}^{m} b_j\,(dx_j)\Big) = \sum_{i,j=1}^{m} a_i b_j \cdot (dx_i \otimes dx_j).$$

Putting all this together, we obtain the description of ω in x_α-coordinates:

$$\omega = \sum_{i_1,i_2=1}^{m} c^{(\alpha)}_{i_1,i_2}(u) \cdot \sum_{j_1,j_2=1}^{m} \Big(\frac{\partial Y_{i_1}}{\partial x_{j_1}}(u)\,\frac{\partial Y_{i_2}}{\partial x_{j_2}}(u)\Big) \cdot (dx_{j_1} \otimes dx_{j_2})_u. \tag{4.55}$$

The coefficients $(\cdots)$ in this sum are clearly $\mathcal{C}^\infty$ functions of $u \in M$, so smoothness of the tensor field is independent of any particular choice of local coordinates. With sufficient patience, the same sort of calculation can be done for higher rank tensor fields on a manifold, but we will not pursure the general change of variable formula here. It uses the same ideas, but the notation becomes more baroque as $r = \text{rank}(\omega)$ increases.

15. The standard Euclidean metric tensor on $M = \mathbb{R}^2$ is

$$g = (dx_1 \otimes dx_1)_u + (dx_2 \otimes dx_2)_u$$

when described in the usual Cartesian coordinates $x_\alpha(u) = (x_1, x_2)$ on M. In polar coordinates $y_\beta(u) = (r, \theta)$, this becomes

$$\begin{aligned} g = {} & A_{11}(u) \cdot (dr \wedge dr)_u + A_{12}(u) \cdot (dr \wedge d\theta)_u \\ & + A_{21}(u) \cdot (d\theta \wedge dr)_u + A_{22}(u) \cdot (d\theta \wedge d\theta)_u. \end{aligned}$$

Find the coefficients $A_{ij}(u)$ expressed in terms of the scalar components of the chart map $y_\beta(u) = (R(u), \Theta(u))$.

REMARK 4.69 (Differential Forms as "Infinitesimal Volume Elements"). In Example 4.50, we computed the relation between the differential forms $dr \wedge d\theta$ and $dx \wedge dy$ on $M = \mathbb{R}^2$ that describes the effect of a transition from Cartesian to polar coordinates in $M = \mathbb{R}^2$:

$$(dx \wedge dy) = r(dr \wedge d\theta).$$

Expressions of this sort should also be familiar from calculus discussions of integrals $\iint_R f(\mathbf{x})\,dxdy$ over regions $R \subseteq \mathbb{R}^2$. When such an integral is computed in Cartesian or polar coordinates, one obtains an identity that can be found in any calculus text:

$$\iint f(x,y)dx\,dy = \iint f(r,\theta)r\,dr\,d\theta.$$

We won't have time to go into the theory of integration on manifolds, which is an extension of integration of functions over hypersurfaces in $\mathbb{R}^n$ (but without the "$\mathbb{R}^n$"). In more advanced accounts you would discover that expressions like $dxdy$ and $rdrd\theta$ in calculus get replaced by the differential forms $(dx \wedge dy)$ and $r(dr \wedge d\theta)$ on the manifold $M = \mathbb{R}^2$; and more generally, functions $f(u)$ on a k-dimensional manifold get integrated against some rank-k (maximal rank) differential form on M to arrive at a numerical value for the resulting integral. When $M = \mathbb{R}^2$, this leads to such identities as

$$\iint_M f(p)\,(dx \wedge dy) = \iint_M f(p)\,r\,(dr \wedge d\theta)$$

involving integration of functions against maximal rank differential forms on the manifold.

With this in mind, mathematicians have often regarded expressions such as $dx \wedge dy$ and $r\,dr \wedge d\theta$ as *infinitesimal elements of oriented volume* in the manifold over which the integration is being performed. That is why change of variable formulas such as $(dx \wedge dy) = r\,(dr \wedge d\theta)$ worked out in Example 4.50 become important when attention turns to integration on manifolds. For example, if volumes computed using integrals over regions in $\mathbb{R}^2$ are to make geometric sense, each small region of the form $dr \times d\theta$ in polar coordinate space (r, θ) has to be assigned a two-dimensional volume $\approx r\,dr \cdot d\theta$, while small regions $dx \times dy$ in Cartesian coordinate space (x, y) are assigned volumes $= dx \cdot dy$, if the calculations are to agree. ○

Section 4.4. Div, Grad, Curl, and All That

1. TRUE/FALSE QUESTIONS ("True" if the statement is *always* true.)
   (a) The triple product $(\mathbf{a} \times \mathbf{b}) \bullet \mathbf{c}$ of vectors in $\mathbb{R}^3$ is the volume of the parallelopiped

   $$P = \{c_1\mathbf{a} + c_2\mathbf{b} + c_3\mathbf{c} \,:\, 0 \le c_i \le 1\}$$

   with one corner at the origin and $\mathbf{a}, \mathbf{b}, \mathbf{c}$ as its edges emanating from the origin.

   (b) We have $\mathbf{a} \times \mathbf{b} \neq \mathbf{0}$ for vectors in $\mathbb{R}^3$ unless they are collinear.

   (c) We have $\mathbf{a} \times \mathbf{b} = \mathbf{0}$ for vectors in $\mathbb{R}^3$ if and only if they are perpendicular.

2. Some calculus texts describe the cross product $\mathbf{a} \times \mathbf{b}$ as an operation whose input is an ordered pair of vectors in $\mathbb{R}^3$ and output is another vector in $\mathbb{R}^3$.
   (a) Verify that the cross product is a vector-valued *bilinear* map from $V \times V \to V$: a linear function of each input vector when the other is held fixed.

There seems to be no reason that we could not then compute iterated products in $\mathbb{R}^3$ such as $\mathbf{a}\times\mathbf{b}\times\mathbf{c}$. Or is there?

(b) Try it, taking

$$\mathbf{a}=(1,2,0),\quad \mathbf{b}=(1,-1,1),\quad \mathbf{c}=(-1,2,3).$$

Did you encounter any cognitive dissonance in carrying out this calculation based on the definition of $\mathbf{a}\times\mathbf{b}$ in (4.53)?

(c) Explain why it is not possible to make a consistent definition of products like $\mathbf{a}\times\mathbf{b}\times\mathbf{c}$ or $\mathbf{a}\times\mathbf{b}\times\mathbf{c}\times\mathbf{d}$.

HINT. The cross product operation only accepts two inputs. There are two different ways to insert parentheses into $\mathbf{a}\times\mathbf{b}\times\mathbf{c}$ to get expressions that only involve multiplying two vectors at a time. Write out and evaluate both expressions using the vectors listed in (b). Do the outcomes agree?

3. Given nonzero vectors $\mathbf{a}$ and $\mathbf{b}$ in $\mathbb{R}^3$,

(a) Explain why the cross product of two vectors in $\mathbb{R}^3$ has Euclidean length

$$||\mathbf{a}\times\mathbf{b}|| = ||\mathbf{a}||\,||\mathbf{b}||\cdot\cos\theta(\mathbf{a},\mathbf{b})$$

where θ is the angle between $\mathbf{a}$ and $\mathbf{b}$, measured in the plane they span in $\mathbb{R}^3$.

(b) Explain why $||\mathbf{a}\times\mathbf{b}|| = \text{Area}(Q)$ where Q is the parallelogram

$$Q=\{c_1\mathbf{a}+c_2\mathbf{b} \,:\, 0\le c_1,c_2\le 1\}\subseteq\mathbb{R}^3$$

determined by $\mathbf{a}$ and $\mathbf{b}$. What determines whether a $(+)$ or $(-)$ sign should be applied to get the "signed area" of Q?

4. Prove that the cross product $\mathbf{a}\times\mathbf{b}$ of two vectors in $\mathbb{R}^3$ is perpendicular to the plane Q determined by $\mathbf{a}$ and $\mathbf{b}$ (orthogonal with respect to the Euclidean inner product in $\mathbb{R}^3$).

5. Why isn't

$$\nabla(\nabla\circ\mathbf{F}) = \mathbf{grad}(\mathbf{div}\,\mathbf{F})\equiv 0$$

a valid identity for smooth vector fields $\mathbf{F}(\mathbf{x})$ on $\mathbb{R}^3$?

6. Mathematicians often use the symbol "Δ" to indicate the *Laplace operator*, the second order partial differential operator

$$\Delta(F) = D^2_{x_1}F+\cdots+D^2_{x_n}F = \frac{\partial^2 F}{\partial x_1^2}+\cdots+\frac{\partial^2 F}{\partial x_n^2}$$

that acts on smooth scalar-valued functions $f(\mathbf{x})$ on $\mathbb{R}^n$. A more archaic notation, still widely used when $n=3$, writes the Laplacian as

$$\nabla^2 F = \nabla\circ(\nabla F) = \mathbf{div}(\mathbf{grad}\,F).$$

(a) Verify that this is correct for $n=3$. (It is even true when $n=2$ using our modified definitions of **div** and **grad** on two-dimensional space.)

(b) What goes wrong when the dimension is ≥ 4?

CHAPTER 5

Matrix Lie Groups

An Overview of This Chapter. The subject of *Lie groups* is the nexus where three major areas of mathematics come into play — calculus-style analysis, modern algebra, and differential geometry are all intertwined in this field, underpinned by the tools linear algebra provides. That interplay among so many disciplines can be daunting, but it is also what makes Lie groups such an interesting subject. [1] This brief chapter cannot be a treatise on the subject but is rather an introduction to the players involved, a few of the most important concepts and examples illustrating the techniques involved in working with them. Some knowledge of differential geometry will be assumed, particularly the concept of a differentiable manifold (which will also be described below), but familiarity with Section 4.1 in Chapter 4 of these *Notes* should suffice.

Many Lie groups of great interest in physics and geometry were originally modeled as curvilinear smooth hypersurfaces M embedded in higher dimensional Euclidean spaces $\mathbb{R}^n$ (and sometimes $\mathbb{C}^n$), for example the "classical matrix groups" – the orthogonal groups $\mathrm{O}(n)$, unitary groups $\mathrm{U}(n)$, and symplectic groups $\mathrm{Sp}(2n)$ introduced in Chapter 3 as the symmetry groups associated with various bilinear forms that arise throughout mathematics. They will often be used as concrete examples.

Beginning in the 20th century, Lie theory (and most of differential geometry as well) moved away from the notion of hypersurfaces embedded in Euclidean spaces, and developed an "intrinsic" theory of hypersurfaces — now called *manifolds* — that viewed Lie groups and manifolds as self-contained universes in their own right. Although we will speak of Lie groups as manifolds in these *Notes*, to keep things simple we will focus on *matrix Lie groups* which can be modeled as subsets of matrix space $\mathrm{M}(n, \mathbb{K})$ taking $\mathbb{K} = \mathbb{R}$ or $\mathbb{C}$. Besides, almost everything said for matrix Lie groups carries over to the general theory of Lie groups, although some proofs get harder in that setting.

In this chapter we will

- Show how differentiable structure can be defined on a matrix group G using the implicit function theorem to define Euclidean coordinates on a family of open sets U_α that cover G. Equipped with these "coordinate charts," G becomes a "Lie group," in which the group multiplication operation $G \times G \to G$ becomes a differentiable map.

[1] The "Lie" in "Lie groups" is pronounced "*Lee.*" Sophus Lie was a Norwegian mathematician who pioneered the study of these structures in the late 1800's. Esoteric concepts then, they are ubiquitous and indispensible in modern physics and differential geometry.

- Discuss what "tangent vectors," "tangent spaces," and "derivatives of smooth functions" $f : G \to \mathbb{R}$ might mean on a group G equipped with differentiable structure, which might be realized as curvilinear hypersurface of high dimension. This is the realm of differential geometry.
- The result is a construct G that has both geometric and algebraic aspects. Lie groups can be quite complicated, but deal with them we must because they turn up at the heart of modern physics (and other fields), for instance as the symmetry groups that govern the interactions of subatomic particles, the behavior of the cosmos according to special and general relativity, and the seemingly paradoxical rules of quantum mechanics. Even the periodic table of chemistry can in the end be deduced as a consequence of the mathematics associated with that classic example of a Lie group — the "special orthogonal group" SO(3) of rotations in 3-dimensional space.
- Finally, we will show that the tangent space to a Lie group G at its identity element, which at first sight is just a vector space of the same dimension as the coordinate charts that cover G, acquires algebraic structure induced by the multiplication law in G and becomes what is known as a "Lie algebra." The important point is that this Lie algebra, which is a *linear* structure, is much easier to deal with than G, and yet it encodes almost enough information to completely reconstruct and understand the Lie group from which it was derived.

5.1. Matrix Groups and the Implicit Function Theorem

The rank of a linear operator $T : V \to W$ is

$$\text{rank}(T) = \dim(\text{range}(T)) = \dim(V) - \dim(\ker(T)).$$

In this chapter, we will often abbreviate rank(T) to "rk(T)" for linear operators, and likewise for matrices.

If $\mathfrak{X}, \mathfrak{Y}$ are bases in V, W, the rank of T can be determined from the matrix $A = [T]_{\mathfrak{Y}\mathfrak{X}}$ as follows. A $k \times k$ submatrix is obtained by designating k rows and k columns and extracting from A the $k \times k$ array where these meet. To describe the outcome, we must specify the row indices $I = \{i_1 < \cdots < i_k\}$ and column indices $J = \{j_1 < \cdots < j_k\}$; the resulting submatrix is indicated by writing A_{IJ} (see Figure 5.1). The matrix A itself is not necessarily square; it is $n \times m$ if $\dim(V) = m$, $\dim(W) = n$.

LEMMA 5.1. *Given a nonzero $n \times m$ matrix A its rank* rk(A) *is*

$$k_{max} = \max\{k \in \mathbb{N} : A \textit{ has a nonsingular } k \times k \textit{ submatrix } A_{IJ}\}.$$

PROOF. We have rk$(A) \geq k_{\max}$ because if A_{IJ} is nonsingular, its columns $\{C'_{j_1}, \dots, C'_{j_k}\}$ are linearly independent, and are truncated versions of the corresponding columns $\{C_{j_1}, \dots, C_{j_k}\}$ of A, which forces the latter to be linearly independent. Hence, $|J| \leq k_{max} \leq$ rk(A).

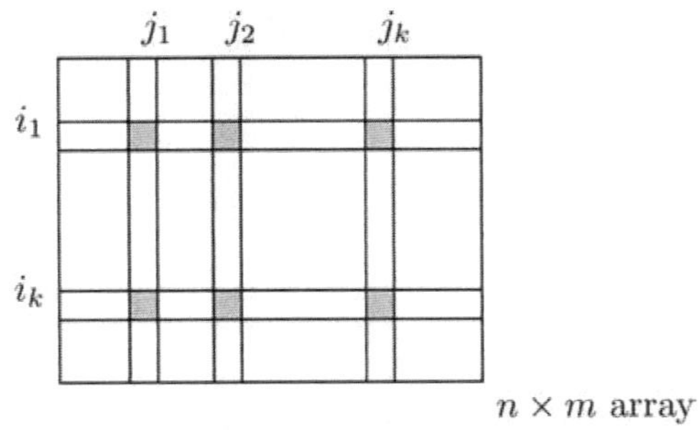

FIGURE 5.1. A square $k \times k$ submatrix A_{IJ} is extracted from an $n \times m$ matrix by specifying row indices $I = \{i_1 < \cdots < i_k\}$ and column indices $J = \{j_1 < \cdots < j_k\}$. The rank $\text{rk}(A)$ is equal to k if A_{IJ} is nonsingular, and all nonsingular square submatrices have size $r \leq k$.

We also have $\text{rk}(A) \leq k_{max}$, for if $\text{rk}(A) = k$, there is some set of column indices with $|J| = k$ such that $\{C_j : i \in J\}$ are linearly independent. If B is the $n \times k$ matrix $[C_{j_1}; \dots; C_{j_k}]$, it is well known that

$$\text{row rank}(B) = \text{ column rank}(B)$$

(see Corollary 4.47 of LA I), so we can find a set I of row indices with $|I| = |J| = k$ such that the rows $\{R_i(B) : i \in I\}$ are linearly independent. The rows in the $n{\times}k$ matrix $B = [C_{j_1}; \dots; C_{j_k}]$ are truncated versions of the corresponding rows in A, and those with row indices in J are precisely the rows of the $k{\times}k$ submatrix A_{IJ}. Obviously, this submatrix is nonsingular, so $k_{max} \geq k = \text{rk}(A_{IJ}) = \text{rk}(A)$. $\square$

Note that various choices I, J of row and column indices may yield nonsingular square submatrices A_{IJ} of maximal size.

Smooth Mappings and Their Differentials. Now consider a mapping $\mathbf{y} = \phi(\mathbf{x}) = \big(\phi_1(\mathbf{x}), \dots, \phi_n(\mathbf{x})\big)$ from $\mathbb{K}^m \to \mathbb{K}^n$ ($\mathbb{K} = \mathbb{R}$ or $\mathbb{C}$, but mostly $\mathbb{R}$ in our discussion). We say that ϕ is a $\mathcal{C}^\infty$ *map* (or *smooth map*) if the scalar components $\phi_k(\mathbf{x})$ have continuous partial derivatives of all orders.

The *Jacobian matrix* for ϕ at base point p is the $n \times m$ matrix

$$(D\phi)_p = \begin{pmatrix} \frac{\partial y_1}{\partial x_1}(p) & \cdots & \frac{\partial y_1}{\partial x_m}(p) \\ \vdots & & \vdots \\ \frac{\partial y_n}{\partial x_1}(p) & \cdots & \frac{\partial y_n}{\partial x_m}(p) \end{pmatrix}_{n \times m}$$

whose entries are smooth scalar-valued functions of $\mathbf{x} \in \mathbb{K}^m$. We will be concerned with the rank $\text{rk}(d\phi)_\mathbf{x}$ of the Jacobian matrix at and near various base points. We assign a linear operator, the *differential of ϕ at p*,

$$(d\phi)_p : \mathbb{K}^m \to \mathbb{K}^n \quad \text{such that} \quad (d\phi)_p(\mathbf{v}) = (D\phi)_p \cdot \text{col}(v_1, \dots, v_m)$$

at each base point $v \in \mathbb{K}^m$ where ϕ is smooth. The operator $(d\phi)_p$ is the unique *linear* operator $\mathbb{K}^m \to \mathbb{K}^n$ that "closely approximates" the behavior of the (nonlinear) map $\phi : \mathbb{K}^m \to \mathbb{K}^n$ near p in the sense that

$$\Delta\phi = \phi(p + \Delta\mathbf{x}) - \phi(p) = (d\phi)_p \cdot (\Delta\mathbf{x}) + E(\Delta\mathbf{x}),$$

in which the "error term" $E(\Delta\mathbf{x})$ becomes very small compared to $\Delta\mathbf{x}$ for small increments away from the base point p:

$$\text{(5.1)} \quad \text{ERROR ESTIMATE:} \quad \frac{\|E(\Delta\mathbf{x})\|}{\|\Delta\mathbf{x}\|} \longrightarrow 0 \text{ in } \mathbb{K}^n \quad \text{as} \quad \|\Delta\mathbf{x}\| \to 0 \text{ in } \mathbb{K}^m.$$

We indicate this by writing "$E(\Delta\mathbf{x}) = o(\Delta\mathbf{x})$." As a function of the base point $p \in \mathbb{K}^m$, the linear operator $(d\phi)_p$ is a $\mathcal{C}^\infty$ map from $\mathbb{K}^m$ into $\mathbb{K}^n$.

The *rank at p* of a map ϕ is the rank $r = \text{rk}(d\phi)_p$ of its Jacobian matrix. As above, $\text{rk}(d\phi)_p = r \Leftrightarrow \dim(\text{range}(d\phi)_p) = r \Leftrightarrow m - \dim(\ker(d\phi)_p) = m - r \Leftrightarrow$ there are r row indices $I = \{i_1 < \cdots < i_r\}$ and column indices $J = \{j_1 < \cdots < j_r\}$ such that

1. The $r \times r$ submatrix $(d\phi_p)_{IJ}$ is nonsingular, and
2. No larger square submatrix (with $k > r$) can be nonsingular, so $r \times r$ is the maximum size of any nonsingular square submatrix.

Note the following points:

- Various choices of I, J may yield nonsingular submatrices $(d\phi)_{IJ}$ of maximal size $r \times r$. The valid choices may vary with the base point p.
- For fixed I, J the entries in the $r \times r$ matrix $(d\phi_{\mathbf{x}})_{IJ}$ vary smoothly with $\mathbf{x}$, and so does the determinant $\det(d\phi)_{\mathbf{x}}$ since the entries $\partial\phi_i/\partial x_j$ are $\mathcal{C}^\infty$ on any chart domain. Thus, if the determinant is nonzero at p, it must also be nonzero for all x near p by continuity of the determinant $\det : \mathrm{M}(k, \mathbb{K}) \to \mathbb{K}$. Hence, for fixed choice of I, J, we have

  $$\text{rk}(d\phi_{\mathbf{x}})_{IJ} \geq r \quad \text{for all } \mathbf{x} \text{ near } p \text{ if} \quad \text{rk}(d\phi_p)_{IJ} = r.$$

  If r_{max} is the largest value of $\text{rk}(d\phi)_p$ on $\mathbb{K}^m$, then $\text{rk}(d\phi)_{\mathbf{x}} \equiv r_{max}$ (constant rank) on some open neighborhood of p in $\mathbb{K}^m$. Often, as when $\phi : \mathbb{K}^m \to \mathbb{K}^n$ has scalar components $\phi = (\phi_1(x), \ldots, \phi_n(x))$ that are polynomials in $\mathbf{x} = (x_1, \ldots, x_m)$, this open set is dense in $\mathbb{K}^m$, and its complement has Lebesgue measure zero — i.e., maximal rank is achieved at "almost all" points in $\mathbb{K}^m$.
- For any choice of an index set $J \subset [1, m]$ with $|J| = k \leq \min\{m, n\}$, define $J' = [1, m] \sim J$ and let $\mathbb{K}^J, \mathbb{K}^{J'} \subseteq \mathbb{K}^m$ be the subspaces

  $$\mathbb{K}^J = \mathbb{R}\text{-span}\{e_i : i \in J\} \quad \text{and} \quad \mathbb{K}^{J'} = \mathbb{R}\text{-span}\{e_i : i \in J'\}$$

  where $\{e_i\}$ is the standard basis in $\mathbb{K}^m$. Then $\mathbb{K}^m$ splits as a direct sum $\mathbb{K}^{J'} \oplus \mathbb{K}^J$, and this decomposition determines projections

  $$\pi^{J'} : \mathbb{K}^m \to \mathbb{K}^{J'} \quad \text{and} \quad \pi^J : \mathbb{K}^m \to \mathbb{K}^J$$

  onto these subspaces.

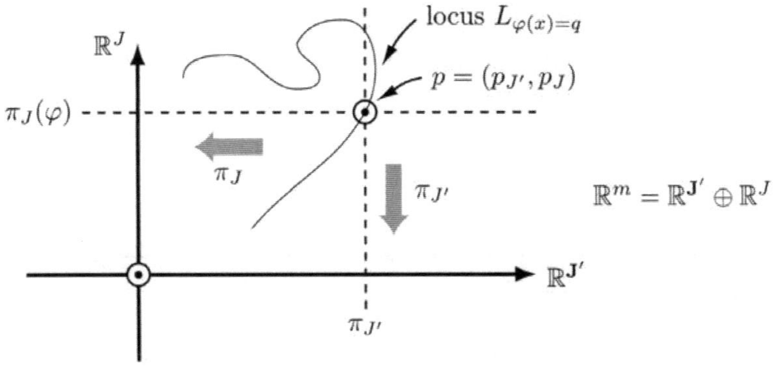

FIGURE 5.2. The projection maps $\pi^{J'}, \pi^J$ and direct sum decomposition $\mathbb{K}^m = \mathbb{K}^{J'} \oplus \mathbb{K}^J$ are associated with a partition of column indices $[1, m] = J' \cup J$, $J' \cap J = \emptyset$. Near any base point $p \in M$, this splits the variables in $\mathbf{x} = (x_1, \dots, x_m)$ into two groups so $\mathbf{x} = (\mathbf{x}_{J'}, \mathbf{x}_J)$ with $x_J = (x_{k_1}, \dots, x_{k_r})$ and $x_{J'} = (x_{\ell_1}, \dots, x_{\ell_s})$, where $r = |J|, s = |J'|$, and $r + s = m = \dim(\mathbb{K}^m)$. In our narrative, such partitions of variables arise in discussing the rank $r = \text{rk}(d\phi)_p$ of the the $n \times m$ Jacobian matrix $[\partial f_i / \partial x_j]$ of a differentiable map $\phi : \mathbb{K}^m \to \mathbb{K}^n$ at points on a typical level set $L_{\phi(\mathbf{x})=q}$.

By composing ϕ with translations in $\mathbb{K}^m$ and $\mathbb{K}^n$ we can assume ϕ maps the origin in $\mathbb{K}^m$ to the origin in $\mathbb{K}^n$. This will not change $\text{rk}(d\phi_p)$, but it does simplify the notation. The following exercise shows that this maneuver does not affect the Jacobian matrices or their determinants.

EXERCISE 5.2. Let $p \in \mathbb{R}^m$.

(a) Let $\mathbf{y} = \phi(\mathbf{x}) = (x_1 + p_1, \dots, x_m + p_m)$ be a translation operator from $\mathbb{R}^m \to \mathbb{R}^m$. Prove that $(d\phi)_p = I_{m \times m}$ at every base point p.

(b) Given smooth maps $\mathbb{R}^m \xrightarrow{\phi} \mathbb{R}^n \xrightarrow{\psi} \mathbb{R}^k$ and base points $p \in \mathbb{R}^m, q = \phi(p) \in \mathbb{R}^n$, explain why the differential of $\psi \circ \phi$ from $\mathbb{R}^m \to \mathbb{R}^k$ is the matrix product of their differentials

$$d(\psi \circ \phi)_p = (d\psi)_{\phi(p)} \cdot (d\phi)_p.$$

Smooth Hypersurfaces and the Implicit Function Theorem (IFT). The implicit function theorem (IFT) concerns itself with *level sets*

$$L_{\phi(\mathbf{x})=q} = \{\mathbf{x} \in \mathbb{K}^m : \phi(\mathbf{x}) = q\} \subseteq \mathbb{K}^m,$$

on which a smooth mapping $\phi : \mathbb{K}^m \to \mathbb{K}^n$ has constant (vector) value $\phi(\mathbf{x}) = q \in \mathbb{K}^n$. In essence, the IFT says that if p is in a level set $L_{\phi(\mathbf{x})=\mathbf{q}}$, then the locus $L_{\phi(\mathbf{x})=\mathbf{q}}$ is carved out of $\mathbb{K}^m$ by imposing n scalar constraints

$$\phi_1(\mathbf{x}) = c_1, \dots, \phi_n(\mathbf{x}) = c_n$$

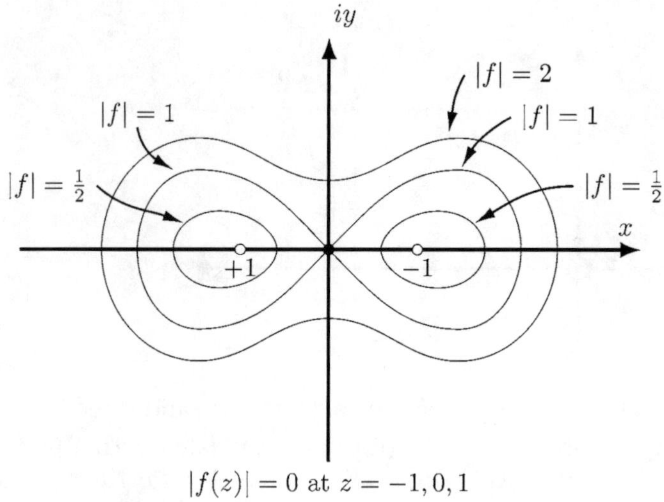

FIGURE 5.3. Level sets for the map $f : \mathbb{R}^2 \to \mathbb{R}$ with $f(x, y) = |z^2 - 1|^2$, identifying $z = x + iy$ with $(x, y) \in \mathbb{R}^2$. If $c < 0$ the level set $L_c = L_{f(z)=c}$ is empty; when $c = 0$ it consists of three isolated points $z = -1, 0, +1$; and for $c > 0$ the locus is usually a smooth curve (perhaps with more than one connected component, as when $0 < c < 1$). But when $c = 1$ the locus has a singularity at the origin. It cannot be represented near the origin as the graph of any smooth function $y = h(x)$ or $x = g(y)$. The origin is a "branch point" for the locus.

on the components of ϕ, and when the hypotheses of the IFT are satisfied at p, this level set can be described locally as a smooth hypersurface of dimension $m - r$ in $\mathbb{K}^m$, where $r = \text{rk}(d\phi)_p$. Thus, the level set is the graph

$$\Gamma = \{(\mathbf{x}, f(\mathbf{x})) : \mathbf{x} \in \mathbb{K}^{m-r}\} \subseteq \mathbb{K}^m = \mathbb{K}^{m-r} \times \mathbb{K}^r$$

of a smooth map $f : \mathbb{K}^{m-r} \to \mathbb{K}^r$. The idea is illustrated in the following example (see also Figure 5.2.) The map $\mathbf{y} = f(\mathbf{x})$ is the "implicit function" of the IFT.

EXAMPLE 5.3. Define $\phi(z) = |z^2 - 1|^2$ for $z \in \mathbb{C}$ and regard it as a map $\mathbb{R}^2 \to \mathbb{R}$ by identifying $z = x + iy \in \mathbb{C}$ with $\mathbf{x} = (x, y) \in \mathbb{R}^2$. Then ϕ becomes a 4th degree polynomial in x and y:

$$\phi(x, y) = x^4 + 2x^2y^2 + y^4 - 2x^2 + 2y^2 + 1.$$

The level sets $L_c = L_{\phi(\mathbf{x})=c}$ are empty if $c < 0$; reduce to the isolated points $\{-1, 0, +1\}$ if $c = 0$; and for $c > 0$ are smooth curves (with more than one connected component when $0 < c < 1$). However there is one exception. When $c = 1$, the locus $L_{\phi=1}$, shown in Figure 5.3, has a singularity at the origin. Near $z = 0 + i0$, it cannot be described locally as the graph of a smooth function $y = h(x)$ or $x = g(y)$. The 1×2 Jacobian matrix $D\phi(z) = [\partial\phi/\partial x(z), \partial\phi/\partial y(z)] = (d\phi)_z$ has $\text{rk}(d\phi)_z \equiv 1$ (constant) throughout $\mathbb{R}^2$ except at $z = 0$, $z = -1$, and

$z = +1$ on the real axis (the "critical points" where both partial derivatives of f are zero).

EXERCISE 5.4. Let $\phi : \mathbb{R}^2 \to \mathbb{R}$ be the function in Example 5.3.

(a) Verify that the Jacobian matrix

$$(d\phi)_{\mathbf{x}} = \left[\frac{\partial \phi}{\partial x}, \frac{\partial \phi}{\partial y}\right]$$

has rank zero (both components $= 0$) if and only if $\mathbf{x} = (-1, 0), (0, 0)$, or $(+1, 0)$ in $\mathbb{R}^2$ by solving the system of equations

$$\frac{\partial \phi}{\partial x}(\mathbf{x}) = 0 \qquad \frac{\partial \phi}{\partial y}(\mathbf{x}) = 0.$$

(b) At which points $\mathbf{x}$ is one of the derivatives $\partial\phi/\partial x$ and $\partial\phi/\partial y$ zero while the other is nonzero? Draw pictures of the sets

$$S_1 = \{\mathbf{x} : \partial\phi/\partial x = 0 \text{ and } \partial\phi/\partial y \neq 0\}$$
$$S_2 = \{\mathbf{x} : \partial\phi/\partial x \neq 0 \text{ and } \partial\phi/\partial y = 0\}.$$

(c) Verify that $\text{rk}(d\phi)_{\mathbf{x}} = 1$ near every point in the sets S_1 and S_2 identified in (b). How are these points related to the pattern of level curves shown in Figure 5.3?

In discussing the IFT, we will discover that a level set $L_c = L_{\phi(\mathbf{x})=c}$ in Example 5.3 can be described as the graph of a smooth function $y = f(x)$ near any point p on the locus where $\partial\phi/\partial x(p) \neq 0$, and similarly we can write $x = g(y)$ if $\partial\phi/\partial y(p) \neq 0$. In Example 5.3, at least one of these conditions is satisfied at every base point except the origin $\mathbf{x} = (0, 0)$, which lies on the locus $L_{\phi=1}$, and the points $\mathbf{x} = (-1, 0)$ and $(1, 0)$, which make up the degenerate locus $L_{\phi=0}$. Consequently, for $c \neq 0$ or 1, the nonempty level sets $L_c = L_{\phi(\mathbf{x})=c}$ can be described locally as smooth curves (the graphs of smooth functions $y = f(x)$ or $x = g(y)$). Furthermore, L_c can be described *both* ways (with $y = f(x)$ or with $x = g(y)$) near most points $p \in L_c$, but at a few points, only one such description is possible — these are the points on the curves in Figure 5.3 at which L_c has either a horizontal or vertical tangent line.

We will apply the IFT to show that the "classical matrix groups" $\text{O}(n)$, $\text{SO}(n)$, $\text{U}(n)$, $\text{SO}(n, \mathbb{C})$, etc., are "matrix Lie groups," smooth hypersurfaces in matrix space $\text{M}(n, \mathbb{K}) \simeq \mathbb{K}^{n^2}$ that are also groups under matrix multiplication. This combination of algebraic and geometric aspects, not to mention their appearance in many aspects of modern physics. makes them objects of considerable interest.

Given a $\mathcal{C}^\infty$ map $\phi : \mathbb{K}^m \to \mathbb{K}^n$ with $m \geq n$, we write $(d\phi)_{\mathbf{x}}$ for the Jacobian matrix $[\partial\phi_i/\partial x_j(\mathbf{x})]_{n\times m}$ at $\mathbf{x} \in \mathbb{R}^m$ and $\text{rk}(d\phi)_{\mathbf{x}}$ for its rank. The simplest version of the IFT asserts that if $\text{rk}(d\phi)_p = n$ (the maximum possible rank), then the level set

$$L_{\mathbf{c}} = L_{\phi(\mathbf{x})=\mathbf{c}} \quad \text{with} \quad \mathbf{c} = (c_1, \dots, c_n) = \phi(p) \text{ in } \mathbb{R}^n$$

can be described near p as a smooth hypersurface of dimension $m - n$ in $\mathbb{K}^m$. If the rank condition holds near each base point in $L_{\mathbf{c}}$, the whole level set is a smooth surface embedded in $\mathbb{R}^m$. We will occasionally invoke a stronger version which says this: if we only know that $\text{rk}(d\phi)_{\mathbf{x}} \equiv r$ (constant) on $L_{\mathbf{c}}$, without requiring maximal rank $r = n$, the conclusion is still true. This will show that various level sets are in fact differentiable manifolds.

For simplicity, we shall focus on these results taking $\mathbb{K} = \mathbb{R}$, although most remain true almost verbatim for $\mathbb{K} = \mathbb{C}$.

THEOREM 5.5 (Implicit Function Theorem (IFT)). *Let $\phi : \mathbb{R}^m \to \mathbb{R}^n$ with $m \geq n$ be a $\mathcal{C}^\infty$ map defined near $p \in \mathbb{R}^m$, let $\mathbf{c} = \phi(p)$, and let $M = L_{\phi(\mathbf{x})=\mathbf{c}}$ be the level set in $\mathbb{R}^m$ containing p, which is determined by n scalar constraint identities*

$$\phi_1(\mathbf{x}) = c_1, \dots, \phi_n(\mathbf{x}) = c_n.$$

Assume $\text{rk}(d\phi)_{\mathbf{x}} \equiv r$ for $\mathbf{x}$ near p. Then there exist index sets

$$I = \{i_1 < \cdots < i_r\} \subseteq [1, n] \quad \text{and} \quad J = \{j_1 < \cdots < j_r\} \subseteq [1, m]$$

such that $[(d\phi)_p(\mathbf{x})]_{IJ}$ is a nonsingular square submatrix of maximal size. The column indices J determine a partition of indices

$$[1, m] = J \cup J' \quad \text{with} \quad J' = [1, m] \sim J,$$

from which we get the following: a direct sum decomposition $\mathbb{R}^m = \mathbb{R}^{J'} \oplus \mathbb{R}^J$ with $|J| = r$ and $|J'| = m - r$, and linear projections $\pi^{J'}, \pi^J$ onto the summands. Then there is an open rectangular neighborhood $B_1 \times B_2$ of p aligned with the decomposition $\mathbb{R}^m = \mathbb{R}^{J'} \oplus \mathbb{R}^J$ such that:

1. *On the relatively open neighborhood $U_p = (B_1 \times B_2) \cap M$ of p in M, the restriction $\pi^{J'}|_{U_p} : U_p \to B_1$ of the linear projection $\pi^{J'}$ from $\mathbb{R}^m$ to $\mathbb{R}^{J'} \cong \mathbb{R}^{m-r}$ is a bicontinuous bijection between the relatively open set $U_p \subseteq M$ and the open set $B_1 \subseteq \mathbb{R}^{m-r}$. It assigns Euclidean coordinates $\mathbf{x} = (x_1, \dots, x_{m-r})$ to every point in U_p.*
2. *The inverse of this map,*

$$\Psi = \left(\pi^{J'}|_{U_p}\right)^{-1} : B_1 \to U_p \subseteq \mathbb{R}^m,$$

is a $\mathcal{C}^\infty$ map from the open set $B_1 \subseteq \mathbb{R}^{J'}$ into all of $\mathbb{R}^m$ that maps B_1 onto the "relatively open neighborhood" U_p of p (the intersection of the level set M with an open set in $\mathbb{R}^m$).

We obtain a $\mathcal{C}^\infty$ map $f : B_2 \to B_1 \subseteq \mathbb{R}^J \cong \mathbb{R}^n$ by following Ψ with the "horizontal" projection π^J shown in Figure 5.4. Then

$$f = \pi^J \circ \Psi \text{ maps } B_1 \subseteq \mathbb{R}^{m-r} \text{ to } B_2 \subseteq \mathbb{R}^r,$$

and Ψ is the graph map of f because

$$\Psi(x) = \left(\pi^{J'}(\Psi(x)), \pi^J(\Psi(x))\right) = \left(x, \pi^J \circ \Psi(x)\right) = (x, f(x)) \quad \text{for } x \in B_1.$$

In particular, the open neighborhood U_p in M is precisely the graph of the $\mathcal{C}^\infty$ map $f : \mathbb{R}^{J'} \to \mathbb{R}^J$.

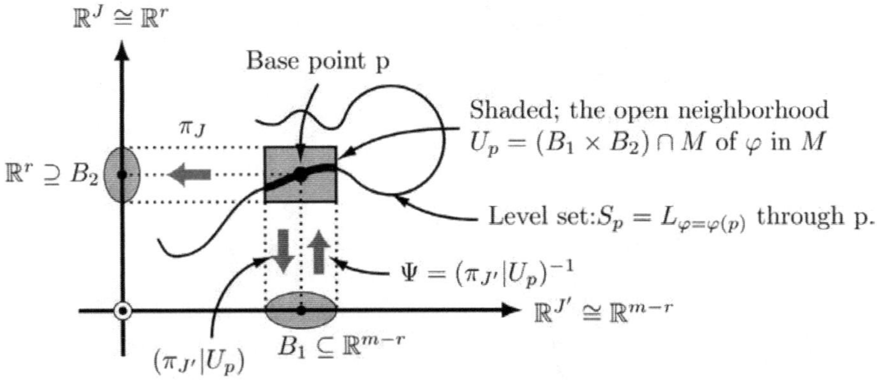

FIGURE 5.4. A diagram showing the players in the implicit function theorem. Here, S_p is the level set passing through p for a $\mathcal{C}^\infty$ map $\phi : \mathbb{R}^m \to \mathbb{R}^n$. By splitting coordinates in $\mathbb{R}^m$ into two groups, we get a decomposition $\mathbb{R}^m = \mathbb{R}^{J'} \oplus \mathbb{R}^J$ and associated projections $\pi^{J'}, \pi^J$ from $\mathbb{R}^m$ to $\mathbb{R}^{J'}$ or $\mathbb{R}^J$ such that (i) the restriction $\pi^{J'}|_{U_p}$ becomes a *bijective* bicontinuous map to an open set B_1 in $\mathbb{R}^{J'}$ for a suitably chosen open neighborhood U_p of p in S_p, and (ii) the inverse $\Psi = \left(\pi^{J'}|_{U_p}\right)^{-1} : B_1 \to U_p$ is a $\mathcal{C}^\infty$ map from the open set $B_1 \subseteq \mathbb{R}^{J'}$ into the entire Euclidean space $\mathbb{R}^m$ in which the level set S_p lives. The set $U_p = S_p \cap (B_1 \times B_2)$ (the shaded rectangular region containing p) is a relatively open neighborhood of p in the level set $L_{\mathbf{c}}$ where $\phi(\mathbf{x}) = \mathbf{c}$.

Conclusion: Near p, the level set $M = L_{\phi(\mathbf{x})=\mathbf{c}}$ looks like part of a smooth hypersurface in $\mathbb{R}^m$ of dimension $d = m-r$. The situation described in the IFT is shown in Figure 5.4.

Rank vs Dimension of Level Sets. A rough general principle is at work here. If $\phi : \mathbb{R}^m \to \mathbb{R}$ is a scalar valued $\mathcal{C}^\infty$ function, one often finds that the solution set $L_{\phi(\mathbf{x})=c}$, $c \in \mathbb{R}$, is a smooth hypersurface of dimension $m-1$. For a vector valued map $\phi : \mathbb{R}^m \to \mathbb{R}^n$ with $\phi = (\phi_1, \dots, \phi_n)$, a typical level set $L_{\phi(\mathbf{x})=\mathbf{c}}$ with $\mathbf{c} = (c_1, \dots, c_n)$ is the intersection of the solution sets for the system of scalar constraints

$$\phi_1(\mathbf{x}) = c_1, \dots, \phi_r(\mathbf{x}) = c_n \quad \text{if} \quad \mathbf{c} = (c_1, \dots, c_n).$$

The solution set tends to lose one degree of freedom for each imposed constraint, so the outcome is usually a smooth hypersurface in $\mathbb{R}^m$ of dimension $m-r$. But that is not always the case, and the point of the IFT is to make clear when it is true. This principle also suggests why it is natural to restrict attention to the case $m \geq n$, in which "maximal rank" means $\operatorname{rk}(d\phi)_p = n$, the dimension of the "target space" $\mathbb{R}^n$ for ϕ. When the number of constraints n exceeds the dimension m of the space $\mathbb{R}^m$ in which the level set lives, the locus may be degenerate with no solutions at all, or it may reduce to a set of isolated points in $\mathbb{R}^m$.

It is worth asking "*What's going on if* $r = \text{rk}(d\phi)_p < n$?" The short answer is that the imposed constraints

$$\phi_1(\mathbf{x}) = c_1, \ldots, \phi_n(\mathbf{x}) = c_n$$

(one for each degree of freedom in the target space $\mathbb{R}^n$) include some redundant constraints whose elimination would not change the solution set $L_{\phi(\mathbf{x})=\mathbf{c}}$ but would make n equal to $r = \text{rk}(d\phi)_p$. Thus, we may as well assume the dimension n of the target space for ϕ is equal to $r = \text{rk}(d\phi)_p$. Unfortunately, ferreting out the redundant constraints could prove an unpleasant chore.

The "Maximal Rank" Case. As a particular example, if ϕ maps $\mathbb{K}^m \to \mathbb{K}^n$ and $m \geq n$, the maximum possible rank of $(d\phi_p)$ is n. If this is achieved at some base point $p \in \mathbb{K}^m$, ϕ will automatically have the same (maximal) rank at all points $\mathbf{x}$ near p in $\mathbb{K}^m$ because entries in the Jacobian matrix $[D\phi(\mathbf{x})]$ vary continuously with $\mathbf{x}$ and $\det[D\phi(p)] \neq 0$. The maximal rank case $r = n$ is the one most often encountered, but sometimes one needs the stronger "constant rank" version in which $r \leq n$.

EXERCISE 5.6. Consider the $\mathcal{C}^\infty$ map $\phi : \mathbb{R}^4 \to \mathbb{R}^2$ given by

$$\mathbf{y} = \big(y_1(\mathbf{x}), y_2(\mathbf{x})\big) = \phi(x_1, x_2, x_3, x_4) = \big(x_1^2 + x_2^2,\, x_3^2 - x_4^2 + x_1x_4\big).$$

(a) Show that the locus $M = L_{\phi(\mathbf{x})=\mathbf{q}}$ can be described as a smooth two-dimensional hypersurface in $\mathbb{R}^4$ near $p = (1, 2, -1, 3)$, at which $q = \phi(p) = (5, -5)$. Identify all pairs of variables x_i, x_j $(1 \leq i < j \leq 4)$ that can be used to smoothly parametrize this hypersurface near p.

(b) Is this locus smooth near *all* of its points?

HINT. In (b), start by showing $x_4 \neq 0$ for every point $\mathbf{x} \in M$, so you can assume $x_4 \neq 0$ in calculations involving points on M, even if you cannot draw a picture. In answering (b), you will have to compute $\text{rk}(A_{IJ})$ for various square submatrices of the 2×4 Jacobian matrix $[\,\partial y_i/\partial x_j\,]$, which has variable coefficients. Do this using symbolic row operations.

EXERCISE 5.7. Consider the $\mathcal{C}^\infty$ scalar-valued function $\phi : \mathbb{R}^2 \to \mathbb{R}$,

$$\phi(x, y) = x^3y + 2e^{xy}.$$

(a) Find all *critical points*, where both partial derivatives $\partial\phi/\partial x$ and $\partial\phi/\partial y$ are zero.

At a critical point p there is *no* way to represent the level set $S_p = L_{\phi(\mathbf{x})=\phi(\mathbf{p})}$ as the graph of a smooth function $y = f(x)$ or $x = g(y)$.

(b) Locate all points p where one of the partial derivatives is zero but the other is not (two cases to consider). Find the value of ϕ at each such base point to determine which level sets $L_{\phi(\mathbf{p})=c}$ contain such points.

(c) The locus $M = L_{\phi(\mathbf{x})=2}$ obviously contains the horizontal and vertical axes. Prove that there are no other points on this locus. (Thus the origin is a singularity for the locus M, and there are no others.)

HINT. In (c): quadrant-by-quadrant, what is the sign of $\partial\phi/\partial y$ off of the x- and y-axes?

EXERCISE 5.8. Let $\mathbf{y} = \phi(\mathbf{x}) = \big(y_1(\mathbf{x}), y_2(\mathbf{x})\big)$ be a $\mathcal{C}^\infty$ map from $\mathbb{R}^3 \to \mathbb{R}^2$ such that

1. $\phi(p) = q = (0,0)$ at base point $p = (3,-1,2)$.
2. At p, the Jacobian matrix is $\big[\partial y_i/\partial x_j\big] = \begin{pmatrix} 1 & 2 & 1 \\ 1 & -1 & 1 \end{pmatrix}$.

Answer the following questions without knowing anything more about ϕ:

(a) Can the level set $M = L_{\phi(\mathbf{x})} = q$ be described near $p = (3,-1,2$ as a smooth hypersurface in $\mathbb{R}^3$? Of what dimension k?

(b) Which of the variables x_1, x_2, x_3 can legitimately be used to parametrize M near $p = (3,-1,2)$ as the graph of a smooth map f from $\mathbb{R}^k$ to $\mathbb{R}^3$? List all valid choices of the parametrizing variables $x_{i_1}, \dots, x_{i_k}$.

EXERCISE 5.9. If $f : \mathbb{K}^r \to \mathbb{K}^s$ is a $\mathcal{C}^\infty$ map defined on open set $B \subseteq \mathbb{K}^r$, its *graph*

$$\Gamma = \{(\mathbf{x}, f(\mathbf{x})) \in \mathbb{K}^r \times \mathbb{K}^s : \mathbf{x} \in B\}$$

is the range of the *graph map* $F(\mathbf{x}) = (\mathbf{x}, f(\mathbf{x}))$ from $\mathbb{K}^r \to \mathbb{K}^{r+s}$. Show that the graph map is $\mathcal{C}^\infty$ for $\mathbf{x} \in B \subseteq \mathbb{K}^r$ and that $\mathrm{rk}(df)_{\mathbf{x}} \equiv r$ (constant) for all $\mathbf{x} \in B$.

The Inverse Mapping Theorem (IMT). A result closely related to the implicit function theorem tells us when a differentiable map $\phi : \mathbb{K}^n \to \mathbb{K}^n$ ($\mathbb{K} = \mathbb{R}$ or $\mathbb{C}$) has a locally defined inverse near a base point p. This variant of the IFT will be central in our discussion of matrix Lie groups.

THEOREM 5.10 (Inverse Mapping Theorem). *Let $\phi(\mathbf{x}) = (\phi_1(\mathbf{x}), \dots, \phi_n(\mathbf{x}))$ be a map from $\mathbb{R}^n \to \mathbb{R}^n$ that is differentiable of class $\mathcal{C}^{(r)}$, so all partial derivatives $D^\alpha\phi_i$ of order $|\alpha| = \alpha_1 + \cdots + \alpha_n \leq r$ exist and are continuous near a base point p in $\mathbb{R}^n$. If the Jacobian matrix at p*

$$D\phi(p) = \left[\frac{\partial\phi_i}{\partial x_j}(p)\right]$$

is invertible, so $\det(D\phi(p)) \neq 0$, there exist open neighborhoods U about p and V about $q = \phi(p)$ such that

1. *The map $\mathbf{y} = \phi(\mathbf{x})$ is of class $\mathcal{C}^{(r)}$ on U and is a bijective map from $U \to V$.*
2. *The inverse map $\mathbf{x} = \psi(\mathbf{y})$ from $V \to U$ is differentiable of class at least $\mathcal{C}^{(r)}$ on V.*

REMARK 5.11. Obviously, the inverse map will be of class $\mathcal{C}^\infty$ if the forward map ϕ is $\mathcal{C}^\infty$; the theorem remains true for multivariate maps $\phi : \mathbb{C}^n \to \mathbb{C}^n$ involving complex variables. Moreover, it even applies to *analytic* functions of several real or complex variables. Analyticity at p means each scalar component ϕ_k has an absolutely convergent multivariable Taylor series expansion

$$\phi_k(\mathbf{x}) = \sum_{\alpha \in \mathbb{Z}_+^n} \frac{D^\alpha\phi_k(\mathbf{p})}{\alpha_1! \cdots \alpha_n!} \cdot (\mathbf{x} - \mathbf{p})^\alpha \quad \big(\mathbf{x}^\alpha = x_1^{\alpha_1} \cdots x_n^{\alpha_n}\big)$$

on some open neighborhood of p. The class of real- or complex-analytic functions on a domain U is indicated by the special symbol $\mathcal{C}^\omega(U)$.

As an example, the exponential map $\text{Exp}(A) = e^A$ is a $\mathcal{C}^\omega$ map on $\mathbb{K}^{n^2}$ when we identify $A \in \text{M}(n, \mathbb{K})$ with

$$\mathbf{a} = (a_{11}, \dots, a_{1n}, \dots, a_{n1}, \dots, a_{nn}) \text{ in } \mathbb{K}^{n^2}$$

for $\mathbb{K} = \mathbb{R}$ (or in $\mathbb{R}^{2n^2}$ when $\mathbb{K} = \mathbb{C}$). Even in one variable, analyticity of ϕ is a much stronger condition than $\phi \in \mathcal{C}^\infty$. ○

Differentiable Manifolds (General Definition). A space M is *locally Euclidean* of dimension d if it can be covered by a family of "charts" $\{(x_\alpha, U_\alpha) : \alpha \text{ in some index set } I\}$, where $x_\alpha : U_\alpha \to V_\alpha \subseteq \mathbb{R}^d$ is a bicontinuous map from an open subset $U_\alpha \subseteq M$ to an open set $V_\alpha = x_\alpha(U_\alpha)$ in the Euclidean space $\mathbb{R}^d$. The "chart maps" x_α assign locally defined Euclidean coordinates $x_\alpha(u) = (x_1^{(\alpha)}(u), \dots, x_k^{(\alpha)}(u))$ for $u \in U_\alpha$. Thus, M looks locally like Euclidean coordinate space $\mathbb{R}^d$.

Where the domains of two charts $(x_\alpha, U_\alpha), (x_\beta, U_\beta)$ overlap, we have the situation shown in Figure 5.5. The intersection $U_\alpha \cap U_\beta$ is an open set in M, the images $N_\alpha = x_\alpha(U_\alpha \cap U_\beta)$, $N_\beta = x_\beta(U_\alpha \cap U_\beta)$ are open sets in coordinate space $\mathbb{R}^k$, and we have induced *coordinate transition maps* that tell us how the coordinates $\mathbf{x} = x_\alpha(u)$ and $\mathbf{y} = x_\beta(u)$ assigned to $u \in M$ by the chart maps are related. These transition maps

$$\mathbf{x} = (x_\alpha \circ x_\beta^{-1})(\mathbf{y}) \quad \text{from } N_\beta \to N_\alpha$$
$$\mathbf{y} = (x_\beta \circ x_\alpha^{-1})(\mathbf{x}) \quad \text{from } N_\alpha \to N_\beta$$

are bicontinuous bijections between the open sets N_α and N_β in $\mathbb{R}^d$.

DEFINITION 5.12. A locally Euclidean space M is a *differentiable manifold* of dimension $\dim(M) = d$ if the charts (U_α, x_α) that cover M each map M into $\mathbb{R}^d$ and are $\mathcal{C}^\infty$-related, so the coordinate transition maps are $\mathcal{C}^\infty$ between the open sets $N_\alpha, N_\beta \subseteq \mathbb{R}^d$ that correspond to the (open) intersection $U_\alpha \cap U_\beta$ of chart domains in M.

This allows us to make sense of such "smooth manifolds" without requiring that they be embedded in some surrounding Euclidean space. It is the starting point for modern differential geometry.

Once we make M a $\mathcal{C}^\infty$ manifold by introducing $\mathcal{C}^\infty$ related covering charts, we can begin to do calculus on M. The following concepts now make sense:

1. SMOOTH FUNCTIONS ON M. A map $f : M \to \mathbb{R}$ is a $\mathcal{C}^\infty$ *function on* M if the scalar-valued function

$$y = F(\mathbf{x}) = (f \circ x_\alpha^{-1})(\mathbf{x}),$$

defined on the open set $V_\alpha = x_\alpha(U_\alpha)$ in coordinate space $\mathbb{K}^m$, has continuous partial derivatives of all orders for each of the covering charts that determine the manifold structure of M.

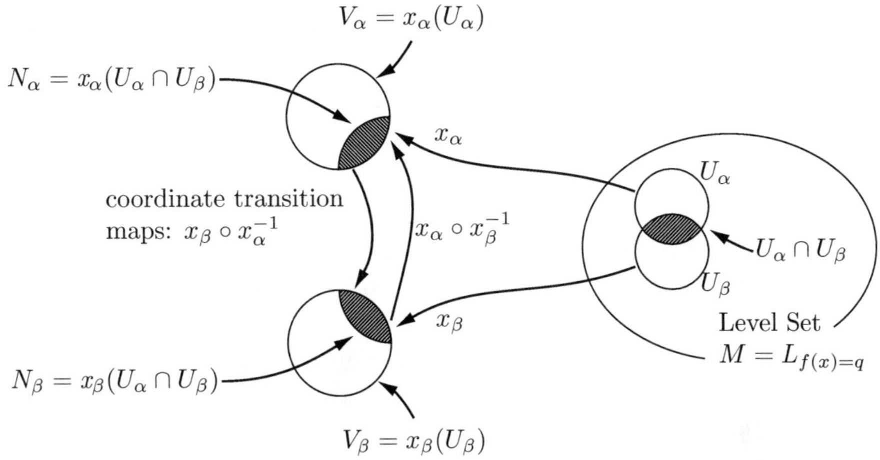

FIGURE 5.5. The coordinate transition maps $x_\alpha \circ x_\beta^{-1}$ and $x_\beta \circ x_\alpha^{-1}$ between two charts $(x_\alpha, U_\alpha), (x_\beta, U_\beta)$ and their (shaded) domains of definition in $\mathbb{R}^r$ are shown. Both shaded domains N_α, N_β in $\mathbb{R}^r$ correspond to the intersection $U_\alpha \cap U_\beta$ of the chart domains, which is an open set in the locally Euclidean space M.

2. SMOOTH MAPS BETWEEN MANIFOLDS. A map $\phi : M \to N$ is a C^∞ *mapping* between manifolds M and N of dimensions m and n if it becomes a $\mathcal{C}^\infty$ map from $\mathbb{R}^m \to \mathbb{R}^n$ when described in local coordinates on M and N. Thus, if (U_α, x_α) and (U_β, y_β) are charts on M and N respectively, the composite
$$\mathbf{y} = \Phi(\mathbf{x}) = y_\beta \circ \phi \circ x_\alpha^{-1}(\mathbf{x})$$
is a $\mathcal{C}^\infty$ map from $\mathbb{R}^m \to \mathbb{R}^n$ wherever it is well-defined.
3. PARAMETRIC CURVES IN M. A *parametric curve* in M is any continuous map $y = \gamma(t)$ from some interval $[a, b] \subseteq \mathbb{R}$ into M. It is a C^∞*-curve* in M if it becomes a $\mathcal{C}^\infty$ vector-valued map
$$\mathbf{x} = \big(y_1(t), \dots, y_m(t)\big) = x_\alpha \circ \gamma(t) \quad \text{for } t \in [a, b]$$
from $[a, b] \to \mathbb{R}^m$ for every chart (U_α, x_α) on M.

EXERCISE 5.13. Suppose $M \xrightarrow{h} N \xrightarrow{f} Q$ are $\mathcal{C}^\infty$ maps between $\mathcal{C}^\infty$ manifolds. In terms of the preceding definitions, explain why the composite $f \circ h$ is a $\mathcal{C}^\infty$ map from $M \to Q$ wherever it is well-defined.

HINT. If M, N, Q are Euclidean coordinate spaces $\mathbb{R}^m, \mathbb{R}^n, \mathbb{R}^q$, this follows by the chain rule of multivariate calculus.

Differentiable Manifolds and the IFT. One consequence of the IFT is this:

If ϕ is a $\mathcal{C}^\infty$ map $\mathbb{R}^m \to \mathbb{R}^n$ and if $\text{rk}(d\phi)_\mathbf{x} \equiv r$ (constant) near *every* point in a level set $M = L_{\phi(\mathbf{x})=\mathbf{c}}$, $\mathbf{c} \in \mathbb{R}^n$, we can use the IFT to create a family of $\mathcal{C}^\infty$-related charts (x_α, U_α) whose domains U_α cover M.

The resulting *standard $\mathcal{C}^\infty$ structure* makes M into a smooth $m-r$-dimensional manifold. The crucial fact that the charts are $\mathcal{C}^\infty$-related follows directly from the way the standard charts are constructed using the IFT, as in the following example.

EXAMPLE 5.14 (Constructing "Standard Charts" on Level Sets). The IFT and the "constant rank" condition allow us to construct a chart (x_α, U_α) about a typical base point $p \in M$.

STEP 1. Write ϕ as $\mathbf{v} = \phi(\mathbf{u})$ in terms of the standard coordinates $\mathbf{u} = (u_1, \dots, u_m)$ and $\mathbf{v} = (v_1, \dots, v_n)$ in $\mathbb{R}^m$ and $\mathbb{R}^n$. By Lemma 5.1 and the "constant rank" condition, we can, for each $\mathbf{u} \in M$, choose row and column indices $I \subseteq [1, n]$ and $J \subseteq [1, m]$ with $|I| = |J| = r = \mathrm{rk}(d\phi)_\mathbf{u}$ such that the square submatrices $\left[\partial v_i / \partial u_j(\mathbf{u})\right]_{IJ}$ are nonsingular for $\mathbf{u}$ near p in $\mathbb{R}^m$. By Lemma 5.1, this cannot be done for any larger square submatrix.

STEP 2. Using the column indices determined in Step 1, let $J' = [1, m] \sim J$, split $\mathbb{R}^m = \mathbb{R}^{J'} \oplus \mathbb{R}^J$, and let $\pi^{J'}, \pi^J$ be the projection maps from $\mathbb{R}^m$ to $\mathbb{R}^{J'}$ or $\mathbb{R}^J$. By the IFT, there is a rectangular open neighborhood $B_1 \times B_2$ of p in $\mathbb{K}^m$ such that the projection

$$\pi^{J'} : (B_1 \times B_2) \to B_1$$

maps the relatively open neighborhood $U_p = (B_1 \times B_2) \cap M$ in M bijectively to the open set $B_1 \subseteq \mathbb{R}^{J'} \cong \mathbb{R}^{m-r}$. To get a chart (x_α, U_α) that imposes Euclidean coordinates on M near p, we take $U_\alpha = U_p$ and bijective chart map

$$x_\alpha = \left(\pi^{J'}|_{U_p}\right) : U_p \to B_1 \quad \text{(an open set in } \mathbb{R}^{m-r}\text{)}.$$

Such charts (x_α, U_α) obviously cover M, owing to constancy of $\mathrm{rk}(d\phi)$ near every point in M.

STEP 3. The inverse map $\Psi = \left(\pi^{J'}|_{U_p}\right)^{-1} : B_1 \to U_p \subseteq M \subseteq \mathbb{R}^m$ is $\mathcal{C}^\infty$ from $B_1 \subseteq \mathbb{R}^{J'}$ into all of $\mathbb{R}^m$, and its range is precisely the chart domain U_p.

Now we must show that charts created this way, perhaps about different base points, are always $\mathcal{C}^\infty$-related wherever they overlap.

THEOREM 5.15. *If $\phi : \mathbb{R}^m \to \mathbb{R}^n$ is a $\mathcal{C}^\infty$ map, let $M = L_{\phi(\mathbf{x})=\mathbf{c}}$ be a level set such that $\mathrm{rk}(d\phi)_x \equiv r$ (constant) on an open neighborhood of every point in M. Then all standard charts on M obtained by the preceding construction are $\mathcal{C}^\infty$ related where they overlap. This determines the standard $\mathcal{C}^\infty$ structure on M. The dimension of the resulting $\mathcal{C}^\infty$ manifold is $d = m - r$.*

PROOF. Consider two standard charts (x_α, U_α) and (x_β, U_β) about a typical point p in $U_\alpha \cap U_\beta$. The chart (x_α, U_α) is determined by a partition of column indices $[1, m] = J'(\alpha) \cup J(\alpha)$ and a choice of row indices $I(\alpha) \subseteq [1, n]$ with $|I(\alpha)| = |J(\alpha)| = r$; then $|J'(\alpha)| = m - r$ and $[(d\phi)_p]_{IJ}$ is nonsingular. In the notation of the IFT, we then have

$$U_\alpha = (B_1^\alpha \times B_2^\alpha) \cap M \quad x_\alpha = \left(\pi_{J'(\alpha)}|_{U_\alpha}\right)$$
$$V_\alpha = x_\alpha(U_\alpha) = B_1^\alpha, \text{ an open set in } \mathbb{R}^{m-r}.$$

The chart map x_α is the restriction to U_α of a *linear* (hence $\mathcal{C}^\infty$) projection map $\pi^{J'}(\alpha) : \mathbb{R}^m \to \mathbb{R}^{J'(\alpha)} \cong \mathbb{R}^{m-r}$,

$$x_\alpha = \left(\pi_{J'(\alpha)}|_{U_\alpha}\right) : U_\alpha \to B_1^\alpha \subseteq \mathbb{R}^{m-r}.$$

Its inverse $\Psi = x_\alpha^{-1}$ is the graph map

$$\Psi_\alpha = \left(\pi_{J'(\alpha)}|_{U_\alpha}\right)^{-1} : V_\alpha \to U_\alpha,$$

which is actually a $\mathcal{C}^\infty$ map from B_1^α into *all of* $\mathbb{R}^m$.

The chart (x_β, U_β) corresponds to some other choice of column and row indices $[1,m] = J'(\beta) \cup J(\beta)$ and $I(\beta) \subseteq [1,n]$, and a corresponding rectangular open neighborhood $B_1^\beta \times B_2^\beta$ of p in $\mathbb{R}^m$. The chart map on $U_\beta = (B_1^\beta \times B_2^\beta) \cap M$ is just the restriction to U_β of a linear projection $\pi_{J'(\beta)}$, and by the IFT its inverse is a $\mathcal{C}^\infty$ map from B_1^β into all of $\mathbb{R}^m$. Therefore the coordinate transition map

$$x_\beta \circ x_\alpha^{-1} = \pi_{J'(\beta)} \circ \left(\pi_{J'(\alpha)}|_{U_\alpha}\right)^{-1} = \pi_{J'(\beta)} \circ \Psi_\alpha$$

is the composite of a linear map and a $\mathcal{C}^\infty$ map

$$\mathbb{R}^{m-r} \cong \mathbb{R}^{J'(\alpha)} \xrightarrow{x_\alpha^{-1}} \mathbb{R}^m \xrightarrow{x_\beta} \mathbb{R}^{J'(\beta)} \cong \mathbb{R}^{m-r}$$

and is certainly $\mathcal{C}^\infty$. Likewise for the reverse transition map. □

The preceding proof is burdened by the complicated notation used to label all the players. A stripped-down outline emphasizes the intuition behind the proof. To show $\mathbf{y} = x_\beta \circ x_\alpha^{-1}(\mathbf{x})$ is $\mathcal{C}^\infty$ near $\mathbf{x}_0$, we observe that

- Near $\mathbf{x}_0$, the map x_α^{-1} coincides with the map $\left(\pi_{J'(\alpha)}|_{U_\alpha}\right)^{-1}$, which by the IFT is a $\mathcal{C}^\infty$ map from an open set in $\mathbb{R}^{m-r}$ into all of $\mathbb{R}^m$ that sends $\mathbf{x}_0 \to p$ and whose range is contained in M.
- Near p, the chart map x_β coincides with the globally defined linear projection map $\pi_{J'(\beta)}$, which is certainly $\mathcal{C}^\infty$.

Therefore the transition map $x_\beta \circ x_\alpha^{-1}$ is the composite of a linear map and a $\mathcal{C}^\infty$ map

$$\mathbb{R}^{m-r} \cong \mathbb{R}^{J'(\alpha)} \xrightarrow{x_\alpha^{-1}} \mathbb{R}^m \xrightarrow{x_\beta} \mathbb{R}^{J'(\beta)} \cong \mathbb{R}^{m-r}$$

and is itself $\mathcal{C}^\infty$. Likewise for the transition map in the reverse direction.

EXAMPLE 5.16. Let $\phi : \mathbb{R}^3 \to \mathbb{R}^1$ with $\phi(\mathbf{x}) = x_3^2 - x_1^2 - x_2^2$. At any $p = (x_1, x_2, x_3)$, the 1×3 Jacobian matrix

$$(d\phi)_p = \left[\frac{\partial \phi}{\partial x_1}, \dots, \frac{\partial \phi}{\partial x_3}\right] = [-2x_1\,,\, -2x_2\,,\, 2x_3]$$

is essentially the classical "gradient" vector $\nabla\phi(p)$. Then $\mathrm{rk}(d\phi)_\mathbf{x}$ is constant $\equiv 1$ unless all three entries are zero, which happens only at the origin $p = (0,0,0)$.

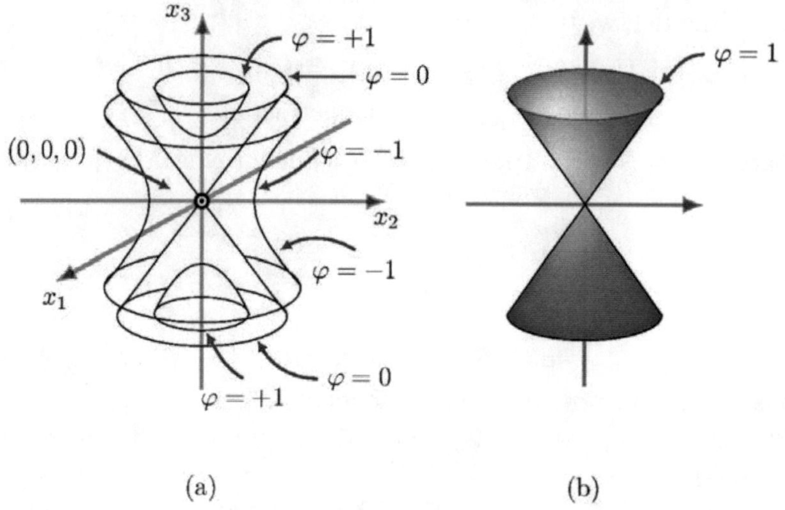

FIGURE 5.6. In (a) we show some level sets $L_c = L_{\phi(\mathbf{x})=c}$ for the map $\phi(\mathbf{x}) = x_3^2 - x_2^2 - x_1^2$ from $\mathbb{R}^3 \to \mathbb{R}^1$. For $c = 0$ the level set where $x_1^2 - x_2^2 - x_3^2 = 0$, shown in (b), is a double cone with a singularity at the origin, where it fails to be locally Euclidean. All other level sets are smooth two-dimensional hypersurfaces in $\mathbb{R}^3$, but the geometry of L_c changes as we pass from $c < 0$ to $c > 0$. For $c < 0$, we get a single connected surface; for $c > 0$ there are two isolated pieces, both smooth.

The level set $M_0 = L_{\phi(\mathbf{x})=0}$ is the double cone shown in Figure 5.6(b). This two-dimensional hypersurface has a singularity at the origin in $\mathbb{R}^3$, where it fails to be locally Euclidean, so $L_{\phi(\mathbf{x})=0}$ cannot be made into a smooth manifold by covering it with suitably defined coordinate charts. All other level surfaces $L_{\phi(\mathbf{x})=c}$, $c \neq 0$, are smooth two-dimensional manifolds, a few of which are shown in Figure 5.6(a).

Consider the possible charts we might impose near the point $p = (1, 1, \sqrt{3})$ on the particular level set $L_{\phi(\mathbf{x})=1}$. Entries in $(d\phi)_p$ at p,

$$(d\phi)_p = [-2x_1, -2x_2, 2x_3],$$

are nonzero at and near p and $\text{rk}(d\phi)_{\mathbf{x}}$ is constant $\equiv 1$ near p, so we may apply the IFT to define standard charts about p. Each nonzero entry in $(d\phi)_p$ corresponds to a nonsingular 1×1 submatrix, so several legitimate groupings of variables are available to parametrize M near p:

$$J = \{1\}, J' = \{2, 3\} \quad J = \{2\}, J' = \{1, 3\} \quad J = \{3\}, J' = \{1, 2\}. \tag{5.2}$$

Thus, $L_{\phi(\mathbf{x})=1}$ can be described as the graph in $\mathbb{R}^3$ of various smooth functions $x_k = f_k(x_i, x_j)$ by solving

$$1 = \phi(\mathbf{x}) = x_3^2 - x_2^2 - x_1^2$$

for one variable in terms of the other two.

1. $x_1 = f_1(x_2, x_3) = +\sqrt{x_3^2 - x_2^2 - 1}$ near $(1, \sqrt{3})$ in the (x_2, x_3)-plane.
2. $x_2 = f_2(x_1, x_3) = +\sqrt{x_3^2 - x_1^2 - 1}$ near $(1, \sqrt{3})$ in the (x_1, x_3)-plane.
3. $x_3 = f_3(x_1, x_2) = +\sqrt{1 + (x_1^2 + x_2^2)}$ near $(1, 1)$ in the (x_1, x_2)-plane.

The coordinate transition map $(x_1, x_3) = y_\beta \circ x_\alpha^{-1}(x_2, x_3)$ can be computed directly by writing $x_1 = f_1(x_2, x_3)$ to get (x_1, x_3) in terms of (x_2, x_3). The resulting transition map

$$\begin{aligned}(x_1, x_3) = \Phi(x_2, x_3) &= y_\beta \circ x_\alpha^{-1}(x_2, x_3)\\ &= (x_1, x_3)\Big|_{x_1 = f_1(x_2, x_3)}\\ &= \left(\sqrt{x_3^2 - x_2^2 - 1}\,,\; x_3\right)\end{aligned}$$

is clearly a $\mathcal{C}^\infty$ map from (x_2, x_3) to (x_1, x_3). So is its inverse.

EXERCISE 5.17. In Example 5.16,

(a) Compute the inverse map $(x_2, x_3) = \Phi^{-1}(x_1, x_3)$.

(b) One of the valid splittings $[1, 3] = J' \cup J$ of column indices listed in (5.2) is $J' = \{1, 2\}$, $J = \{3\}$. Find an explicit formula for the corresponding projection map $(x_1, x_2) = \pi^{J'}(x_1, x_2, x_3)$ that assigns Euclidean coordinates to points $\mathbf{x} = (x_1, x_2, x_3)$ on M near $p = (1, 1, \sqrt{3})$.

(c) Give an explicit formula for the inverse

$$(x_1, x_2, x_3) = \left(\pi^{J'}|_{\mathbb{R}^{J'}}\right)^{-1}(x_1, x_3)$$

of the projection map in (b).

EXERCISE 5.18. Consider the points

$$\text{(a)}\ \ p = (1, 0, \sqrt{2}) \qquad \text{(b)}\ \ p = (0, 0, -1)$$

on the two-dimensional hypersurface $M = L_{\phi(\mathbf{x})=1}$ of Example 5.16. In each case, determine all groups of coordinates $x_{J'} = (x_i, x_j)$ that parametrize M near the base point p.

EXERCISE 5.19. The unit sphere $S^2 = L_{\phi(\mathbf{x})=1}$ in $\mathbb{R}^3$ is determined by $\phi(\mathbf{x}) = x_1^2 + x_2^2 + x_3^2 = 1$. Verify that it is a $\mathcal{C}^\infty$ manifold in $\mathbb{R}^3$ by showing that $\text{rk}(d\phi)_\mathbf{x} \equiv 1$ near *every* point $\mathbf{x} \in S^2$.

EXERCISE 5.20. Describe a set of standard charts covering the unit sphere $S^2 = L_{\phi(\mathbf{x})=1}$ where $\phi(\mathbf{x}) = x_1^2 + x_2^2 + x_3^2$, taking for your chart domains the relatively open hemispheres (boundary circles excluded)

$$U_k^+ = \{\mathbf{x} \in S^2 : x_k > 0\} \qquad U_k^- = \{\mathbf{x} \in S^2 : x_k < 0\}$$

for $k = 1, 2, 3$. (All six hemispheres are required to fully cover the sphere S^2.)

The chart maps $x_1^\pm : U_1^\pm \to \mathbb{R}^2$ project points $\mathbf{x} \in U_1^\pm$ onto the open unit disc $x_2^2 + x_3^2 < 1$ in the (x_2, x_3)-plane when $k = 1$; they project $U_2^\pm$ onto the open disc in the (x_1, x_3)-plane when $k = 2$; and they project onto the disc in the (x_1, x_2)-plane when $k = 3$.

(a) Give explicit formulas for the chart maps on the particular domains U_1^+ and U_3^-.

(b) Compute the coordinate transition maps in both directions for this pair of charts, noting that they have the form

$$(x_i, x_j) = x_\alpha(\mathbf{x}) = x_\alpha(x_1, x_2, x_3)$$

for $\mathbf{x} \in M$.

These are examples of standard charts on the level set $L_{\phi(\mathbf{x})=1}$.

EXERCISE 5.21 (Stereographic Projection on the 2-Sphere S^2). Let H^+ be the two-dimensional hyperplane in $\mathbb{R}^3$ that is tangent to the unit sphere $M = S^2$ at its "north pole" $N = (0, 0, +1)$, and consider the "punctured sphere" $U_\alpha = S^2 \sim \{S\}$ obtained by deleting the south pole $S = (0, 0, -1)$ from the sphere. Each point $\mathbf{u} \in U_\alpha$ determines a straight line in $\mathbb{R}^3$ that emanates from S and passes through $\mathbf{u}$; continuing along this line, we meet the hyperplane H^+ in a unique point with coordinates $\big(x(\mathbf{u}), y(\mathbf{u}), +1\big)$. The resulting bijection Φ^+ from U_α to H^+ is an example of *stereographic projection*. Dropping the redundant coordinate entry "1", we obtain the *stereographic projection map* $x_\alpha : U_\alpha \to \mathbb{R}^2$,

$$x_\alpha(\mathbf{u}) = \big(x_1(\mathbf{u}), x_2(\mathbf{u})\big) \in \mathbb{R}^2,$$

which is bicontinuous from the open subset $U_\alpha^+ \subseteq S^2$ onto all of coordinate space $\mathbb{R}^2$.

Similarly, we may project the punctured sphere $U_\beta = S^2 \sim \{N\}$ along lines radiating from N onto the hyperplane H^- tangent to the sphere at the south pole $S = (0, 0, -1)$ to define a second chart map $x_\beta(\mathbf{v}) = (x', y') \in \mathbb{R}^2$ for $\mathbf{v} \in U_\beta$.

(a) Give an explicit formula for the stereographic projection map $(x_1, x_2) = x_\alpha(\mathbf{u}) = x_\alpha(u_1, u_2, u_3)$. Note carefully that x_β maps triples $\mathbf{u}$ with $u_1^2 + u_2^2 + u_3^3 = 1$ to pairs $(x_1, x_2) \in \mathbb{R}^2$.

(b) Compute the coordinate transition map $\mathbf{v} = x_\beta \circ x_\alpha^{-1}(\mathbf{u})$ and its inverse, and check that it is $\mathcal{C}^\infty$ where defined.

Stereographic projection allows us to cover the sphere S^2 with just two $\mathcal{C}^\infty$-related charts, the minimum number possible since it is well known that S^2 cannot be mapped bicontinuously to an open set in the plane $\mathbb{R}^2$. But note: for many purposes, the covering with hemispheres leads to simpler computations than stereographic projection.

HINT. In (a), use similar triangles and rotational symmetry.

NOTE. The charts H^+ and H^- are not among the founding charts for the differentiable structure on S^2 described in Example 5.14, but they are $\mathcal{C}^\infty$-related to all founding charts (which, by Theorem 5.15, are $\mathcal{C}^\infty$-related to each other). ○

5.2. Matrix Lie Groups

The classical groups, first discussed in Chapter 3, are level sets of certain polynomial maps $\mathbb{K}^m \to \mathbb{K}^n$, except for the general linear group $\mathrm{GL}(n, \mathbb{K}) = \{A \in \mathrm{M}(n, \mathbb{K}) : \det(A) \neq 0\}$, often abbreviated to "GL." It is an open, dense subset in matrix space $\mathrm{M}(n, \mathbb{K}) \cong \mathbb{K}^{n^2}$, and so can be regarded as a smooth manifold covered by a single chart (U_α, x_α) with $U_\alpha = \mathrm{GL}$ and a chart map $x_\alpha : \mathrm{GL}(n, \mathbb{K}) \to \mathbb{K}^{n^2}$, which we shall write as

$$x_\alpha(A) = (a_{11}, \dots, a_{1n}; a_{21}, \dots, a_{2n}; \dots; a_{n1}, \dots, a_{nn})$$

in what follows. Obviously, $\dim_{\mathbb{K}}(\mathrm{GL}) = n^2$ since GL is an open set in the n^2-dimensional matrix space. All other classical groups are closed subsets of lower dimension in $\mathrm{M}(n, \mathbb{K}) \cong \mathbb{K}^{n^2}$.

DEFINITION 5.22. A smooth manifold G is a *Lie group* if

1. It is a group under some product operation $P : G \times G \to G$, indicated by writing $P(x, y) = x \cdot y$, and under the inversion map $J : G \to G$ that sends $x \to x^{-1}$.
2. The product operation P and inversion operation J are $\mathcal{C}^\infty$ maps of the underlying differentiable manifold.

The dimension $d = \dim_{\mathbb{K}}(G)$ is the common dimension of the coordinate charts that cover G and provide its differentiable structure.

In particular, if (U_α, x_α), (U_β, y_β) are coordinate charts, the product operation becomes a $\mathcal{C}^\infty$ map $P : \mathbb{K}^d \times \mathbb{K}^d \to \mathbb{K}^d$ in these coordinates, where $d = \dim(G)$. Thus, if $x \in U_\alpha$, $y \in U_\beta$, and (U_γ, x_γ) is a chart containing the product $z = x \cdot y$, the composite map

$$x_\gamma \circ P \circ (x_\alpha^{-1} \times y_\beta^{-1}) : \mathbb{K}^d \times \mathbb{K}^d \to G \times G \xrightarrow{P} G \xrightarrow{x_\gamma} \mathbb{K}^d$$

is a $\mathcal{C}^\infty$ map. Similarly, if $z \in U_\alpha$ and $z^{-1} \in U_\beta$, then

$$x_\beta \circ J \circ x_\alpha^{-1} : \mathbb{K}^d \to \mathbb{K}^d \quad \text{is a } \mathcal{C}^\infty \text{ map}.$$

The general theory of Lie groups has become a vast subject. In general, we will restrict attention to *matrix Lie groups*, which comprise $\mathrm{GL}(n, \mathbb{K})$ itself and its "closed" subgroups, the subsets $G \subseteq \mathrm{GL}(n, \mathbb{K})$ such that G is :

1. A group under matrix multiplication, as in Definition 5.22.
2. A $\mathcal{C}^\infty$ manifold (smooth hypersurface) in matrix space, as in Definition 5.12.
3. A *closed subgroup* in $\mathrm{GL}(n, \mathbb{K})$, which means that if a sequence of matrices in G converges to a limit A in $\mathrm{M}(n, \mathbb{K})$, so that $A_n \to A$ as $n \to \infty$, the limit A is also in G. Notice that $\mathrm{GL}(n, \mathbb{K})$ itself is not a closed subset of matrix space because the upper triangular matrices $A_n = [\frac{1}{n}, 1; 0, 1]$ converge to $B = [0, 1; 0, 1]$ which has $\det(B) = 0$ and is not in GL.

But this point of view soon becomes cumbersome with its continual need to refer to the external matrix space in which G is embedded.

The general theory, in which G is a differentiable manifold not embedded in any external Euclidean space, becomes accessible if you have read Section 4.1 of Chapter 4. The following discussion is pretty much self-contained but does presume some familiarity with that introductory discussion of manifolds, tangent vectors, vector fields, etc.

Real vs Complex Matrix Groups. The bijection $J : \mathbb{C} \to \mathbb{R}^2 = \mathbb{R} \oplus \mathbb{R}$ given by $z = x + \sqrt{-1}\,y \mapsto J(z) = (x, y) \in \mathbb{R}^2$ identifies $\mathbb{C} \cong \mathbb{R}^2$ as vector spaces over $\mathbb{R}$ since J is an $\mathbb{R}$-linear bijection; however, J is not a $\mathbb{C}$-linear map. Similarly, if we view $\mathrm{M}(n, \mathbb{C})$ as a vector space over $\mathbb{R}$, it becomes a direct sum $\mathrm{M}(n, \mathbb{C}) = \mathrm{M}(n, \mathbb{R}) \oplus \sqrt{-1}\,\mathrm{M}(n, \mathbb{R})$ of vector subspaces over $\mathbb{R}$,

$$\mathrm{M}(n, \mathbb{C}) \cong \mathrm{M}(n, \mathbb{R}) \oplus \mathrm{M}(n, \mathbb{R}) \cong \mathbb{R}^{2n^2},$$

via the $\mathbb{R}$-linear bijection that sends

$$Z = \{z_{ij}\} = \{x_{ij} + \sqrt{-1}y_{ij}\} \to (X, Y) = \big(\{x_{ij}\}, \{y_{ij}\}\big).$$

In particular, we have

$$\dim_{\mathbb{R}}(\mathbb{C}) = 2 \cdot \dim_{\mathbb{C}}(\mathbb{C}) = 2 \cdot 1 = 2$$
$$\dim_{\mathbb{R}} \mathrm{M}(n, \mathbb{C}) = 2 \cdot \dim_{\mathbb{R}} \mathrm{M}(n, \mathbb{R}) = 2n^2.$$

In these *Notes*, we will generally view $\mathrm{M}(n, \mathbb{C})$ as a vector space over $\mathbb{R}$ with dimension $2n^2$ and regard all classical groups mentioned above as real hypersurfaces in $\mathbb{R}^{n^2}$ or $\mathbb{R}^{2n^2}$, even if they are carved out of complex matrix space by real- or complex-valued constraint equations. The hypotheses of the IFT can be shown to hold for all the classical groups, and we shall regard these matrix groups as $\mathcal{C}^\infty$ manifolds equipped with (real) coordinate charts the IFT provides.

For instance in $\mathrm{M}(n, \mathbb{R}) \cong \mathbb{R}^{n^2}$, once the conditions of the IFT are verified, the single real-valued constraint equation $\det(A) = 1$ yields a $\mathcal{C}^\infty$ hypersurface $G_{\mathbb{R}} = \mathrm{SL}(n, \mathbb{R})$ with $\dim_{\mathbb{R}}(G_{\mathbb{R}}) = n^2 - 1$.

If we identify complex matrix space $\mathrm{M}(n, \mathbb{C}) \cong \mathbb{C}^{n^2}$, the matrix group

$$G_{\mathbb{C}} = \mathrm{SL}(n, \mathbb{C}) = \{A \in \mathrm{M}(n, \mathbb{C}) : \det(A) = 1 + i0\}$$

has complex dimension $\dim(G_{\mathbb{C}}) = n^2 - 1$ because it is carved out of $\mathrm{M}(n, \mathbb{C}) \cong \mathbb{C}^{n^2}$ by a single complex constraint. By the complex-variable version of the (IFT), $G_{\mathbb{C}}$ can be described (at least locally) by imposing $n^2 - 1$ independent complex coordinates on the group.

On the other hand, if we view $\mathrm{M}(n, \mathbb{C})$ as a real vector space

$$\mathrm{M}(n, \mathbb{R}) \oplus \mathrm{M}(n, \mathbb{R}) \cong \mathbb{R}^{n^2} \oplus \mathbb{R}^{n^2} \cong \mathbb{R}^{2n^2},$$

then as a subset of $\mathbb{R}^{2n^2}$, the group $\mathrm{SL}(n, \mathbb{C})$ becomes a hypersurface of real dimension $2n^2 - 2 = 2 \cdot \dim_{\mathbb{C}} G_{\mathbb{C}}$ because the complex constraint identity

$$\det(A) = \mathrm{Re}(\det(A)) + i\,\mathrm{Im}(\det(A)) = F_1(A) + iF_2(A) = 1 + i0$$

involves *two* independent real-valued constraints, $F_1(A) = 1$ and $F_2(A) = 0$, for matrices with complex entries.

Geometrically, $\mathrm{SL}(n,\mathbb{R})$ can be viewed as as a subset of $\mathrm{SL}(n,\mathbb{C})$ since

$$\mathrm{SL}(n,\mathbb{R}) = \mathrm{SL}(n,\mathbb{C}) \cap \big(\mathrm{M}(n,\mathbb{R}) + i0\big)$$

is the set of complex matrices with $\det(A) = 1$ whose entries happen to be *real*. We can also think of this intersection as the set of "real points" in the complex matrix group $\mathrm{SL}(n,\mathbb{C})$ when we identify $\mathrm{M}(n,\mathbb{C}) = \mathrm{M}(n,\mathbb{R}) + \sqrt{-1}\,\mathrm{M}(n,\mathbb{R})$ as the complexification of the real vector space $\mathrm{M}(n,\mathbb{R})$.

EXAMPLE 5.23 (Unitary Groups $\mathrm{U}(n)$ and $\mathrm{SU}(n)$). The *unitary groups* are another important family of matrix groups in $\mathrm{M}(n,\mathbb{C})$:

$$\mathrm{U}(n) = \{A : AA^* = I\} \quad \text{and} \quad \mathrm{SU}(n) = \mathrm{U}(n) \cap \mathrm{SL}(n,\mathbb{C}).$$

DISCUSSION. In Chapter 6 of LA I we saw that a complex matrix A is unitary if and only if its rows form an orthonormal basis in $\mathbb{C}^n$, with inner products

$$(R_i, R_j) = \delta_{ij} \quad \text{(Kronecker delta)} \quad \text{for } 1 \le i \le j \le n.$$

This involves $\frac{1}{2}(n^2+n) = \frac{1}{2}(n^2-n) + n$ independent *complex* constraint identities on row vectors in $\mathbb{C}^n$.

If we regard $\mathrm{M}(n,\mathbb{C}) \cong \mathbb{R}^{2n^2}$, the expressions $(R_i,R_i) = \|R_i\|^2 = 1$ are automatically real-valued (and positive) and impose n real constraints, while the $(n^2-n)/2$ complex off-diagonal constraints $(R_j,R_k) = 0 + i0$ for $j < k$ each impose two real constraints. None of of these can be dropped if we want to ensure $AA^* = I$. Thus, when we identify $\mathrm{M}(n,\mathbb{C}) \cong \mathbb{R}^{2n^2}$, $\mathrm{U}(n)$ is determined by $n + 2\cdot(\frac{n^2-n}{2}) = n^2$ real constraints. If the conditions of the IFT can be verified, we would conclude that $\mathrm{U}(n)$ is a matrix Lie group with

$$\begin{aligned}\dim_{\mathbb{R}} \mathrm{U}(n) &= \dim_{\mathbb{R}} \mathrm{M}(n,\mathbb{C}) - \#(\text{irredundant real constraints})\\ &= 2n^2 - n^2 = n^2.\end{aligned}$$

REMARK 5.24 (Dimension of Level Sets in IFT). If $\phi : \mathbb{R}^m \to \mathbb{R}^n$ is a $\mathcal{C}^\infty$ map such that $r = \mathrm{rk}(d\phi)_{\mathbf{x}}$ is constant near each point in a level set $M = L_{\phi(\mathbf{x})=\mathbf{c}}$ ($\mathbf{c} \in \mathbb{R}^n$), the IFT provides a covering of M by $\mathcal{C}^\infty$-related coordinate charts that make it into a differentiable manifold. A careful reading of the implicit function theorem, Theorem 5.5, reveals the connection between the dimension of M and the rank $r = \mathrm{rk}(d\phi)$:

$$\begin{aligned}\dim_{\mathbb{R}}(M) &= \dim(\mathbb{R}^m) - \#(\text{irredundant constraints})\\ &= \dim(\mathbb{R}^m) - \mathrm{rk}(d\phi) = m - r \qquad (5.3)\end{aligned}$$

because this is also the dimension of the coordinate charts that cover M. We would have $r = \mathrm{rk}(d\phi) < n$ if ϕ involves some redundant constraints. ○

As for $\mathrm{SU}(n)$, the additional condition $\det(A) = 1+i0$ might seem to impose two more real constraints. However, the real and imaginary parts of $\det(A)$ are not independent when $A \in \mathrm{U}(n)$, as they were for $A \in \mathrm{SL}(n,\mathbb{C})$. The very definition of $\mathrm{U}(n)$ imposes a constraint on values of $\det(A)$,

$$AA^* = I \Rightarrow \det(A)\det(A^*) = |\det(A)|^2 = 1 \Rightarrow |\det(A)| = 1,$$

which must be taken into account. The value of $\det(A) = e^{i\theta(A)}$ is determined by a single *real-valued* "angle variable" $\theta(A)$. Requiring that $|\det(A)| = 1$ imposes *just one* real constraint on $\mathbb{R}^{2n^2}$, so that

$$\dim_{\mathbb{R}} \mathrm{SU}(n) = \dim_{\mathbb{R}} \mathrm{U}(n) - 1 = n^2 - 1.$$

In particular when $n = 2$, we have $\dim_{\mathbb{R}} \mathrm{M}(2, \mathbb{C}) = 8$, $\dim_{\mathbb{R}} \mathrm{U}(2) = 4$, and $\dim_{\mathbb{R}} \mathrm{SU}(2) = 3$.

The classical groups will mostly be viewed as smooth real hypersurfaces in $\mathbb{R}^{n^2}$ (or $\mathbb{R}^{2n^2}$ if they are defined as subsets of complex matrix space $\mathrm{M}(n, \mathbb{C})$). But a few actually are intrinsically complex manifolds that can be equipped with differentiably related complex coordinate charts. For instance, we have

$$\dim_{\mathbb{C}} \mathrm{GL}(n, \mathbb{C}) = n^2 \quad \text{and} \quad \dim_{\mathbb{C}} \mathrm{SL}(n, \mathbb{C}) = n^2 - 1$$

even though we will mostly regard them as real manifolds with $\dim_{\mathbb{R}} = 2n^2$ or $2n^2 - 2$.

EXAMPLE 5.25 (Complex Orthogonal Groups $\mathrm{O}(n, \mathbb{C})$). Another example of a complex matrix Lie group is $\mathrm{O}(n, \mathbb{C}) = \{A \in \mathrm{M}(n, \mathbb{C}) : AA^{\mathrm{T}} = 1\}$, on which $\det(A)$ can only take the values ± 1 because $AA^{\mathrm{T}} = I \Rightarrow$

$$1 = \det(I) = \det(AA^{\mathrm{T}}) = (\det(A))^2 \Rightarrow \det(A) = \pm 1.$$

Since $\det(A) = \pm 1$, yet varies continuously with A, its value must be constant on some open neighborhood for each $A \in \mathrm{O}(n, \mathbb{C})$. Therefore, the sets of constancy O^+ and O^- for $\det(A)$ are disjoint open sets in $\mathrm{O}(n, \mathbb{C})$, and $\mathrm{O}^+ = \mathrm{SO}(n, \mathbb{C})$ is an open subgroup in the full orthogonal group. This means $\mathrm{SO}(n, \mathbb{C}) \subseteq \mathrm{O}(n, \mathbb{C})$ are complex matrix groups having the same dimension over $\mathbb{C}$ (which gets multiplied by 2 when they are viewed as real hypersurfaces in $\mathbb{R}^{2n^2}$). In particular $\det(A) \equiv +1$ for all A close to the identity in $\mathrm{O}(n, \mathbb{C})$ because $\det(I_{n \times n}) = 1$ and varies continuously with A.

The group $\mathrm{O}(n, \mathbb{C})$ is the the matrix realization of the automorphism group $\mathrm{Aut}(B)$ of the canonical nondegenerate symmetric bilinear form on on $\mathbb{C}^n$,

$$B(\mathbf{z}, \mathbf{w}) = \mathbf{z}^{\mathrm{T}} I_{n \times n} \mathbf{w} = z_1 w_1 + \cdots + z_n w_n \quad \text{for } \mathbf{z}, \mathbf{w} \in \mathbb{C}^n,$$

with respect to the standard basis in $\mathbb{C}^n$ (recall the discussion following Definition 3.17). Although B is bilinear and symmetric on $\mathbb{C}^n$ it is *not* an inner product and should not be confused with the standard Euclidean inner product on $\mathbb{C}^n$,

$$(\mathbf{z}, \mathbf{w}) = \sum_{j=1}^{n} z_j \overline{w_j}.$$

For instance, the values of $B(\mathbf{z}, \mathbf{z})$ need not be real or ≥ 0.

DISCUSSION. By definition of matrix product, we have $AA^{\mathrm{T}} = I \Leftrightarrow$ the rows (and columns) of A are B-orthonormal with

$$B(R_r, R_s) = \delta_{rs} \quad \text{(Kronecker delta)}.$$

Thus $\frac{1}{2}(n^2 + n)$ irredundant complex constraints on $\mathrm{M}(n, \mathbb{C})$ carve out $\mathrm{O}(n, \mathbb{C})$, so $\dim_{\mathbb{C}} \mathrm{O}(n, \mathbb{C}) = \frac{1}{2}(n^2 + n)$. As indcated above, $\mathrm{O}(n, \mathbb{C})$ is made up of two open, disjoint pieces $O^+ \cup O^-$, one being the special orthogonal group $SO(n, \mathbb{C})$. Both have the same (complex) dimension; when they are regarded as real Lie groups, their dimensions are $\dim_{\mathbb{R}} G = n^2 + n$.

It remains to verify the hypotheses of the IFT to show that these are in fact matrix Lie groups, but we omit these details. The calculations would be tedious, and we will soon cite a general theorem (Theorem 5.37 below) that circumvents such issues. ○

Examples of Lie Groups.

EXAMPLE 5.26. The simplest example of a Lie group is $G = (\mathbb{R}^n, +)$ whose differentiable structure is given by the trivial chart $(U_\alpha, x_\alpha) = (\mathbb{R}^n, \text{id})$. Clearly, the $(+)$ operation is a $\mathcal{C}^\infty$ map $\mathbb{R}^n \times \mathbb{R}^n \to \mathbb{R}^n$, as is the inversion map $J(x) = -x$ on $\mathbb{R}^n$. This is a commutative (or *abelian*) Lie group, but strictly speaking it is not a matrix Lie group because its elements are not matrices. However, $(\mathbb{R}^2, +)$ can be realized as a matrix group $G \subseteq M(3, \mathbb{R})$ via the bijective isomorphism

$$\phi(t) = \begin{pmatrix} 1 & 0 & t_1 \\ 0 & 1 & t_2 \\ 0 & 0 & 1 \end{pmatrix} \qquad \mathbf{t} = (t_1, t_2)$$

such that $\phi(\mathbf{s}+\mathbf{t}) = \phi(\mathbf{s}) \cdot \phi(\mathbf{t})$, and this works in all dimensions $n \geq 2$. Similarly $G = \mathbb{C}^n$ with $(+)$ as its group operation is a Lie group with $\dim_{\mathbb{C}} \mathbb{C}^n = n$ and $\dim_{\mathbb{R}} \mathbb{C}^n = 2n$; it can be modeled as a matrix group in the same way as $\mathbb{R}^n$.

The IFT shows that the unit circle $S^1 = \{x + iy \in \mathbb{C} : |z| = x^2 + y^2 = 1\}$ in the complex plane is a $\mathcal{C}^\infty$ manifold with $\dim_{\mathbb{R}}(S^1) = 1$ because S^1 can be covered by two real $\mathcal{C}^\infty$-related charts (recall stereographic projection discussed in Example 5.19).

However, S^1 is also an abelian group under the multiplication operation in $\mathbb{C}$, and the operations $P(z, w) = z \cdot w$, $J(z) = 1/z$ are differentiable when described in local coordinates. There is a natural bijective correspondence between the circle S^1 equipped with complex multiplication and the group $SO(2)$ of rotation matrices on the plane via a diffeomorphism that intertwines the two group operations. This makes S^1 and G isomorphic as Lie groups. Further details about S^1 as a Lie group are provided by Exercises 2 - 3 in Section 5.2 of the Additional Exercises.

EXAMPLE 5.27 ($G = GL(n, \mathbb{K})$). The general linear group $GL = GL(n, \mathbb{K})$ is a noncommutative matrix Lie group with $\dim_{\mathbb{K}}(G) = n^2$ for $\mathbb{K} = \mathbb{R}$ or $\mathbb{C}$. We can assign chart coordinates to elements $A \in GL$, letting

$$x_\alpha(A) = (A_{11}, \dots, A_{1n}; \dots; A_{n1}, \dots, A_{nn}) \in \mathbb{K}^{n^2},$$

which is the restriction to GL of a bijective $\mathbb{K}$-linear map of $M(n, \mathbb{K}) \to \mathbb{K}^{n^2}$. The product operation in GL is given by matrix multiplication,

$$(AB)_{ij} = \sum_{k=1}^{n} A_{ik} B_{kj}, \tag{5.4}$$

and inversion is effected by Cramer's rule (see LA I, Theorem 4.48),

$$A^{-1} = \frac{1}{\det(A)}\mathrm{Cof}(A)^{\mathsf{T}}.$$

This rational map involves the transpose of the cofactor matrix $\mathrm{Cof}(A)$, whose entries are polynomials in the entries of A; $\det(A)$ is also a polynomial in the entries of A and is nonvanishing on GL.

Other matrix Lie groups are level sets for various $\mathcal{C}^\infty$ maps :

$$\phi : \mathrm{M}(n, \mathbb{K}) \to \mathbb{K}^k \text{ with } k \le n^2 = \dim_{\mathbb{K}} \mathrm{M}(n, \mathbb{K}).$$

It follows from the IFT that they are all smooth manifolds, but to verify that they are also Lie groups, we must also prove that the multiplication and inversion operations are *differentiable* maps between manifolds. The next observation takes care of that.

THEOREM 5.28. *Suppose M is the level set through $A \in \mathrm{M}(n, \mathbb{R}^n)$ for some $\mathcal{C}^\infty$ map $\phi : \mathrm{M}(n, \mathbb{R}) \to \mathbb{R}^k$ and that $\mathrm{rk}(d\phi)_A$ is constant on some open neighborhood of each point in M. If M is also a subgroup of $\mathrm{GL}(n, \mathbb{R})$ under matrix multiplication, then M is automatically a Lie group in the standard $\mathcal{C}^\infty$ structure imposed via the IFT as a smooth submanifold in matrix space (identifying $\mathrm{M}(n, \mathbb{R}) \cong \mathbb{R}^{n^2}$).*

PROOF. If $a, b \in M$, let $(U_\alpha, x_\alpha), (U_\beta, x_\beta)$ be standard charts about a, b and let (U_γ, x_γ) be a chart about $c = a \cdot b$. By the IFT, x_γ is the restriction to U_γ of a linear projection $\pi_{J'(\gamma)}$ from $\mathbb{K}^{n^2}$ onto $\mathbb{K}^d \cong \mathbb{K}^{J'(\gamma)}$. The matrix product sending $\mathrm{M}(n, \mathbb{K}) \times \mathrm{M}(n, \mathbb{K}) \to \mathrm{M}(n, \mathbb{K})$, is obviously $\mathcal{C}^\infty$ on matrix space (in fact it is a polynomial map). Then the group operation $P(x, y) = x \cdot y$ is a $\mathcal{C}^\infty$ map with respect to the manifold structure on M because it becomes a $\mathcal{C}^\infty$ map from $\mathbb{R}^d \times \mathbb{R}^d \to \mathbb{R}^d$ when written in local coordinates:

$$x_\gamma \circ P \circ \left(x_{\overline{\alpha}}^{-1} \times y_{\overline{\beta}}^{-1}\right) = \left(\pi_{J'(\gamma)}|_{U_\gamma}\right) \circ P \circ \left(x_{\overline{\alpha}}^{-1} \times y_{\overline{\beta}}^{-1}\right)$$

where $d = \dim_{\mathbb{K}}(G) = n^2 - r$ and $r = \mathrm{rk}(d\phi)$. Breaking this composite into steps, we have

$$\mathbb{K}^{d^2} \cong \mathbb{K}^d \times \mathbb{K}^d \xrightarrow{x_{\overline{\alpha}}^{-1} \times y_{\overline{\beta}}^{-1}} U_\alpha \times U_\beta \xrightarrow{P} U_\gamma \xrightarrow{x_\gamma = \pi_{J'(\gamma)}} \mathbb{K}^d.$$

The matrix product operation $P : \mathbb{K}^{n^2} \times \mathbb{K}^{n^2} \to \mathbb{K}^{n^2}$ is defined and $\mathcal{C}^\infty$ on all of $\mathrm{M}(n, \mathbb{K})$, and $P\big(U_\alpha \times U_\beta\big)$ is contained in G because G is a group. As in the IFT, the map $x_\gamma : U_\gamma \to \mathbb{K}^d$ is the restriction to $U_\gamma \subseteq G$ of a globally defined linear projection map $\pi_{J'(\gamma)}$ from $\mathbb{K}^{n^2} \to \mathbb{K}^{J'(\gamma)} \cong \mathbb{K}^d$, so it is the restriction to U_γ of a globally $\mathcal{C}^\infty$ map. Clearly then, the description of P in local coordinates

$$x_\gamma \circ P \circ (x_{\overline{\alpha}}^{-1} \times y_{\overline{\beta}}^{-1}) = \pi_{J'(\gamma)} \circ P \circ (x_{\overline{\alpha}}^{-1} \times y_{\overline{\beta}}^{-1}) : \mathbb{K}^{d^2} \to \mathbb{K}^d$$

is $\mathcal{C}^\infty$ too. Similarly, the inversion map $J : G \to G$ is $\mathcal{C}^\infty$ when expressed in standard chart coordinates. $\square$

Below is the first example in which we need the implicit function theorem.

EXAMPLE 5.29 (Special Linear Groups $\mathrm{SL}(n, \mathbb{C})$ and $\mathrm{SL}(n, \mathbb{R})$). By appealing to the IFT, we will show that $G = \mathrm{SL}(n, \mathbb{R})$ is a matrix Lie group with $\dim_{\mathbb{R}}(G) = n^2 - 1$. Similar arguments would show that $\mathrm{SL}(n, \mathbb{C})$ is a smooth complex hypersurface in $\mathrm{M}(n, \mathbb{C}) \cong \mathbb{C}^{n^2}$ with

$$\dim_{\mathbb{C}} \mathrm{SL}(n, C) = n^2 - 1 \quad \text{and} \quad \dim_{\mathbb{R}} \mathrm{SL}(2, \mathbb{C}) = 2(n^2 - 1).$$

To keep things simple, we will focus on the case $\mathbb{K} = \mathbb{R}$, which can be handled by the familiar methods of real analysis.

DISCUSSION. $\mathrm{SL}(n, \mathbb{R})$ is the level set $L_{\phi(A)=1}$ where $\phi(A) = \det(A)$ on $\mathbb{R}^{n^2} \cong \mathrm{M}(n, \mathbb{R})$. By the IFT, this is a smooth manifold in $\mathbb{R}^{n^2}$ if the $1 \times n^2$ Jacobian matrix has $\mathrm{rk}(d\phi)_A = 1$ near every $A \in \mathrm{SL}(n, \mathbb{R})$. This can be proved in all dimensions. In fact, if we list the entries in A as $(a_{11}, \dots, a_{1n}; \dots; a_{n1}, \dots, a_{nn})$ in $\mathbb{R}^{n^2}$, then by definition of determinant,

$$\det(A) = \sum_{\sigma \in S_n} \mathrm{sgn}(\sigma) \cdot a_{1,\sigma(1)} \cdots a_{n,\sigma(n)}$$

is a polynomial in the n^2 entries a_{ij}. By the product formula for derivatives we get

$$\begin{aligned} \frac{\partial \phi}{\partial a_{k,\ell}} &= \sum_{\sigma \in S_n} \mathrm{sgn}(\sigma) \cdot \left[\sum_{j=1}^{n} a_{1,\sigma(1)} \cdot \dots \cdot \frac{\partial a_{j,\sigma(j)}}{\partial a_{k,\ell}} \cdot \dots \cdot a_{n,\sigma(n)} \right] \\ &= \sum_{\sigma : \sigma(k) = \ell} \mathrm{sgn}(\sigma) \cdot a_{1,\sigma(1)} \cdot \dots \cdot \widehat{a_{k,\ell}} \cdot \dots \cdot a_{n,\sigma(n)}, \end{aligned}$$

because

$$\partial a_{j,\sigma(j)} / \partial a_{k,\ell} = \delta_{jk} \delta_{\sigma(j),\ell} \quad \text{(product of Kronecker deltas).}$$

(Here we use the standard math notation $b_1 \cdots \widehat{b_i} \cdots b_n = \prod_{j \neq i} b_j$ to indicate a deleted factor in a product.) Indeed, $\partial a_{j,\sigma(j)} / \partial a_{k,\ell} = 0$ unless $(j, \sigma(j)) = (k, \ell)$ which happens if and only if $j = k$ and $\sigma(k) = \ell$, in which case it is equal to 1. Therefore, we have

$$a_{k\ell} \frac{\partial \phi}{\partial a_{k,\ell}} = \sum_{\sigma : \sigma(k) = \ell} \mathrm{sgn}(\sigma) \cdot a_{1,\sigma(1)} \cdot \dots \cdot a_{k,\ell} \cdot \dots \cdot a_{n,\sigma(n)}.$$

But the $1 \times n^2$ matrix $[\partial \phi / \partial a_{k\ell}]$ has rank $= 1$ (maximal rank) unless all entries are zero, in which case we would have

$$\begin{aligned} \det(A) &= \sum_{\sigma(1)=1} (\cdots) + \sum_{\sigma(1)=2} (\cdots) + \cdots + \sum_{\sigma(1)=n} (\cdots) \\ &= \sum_{\sigma \in S_n} \mathrm{sgn}(\sigma) \cdot a_{1,\sigma(1)} \cdots a_{n,\sigma(n)} = 0. \end{aligned}$$

That contradicts the hypothesis $\det(A) = 1$ and cannot occur. Thus $(d\phi)_A$ has maximal rank ($= 1$) at each point in $\mathrm{SL}(n, \mathbb{R})$, so $\dim_{\mathbb{R}} \mathrm{SL}(n, \mathbb{R}) = n^2 - 1$ as claimed. ○

EXAMPLE 5.30 (Real Orthogonal Groups $\mathrm{O}(n)$ and $\mathrm{SO}(n)$). The *(real) orthogonal groups* are defined as hypersurfaces in $\mathrm{M}(n,\mathbb{R})$, and both have the same dimension $\dim_{\mathbb{R}} G = \frac{1}{2}(n^2 - n)$ for $n \geq 2$. In particular $\dim_{\mathbb{R}} \mathrm{SO}(3) = 3$.

DISCUSSION. The group $\mathrm{O}(n)$ is a closed set (hence a *closed subgroup*) in $\mathrm{M}(n,\mathbb{R})$ because if a sequence $\{A_k\}$ in $\mathrm{O}(n)$ converges to some limit A in matrix space, so $A_k \to A$ as $k \to \infty$, then $A_k^{\mathsf{T}} \to A^{\mathsf{T}}$ (obvious) and

$$I = A_k A_k^{\mathsf{T}} \to AA^{\mathsf{T}} \quad \text{as } k \to \infty$$

by joint continuity of the matrix product operation (see LA I, Exercise 5.36). Thus, A is in $\mathrm{O}(n)$ and is in $\mathrm{SO}(n)$ if $A_k \in \mathrm{SO}(n)$ since $\det(A_k) \to \det(A)$.

On $\mathrm{O}(n)$ we have

$$1 = \det(I) = \det(AA^{\mathsf{T}}) = (\det(A))^2,$$

so $\det(A)$ can only achieve values ± 1, and -1 is actually achieved, for instance at $J = \mathrm{diag}(1, \dots, 1, -1)$. Thus $\mathrm{O}(n)$ consists of two disjoint pieces — one, the subgroup $\mathrm{SO}(n)$ where $\det(A) = +1$ by definition, and the other a multiplicative coset $J \cdot \mathrm{SO}(n)$ on which $\det(A) \equiv -1$. Both are open subsets of $\mathrm{O}(n)$, so they have the same dimension as real manifolds.

These two sets have geometric significance. Elements $A \in \mathrm{SO}(n)$ yield the orientation-preserving orthogonal linear maps $L_A : \mathbb{R}^n \to \mathbb{R}^n$, while matrices in $J \cdot \mathrm{SO}(n)$ correspond to orientation-reversing linear operators on $\mathbb{R}^n$. The latter do not form a subgroup in $\mathrm{O}(n)$ because the product of two orientation-reversing maps is orientation-preserving.

We show $\mathrm{O}(n)$ is a $\mathcal{C}^\infty$ manifold by directly verifying the hypotheses of the IFT. Recall that $A \in \mathrm{O}(n) \Leftrightarrow AA^{\mathsf{T}} = I \Leftrightarrow$ the rows R_i are an orthonormal basis in $\mathbb{R}^n$, so their innner products in $\mathbb{R}^n$ are

$$(R_i, R_j) = \delta_{ij} \quad (\text{Kronecker delta for } 1 \leq i, j \leq n).$$

Eliminating redundant entries in this list of n^2 identities by requiring $i \leq j$, we define the map

$$\phi : \mathbb{R}^{n^2} \cong \mathrm{M}(n,\mathbb{R}) \to \mathbb{R}^{(n^2+n)/2} = \mathbb{R}^d$$

whose entries are inner products (R_i, R_j) of the rows of A such that $i \leq j$,

$$\phi(A) = \big((R_1,R_1), \dots, (R_1,R_n); (R_2,R_2), \dots, (R_2,R_n); \cdots; (R_n,R_n)\big).$$

Then $\mathrm{O}(n)$ is the level set $L_{\phi(A)=\mathbf{c}}$, in which

$$\mathbf{c} = (1, 0, \dots, 0; \dots; 1, 0, 0, ; 1, 0; 1),$$

and we must show $\mathrm{rk}(d\phi)_A = d = \frac{1}{2}(n^2 + n)$ for all $A \in \mathrm{O}(n)$. The idea is clearly illustrated by the case $n = 3$, where $\phi(A) = (\phi_1, \dots, \phi_6)$ with

$$\phi(A) = \Big(\sum_{i=1}^{3} a_{1i}^2, \sum_i a_{1i}a_{2i}, \sum_i a_{1i}a_{3i}; \sum_i a_{2i}^2, \sum_i a_{2i}a_{3i}; \sum_i a_{3i}^2\Big).$$

Dropping duplicate constraints, the 6×9 Jacobian matrix $(d\phi)_A$ is

$$\left[\frac{\partial \phi_i}{\partial a_{ij}}\right] = \left(\begin{array}{ccc|ccc|ccc} 2a_{11} & 2a_{12} & 2a_{13} & 0 & 0 & 0 & 0 & 0 & 0 \\ a_{21} & a_{22} & a_{23} & a_{11} & a_{12} & a_{13} & 0 & 0 & 0 \\ a_{31} & a_{32} & a_{33} & 0 & 0 & 0 & a_{11} & a_{12} & a_{13} \\ 0 & 0 & 0 & 2a_{21} & 2a_{22} & 2a_{23} & 0 & 0 & 0 \\ 0 & 0 & 0 & a_{31} & a_{32} & a_{33} & a_{21} & a_{22} & a_{23} \\ 0 & 0 & 0 & 0 & 0 & 0 & 2a_{31} & 2a_{32} & 2a_{33} \end{array}\right).$$

Recall that row and column rank of any matrix are equal. Symbolic row/column operations show that the row rank $\mathrm{rk}(d\phi)_A = 6 = \frac{1}{2}(n^2 + n)$, so $\dim_{\mathbb{R}} \mathrm{O}(3) = \dim_{\mathbb{R}} \mathrm{SO}(3) = 9 - 6 = 3$ (see next exercise for hints on this calculation). ○

EXERCISE 5.31. For the real-orthogonal group SO(3) in $\mathrm{M}(3, \mathbb{R})$,

(a) In terms of the matrix elements a_{ij} for $A \in \mathrm{SO}(3)$, write out the constraints

$$\phi_1(A) = c_1, \dots, \phi_r(A) = c_6$$

that determine SO(3) as a hypersurface in $\mathrm{M}(3, \mathbb{R})$, and verify that the entries in the preceding Jacobian matrix are correct.

(b) Verify that the "constant rank" condition $\mathrm{rk}(d\phi)_A = 6$ of the IFT is satisfied by all A *close to the identity* I in SO(3).

HINT. Row rank is the number of linearly independent rows or columns. After scaling the first row of $(d\phi)_A$ by $1/2$, the original matrix A appears as a 3×3 submatrix in the upper left corner with zeros below it. The original rows and columns of A are linearly independent, so we can transform the submatrix A to $I_{3\times3}$ by column operations. What happens when these column operations are applied to the full Jacobian matrix? (Additional steps, using both row and column operations, will be needed to finish the calculation.)

REMARK 5.32. In the next section, we will show that if a matrix Lie group $G \subseteq \mathrm{M}(m, \mathbb{R})$ is determined by constraints $\phi(A) = \mathbf{c}$, then

The rank $\mathrm{rk}(d\phi)_A$ *is constant near every* $A \in G$ *if it is constant for all matrices* $A \in G$ *close to the identity element* $I_{m\times m}$.

This is entirely due to the fact that G is a *group* of matrices, which forces the behavior of the rank $(d\phi)_A$ near any $A \in G$ to be the same as that near the identity element. Constancy of $\mathrm{rk}(d\phi)_A$ near the identity is a "local" condition on G, much easier to verify than constancy throughout G. ○

EXERCISE 5.33. In Example 5.30 we described the orthogonal group $G = \mathrm{SO}(3)$ as the level set $M = L_{\phi(x)=\mathbf{c}}$ for a $\mathcal{C}^\infty$ map $\phi : \mathrm{M}(3, \mathbb{R}) \cong \mathbb{R}^9 \to \mathbb{R}^6$ and computed the 6×9 Jacobian matrix for arbitrary $A \in G$.

(a) Write out the Jacobian matrix of $(d\phi)_A$ at the identity element $A = I_{3\times3}$ in G and show that $\mathrm{rk}(d\phi)_I = 6$. (Notice how the rank calculations simplify in this special case.)

(b) Explain why we must then have $\mathrm{rk}(d\phi)_A = 6$ for all A in SO(3) that lie *close to* I.

HINT. What is the maximum possible rank for $d\phi$ in this example?

EXERCISE 5.34. Verify that the special orthogonal group SO(2) in $\mathrm{M}(2,\mathbb{R}) \cong \mathbb{R}^4$ is a commutative matrix Lie group with $\dim_{\mathbb{R}} \mathrm{SO}(2) = 1$.

The following questions outline the details needed to fully verify that the unitary groups

$$\mathrm{U}(2) = \{A \in \mathrm{M}(2,\mathbb{C}) : AA^* = I\}$$
$$\mathrm{SU}(2) = \{A \in \mathrm{M}(2,\mathbb{C}) : AA^* = I \text{ and } \det(A) = 1\}$$

satisfy the conditions of the IFC, so they are matrix Lie groups in $\mathbb{R}^8$ with $\dim_{\mathbb{R}} \mathrm{U}(2) = 4$ and $\dim_{\mathbb{R}} \mathrm{SU}(2) = 3$ when we identify $\mathrm{M}(2,\mathbb{C}) \cong \mathbb{R}^8$. To do so, we shall identify complex matrices

$$A = \begin{pmatrix} z_1 & z_2 \\ z_3 & z_4 \end{pmatrix} \quad \text{with entries } z_k = x_k + \sqrt{-1}\, y_k,\ (x_k, y_k \in \mathbb{R})$$

with real 8-tuples $\mathbf{x} = (x_1, \dots, x_4; y_1, \dots, y_4)$, regarding these groups as subsets of $\mathbb{R}^8$. Our comments will focus on verifying the hypotheses of IFT for SU(2). Analysis of U(2) is more complicated, but that case can be handled directly once we know SU(2) is a matrix Lie group by noting that U(2) is a product $S^1 \cdot \mathrm{SU}(2)$ of familar subgroups. Here S^1 is the unit-circle group realized as the set of complex matrices $S^1 = \{e^{i\theta} I_{2\times 2} : \theta \in \mathbb{R}\}$ in $\mathrm{M}(2,\mathbb{C})$.

EXERCISE 5.35. Let

$$A = \begin{pmatrix} a & b \\ c & d \end{pmatrix} \quad (a, b, c, d \in \mathbb{C})$$

be a matrix in $\mathrm{M}(2,\mathbb{C})$.

(a) Write out the complex constraint identities on a, b, c, d that hold when $AA^* = I_{2\times 2}$ and $\det(A) = 1$. Then rewrite this system to eliminate the variables c and d and prove that $A \in \mathrm{SU}(2) \Leftrightarrow A$ has the form

$$A = \begin{pmatrix} a & b \\ -\bar{b} & \bar{a} \end{pmatrix} \quad \text{with} \quad |a|^2 + |b|^2 = 1$$

for $a, b \in \mathbb{C}$.

(b) Recast these as identities in the real variables $x_1, \dots, x_4; y_1, \dots, y_4$ to describe SU(2) as the level set $L_{\phi(\mathbf{x})=\mathbf{c}}$ for a suitably chosen map ϕ from $\mathbb{R}^8 \to \mathbb{R}^5$. What is the appropriate value $\mathbf{c} = (c_1, \dots, c_5)$?

(c) Once SU(2) has been described as a level set as in (b), compute the 5×8 Jacobian matrix and prove that $\mathrm{rk}(d\phi)_{\mathbf{p}} = 6$ for all $p \in \mathrm{SU}(2)$. (Then rank $= 6$ *near* every p in $\mathrm{SU}(2) \subseteq \mathrm{M}(2,\mathbb{C})$ because this is a maximal rank case.)

HINT. We have $AA^* = I \Leftrightarrow$ the rows in A are an orthonormal basis in $\mathbb{C}^2$; eliminating redundant (complex) identities leaves us with three identities $(R_i, R_j) = \delta_{ij}$ (Kronecker delta) plus the constraint $\det(A) = ad - bc = 1$. Recall that $|\det(A)| = 1$ for all unitary matrices, so $\det(A) = e^{i\theta}$ for some $\theta \in \mathbb{R}$.

Continuing the discussion of Exercise 5.35, we pose a different question:

Which choices of d coordinates $x_{i_1}, \dots, x_{i_d}$ from $x_1, \dots, x_8$ give valid parametrizations of SU(2) *near the identity element?*

As in the IFT, we must find a partition of coordinate indices $[1, 8] = J \cup J'$ such that the restriction $(\pi_{J'}|_M) : \mathrm{SU}(2) \to \mathbb{R}^{J'}$ is an admissible coordinate chart near I. (Not all choices work.)

EXERCISE 5.36. Consider the identity element $I_{2\times 2}$ in $G = \mathrm{SU}(2)$.

(a) Produce a partition of $[1, 8] = J' \cup J$ such that $\pi_{J'} : \mathbb{R}^8 \to \mathbb{R}^{J'}$ restricts to SU(2) to give a standard chart defined near $I_{2\times 2}$.

It will follow that $\dim_{\mathbb{R}} \mathrm{SU}(2) = |J'| = 3$.

(b) Determine *all* partitions $[1, 8] = J' \cup J$ that produce admissible chart maps $\pi^{J'} : U_\alpha \to \mathbb{R}^3$ near the identity in SU(2).

(c) Produce an explicit partition $[1, 8] = J' \cup J$ with $|J'| = 3$ that *does not* yield valid coordinates describing SU(2) near the identity I.

In practice, the need for such intricate calculations to prove that a group G in $\mathrm{M}(n, \mathbb{K})$ is a differentiable manifold (and hence a Lie group) can often be circumvented by appeal to a theorem due to von Neumann and developed further by Elie Cartan.

THEOREM 5.37 (J. von Neumann/E. Cartan). *Every closed subgroup G in $\mathrm{M}(n, \mathbb{R})$ is a matrix Lie group: it can be covered with $\mathcal{C}^\infty$-related charts (x_α, U_α) that make it into a Lie group over $\mathbb{R}$.*

This result does not identify the dimension $d = \dim(G)$ of the charts that cover G, which must be determined separately. On the other hand, it does not require that G be a level set for some map $\phi : \mathbb{R}^m \to \mathbb{R}^n$, and even if it is, there is no need to investigate $\mathrm{rk}(d\phi)_A$ on G. The subgroup G could even be a closed *discrete* subgroup (a zero-dimensional Lie group consisting of isolated points in matrix space) such as $G = \mathbb{Z} \subseteq \mathbb{R}$ or

$$\mathrm{SL}(n, \mathbb{Z}) = \{A \in \mathrm{M}(n, \mathbb{R}) : \text{ all } a_{ij} \in \mathbb{Z} \text{ and } \det(A) = 1\},$$

for which differentiability considerations are irrelevant.

As an example, consider $G = \mathrm{SO}(n, \mathbb{C})$. The constant rank conditions of the IFT could be verified by brute force calculations, but it is easily seen that this matrix group is a closed subset in matrix space. Indeed, if $A_k \to A$ as $k \to \infty$ in matrix space and the A_k are in $\mathrm{SO}(n, \mathbb{C})$, the limit A must be in $\mathrm{SO}(n, \mathbb{C})$ because $I = A_k A_k^{\mathrm{T}} \to AA^{\mathrm{T}}$ and $\det(A) = \lim_{k\to\infty} \det(A_k) = 1$. The same is true for the groups $\mathrm{U}(n)$ and $\mathrm{SU}(n)$ because $A_k \to A$ in $\mathrm{M}(n, \mathbb{C}) \Rightarrow A_k^* \to A^*$ with $A_k \in \mathrm{SU}(n) \Rightarrow AA^* = \lim_{k\to\infty} A_k A_k^* = I$ so $A \in \mathrm{U}(n)$, and if $\det(A_k) = 1$ then $A \in \mathrm{SU}(n)$ because $\det(A) = \lim_{k\to\infty} \det(A_k) = 1$.

Translations and Automorphisms on a Lie Group. Many algebraic symmetry operations on matrix Lie groups are actually $\mathcal{C}^\infty$ maps on the underlying manifolds. If $g \in G$, we can define *left and right translation operators* $\lambda_x, \rho_x : G \to G$, letting

$$\lambda_x(g) = x \cdot g \quad \text{and} \quad \rho_x(g) = g \cdot x.$$

These are continuous and invertible maps with the properties

$$\lambda_e = \mathrm{id}_G \quad \lambda_{x\cdot y} = \lambda_x \circ \lambda_y \quad \lambda_{x^{-1}} = \left(\lambda_x\right)^{-1},$$

and likewise for right translations, except that

$$\rho_{x\cdot y} = \rho_y \circ \rho_x \quad \text{(note reversal of factors).}$$

These maps are *diffeomeorphisms* on G: they are bijections, and when λ_x or ρ_x are described in terms of chart coordinates, they (and their inverses) become $\mathcal{C}^\infty$ maps from $\mathbb{R}^d \to \mathbb{R}^d$, where $d = \dim_{\mathbb{R}}(G)$. In fact, λ_x is obtained by holding fixed the first variable in the product map $P(x, y) = xy$ while the other is allowed to vary. Since $P : G \times G \to G$ is, by definition, jointly differentiable in all its variables, fixing one variable or the other yields $\mathcal{C}^\infty$ maps λ_x and ρ_g on G. There is a related action $G \times \mathcal{C}^\infty(G) \to \mathcal{C}^\infty(G)$ of G on globally smooth functions f on G:

$$L_g f(x) = f(g^{-1}x) = f \circ \lambda_{g^{-1}}(x) \quad (g, x \in G).$$

The inverse g^{-1} appearing here makes this a *left action* on $\mathcal{C}^\infty(G)$ in the sense that

$$L_e = I \quad L_{g_1 g_2} = L_{g_1} \circ L_{g_2} \quad L_{g^{-1}} = (L_g)^{-1}.$$

(The action $f(x) \mapsto L'_g f(x) = f(g \cdot x) = f \circ \lambda_g(x)$ is a *right action*, with $L'_{g_1 g_2} = L'_{g_2} \circ L'_{g_1}$, in which the order of the factors is reversed.)

DEFINITION 5.38 (The Automorphism Group Aut(G)). An *automorphism* of a group G is a bijection $\alpha : G \to G$ that preserves all group operations, so

$$\alpha(e) = e \quad \alpha(x \cdot y) = \alpha(x) \cdot \alpha(y) \quad \alpha(x^{-1}) = (\alpha_x)^{-1}$$

for $x, y \in G$. The automorphisms Aut(G) of a matrix Lie group G are the bijections that preserve both the algebraic and the differentiable manifold structure of G. They are the $\mathcal{C}^\infty$ diffeomorphisms of G that are also group automorphisms.

EXERCISE 5.39. Show that the inverse α^{-1} of an automorphism is also an automorphism and that the set of automorphisms Aut(G) is itself a group under composition of operators,

$$(\alpha_1 \circ \alpha_2)(x) = \alpha_1(\alpha_2(x)) \quad \text{for all } x \in G.$$

EXAMPLE 5.40. The group G acts on itself by "conjugation," yielding the *inner automorphisms* $i_g : x \to gxg^{-1}$ of G.[2] These are diffeomorphisms from $G \to G$ because $i_g = \lambda_g \circ \rho_{g^{-1}}$ is a composite of invertible $\mathcal{C}^\infty$ translations. The inverse of i_g is the inner automorphism $(i_g)^{-1} = i_{g^{-1}}$, and the correspondence $\Phi : g \to i_g$ is a homomorphism from G into the full group Aut(G) of automorphisms. The set Inn(G) of inner auromorphisms is a subgroup of Aut(G). Notice that λ_x, ρ_y commute for all $x, y \in G$ since they act on opposite sides of each group element.

[2]There may be other automorphisms of G, called *outer automoprhisms*, that do not arise as conjugations by elements of G. For instance, the identity map *id* is the only inner automorphism on a commutative group such as $(\mathbb{R}^n, +)$, but this group has plenty of outer automorphisms: every invertible linear operator is an automorphism of $\mathbb{R}^n$.

EXERCISE 5.41. Verify that conjugations $i_g(x) = gxg^{-1}$ have the properties that make $\Phi(g) = i_g$ a group homomorphism from $G \to \mathrm{Aut}(G)$:

$$\text{(a) } i_e = \mathrm{id}_G \quad \text{(b) } i_{g_1 g_2} = i_{g_1} \circ i_{g_2} \quad \text{(c) } (i_g)^{-1} = i_{g^{-1}}$$

for all g, g_1, g_2 in G.

The $\mathcal{C}^\infty$ structure on a matrix Lie group is often obtained by identifying G as a subset in $\mathrm{M}(n, \mathbb{R}) \cong \mathbb{R}^{n^2}$ and then covering G with standard charts (x_α, U_α) constructed via the IFT. The chart maps x_α are just restrictions to G of various projection maps $\pi_{J'} : \mathbb{R}^{n^2} \to \mathbb{R}^{J'}$ that correspond to splittings $\mathbb{R}^{n^2} = \mathbb{R}^J \oplus \mathbb{R}^{J'}$ with $|J'| = d = \dim(G)$. Because of the way standard charts are defined, the next result makes it easy to check whether an arbitrary map $\psi : \mathbb{R}^k \to G$ into a matrix Lie group is a $\mathcal{C}^\infty$ map with respect to the standard manifold structure on G.

PROPOSITION 5.42. *Suppose $M = L_{\phi(\mathbf{x})=\mathbf{c}}$ is a smooth d-dimensional manifold in $\mathbb{R}^m$, a level set for some $\mathcal{C}^\infty$ map ϕ from $\mathrm{M}(m, \mathbb{R}) \cong \mathbb{R}^{m^2} \to \mathbb{R}^n$. If $\psi : \mathbb{R}^k \to \mathrm{M}(m, \mathbb{R})$ is a $\mathcal{C}^\infty$ map whose range happens to lie within M, then $\psi : \mathbb{R}^k \to M$ is automatically a $\mathcal{C}^\infty$ map with respect to the standard $\mathcal{C}^\infty$ manifold structures on M and $\mathbb{R}^k$.*

PROOF. If $\mathbf{x}_0 \in \mathbb{R}^k$, $p_0 = \psi(\mathbf{x}_0)$, and (U_α, x_α) is a standard chart on M about p_0, then $x_\alpha \circ \psi$ is a composite $\pi^{J'} \circ \psi$ of a $\mathcal{C}^\infty$ map $\psi : \mathbb{R}^k \to M \subseteq \mathbb{R}^{m^2}$, and a linear map $\pi^{J'} : \mathbb{R}^{m^2} \to \mathbb{R}^{J'} \cong \mathbb{R}^d$ with $d = \dim(M)$, so ψ is a $\mathcal{C}^\infty$ map between Euclidean spaces when described in local coordinates on M. □

Tangent Spaces TM_p of a Manifold. The tangent space TM_p to a smooth hypersurface $M \subseteq \mathbb{R}^m$ at a base point p is often thought of as the set of (vector-valued) derivatives $\gamma'(0)$ of smooth parametrized curves $\gamma(t)$ that pass through p when $t = 0$ and remain within M at all times. If $\gamma(t) = \sum_{j=1}^m x_j(t)\,\mathbf{e}_j$ has differentiable coefficients, the classical derivative $d\gamma/dt$ is given by

$$\begin{aligned}\gamma'(0) = \frac{d\gamma}{dt}(0) &= \lim_{\Delta t \to 0} \frac{\Delta\gamma}{\Delta t} = \lim_{\Delta t \to 0} \frac{\gamma(t + \Delta t) - \gamma(t)}{\Delta t} \\ &= \frac{dx_1}{dt}(0)\,\mathbf{e}_1 + \cdots + \frac{dx_m}{dt}(0)\,\mathbf{e}_n,\end{aligned}$$

where the $\mathbf{e}_j$ are the standard basis vectors in $\mathbb{R}^m$. The resulting *tangent vectors* $\gamma'(0)$ at p form a vector space, the *tangent space* $\mathrm{TM}_p \subseteq \mathbb{R}^m$, whose elements are generally denoted by capital letters X_p, Y_p, etc. Here, $d = \dim(M) \leq m = \dim(\mathbb{R}^m)$.

However, this classical interpretation of tangent vectors for hypersurfaces embedded in Euclidean space has proved inadequate for modern differential geometry and is antithetical to its spirit, which focuses on things intrinsic to the manifold and not hypothetical spaces in which they might (or might not) be embedded. Section 4.1 of these *Notes* provides a fast introduction to this point of view, which regards tangent vectors at p as *operators* $X_p : \mathcal{C}^\infty(p) \to \mathbb{R}$ that act on the *local algebra* $\mathcal{C}^\infty(p)$, the space of smooth scalar-valued functions defined near p. Instead of regarding tangent vectors as derivatives $\gamma'(0)$ along differentiable curves passing through p when $t = 0$, we are instead using $\gamma'(0)$ to define a *directional derivative* X_p, a linear operator whose action on functions in $\mathcal{C}^\infty(p)$ is defined to be

$$\langle X_p, f\rangle = \langle \gamma'(0), f\rangle = \frac{d}{dt}\{f(\gamma(t))\}\Big|_{t=0} \qquad \text{for } f \in \mathcal{C}^\infty(p).$$

Thus, $\langle X_p, f\rangle$ is the instantaneous rate of change in the values of f, seen as $\gamma(t)$ passes through p. The operators $X_p : \mathcal{C}^\infty(p) \to \mathbb{R}$ are called *derivations at p* which are completely characterized by two simple algebraic properties:

1. LINEARITY. They are linear functionals on the infinte dimensional local algebra $\mathcal{C}^\infty(p)$, so
$$\langle X_p, c_1 f_1 + c_2 f_2\rangle = c_1\langle X_p, f_1\rangle + c_2\langle X_p, f_2\rangle$$
for $f_1, f_2 \in \mathcal{C}^\infty(p), c_1, c_2 \in \mathbb{R}$.
2. DERIVATION PROPERTY. If $f \cdot h$ is the pointwise product of functions in $\mathcal{C}^\infty(p)$, then
$$\langle X_p, f \cdot h\rangle = \langle X_p, f\rangle \cdot h(p) + f(p) \cdot \langle X_p, h\rangle$$
(as in the "product rule" for derivatives in calculus).

The space of derivations *is* the modern interpretation of the tangent space TM_p. It is a vector space because sums and scalar multiples of derivations are again derivations.

The Differential of a Map. A $\mathcal{C}^\infty$ map between manifolds induces linear maps between the various tangent spaces.

DEFINITION 5.43 (Differential $(d\phi)_p$ of a Map ϕ). A $\mathcal{C}^\infty$ map $\phi : M \to N$ between manifolds that sends $p \in M$ to $q = \phi(p) \in N$ induces a natural map between tangent spaces, a linear map

$$(d\phi)_p : \mathrm{TM}_p \to \mathrm{TN}_{\phi(p)}$$

called the *differential of ϕ at p*. Notice that ϕ moves points forward from M to N and "pulls back" functions from N to M via $f \mapsto f \circ \phi$, while the differential $(d\phi)_p$ "pushes tangent vectors forward " from $\mathrm{TM}_p \to \mathrm{TN}_{\phi(p)}$.

If we regard tangent vectors in $\mathrm{TM}_p, \mathrm{TN}_q$ as derivation operators on the local algebras $\mathcal{C}^\infty_M(p), \mathcal{C}^\infty_N(q)$, the action of the differential $(d\phi)_p$ can be described by specifying its action on elements $X_p \in \mathrm{TM}_p$, letting

$$\langle (d\phi)_p X_p, f\rangle = \langle X_p, f \circ \phi\rangle \quad \text{for } f \in \mathcal{C}^\infty_N(q). \tag{5.5}$$

On the other hand, if $X_p \in \mathrm{TM}_p$ is viewed as a vector derivative $\gamma'(0)$ along some curve $\gamma(t)$ in M passing through p, then ϕ maps $\gamma(t)$ to a smooth curve

$\eta(t) = \phi(\gamma(t))$ in N that passes through $q = \phi(p)$ when $t = 0$. The derivative $Y_q = \eta'(0)$ of this curve is precisely the image $(d\phi)_p X_p$. The formal version of this statement is:

LEMMA 5.44. *If* $X_p = \gamma'(0)$ *in* TM_p *and* $\phi : M \to N$ *is a* $\mathcal{C}^\infty$ *map, then in* $\mathrm{TN}_{\phi(p)}$ *we have*

$$(d\phi)_p X_p = (\phi \circ \gamma)'(0) = \eta'(0) \quad \textit{where} \quad \eta(t) = \phi(\eta(t)).$$

PROOF. By definition of the vector derivative $d\gamma/dt$, we have

$$\begin{aligned} \langle (d\phi)_p X_p, f\rangle = \langle X_p, f \circ \phi\rangle &= \frac{d}{dt}\{f(\phi \circ \gamma(t))\}\Big|_{t=0} \\ &= \frac{d}{dt}\{f(\eta(t))\}\Big|_{t=0} = \langle \eta'(0), f\rangle \end{aligned}$$

for all $f \in \mathcal{C}^\infty(q)$, so $X_p = \eta'(0)$. □

EXERCISE 5.45. Given $\mathcal{C}^\infty$ maps $M \overset{\psi}{\to} N \overset{\phi}{\to} R$ and points $p \in M$, $q = \psi(p) \in N$, prove that

$$d(\phi \circ \psi)_p = (d\phi)_q \circ (d\psi)_p$$

as maps $\mathrm{TM}_p \to \mathrm{TN}_q \to \mathrm{TR}_{\phi(q)} = \mathrm{TR}_{\phi(\psi(p))}$.

Smooth Vector Fields on a Manifold. A *vector field* $\widetilde{X}$ on a manifold assigns a tangent vector $X_p \in \mathrm{TM}_p$ to each base point in M.

EXAMPLE 5.46 (Translation-Invariant Vector Fields). For instance if M is a matrix Lie group, we can move a tangent vector $X \in \mathfrak{g} = \mathrm{TG}_e$ at the identity e to a tangent vector $(d\lambda_g)_e X$ in TG_g, obtaining a uniquely defined field of tangent vectors $\widetilde{X}_g = (d\lambda_g)_e X$ on G. This is a *left-invariant vector field* because

$$X_{g\cdot p} = (d\lambda_g)_p X_p \text{ in } \mathrm{TG}_{g\cdot p} \quad \text{for all } g, p \text{ in } G \text{ and } X_p \in \mathrm{TG}_p. \tag{5.6}$$

Such fields are uniquely determined by their value $\widetilde{X}_e = X$ at the identity $e \in G$. Similarly, *right-invariant vector fields* are obtained by applying right translations $(d\rho_g)_e$ to an element $X \in \mathrm{TG}_e$. But left- or right-invariant vector fields are quite special. For arbitrary vector fields, the values $\widetilde{X}_p$ are not correlated as we move from one base point to another.

Our interest here will be in "smoothly varying vector fields," but the notion of smoothness can be defined in various ways, and it requires some preliminary discussion. For matrix Lie groups, one (not very elegant) approach is to argue that

> *When we view a manifold M as a d-dimensional hypersurface embedded in some higher-dimensional space* $\mathbb{R}^m$ *(perhaps a matrix space), tangent vectors are described as vectors in the larger space* $\mathbb{R}^m$
>
> $$\widetilde{X}_p = \gamma'(0) = (c_1(p), \dots, c_m(p)) \in \mathbb{R}^m$$

> *attached to points $p \in M$. Why not define smoothness of $\widetilde{X}$ to mean that the entries $c_j(p)$, $1 \leq j \leq m$, are $\mathcal{C}^\infty$ functions on U_α for all charts?*

That works for some purposes but is in many ways unsatisfactory. For one thing, the dimension $d = \dim(M)$ is usually $< m =$ the dimension of the larger space $\mathbb{R}^m$ in which M is embedded. All those extra dimensions lie outside of M and have little to do with its geometry. One would like to define smoothness of a vector field on M in terms of constructs intrinsic to M.

A better way to describe vector fields and perform calculations is to systematically prescribe basis vectors in the individual tangent spaces TM_p attached to M and use the fact that the tangent spaces are permuted by left translations $(d\lambda_g)$. The charts (U_α, x_α) that cover M provide a way to do this. Given a chart, we can construct correlated bases

$$\mathfrak{X}_p = \left\{\left(\frac{\partial}{\partial x_1}\bigg|_p\right), \dots, \left(\frac{\partial}{\partial x_d}\bigg|_p\right)\right\} \tag{5.7}$$

throughout U_α. (We will soon explain this odd notation for the basis vectors.) If $\mathbf{u} = x_\alpha(u) = (x_1(u), \dots, x_d(u))$ in coordinate space $\mathbb{R}^d$ for $u \in U_\alpha$, we can define smooth straight-line curves in $\mathbb{R}^d$ emanating from $\mathbf{p} = x_\alpha(p)$:

$$\gamma_i^*(t) = \mathbf{p} + t\mathbf{e}_i \quad (\mathbf{e}_i \text{ the standard basis vectors in } \mathbb{R}^d, 1 \leq i \leq d).$$

The inverse $x_\alpha^{-1} : \mathbb{R}^d \to M$ of the chart map transfers these curves back to M, and the resulting smooth curves $\gamma_i(t) = x_\alpha^{-1}(\gamma_i^*(t))$ in M determine tangent vectors $(\partial/\partial x_i|_p)$ in TM_p whose action as derivations on $\mathcal{C}^\infty(p)$ is

$$\begin{aligned}\left\langle \frac{\partial}{\partial x_i}\bigg|_p, f\right\rangle &= \langle \gamma_i'(0), f\rangle = \frac{d}{dt}\{f(\gamma_i(t))\}\bigg|_{t=0} \\ &= \frac{d}{dt}\{f \circ x_\alpha^{-1}(\mathbf{p} + t\mathbf{e}_i)\bigg|_{t=0}\} \\ &= D_{x_i}(f \circ x_\alpha^{-1})(\mathbf{p}),\end{aligned}$$

where D_{x_i} is the classical i^{th} partial derivative at $\mathbf{p} = x_\alpha(p)$ of functions on $\mathbb{R}^d$. The corresponding derivation $\gamma_i'(0)$ on $\mathcal{C}^\infty(p)$ is just the partial derivative D_{x_i} on $\mathbb{R}^d$ lifted back to M — which explains the suggestive notation (5.6) for the basis vectors associated with a chart. With this in mind, we can also define partial derivatives

$$\frac{\partial f}{\partial x_i}(p) = \left\langle \frac{\partial}{\partial x_i}\bigg|_p, f\right\rangle \quad (\text{for } p \in U_\alpha; 1 \leq i \leq d)$$

of smooth functions $f : M \to \mathbb{R}$ on the chart domain. Note carefully that $f(u)$ and its partial derivatives $(\partial f/\partial x_i)(u)$ all live *on the manifold* and not on coordinate space $\mathbb{R}^d$. Furthermore, we have simultaneously assigned basis vectors $(\partial/\partial x_1\big|_p), \dots, (\partial/\partial x_d\big|_p)$ in TM_p for all $p \in U_\alpha$.

EXERCISE 5.47. The chart map $x_\alpha : U_\alpha \to \mathbb{R}^d$ has scalar components

$$x_\alpha(u) = \big(X_1(u), \dots, X_d(u)\big),$$

where each $X_k : U_\alpha \to \mathbb{R}$ is a scalar-valued $\mathcal{C}^\infty$ function on M.

(a) Prove from the definitions that

$$\frac{\partial X_i}{\partial x_j}(u) = \Big\langle \frac{\partial}{\partial x_j}\Big|_u , X_i(u)\Big\rangle = \delta_{ij} \quad \text{(Kronecker delta)}$$

on all of U_α.

(b) Use this to show that the vectors $(\partial/\partial x_i|_p)$ are linearly independent in TM_p for each $p \in U_\alpha$.

(c) Explain why these vectors also span the tangent space TM_p, so $\mathfrak{X}_p$ is a basis.

HINT. Proving (a) is easy once you get comfortable with the notation; then (a) ⇒ (b) follows easily. In (c), consider the tangent vectors you get by taking directional derivatives along curves $\gamma(t)$ through $p \in M$ that correspond to straight lines through $\mathbf{p} = x_\alpha(p)$ in coordinate space.

We shall define smoothness of vector fields in terms of the bases (5.6) in TM_p determined by charts on M. This definition is intrinsic to the manifold because it makes no reference to any "surrounding space" $\mathbb{R}^m$. The following lemma, proved as Corollary 4.16 of Chapter 4, makes it easy to write tangent vectors $\gamma'(0)$ to smooth curves through p in terms of the basis $\mathfrak{X}_p$.

LEMMA 5.48. *If $p \in M$ lies in a chart (U_α, x_α) and $\gamma(t)$ is a $\mathcal{C}^\infty$ curve passing through p when $t = 0$, then in terms of the basis (5.6) we have*

$$\gamma'(0) = \sum_{j=0}^{d} \frac{dx_j}{dt}(0) \cdot \Big(\frac{\partial}{\partial x_j}\Big|_p\Big),$$

where

$$x_\alpha \circ \gamma(t) = \big(x_1(t), \dots, x_d(t)\big)$$

is the description of $\gamma(t)$ in local chart coordinates.

PROOF. Apply both sides to a typical $\mathcal{C}^\infty$ function defined near p and compare. □

With all this in mind, we make the following definition of smoothness.

DEFINITION 5.49 (Smooth Vector Fields on a Manifold M). On any chart (U_α, x_α), a vector field $\widetilde{X}$ on M can be uniquely represented in terms of the basis vectors $\mathfrak{X}_u$,

$$X_u = \sum_{j=1}^{d} c_k^\alpha(u) \cdot \Big(\frac{\partial}{\partial x_j}\Big|_u\Big) \quad \text{for } u \in U_\alpha. \tag{5.8}$$

We say $\widetilde{X}$ is a *smooth vector field* on M if the coefficients $c_j^\alpha(u)$ are $\mathcal{C}^\infty$ functions of $u \in U_\alpha$ for all charts (so $c_k^\alpha \circ x_\alpha^{-1} : \mathbb{R}^d \to \mathbb{R}$ is $\mathcal{C}^\infty$).

This definition is independent of the choice of local chart coordinates: in Lemma 4.18 and Corollary 4.16 of Chapter 4 we proved that if the coefficients are $\mathcal{C}^\infty$ near p for one chart, they are $\mathcal{C}^\infty$ for any other chart containing p.

EXERCISE 5.50 (Matrix Lie Groups Versus Lie Groups). Elements of a (real) matrix Lie group G are matrices $[A_{ij}] \in \mathrm{M}(n, \mathbb{R})$; tangent vectors $X \in \mathrm{TG}_p$ are also matrices in $\mathrm{M}(n, \mathbb{R})$, being derivatives $\gamma'(0) = [X_{ij}(p)]$ of smooth matrix-valued curves in $\mathrm{M}(n, \mathbb{R})$ that remain within G at all times. Vector fields on G are likewise matrix-valued since they are derivatives of matrix-valued curves in matrix space.

Suppose we define smoothness of such a vector field $\widetilde{X}$ to mean the individual matrix entries $X_{ij}(p)$ are $\mathcal{C}^\infty$ functions with respect to the charts that determine the differentiable structure of G, so $X_{ij} \circ x_\alpha^{-1} : \mathbb{R}^d \to \mathbb{R}$ is $\mathcal{C}^\infty$ in the usual sense. Prove that $\widetilde{X}$ is a smooth vector field as in Definition 5.49.

HINT. This is a matter of reconciling various definitions, but it is a good check to whether you've got the definitions straight.

For instance, when G is a matrix Lie group, every left-invariant vector field on G is obtained by transporting a fixed tangent vector $X \in \mathrm{TG}_e$ at the identity to other base points in G by left translation, letting $\widetilde{X}_g = (d\lambda_g)_e X$. In this situation, $\widetilde{X}$ is automatically smooth, essentially because group operations are differentiable. Although this seems intuitively obvious, proving it from the preceding definition is a bit messy and not particularly instructive, so we won't go into those details here.

EXERCISE 5.51. If G is a Lie group as in Definition 5.22 and $X \in \mathrm{TG}_e$, we get a well-defined field of vectors $\widetilde{X}$ on G if we take $\widetilde{X}_g = (d\lambda_g)_e X$.

(a) Verify that $\widetilde{X}$ is left-invariant, so that $\widetilde{X}_{g \cdot p} = \widetilde{X}_p$ for all $p, g \in G$.

(b) If we define left translates of functions on G, letting $L_g f(x) = f(g^{-1}x)$ for all $f \in \mathcal{C}^\infty(G)$ and $g, x \in G$, verify that $\widetilde{X}$ is left-invariant $\Leftrightarrow$ $L_g(\widetilde{X}f) = \widetilde{X}(L_g(f))$ for all $f \in \mathcal{C}^\infty(G)$ — i.e., $\widetilde{X}$ commutes with all translation operators L_g on $\mathcal{C}^\infty(G)$.

(c) Using the fact that left translations λ_g are $\mathcal{C}^\infty$ diffeomorphisms of G, verify the claim that $\widetilde{X}$ is always a smooth vector field with respect to the manifold structure of G for any $X \in \mathfrak{g}$ using the fact that translations λ_g are diffeomorphisms of the manifold G.

Smooth Vector Fields as Differential Operators on M**.** The set $\mathcal{D}(M)$ of all smooth vector fields on a manifold M is an infinite-dimensional vector space if we define the vector space operations, letting

$$(c\widetilde{X})_p = c \cdot \widetilde{X}_p \text{ for } c \in \mathbb{R} \quad \text{and} \quad (\widetilde{X} + \widetilde{Y})_p = \widetilde{X}_p + \widetilde{Y}_p$$

in TM_p at every base point $p \in M$. Since the local algebras $\mathcal{C}^\infty(p)$ at various base points all include the space $\mathcal{C}^\infty(M)$ of globally $\mathcal{C}^\infty$ functions on M, each smooth vector field $\widetilde{X}$ determines a natural linear operator $\widetilde{X}$ on the space of globally defined smooth functions $\mathcal{C}^\infty(M)$ if we let

$$\widetilde{X}f(p) = \langle \widetilde{X}_p, f\rangle \quad \text{for all} \quad f \in \mathcal{C}^\infty(M), p \in M.$$

The functions $\widetilde{X}f$ are $\mathcal{C}^\infty$ because, by equation (5.8), on any chart domain U_α we have

$$\begin{aligned} \widetilde{X}f(u) &= \left\langle \sum_{j=1}^{d} c_k^\alpha(u) \frac{\partial}{\partial x_j}\Big|_u \, , \, f(u)\right\rangle \\ &= \sum_{j=1}^{d} c_k^\alpha(u) \frac{\partial f}{\partial x_j}(u) \in \mathcal{C}^\infty(U_\alpha) \end{aligned} \tag{5.9}$$

for $f \in \mathcal{C}^\infty(U_\alpha)$. In particular, a vector field $\widetilde{X}$ on a matrix Lie group G is left-invariant $\Leftrightarrow$ it commutes with left translations, so that

$$L_g(\widetilde{X}f) = \widetilde{X}(L_g(f)) \quad \text{or} \quad L_g \circ \widetilde{X} \circ L_g^{-1}(f) = f \tag{5.10}$$

for all $g \in G$ and $f \in \mathcal{C}^\infty(G)$, [3] where $L_g f(x) = f(g^{-1}x)$.

Thus, in local coordinates on a chart (U_α, x_α), the action of $\widetilde{X}$ on smooth functions $f \in \mathcal{C}^\infty(M)$ is the same as that of the partial differential operator with variable coefficients

$$L = \sum_{j=1}^{d} c_j^\alpha(u) \cdot \left(\frac{\partial}{\partial x_j}\Big|_u\right) \quad \text{for } u \in U_\alpha. \tag{5.11}$$

Notice that L is described on U_α as a first-order partial differential operator with $\mathcal{C}^\infty$ coefficients and is *homogeneous* in the sense that it has no degree-0 scalar term $c_0(u)I$. This representation of $\widetilde{X}$ will change if we pass to a different local chart (U_β, y_β), but the new description is easily found using the change of variables formula (Corollary 4.16), and its coefficients are again $\mathcal{C}^\infty$ on U_β.

It is not so clear what one might mean by a "partial differential operator with smooth coefficients" on a manifold M, since M has no preferred system of global coordinates. A natural definition, then, is to say L is a *smooth differential operator* on M if it is a linear operator $L : \mathcal{C}^\infty(M) \to \mathcal{C}^\infty(M)$ whose action looks like that of a smooth-coefficient partial differential operator whenever it is described in local coordinates, so that on any chart (U_α, x_α) the action of L is described by a finite sum

$$Lf(u) = c_0(u)f(u) + \sum_{j=1}^{m} c_j(u)\frac{\partial f}{\partial x_j}(u) + \sum_{j,k=1}^{m} c_{jk}(u)\frac{\partial^2 f}{\partial x_j \partial x_k}(u) + \cdots \tag{5.12}$$

for all $f \in \mathcal{C}^\infty(U_\alpha)$. The "degree" $\deg(L)$ is that of the highest order partial derivative appearing with a nonzero coefficient, and the order-zero term $c_0(u)I$ is referred to as the "scalar term" (which might be zero).

[3]The "g^{-1}" that appears in the definition of $L_g f(x) = f(g^{-1}x)$ ensures that we have a "left action" of G on smooth vector fields, with $L_{g_1 g_2} = L_{g_1} \circ L_{g_2}$; without it we would have a "right action," in which the factors on the right appear in reverse order.

It is clear from (5.11) that smooth vector fields on M become the smooth homogeneous first-order partial differential operators on M. (It is not hard to check that the notions of "first order" and "homogeneous" persist and make sense in all coordinate systems.) The set $\mathcal{D}(M)$ of all smooth vector fields on a manifold M becomes an infinite-dimensional vector space if we define sums $\widetilde{X} + \widetilde{Y}$ and scalar multiples $c \cdot \widetilde{X}$ base-point by base-point, letting

$$(\widetilde{X} + \widetilde{Y})_p = \widetilde{X}_p + \widetilde{Y}_p \quad (c\widetilde{X})_p = c \cdot \widetilde{X}_p$$

in TM_p for all $p \in M$.

Now let M be a matrix Lie group G. We have noted that left-invariant vector fields on G are automatically smooth (essentially because group multiplication is a differentiable map $G \times G \to G$). They form a vector subspace $\mathcal{L}(G)$ in the larger space $\mathcal{D}(G)$ because two such fields $\widetilde{X}, \widetilde{Y}$ are determined by their values $X = \widetilde{X}_e, Y = \widetilde{Y}_e$ in the tangent space TG_e and $(\widetilde{X} + \widetilde{Y})_e = X + Y = (\widetilde{X + Y})_e$.

Then the map $\Phi : \mathfrak{g} \to \mathcal{L}(G)$ that assigns to $X \in \mathfrak{g}$ the unique left-invariant vector field $\widetilde{X}$ with values

$$\widetilde{X}_p = (d\lambda_p)_e(X) \in \mathrm{TG}_p \quad \text{for } p \in G, X \in \mathfrak{g}$$

is a bijective linear isomorphism between $\mathfrak{g} = \mathrm{TG}_e$ and the vector space $\mathcal{L}(G)$ in $\mathcal{D}(G)$.

This correspondence between tangent vectors X at the identity and left-invariant partial differential operators $\widetilde{X}$ on the whole group leads to many important insights about Lie groups. For one thing, it reveals, in a natural way, that the tangent space $\mathfrak{g} = \mathrm{TG}_e$ of a Lie group has remarkable "Lie algebra structure" that becomes apparent when we realize $\mathfrak{g}$ as a space $\mathcal{L}(G)$ of left-invariant partial differential operators on G.

In order to explain this connection, we shall take time below to describe the general concept of a "Lie algebra" and provide a direct construction of the associated "Lie bracket operation" in TG_e. That direct construction (for matrix Lie groups) will perhaps seem unintuitive, but its techniques are transparent and are elementary enough to allow straightforward matrix calculations.

An alternative approach is needed to construct the Lie bracket that makes $\mathfrak{g} = \mathrm{TG}_e$ a Lie algebra for general Lie groups, and the identification of $\mathfrak{g} \cong \mathcal{L}(G)$ is the key to that construction at the heart of modern Lie theory. In this limited account, we will not be able to present all those details, but here is a meaningful outline of what happens:

1. One begins by identifying $\mathfrak{g}$ with the space $\mathcal{L}(G)$ of left-invariant vector fields $\widetilde{X}$ on G, which in turn can be identified as the finite-dimensional vector space $\mathcal{L}(G)$ of homogeneous first-order partial differential operators defined on all of G. As such, $\mathcal{L}(G)$ is a subspace of the infinite-dimensional vector space $\mathrm{Diff}(G)$ of all linear partial differential operators (of any degree) with $\mathcal{C}^\infty$ coefficients. But in addition to being a vector space, $\mathrm{Diff}(G)$ also has a natural multiplication operation given by composition of operators,

$$(S \cdot T)(f) = S(T(f)) \quad \text{for } f \in \mathcal{C}^\infty(G).$$

This makes Diff(G) a (noncommutative) associative algebra with identity.

2. Any associative algebra $\mathcal{A}$ has a natural bracket operation, the *associative commutator* of two operators

$$[S,T] = ST - TS \quad (S,T \in \mathcal{A}).$$

This bilinear operation on pairs of elements $S,T \in \mathcal{A}$ measures the degree to which they fail to commute, since $[S,T] = 0 \Leftrightarrow S$ and T commute. Equipped with this bracket operation, Diff(G) becomes a "Lie algebra," a concept that will be defined and examined further in Section 5.3.

3. We now ask

Is the finite-dimensional vector subspace $\mathcal{L}(G) \subseteq$ Diff(G) closed under formation of commutators $[\widetilde{X},\widetilde{Y}] = \widetilde{X}\widetilde{Y} - \widetilde{Y}\widetilde{X}$?

There are several reasons this may seem unlikely, but they aren't worth discussing because the answer is "*yes*," and for an interesting reason that we now explain. In any case, once we know that $[\mathcal{L}(G),\mathcal{L}(G)] \subseteq \mathcal{L}(G)$, it follows immediately that $\mathcal{L}(G)$ becomes a finite-dimensional Lie algebra. By transferring this operation over to the tangent space $\mathfrak{g} = \mathrm{TG}_e \cong \mathcal{L}(G)$ we make $\mathfrak{g}$ into a Lie algebra with

$$\dim(\mathfrak{g}) = \dim(G) = \dim \mathcal{L}(G).$$

We close this commentary by showing why $\mathcal{L}(G)$ is closed under formation of comutators. Smooth vector fields $\widetilde{X}$ on G are particular examples of globally defined smooth partial differential operators in Diff(G). If we describe $\widetilde{X}$ in local coordinates on a chart (U_α, x_α) it becomes a homogeneous first-order partial differential operator with variable coefficents (no scalar term):

$$\widetilde{X}_u = \sum_{j=1}^{d} c_j^\alpha(u) \cdot \left(\frac{\partial}{\partial x_j}\bigg|_u\right) \quad \text{for } u \in U_\alpha$$

with coefficients in $\mathcal{C}^\infty(U_\alpha)$. Such operators are determined by what they do to smooth functions $f \in \mathcal{C}^\infty(U_\alpha)$, and within Diff($G$) two partial differential operators on U_α can be multiplied by composition to get

$$\widetilde{X}\widetilde{Y}(f) = \widetilde{X}(\widetilde{Y}(f)) \quad \text{for } f \in \mathcal{C}^\infty(U_\alpha).$$

When we describe $\widetilde{X},\widetilde{Y}$ in local coordinates,

$$\widetilde{X} = \sum_{j=1}^{d} a_j(u) \cdot \frac{\partial}{\partial x_j} \quad \text{and} \quad \widetilde{Y} = \sum_{k=1}^{d} b_k(u) \cdot \frac{\partial}{\partial x_k},$$

we find that their product is a *second order* partial differential operator,

$$\widetilde{X}\widetilde{Y}f(u) = \sum_{j,k=1}^{d} a_j(u)\Big(\sum_{k=1}^{d} \frac{\partial b_k}{\partial x_j}(u) \cdot \frac{\partial f}{\partial x_k}(u)\Big) + a_j(u)b_k(u)\frac{\partial^2 f}{\partial x_j \partial x_k}(u)$$

for $f \in \mathcal{C}^\infty(U_\alpha)$. Similarly for $\widetilde{Y}\widetilde{X}f$. But when we take the difference of these expressions, all terms in the commutator $[\widetilde{X},\widetilde{Y}] = \widetilde{X}\widetilde{Y} - \widetilde{Y}\widetilde{X}$ involving second partial derivatives cancel because the mixed second order partial derivatives of smooth functions are equal. Thus the action of the commutator $[\widetilde{X},\widetilde{Y}]$ on $\mathcal{C}^\infty(M)$ is described, on any chart, as a homogeneous *first-order operator*:

$$[\widetilde{X},\widetilde{Y}]f(u) = \sum_{j=1}^{m}\Big(\sum_{i=1}^{n} a_i\frac{\partial b_j}{\partial x_i} - b_i\frac{\partial a_j}{\partial x_i}\Big)\cdot\Big(\frac{\partial f}{\partial x_j}\Big|_u\Big) \qquad \text{for } u \in U_\alpha.$$

Since this is true on every chart domain U_α, the commutator $[\widetilde{X},\widetilde{Y}]$ is a smooth vector field in $\mathcal{D}(M)$.

Finally, this commutator field is left-invariant (as in (5.10)) and hence is equal to $\widetilde{Z}$ for a unique element $Z \in \mathfrak{g}$. Thus $\mathfrak{g} \cong \mathcal{L}(G)$ is closed under formation of associative commutators in Diff(G) and is a Lie algebra.

5.3. Lie Algebra Structure in Tangent Spaces of Lie Groups

The tangent space TG_e of a matrix Lie group at its identity element $e \in G$ is of special interest to us, and we shall denote such spaces by German "fraktur" letters $\mathfrak{g} = \mathrm{TG}_e$, $\mathfrak{h} = \mathrm{TH}_e$, etc. These are of course vector subspaces in the matrix space $\mathrm{M}(n,\mathbb{R})$ in which G is embedded. But their most important property is that the group operation $x \cdot y$ in G induces a distinctive bilinear *Lie bracket operation* $[X,Y]$ on tangent vectors that sends a pair X, Y in $\mathfrak{g}$ to a new tangent vector $[X,Y]$ in $\mathfrak{g}$. This makes $\mathfrak{g}$ into a "Lie algebra."

DEFINITION 5.52. A *Lie algebra* $\mathfrak{g}$ over a field of scalars $\mathbb{K}$ is a vector space over $\mathbb{K}$, equipped with a bracket operation $B(X,Y) = [X,Y]$ that has the following properties:

1. BILINEARITY. It is a vector-valued bilinear map from $\mathfrak{g}\times\mathfrak{g} \to \mathfrak{g}$ (linear in each input when the other is held fixed).
2. SKEW-SYMMETRY. $[Y,X] = -[X,Y]$ for all $X,Y \in \mathfrak{g}$. In particular $[X,X] = 0$ (the zero vector in $\mathfrak{g}$) for all X.
3. JACOBI IDENTITY. For $X,Y,Z \in \mathfrak{g}$ we have
$$[X,[Y,Z]] + [Y,[Z,X]] + [Z,[X,Y]] = 0.$$

In view of these laws, if a Lie algebra $\mathfrak{g}$ is finite dimensional and $\mathfrak{X} = \{X_1,\dots,X_d\}$ is a vector basis, its algebraic structure is completely determined once we know the brackets of the basis vectors

$$[X_i,X_j] = \sum_{k=1}^{d} c_{ij}^{(k)}X_k \quad (c_{ij}^{(k)} \in \mathbb{K}, 1 \le i,j \le d).$$

The d^3 coefficients appearing here are the *structure constants* of $\mathfrak{g}$ with respect to the basis $\mathfrak{X}$.

The basic example of a Lie algebra is provided by the "commutator bracket" operation

$$[X,Y] = XY - YX \quad \text{(matrix products)}$$

on the associative algebra $M(n, \mathbb{K})$ of $n \times n$ matrices with entries in $\mathbb{K}$; it measures the degree to which two matrices X and Y fail to commute.

EXERCISE 5.53. Verify that the Jacobi identity of Definition 5.52 holds for the associative commutator $[X, Y] = XY - YX$ on the matrix algebra $M(n, \mathbb{K})$

HINT. Expand all commutators in long form, as sums of expressions that are permutations of XYZ. Everything cancels.

Similarly, the space $\text{Hom}_{\mathbb{K}}(V, V)$ of $\mathbb{K}$-linear operators on a vector space V becomes an associative algebra if we take composition of operators $S \circ T$ as its multiplication operator; it becomes a Lie algebra when equipped with the commutator bracket $[S, T] = S \circ T - T \circ S$. No surprise here: if $\dim(V) = n$ and $\mathfrak{X}$ is a basis in V, we saw in Section 2.4 of LA I that the map $T \to [T]_{\mathfrak{X}\mathfrak{X}}$ is a natural isomorphism $\text{Hom}_{\mathbb{K}}(V, V) \to M(n, \mathbb{K})$ of associative algebras. This respects linear combinations and products, so it sends commutators to commutators. Hence it is also an isomorphism between the Lie algebras.

Lie Algebra of a Matrix Lie Group. We can now explain how the multiplication operation $x \cdot y$ in G induces a natural Lie bracket operation in the tangent space $\mathfrak{g} = \text{TG}_e$ at the identity element in G, making $\mathfrak{g}$ into a Lie algebra. Two observations motivate our interest in this Lie algebra structure.

- Being a vector space, the Lie algebra $\mathfrak{g}$ is a *linear structure* to which we can apply all the tools of Linear Algebra. That is a good thing because Lie algebras are much more accessible objects than the nonlinear Lie groups from which they are derived.
- The Lie algebra $\mathfrak{g}$ might have remained a mathematical curiosity, were it not for its most important property:

  The Lie algebra structure of $\mathfrak{g}$, equipped with its Lie bracket, encodes almost all the information needed to completely reconstruct the original Lie group.

  This has been heavily exploited to reduce many investigations of Lie groups to purely algebraic questions about their Lie algebras.

By now, Lie algebras have become a topic as important as the Lie groups from which they arose.

It is not immediately apparent how the multiplication operation in G induces a Lie bracket operation in the tangent space $\mathfrak{g}$. One approach is to get a handle on vectors in the tangent spaces TG_p by regarding tagent vectors at p as derivatives $\gamma'(0)$ along differentiable curves $\gamma(t)$ in G that pass through p at time $t = 0$. This identifies tangent vectors at p with derivation operators X_p on functions f in the local algebra $\mathcal{C}^\infty(p)$, letting

$$\langle X_p, f \rangle = \langle \gamma'(0), f \rangle = \left\langle \frac{d}{dt}\{\gamma(t)\}\Big|_{t=0}, f \right\rangle = \frac{d}{dt}\{f(\gamma(t))\}\Big|_{t=0}$$

for $f \in \mathcal{C}^\infty(p)$. These operators are derivations on $\mathcal{C}^\infty(p)$, and in modern differential geometry *they* are the elements of the tangent space TM_p.

Next, some basic facts about derivatives of algebraic combinations of smooth curves in a Lie group, and the tangent vectors $X_p \in \mathrm{TG}_p$ on local algebras $\mathcal{C}^\infty(p)$ they determine.

LEMMA 5.54. *If $\gamma, \eta : \mathbb{R} \to \mathrm{M}(n, \mathbb{K})$ are $\mathcal{C}^\infty$ curves defined for $a < t < b$, then*

1. SUM RULE. $\frac{d}{dt}\{\gamma(t) + \eta(t)\} = \frac{d}{dt}\{\gamma(t)\} + \frac{d}{dt}\{\eta(t)\}$ *for all* t.
2. SCALING LAW. $\frac{d}{dt}\{c \cdot \gamma(t)\} = c \cdot \frac{d\gamma}{dt}(t)$ *for* $c \in \mathbb{K}$.
3. DERIVATION LAW. *The derivative of the matrix product* $\gamma(t) \cdot \eta(t)$ *is*
$$\frac{d}{dt}\{\gamma(t) \cdot \eta(t)\} = \frac{d\gamma}{dt}(t) \cdot \eta(t) + \gamma(t) \cdot \frac{d\eta}{dt}(t) \quad \text{for all } t.$$
The factors cannot be interchanged unless G is an abelian group.

PROOF. Items (1.) and (2.) are trivial. For (3.) write the appropriate difference quotients as

$$\begin{aligned}
&\frac{\gamma(t+\Delta t)\cdot\eta(t+\Delta t) - \gamma(t)\cdot\eta(t)}{\Delta t} \\
&\quad = \frac{\gamma(t+\Delta t)\cdot\eta(t+\Delta t) - \gamma(t)\cdot\eta(t+\Delta t)}{\Delta t} + \frac{\gamma(t)\cdot\eta(t+\Delta t) - \gamma(t)\cdot\eta(t)}{\Delta t} \\
&\quad = \frac{\Delta\gamma}{\Delta t}\cdot\eta(t+\Delta t) + \gamma(t)\cdot\frac{\Delta\eta}{\Delta t}.
\end{aligned}$$

These converge to $\gamma'(t)\cdot\eta(t)+\gamma(t)\cdot\eta'(t)$ as $\Delta t \to 0$ because the product operation is *jointly continuous*, with $\gamma(s) \cdot \eta(t) \to \gamma(0) \cdot \eta(0)$ as $s, t \to 0$ independently. □

This yields the following description of the vector space operations in the tangent space TG_e of a matrix Lie group.

LEMMA 5.55. *The Lie algebra $\mathfrak{g} = \mathrm{TG}_e$ of a matrix Lie group is a vector space over $\mathbb{R}$, as are the tangent spaces TG_p at arbitrary base points in G.*

PROOF. If $c \in \mathbb{R}$ and $\gamma(t)$ is a $\mathcal{C}^\infty$ curve in G passing through the identity I when $t = 0$, and if $\gamma'(0) = X \in \mathfrak{g}$, then $\eta(t) = \gamma(ct)$ is also a smooth curve, with

$$\frac{d}{dt}\{\eta(t)\} = c \cdot \frac{d\gamma}{dt}(t) \quad \text{for all } t.$$

Hence $\eta'(0) = c \cdot X$ is in $\mathfrak{g} = \mathrm{TG}_e$ for $X \in \mathfrak{g}$.

If $\mathrm{g} \in G$ and γ, η have $\gamma(0) = \eta(0) = e$ and $\gamma'(0) = X$, $\eta'(0) = Y$ in $\mathfrak{g}$, then the product $\Phi(t) = \gamma(t) \cdot \mathrm{g}^{-1} \cdot \eta(t)$ is a $\mathcal{C}^\infty$ curve with $\Phi(0) = \mathrm{g}\mathrm{g}^{-1}\mathrm{g} = \mathrm{g}$, and by the derivation law Lemma 5.54 we get

$$\Phi'(0) = \gamma'(0) \cdot \mathrm{g}^{-1}\eta(0) + \gamma(0)\mathrm{g}^{-1} \cdot \eta'(0) = X + Y$$

in $\mathfrak{g}$. □

LEMMA 5.56. *Let $\gamma(t)$ be a $\mathcal{C}^\infty$ curve in a matrix Lie group G such that $\gamma(0) = I$ and $\gamma'(0) = X$ in* TG_e.

1. *Inverses $\mu(t) = \big(\gamma(t)\big)^{-1}$ exist for sufficiently small values of t, and $\mu(t)$ is a $\mathcal{C}^\infty$ curve in G with $\mu(0) = I$ and $\mu'(0) = -\gamma'(0)$.*
2. *For fixed $g \in G$, $\eta(t) = \lambda_g(\gamma(t)) = g \cdot \gamma(t)$ is a $\mathcal{C}^\infty$ curve in G with $\eta(0) = g$ and vector derivative*
$$\eta'(t) = g \cdot \gamma'(t) \quad (\textit{matrix product in } \mathrm{M}(n, \mathbb{R}))$$
for all $t \in \mathbb{R}$.
3. *The differential $(d\lambda_g)_e$ sends $X \in \mathrm{TG}_e$ to*
$$(d\lambda_g)_e X = \eta'(0) = g \cdot X \quad \big(\textit{matrix product in } \mathrm{M}(n, \mathbb{R})\big).$$
in TG_g. Similarly for right translations we have $(d\rho_g)_e X = X \cdot g$.

PROOF. Since $\gamma(t) \to \gamma(0) = I$ as $t \to 0$, $\det(\gamma(t)) \to \det(I) = 1$, so $\det(\gamma(t))$ is nonzero and $\gamma(t)$ is invertible for small t. Inversion $x \mapsto J(x) = x^{-1}$ is a $\mathcal{C}^\infty$ map (by definition) and we are assuming $\gamma : \mathbb{R} \to G \subseteq \mathrm{M}(n, \mathbb{R})$ is $\mathcal{C}^\infty$, so the composite $\mu(t) = \gamma(t)^{-1} = J(\gamma(t))$ is also $\mathcal{C}^\infty$. To prove $\mu'(0) = -\gamma'(0)$ we differentiate the identity $I_{n\times n} = \gamma(t) \cdot \gamma(t)^{-1} = \gamma(t) \cdot \mu(t)$ to get
$$0 = \frac{d\gamma}{dt} \cdot \mu(t) + \gamma(t) \cdot \frac{d\mu}{dt}.$$
Set $t = 0$ and rearrange terms to conclude that $\mu'(0) = -\gamma'(0)$.

In (2) left translations $\lambda_g : G \to G$ are diffeomorphisms of G so $\eta(t) = \lambda_g(\gamma(t))$ is a $\mathcal{C}^\infty$ curve in G with $\eta(0) = g$. Then $\eta'(t) = g \cdot \gamma'(t)$ because
$$\begin{aligned}\eta'(t) = \frac{d}{dt}\{\lambda_g(\gamma(t))\} &= \lim_{\Delta t \to 0} \frac{g \cdot \gamma(t + \Delta t) - g \cdot \gamma(t)}{\Delta t} \\ &= g \cdot \lim_{\Delta t \to 0} \frac{\Delta\gamma}{\Delta t} = g \cdot \gamma'(t) \quad (\text{product of } n \times n \text{ matrices})\end{aligned}$$
for all t.

For (3) we must wend our way through several definitions. Viewing tangent vectors in TG_e and TG_g as derivations on the local algebras $\mathcal{C}^\infty(e)$ and $\mathcal{C}^\infty(g)$ we have just seen that $\eta'(0) = g \cdot X$, while for all $f \in \mathcal{C}^\infty(g)$ we have
$$\begin{aligned}\langle (d\lambda_g)_e X, f\rangle &= \langle X, (f \circ \lambda_g)\rangle = \langle \gamma'(0), (f \circ \lambda_g)\rangle \\ &= \frac{d}{dt}\{f(\lambda_g \circ \gamma(t))\}\Big|_{t=0} = \frac{d}{dt}\{f(g \cdot \gamma(t))\}\Big|_{t=0} \\ &= \langle \eta'(0), f\rangle = \langle g \cdot X, f\rangle.\end{aligned}$$
Thus $(d\lambda_g)_e X = g \cdot X$ in $\mathrm{M}(n, \mathbb{K})$. □

We can now show how the tangent space space $\mathfrak{g} = \mathrm{TG}_e$ acquires a Lie algebra structure.

PROPOSITION 5.57 (The Lie Bracket in $\mathfrak{g}$). *If $G \subseteq \mathrm{M}(n,\mathbb{R})$ is a matrix Lie group its tangent space at the identity $\mathfrak{g} = \mathrm{TG}_e$ is closed under formation of associative commutators in $\mathrm{M}(n,\mathbb{R})$, then*

$$X, Y \in \mathfrak{g} \Rightarrow [X,Y] = XY - YX \text{ is also in } \mathfrak{g}, \tag{5.13}$$

so $\mathfrak{g}$ is a Lie subalgebra in $\mathrm{M}(n,\mathbb{R})$.

NOTE. Tangent vectors $X, Y \in \mathfrak{g} = \mathrm{TG}_e$ are in $\mathrm{M}(n,\mathbb{R})$ so we can form their matrix products XY and YX, but these *need not* be elements of $\mathfrak{g}$; nevertheless, the associative commutator $[X,Y] = XY - YX$ always ends up in $\mathfrak{g}$. Once we prove that it will be obvious that the bracket operation (5.6) makes $\mathfrak{g}$ into a Lie algebra over $\mathbb{R}$. ○

PROOF. Let $\gamma(t), \eta(t)$ be $\mathcal{C}^\infty$ curves in G that pass through $e = I$ when $t = 0$ and have derivatives $\gamma'(0) = X, \eta'(0) = Y$ in $\mathfrak{g}$. For $s,t \in \mathbb{R}$ the function

$$f(t) = \gamma(s)\eta(t)\gamma(s)^{-1} \quad \text{(product of elements in } G).$$

is $\mathcal{C}^\infty$, $f(0) = I$ for all values of s, and $f(t)$ remains within G for all t (and s) because G is a group. Thus $f'(0)$ is in the Lie algebra $\mathfrak{g}$. But we also have

$$f'(0) = \frac{d}{dt}\{f(t)\}\Big|_{t=0} = \frac{\partial}{\partial t}\{\gamma(s)\eta(t)\gamma(s)^{-1}\}\Big|_{t=0} = \gamma(s)Y\gamma(s)^{-1}$$

for small values of s. Hence $\gamma(s)\cdot \mathrm{TG}_e \cdot \gamma(s)^{-1} \subseteq \mathrm{TG}_e$ for all small s and we get

$$\begin{aligned}
\frac{\partial}{\partial s}\{\gamma(s)Y\gamma(s)^{-1}\}\Big|_{s=0} &= \gamma'(0)\cdot Y\gamma(0)^{-1} + \gamma(0)Y\cdot \frac{d}{ds}\{\gamma(s)^{-1}\}\Big|_{s=0} \\
&= \gamma'(0)\cdot Y\gamma(0)^{-1} + \gamma(0)Y\cdot -\gamma'(0) \\
&= XY - YX = [X,Y] \quad \text{(commutator in } \mathrm{M}(n,\mathbb{R}))
\end{aligned}$$

as required to show $[\mathfrak{g},\mathfrak{g}] \subseteq \mathfrak{g}$. □

Another way to say this is:

$$[X,Y] = \frac{\partial^2 f}{\partial s \partial t}(0) = \frac{\partial^2}{\partial s\, \partial t}\{\gamma(s)\eta(t)\gamma(s)^{-1}\}\Big|_{s=t=0}. \tag{5.14}$$

Adjoint Action of G on Its Lie Algebra $\mathfrak{g}$. The action of a Lie group on itself by inner automorphisms $i_g(x) = gxg^{-1}$ (aka *conjugations*) plays a major role in Lie theory. The next result shows that the differentials $(di_g)_e$ are linear operators on the Lie algebra $\mathfrak{g}$, and the resulting linear action $G\times\mathfrak{g} \to \mathfrak{g}$ is called the *adjoint action* of G on the Lie algebra. The modern literature reflects this now universal terminology by writing the diffential $(di_g)_e$ of a conjugation as $\mathrm{Ad}_g : \mathfrak{g} \to \mathfrak{g}$.

LEMMA 5.58. *An inner automorphisms $i_g : x \to gxg^{-1}$ fixes the identity element in G and its differential $\mathrm{Ad}_g = (di_g)_e$ leaves the Lie algebra $\mathfrak{g} = \mathrm{TG}_e$ invariant for every $g \in G$, so $\mathrm{Ad}_g(\mathfrak{g}) \subseteq \mathfrak{g}$. Furthermore, we have*

$$\mathrm{Ad}_g(Y) = gYg^{-1} \quad (\textit{matrix product})$$

for all $Y \in \mathfrak{g}, g \in G$.

PROOF. If $X \in \mathfrak{g}$, $g \in G$, and γ is a $\mathcal{C}^\infty$ curve in G such that $\gamma(0) = I$ and $\gamma'(0) = X$, then $\eta(t) = i_g(\gamma(t)) = g\gamma(t)g^{-1}$ is a smooth curve in G with $\eta(0) = I$ because products and inverses in G are $\mathcal{C}^\infty$ operations in G. By Lemma 5.54 we have $\eta'(t) = g\gamma'(t)g^{-1}$, and setting $t = 0$ we see that

$$\eta'(0) = gXg^{-1} \quad \text{(matrix product)}$$

is always in $\mathfrak{g}$. By Part (3.) of Lemma 5.56 the inner automorphism $i_g(x) = gxg^{-1}$ can be written $i_g = \lambda_g \circ \rho_g^{-1}$, hence

$$\begin{aligned}\mathrm{Ad}_g(X) &= (di_g)_e(X) = (d\lambda_g)_e \circ (d\rho_{g^{-1}})_e(X)\\ &= (d\lambda_g)_e(X \cdot g^{-1}) = gXg^{-1},\end{aligned}$$

as claimed. □

EXERCISE 5.59. Verify that the adjoint action $\mathrm{Ad} : G \times \mathfrak{g} \to \mathfrak{g}$ of a matrix Lie group on its Lie algebra $\mathfrak{g} = \mathrm{TG}_e$ has the following properties.

(a) $\mathrm{Ad}_g = (di_g)_e$ is an invertible linear operator on $\mathfrak{g}$.

(b) The map "*Ad*" that sends $g \in G$ to the invertible linear operator $\mathrm{Ad}_g = (di_g)_e$ on $\mathfrak{g}$ is a group homomorphism from $G \to \mathrm{Aut}(\mathfrak{g})$.

$$\mathrm{Ad}_e = \mathrm{id}_\mathfrak{g} \quad \mathrm{Ad}_{g_1g_2} = \mathrm{Ad}_{g_1} \circ \mathrm{Ad}_{g_2} \quad (\mathrm{Ad}_g)^{-1} = \mathrm{Ad}_{g^{-1}}$$

for $g_1, g_2 \in G$.

EXERCISE 5.60. For a Lie algebra $\mathfrak{g}$ over any field of scalars $\mathbb{K}$ we can define the linear map *ad* that sends $X \in \mathfrak{g}$ to a linear operator ad_X on $\mathfrak{g}$, letting

$$\mathrm{ad}_X(Y) = [X, Y] \quad \text{for all } X, Y \text{ in } \mathfrak{g}.$$

(a) Prove that ad_X is a *Lie algebra derivation*, so that

$$\mathrm{ad}_X([Y, Z]) = [\mathrm{ad}_X(Y), Z] + [Y, \mathrm{ad}_X(Z)] \quad \text{for all } X, Y, Z \in \mathfrak{g}.$$

(b) Verify that this derivation property of ad_X is equivalent to the Jacobi identity in the definition of Lie Algebras.

We will have more to say about the adjoint action before long.

Lie Algebras of the Classical Groups. There is more to be done if we wish to understand particular Lie algebras $\mathfrak{g}$ and do effective calculations with them. We need concrete models of $\mathfrak{g}$, and vector bases whose pattern of structure constants is simple enough to reveal the properties of $\mathfrak{g}$. Using the facts just developed we will identify the matrices appearing in the Lie algbras of the classical matrix Lie groups, and in some low-dimensional examples we will produce vector bases with particularly illuminating patterns of structure constants.

When $G = \mathrm{GL}(n, \mathbb{K})$ there is nothing to do because $\mathrm{GL}(n, \mathbb{K})$ is an open set that fills most of matrix space, and its Lie algebra $\mathfrak{gl}(n, \mathbb{K})$ is all of $\mathrm{M}(n, \mathbb{K})$ equipped with the usual associative Lie bracket. The standard "matrix units" E_{ij} with (k, ℓ) entry equal to $\delta_{ik} \cdot \delta_{j\ell}$ (Kronecker deltas) are a $\mathbb{K}$-basis for $\mathfrak{gl}(n, \mathbb{K})$. The structure constants are given by the commutator relations

$$[E_{ij}, E_{k\ell}] = E_{ij}E_{k\ell} - E_{k\ell}E_{ij} = \delta_{jk}E_{i\ell} - \delta_{i\ell}E_{kj}.$$

This is admittedly a fairly complicated set of relations. It becomes a little more transparent if you notice that the pairs of addresses $(ij), (k\ell)$ and $(kj), (i,\ell)$ lie at diametrically opposite corners of a rectangle in the $n \times n$ array of matrix addresses.

EXAMPLE 5.61 (Special Linear Group $\mathrm{SL}(n,\mathbb{K})$). The Lie algebra of $G = \mathrm{SL}(n,\mathbb{K}) = \{A \in \mathrm{M}(n,\mathbb{K}) : \det(A) = 1\}$ is

$$\mathfrak{sl}(n,\mathbb{K}) = \{X \in \mathrm{M}(n,\mathbb{K}) : \mathrm{Tr}(X) = 0\},$$

which has $\dim_{\mathbb{K}} \mathfrak{sl}(n,\mathbb{K}) = n^2 - 1$ because it is the kernel of the $\mathbb{K}$-linear trace function $\mathrm{Tr} : \mathfrak{g} \to \mathbb{K}$.

DISCUSSION. Let $\gamma(t)$ be a $\mathcal{C}^\infty$ curve in G with $\gamma(0) = I_{n\times n}$ and $\gamma'(0) = X$ in the Lie algebra $\mathfrak{g}$. By definition of SL,

$$1 \equiv \det\big(\gamma(t)\big) = \sum_{\pi\in S_n} \mathrm{sgn}(\pi) \cdot \prod_{j=1}^{n} \gamma(t)_{j,\pi(j)} \quad \text{for all } t.$$

The product rule implies

$$0 = \frac{d}{dt}\{\gamma(t)\} = \sum_{\pi\in S_n} \mathrm{sgn}(\pi) \cdot \Big(\sum_{k=1}^{n} \gamma(t)_{1,\pi(1)} \cdot \dots \cdot \gamma'(t)_{k,\pi(k)} \cdot \dots \cdot \gamma_{n,\pi(n)}(t) \Big).$$

Since $\gamma(0) = I_{n\times n}$ when $t = 0$, for $j \neq k$ the jth term in $\big(\sum_{k=1}^{n} \cdots\big)$ is zero unless $\pi(j) = j$, but when $j = k$ we must have $\pi(k) = k$ too, so $\pi = \mathrm{id}$ in S_n. We conclude that the entire sum over k is zero, except for the term corresponding to $\pi = \mathrm{id}$ in the outer sum over S_n. Since $\gamma'(0) = X$ in $\mathrm{M}(n,\mathbb{K})$ we get

$$0 = \sum_{k=1}^{n} 1 \cdot \dots \cdot 1 \cdot x_{kk} \cdot 1 \cdot \dots \cdot 1 = \sum_{k=1}^{n} x_{kk} = \mathrm{Tr}(X)$$

where $x_{ij} = [\gamma'(0)]_{ij}$.

Conversely, if $X \in \mathrm{M}(n,\mathbb{K})$ and $\mathrm{Tr}(X) = 0$ the matrix-exponential series

$$\gamma_X(t) = e^{tX} = \sum_{n=0}^{\infty} \frac{t^n}{n!} X^n$$

converges in matrix space to a $\mathcal{C}^\infty$ curve that passes through I when $t = 0$. Furthermore, the exponent law $e^{(s+t)X} = e^{sX} \cdot e^{tX}$ implies that

$$\frac{d}{dt}\{e^{tX}\} = X \cdot e^{tX} \quad \text{for all } X \in \mathrm{M}(n,\mathbb{K}),\ t \in \mathbb{R}, \tag{5.15}$$

so $\gamma_X'(0) = X$. (Matrix exponentials e^{tA} are discussed in Sections 5.4 and 5.5 of LA I, see Application 5.5: Linear Systems of Differential Equations.)

Finally, if $X \in \mathrm{M}(n,\mathbb{K})$ has trace $\mathrm{Tr}(X) = 0$ the exponentials e^{tX} remain within G for all t because

$$\det\big(e^{tX}\big) = e^{t\cdot\mathrm{Tr}(X)} = 1$$

(see Lemma 5.62 below). We conclude that

$$\begin{aligned}\mathfrak{sl}(n,\mathbb{K}) &= \{X \in \mathrm{M}(n,\mathbb{K}) : \mathrm{Tr}(X) = 0\}\\ &= \{X \in \mathrm{M}(n,\mathbb{K}) : e^{tX} \in \mathrm{SL}(n,\mathbb{C}) \text{ for all } t \in \mathbb{R}\}.\end{aligned}$$

When $\mathbb{K} = \mathbb{C}$, $G = \mathrm{SL}(n,\mathbb{C})$ is a complex Lie group and $\mathfrak{sl}(n,\mathbb{C})$ is a Lie algebra over $\mathbb{C}$. Whether $\mathbb{K} = \mathbb{R}$ or $\mathbb{C}$, $\dim_{\mathbb{K}}(G) = \dim_{\mathbb{K}} \mathfrak{sl}(n,\mathbb{K})$ and our identification of the Lie algebra immediately shows that $\dim_{\mathbb{K}} \mathfrak{sl}(n,\mathbb{K}) = n^2 - 1$, being the kernel of the linear trace function $\mathrm{Tr} : \mathfrak{g} \to \mathbb{K}$.

LEMMA 5.62. *When $\mathbb{K} = \mathbb{C}$ or $\mathbb{R}$ the identity*

$$\det(e^A) = e^{\mathrm{Tr}(A)} \tag{5.16}$$

holds for $\mathbb{K} = \mathbb{R}$ or $\mathbb{C}$, or for arbitrary $\mathbb{K}$ if the characteristic polynomial p_A splits over $\mathbb{K}$.

PROOF. If p_A splits this follows because the Jordan canonical form of A exists and is upper triangular. Its diagonal entries are the eigenvalues $\lambda_j \in \mathbb{K}$, appearing according to their multiplicities m_j. By Corollary 1.53 we get

$$\det(A) = \prod_{j=1}^{r} \lambda_j^{m_j} \quad \text{while} \quad \mathrm{Tr}(A) = \sum_{j=1}^{r} m_j\lambda_j.$$

Then e^A is upper triangular with diagonal entries e^{λ_j}, each appearing m_j times, so $\det(A) = \prod_{j=1}^{r} e^{m_j\lambda_j}$. Meanwhile, $e^{\mathrm{Tr}(X)} = e^{(m_1\lambda_1+\cdots+m_r\lambda_r)}$, and the identity follows.

When $\mathbb{K} = \mathbb{R}$ equation (5.16) remains valid even if the characteristic polynomial does not split over $\mathbb{R}$. If $T = L_A : \mathbb{R}^n \to \mathbb{R}^n$ for some A in $\mathrm{M}(n,\mathbb{R})$ (5.16) follows by examining the Jordan canonical form of the complexification $T_{\mathbb{C}}$ on $\mathbb{C}^n$ and observing that

$$\det(T_{\mathbb{C}}) = \det(T) = \det(A) \quad \text{and} \quad \mathrm{Tr}(T_{\mathbb{C}}) = \mathrm{Tr}(T) = \mathrm{Tr}(A).$$

The formulas for $\det(A)$ and $\mathrm{Tr}(A)$ involve all the complex eigenvalues of the real matrix A, but the terms in (5.16) are real-valued owing to cancellations. □

PROPOSITION 5.63 (Lie Algebras of Unitary Groups $\mathrm{U}(n)$ and $\mathrm{SU}(n)$). *When $G = \mathrm{U}(n)$ or $\mathrm{SU}(n)$ the Lie algebras are*

$$\begin{aligned}\mathfrak{u}(n) &= \{X \in \mathrm{M}(n,\mathbb{C}) : X^* = -X\}\\ \mathfrak{su}(n) &= \{X \in \mathrm{M}(n,\mathbb{C}) : X^* = -X \text{ and } \mathrm{Tr}(X) = 0\}\end{aligned}$$

These groups are real, not complex, even though they are carved out of $\mathrm{M}(n,\mathbb{C})$, and we have

$$\dim_{\mathbb{R}} \mathfrak{u}(n) = n^2 \quad \dim_{\mathbb{R}} \mathfrak{su}(n) = n^2 - 1.$$

PROOF. For $(\subseteq)$, let $\gamma(t)$ be a $\mathcal{C}^\infty$ curve in G with $\gamma(0) = I$, $\gamma'(0) = X$ in $\mathfrak{g}$. Since

$$\frac{d}{dt}\{\gamma(t)^*\} = \left(\frac{d\gamma}{dt}\right)^*$$

and

$$0 = \frac{d}{dt}\{\gamma(t)^* \cdot \gamma(t)\} = \frac{d}{dt}\{\gamma(t)\}^* \cdot \gamma(t) + \gamma(t)^* \cdot \frac{d}{dt}\{\gamma(t)\}$$

we get

$$0 = \gamma'(t)^*\gamma(t) + \gamma(t)^*\gamma'(t) \quad \text{for all small } t.$$

Setting $t = 0$ gives $X^* = -X$, so $\mathfrak{u}(n) \subseteq \{X : X^* = -X\}$; similarly $\mathfrak{su}(n) \subseteq \{X : X^* = -X \text{ and } \mathrm{Tr}(X) = 0\}$.

For the converse inclusion ($\supseteq$), if $X^* = -X$ and $\gamma(t) = e^{tX}$ we have

$$\gamma(t)^* = e^{tX^*} = e^{-tX} = \gamma(t)^{-1} \quad \text{for all } t,$$

so $\gamma(\mathbb{R}) \subseteq \mathrm{U}(n)$. Since $\gamma(t)$ is $\mathcal{C}^\infty$, $\gamma(0) = I$, and $\gamma'(0) = X$, the tangent vector X is in $\mathfrak{u}(n)$, proving the inclusion ($\supseteq$). The same argument works for $\mathfrak{su}(n)$. □

EXERCISE 5.64. Review the preceding discussion to explain why

(a) The exponential curve $\gamma_X(t) = e^{tX}$ remains within $\mathrm{U}(n)$ if $X \in \mathfrak{u}(n)$.

(b) A vector $X \in \mathrm{M}(n, \mathbb{C})$ is in the Lie algebra $\mathfrak{u}(n) \Leftrightarrow e^{tX}$ is in $\mathrm{U}(n)$ for all $t \in \mathbb{R}$.

EXAMPLE 5.65 (Complex Orthogonal Groups $\mathrm{O}(n, \mathbb{C})$ and $\mathrm{SO}(n, \mathbb{C})$). The Lie algebras of $\mathrm{O}(n, \mathbb{C})$ and $\mathrm{SO}(n, \mathbb{C})$ are the same. Since $\det(A)$ can only take the values ± 1, $\mathrm{O}(n, \mathbb{C})$ consists of two disjoint open subsets, the one containing the identity I being $\mathrm{SO}(n, \mathbb{C})$. Thus we have

$$\mathfrak{so}(n, \mathbb{C}) = \mathfrak{o}(n, \mathbb{C}) = \{X \in \mathrm{M}(n, \mathbb{C}) : X^\mathsf{T} = -X\}.$$

This automatically implies that diagonal entries of X are zero, so $\mathrm{Tr}(X) = 0$.

DISCUSSION. If $\gamma(s)$ is a smooth curve in $\mathrm{SO}(n, \mathbb{C})$ passing through I when $s = 0$, we have $\gamma(s)\gamma(s)^\mathsf{T} \equiv I$. Applying d/ds we get

$$\gamma'(s)\gamma(s)^\mathsf{T} + \gamma(s)\gamma'(s)^\mathsf{T} = 0 \quad \text{for all } s,$$

which implies that $X^\mathsf{T} = -X$ when we set $s = 0$. Conversely if $X^\mathsf{T} = -X$, the exponential curve $\gamma(s) = e^{sX}$ has $\gamma(0) = I$, $\gamma'(0) = X$, and remains confined within $\mathrm{SO}(n, \mathbb{C})$ for all s because

$$\gamma(s)^\mathsf{T} = \left(e^{sX}\right)^\mathsf{T} = e^{sX^\mathsf{T}} = e^{-sX} = \gamma(s)^{-1}.$$

Thus $\gamma(s)\gamma(s)^\mathsf{T} = I$ for all $s \in \mathbb{R}$.

The subgroup $\mathrm{SO}(n, \mathbb{C})$ is a complex group with

$$\dim_\mathbb{C} \mathrm{SO}(n, \mathbb{C}) = \dim_\mathbb{C} \mathfrak{so}(n, \mathbb{C}) = \frac{1}{2}(n^2 - n),$$

but $\mathrm{SO}(n, \mathbb{C})$ can also be regarded as a Lie group over $\mathbb{R}$ of twice the dimension,

$$\dim_\mathbb{R} \mathfrak{so}(n, \mathbb{C}) = n^2 - n.$$

EXAMPLE 5.66 (Real Orthogonal Groups $\mathrm{O}(n)$ and $\mathrm{SO}(n)$). The groups $\mathrm{O}(n)$ and $\mathrm{SO}(n)$ are the real points in the corresponding complex groups,

$$\mathrm{O}(n) = \mathrm{O}(n, \mathbb{C}) \cap \left(\mathrm{M}(n, \mathbb{R}) + i0\right)$$
$$\mathrm{SO}(n) = \mathrm{SO}(n, \mathbb{C}) \cap \left(\mathrm{M}(n, \mathbb{R}) + i0\right).$$

They have the same Lie algebra because $\mathrm{SO}(n)$ is an open subgroup in $\mathrm{O}(n)$ (because $\det(A)$ varies continuously and $\det(A) = \pm 1$ on $\mathrm{O}(n)$). The same argument used in Example 5.65 shows that

$$\mathfrak{so}(n) = \{X \in \mathrm{M}(n, \mathbb{R}) : X^{\mathsf{T}} = -X\};$$

(and the trace $\mathrm{Tr}(X)$ is automatically zero). It is also clear that

$$\dim_{\mathbb{R}} \mathrm{SO}(n) = \dim_{\mathbb{R}} \mathfrak{so}(n) = \frac{1}{2}(n^2 - n).$$

EXAMPLE 5.67 (Symplectic Group $\mathrm{Sp}(2n, \mathbb{K})$). The Lie algebra of the symplectic group $\mathrm{Sp}(2n, \mathbb{K})$ is

$$\mathfrak{sp}(2n, \mathbb{K}) = \{X \in \mathrm{M}(2n, \mathbb{K}) : X^{\mathsf{T}} J + JX = 0\},$$

where

$$J = \begin{pmatrix} 0 & I_{n\times n} \\ -I_{n\times n} & 0 \end{pmatrix}_{2n\times 2n},$$

and $J^{\mathsf{T}} = -J = J^{-1}$. The dimensions over $\mathbb{K} = \mathbb{R}$ or $\mathbb{C}$ are

$$\dim_{\mathbb{K}} \mathfrak{sp}(2n, \mathbb{K}) = 2n^2 + n.$$

Here $2n$ is needed because, as we saw in Chapter 3, only even-dimensional vector spaces can support nondegenerate skew-symmetric bilinear forms.

Matrices in the Lie algebra $\mathfrak{sp}(2n, \mathbb{K})$ have the form

$$X = \begin{pmatrix} A & B \\ C & -A^{\mathsf{T}} \end{pmatrix}_{2n\times 2n},$$

where A, B, C are arbitrary $n \times n$ matrices such that $B^{\mathsf{T}} = B$ and $C^{\mathsf{T}} = C$.

DISCUSSION. The inclusion ($\subseteq$) is seen by differentiating the relation that defines $A \in \mathrm{Sp}(2n, \mathbb{K})$: A is nonsingular and

$$A^{\mathsf{T}} J A = J \quad (\text{or } JAJ^{-1} = (A^{\mathsf{T}})^{-1})$$

(recall Definition 3.39). Since $\det(J) \neq 0$, it follows that $\big(\det(A)\big)^2 = 1$, so $\det(A) = \pm 1$ for $A \in \mathrm{Sp}(2n, \mathbb{K})$, whether $\mathbb{K} = \mathbb{R}$ or $\mathbb{K} = \mathbb{C}$.

To prove the inclusion ($\supseteq$), we consider a $\mathcal{C}^\infty$ curve $\gamma(t)$ in $G = \mathrm{Sp}(n, \mathbb{K})$ such that $\gamma(0) = I$ and $\gamma'(0) = X$ in $\mathfrak{g}$. Differentiate the identity $\gamma(t)^{\mathsf{T}} J \gamma(t) = J$ to get

$$(\gamma'(t))^{\mathsf{T}} J \gamma(t) + \gamma(t)^{\mathsf{T}} J \gamma'(t) = 0\,;$$

setting $t = 0$ yields

$$X^{\mathsf{T}} J + JX = 0 \quad \text{or} \quad X^{\mathsf{T}} J = -JX.$$

Notice that $X^{\mathsf{T}} = -JXJ^{-1} = JXJ$ since $J^{-1} = -J$, so

$$\mathrm{Tr}(X) = \mathrm{Tr}(X^{\mathsf{T}}) = \mathrm{Tr}(JXJ) = \mathrm{Tr}(XJ^2) = -\mathrm{Tr}(X).$$

Thus $\mathrm{Tr}(X) = 0$ automatically for elements in $\mathfrak{sp}(2n, \mathbb{K})$.

To prove the reverse inclusion ($\supseteq$) consider any X such that $X^{\mathsf{T}} J = -JX$ (so $X^{\mathsf{T}} = JXJ = -JXJ^{-1}$). Then $\gamma(s) = e^{sX}$ is a $\mathcal{C}^\infty$ matrix-valued curve such that $\gamma(0) = I$, $\gamma'(0) = X$, and $\gamma(t)$ remains within $\mathrm{Sp}(n, \mathbb{K})$ because

$$J\gamma(s)J^{-1} = e^{s(JXJ^{-1})} = e^{-sX^{\mathsf{T}}} = (e^{-sX})^{\mathsf{T}} = \big(\gamma(s)^{\mathsf{T}}\big)^{-1} \quad \text{for all } s \in \mathbb{R}.$$

EXERCISE 5.68. Using the symmetry properties of matrices in $\mathfrak{sp}(2n,\mathbb{K})$ to verify that matrices in these Lie algebras have the form shown in Example 5.67. Verify that $X \in \mathrm{M}(2n,\mathbb{K})$ is in $\mathfrak{sp}(2n,\mathbb{K}) \Leftrightarrow e^{tX} \in \mathrm{Sp}(2n,\mathbb{K})$ for all $t \in \mathbb{K}$.

Some Lie Algebras Determined by Commutation Relations. Commutative Lie algebras are determined by the trivial relations $[X_i, X_j] = 0$ for any basis. The simplest noncommutative example is the Lie algebra of the "$ax+b$ *group*,"

EXAMPLE 5.69 (Lie Algebra of "$ax+b$" Group). The *affine transformations* $T_{a,b}(x) = ax + b \ (a \neq 0)$ of the real line form a group of transformations under composition that is isomorphic to the matrix group

$$G = \left\{ \begin{pmatrix} a & b \\ 0 & 1 \end{pmatrix} : a, b \in \mathbb{R}, a \neq 0 \right\}$$

under the map

$$\phi\begin{pmatrix} a & b \\ 0 & 1 \end{pmatrix} = T_{a,b},$$

(for more details see the Additional Exercises for Section 5.3). It is not hard to see this is a two-dimensional matrix Lie group. $\mathcal{C}^\infty$ curves through the identity I that determine basis vectors in the tangent space $\mathfrak{g} = \mathrm{TG}_e$ are provided by the exponential functions $\gamma(t) = e^{tX_k}$ where

$$X_1 = \begin{pmatrix} 1 & 0 \\ 0 & 0 \end{pmatrix} \quad \text{with} \quad e^{tX_1} = \begin{pmatrix} e^t & 0 \\ 0 & 1 \end{pmatrix}$$
$$X_2 = \begin{pmatrix} 0 & 1 \\ 0 & 0 \end{pmatrix} \quad \text{with} \quad e^{tX_2} = \begin{pmatrix} 1 & t \\ 0 & 1 \end{pmatrix}.$$

The tangent vectors $\gamma_k'(0) = X_k$ span a two-dimensional vector space $\mathfrak{m}$ in $\mathrm{M}(2,\mathbb{R})$; since $[X_1, X_2] = X_2$ and the Lie bracket is bilinear, it is easy to check that commutators $[A,B] = AB - BA$ of arbitrary elements in $\mathfrak{m}$ are in $\mathfrak{m}$. Thus $\mathfrak{m}$ is a Lie algebra over $\mathbb{R}$, and is precisely the two-dimensional tangent space of the $ax+b$ group.

NOTE. Although the $ax+b$ group G has been defined as a group of matrices, it is not a "matrix Lie group" in our sense because it is not a *closed* subset in matrix space $\mathrm{M}(2,\mathbb{R}) \cong \mathbb{R}^4$, although it is a Lie group in the more general sense. For instance, the diagonal matrices $A_n = [\frac{1}{n}, 0; 0, 1] \in G$ converge in matrix space to $B = [0,0;0,1]$ which is not invertible and not in G. All the classical groups are closed subgroups of matrices. ○

EXERCISE 5.70. In the Lie algebra $\mathfrak{g}$ of the $ax+b$ group, write

$$X = \begin{pmatrix} 1 & 0 \\ 0 & 0 \end{pmatrix} \quad \text{and} \quad Y = \begin{pmatrix} 0 & 1 \\ 0 & 0 \end{pmatrix}.$$

Prove that

(a) $[X, Y] = Y$.

(b) $[\mathfrak{m}, \mathfrak{m}] \subseteq \mathfrak{m}$ if $\mathfrak{m} = \mathbb{R}\text{-span}\{X, Y\}$.

(c) $\mathfrak{m}$ is contained in the tangent space TG_e of the $ax + b$ group.

(d) $\mathfrak{m}$ is equal to the Lie algebra of the $ax + b$ group.

(e) Compute the one-parameter subgroups e^{sX} and $e^{t}Y$ and their products $e^{sX}e^{tY}$. Does every element A in the $ax + b$ group have such a factorization? (What if $\det(A) < 0$?) Are such factorizations unique when $\det(A) > 0$?

EXAMPLE 5.71 (The Standard Basis in $\mathfrak{sl}(2, \mathbb{R})$). The Lie algebra of $\text{SL}(2, \mathbb{R})$,

$$\mathfrak{sl}(2, \mathbb{R}) = \left\{ \begin{pmatrix} a & b \\ c & -a \end{pmatrix} : a, b, c \in \mathbb{R} \right\},$$

has a distinguished basis

$$H = \begin{pmatrix} 1 & 0 \\ 0 & -1 \end{pmatrix} \quad E = \begin{pmatrix} 0 & 1 \\ 0 & 0 \end{pmatrix} \quad F = \begin{pmatrix} 0 & 0 \\ 1 & 0 \end{pmatrix},$$

whose commutation relations

$$[H, E] = 2E \quad [H, F] = -2F \quad [E, F] = H$$

are easily verified by direct calculation. It is clear that $\text{ad}_H(X) = [H, X]$ leaves the two-dimensional subspace $\mathfrak{m} = \mathbb{R}\text{-span}\{E, F\}$ invariant, and $\text{ad}_H|_{\mathfrak{m}}$ is diagonalizable with eigenvalues $\lambda_{\pm} = \pm 2$. The basis vectors are the derivatives $\gamma'(0)$ for the one-parameter subgroups in $\text{SL}(2, \mathbb{R})$

$$e^{tH} = \begin{pmatrix} e^t & 0 \\ 0 & e^{-t} \end{pmatrix} \quad e^{tE} = \begin{pmatrix} 1 & t \\ 0 & 1 \end{pmatrix} \quad e^{tF} = \begin{pmatrix} 1 & 0 \\ t & 1 \end{pmatrix},$$

which all lie in $\text{SL}(2, \mathbb{R})$. The matrices E, F, H are obviously linearly independent and lie in $\mathfrak{sl}(2, \mathbb{R})$, so they are a basis for this 3-dimensional Lie algebra.

EXERCISE 5.72 (Standard Basis in $\mathfrak{so}(3)$). Verify that:

(a) The matrices in $\text{M}(3, \mathbb{R})$

$$\mathbf{I} = \begin{pmatrix} 0 & 0 & 0 \\ 0 & 0 & 1 \\ 0 & -1 & 0 \end{pmatrix} \quad \mathbf{J} = \begin{pmatrix} 0 & 0 & 1 \\ 0 & 0 & 0 \\ -1 & 0 & 0 \end{pmatrix} \quad \mathbf{K} = \begin{pmatrix} 0 & 1 & 0 \\ -1 & 0 & 0 \\ 0 & 0 & 0 \end{pmatrix}$$

satisfy the commutation relations

$$[\mathbf{I}, \mathbf{J}] = \mathbf{K} \quad [\mathbf{J}, \mathbf{K}] = \mathbf{I} \quad [\mathbf{K}, \mathbf{I}] = \mathbf{J}.$$

(b) Verify that **I**, **J**, **K** are in the Lie algebra $\mathfrak{so}(3)$ of the group $\text{SO}(3)$ of orientation-preserving rotations about axes through the origin in $\mathbb{R}^3$.

(c) Compute one-parameter groups $R_1(t) = e^{t\mathbf{I}}, R_2(t) = e^{t\mathbf{J}}, R_3(t) = e^{t\mathbf{K}}$ in "closed form"– i.e. find explicit formulas for the series sums.

(d) The $R_k(t)$ are rotation operators in $\mathbb{R}^3$. In each case find the axis of rotation ℓ_k and angle of of rotation $\theta_k(t)$ when $t = 1$.

HINT. For (c) you might revisit Euler's Theorem (Theorem 6.84 in LA I).

EXERCISE 5.73 (The Standard Basis in $\mathfrak{su}(2)$). Verify that:

(a) The matrices in $\mathrm{M}(2,\mathbb{C})$

$$\mathbf{I}' = \frac{1}{2}\begin{pmatrix} i & 0 \\ 0 & -i \end{pmatrix} \quad \mathbf{J}' = \frac{1}{2}\begin{pmatrix} 0 & 1 \\ -1 & 0 \end{pmatrix} \quad \mathbf{K}' = \frac{1}{2}\begin{pmatrix} 0 & i \\ i & 0 \end{pmatrix}$$

satisfy the commutation relations

$$[\mathbf{I}',\mathbf{J}'] = \mathbf{K}' \quad [\mathbf{J}',\mathbf{K}'] = \mathbf{I}' \quad [\mathbf{K}',\mathbf{I}'] = \mathbf{J}'.$$

(b) These matrices are basis vectors for the Lie algebra $\mathfrak{su}(2)$ of the special unitary group SU(2).

(c) Find explicit formulas for the matrices in the one parameter subgroups $\gamma(t) = e^{tA}$ of SU(2) for each of the matrices $A = \mathbf{I}',\mathbf{J}',\mathbf{K}'$.

NOTE. Our labeling of the basis vectors for $\mathfrak{so}(3)$ and $\mathfrak{su}(2)$ reveals that they have the same commutation relations and structure constants. This immediately implies there is a unique Lie algebra isomorphism between SO(3) and SU(2) that sends $\mathbf{I} \to \mathbf{I}', \dots, \mathbf{K} \to \mathbf{K}'$. These commutation relations should look familiar. The same relations

$$\mathbf{i}\times\mathbf{j} = \mathbf{k} \quad \mathbf{j}\times\mathbf{k} = \mathbf{i} \quad \mathbf{k}\times\mathbf{i} = \mathbf{j}$$

govern the calculus cross product operation $\mathbf{a}\times\mathbf{b}$ on $\mathbb{R}^3$ when applied to the standard basis vectors $\mathbf{i} = (1,0,0)$, $\mathbf{j} = (0,1,0)$, $\mathbf{k} = (0,0,1)$. This is not an accident. In classical and quantum mechanics the basis vectors $\mathbf{I},\mathbf{J},\mathbf{K}$ in $\mathfrak{so}(3)$ are often described as "infinitesimal generators" of rotations about the x-axis, y-axis, and z-axis in $\mathbb{R}^3$. (See also the discussion of "Differential Forms and the Cross-Product" in Section 4.4 of these *Notes*.) ○

5.4. The Exponential Map for Matrix Lie Groups

For $\mathbb{K} = \mathbb{R}$ or $\mathbb{C}$, a matrix Lie group G is a closed subgroup in $\mathrm{GL}(n,\mathbb{K})$ that is also a smooth differentiable manifold embedded in matrix space $\mathrm{M}(n,\mathbb{K}) \cong \mathbb{K}^{n^2}$, with dimension $d = \dim(G) \le n^2$. Its Lie algebra is the tangent space $\mathfrak{g} = \mathrm{TG}_e$ at the identity in G and the other tangent spaces TG_g are left translates

$$(d\lambda_g)_e(\mathfrak{g}) = g\cdot\mathfrak{g} \quad \text{(matrix product)}.$$

The principal tools for understanding Lie groups G and their Lie algebras $\mathfrak{g}$ are the *exponential map* $\exp : \mathfrak{g} \to G$, and the *one-parameter subgroups* $\gamma_X(t) = \exp(tX)$ it determines for elements $X \in \mathfrak{g}$. Defining *exp* for general Lie groups would require a digression on integral curves (solution curves) for smooth left-invariant vector fields $\widetilde{X}$ on G — essentially a matter of solving partial differential equations on the manifold, with $\widetilde{X}$ playing the role of the differential operator. The exponential is simpler for matrix Lie groups, so they will be the focus of these *Notes*. For such groups, $\mathfrak{g}$ is a subspace in $\mathrm{M}(n,\mathbb{K})$ and *exp* turns out to be the restriction to $\mathfrak{g} = \mathrm{TG}_e$ of the classical exponential map for matrices,

$$\mathrm{Exp}(A) = e^A = \sum_{n=0}^{\infty} \frac{1}{n!}A^n \quad (A \in \mathrm{M}(m,\mathbb{K})). \tag{5.17}$$

Our whole discussion will rest on the fact that the subgroups $\gamma_X(t) = e^{tX}$ generated by elements $X \in \mathrm{M}(n, \mathbb{K})$ *remain within G for all $t \in \mathbb{K}$* if and only if $X \in \mathfrak{g}$. This was true for each of the classical groups, but for general matrix Lie groups this is not evident from the definition of $\mathfrak{g} = \mathrm{TG}_e$, which says:

> $X \in \mathrm{TG}_e \Leftrightarrow X = \gamma'(0)$ *for some $\mathcal{C}^\infty$ curve in* $\mathrm{M}(n, \mathbb{K})$ *defined near $t = 0$, such that $\gamma(0) = I$ and $\gamma(t) \in G$ for all t.*

For any matrix X the curves $\gamma_X(t) = e^{tX}$ from $\mathbb{R}$ into $\mathrm{GL}(n, \mathbb{K})$ are obviously $\mathcal{C}^\infty$ maps defined for all $t \in \mathbb{R}$. Once we know they remain confined to G when $X \in \mathfrak{g}$ we can use *Exp* to impose "exponential coordinates" that identify points near the identity in G with an open subset of vector space $\mathfrak{g}$ with dimension $d = \dim(G)$. Our first task is to resolve this issue for $\mathrm{Exp} : \mathrm{M}(n, \mathbb{K}) \to \mathrm{GL}(n, \mathbb{K})$, and then for the restrictions $\exp = \mathrm{Exp}|_{\mathfrak{g}} : \mathfrak{g} \to G$ for arbitrary matrix Lie groups.

One-Parameter Subgroups in Matrix Lie Groups. For $\mathbb{K} = \mathbb{R}$ or $\mathbb{C}$, the exponential map (5.17) sends $\mathrm{M}(n, \mathbb{K})$ into $\mathrm{GL}(n, \mathbb{K})$, but is not surjective or one-to-one. That presents some problems when we try to define logartihms $\mathrm{Log}(A)$ of matrices in $\mathrm{GL}(n, \mathbb{K})$. As maps $\mathbb{K}^{n^2} \to \mathbb{K}$, matrix entries $(e^A)_{ij}$ are $\mathcal{C}^\infty$ functions of the entries in A, so *Exp* is a $\mathcal{C}^\infty$ map from $\mathrm{M}(n, \mathbb{K}) \to \mathrm{M}(n, \mathbb{K})$ with range in $\mathrm{GL}(n, \mathbb{K})$.[4]

Each matrix $X \in \mathrm{M}(n, \mathbb{K})$ determines a *one-parameter subgroup*

$$\gamma_X(t) = \mathrm{Exp}(tX) = e^{tX} \quad (t \in \mathbb{R}).$$

This $\mathcal{C}^\infty$ map is also a group homorphism $(\mathbb{R}, +) \to (\mathrm{GL}(n, \mathbb{K}), \cdot\,)$ because

$$\gamma_X(s+t) = \gamma_X(s) \cdot \gamma_X(t) \qquad \gamma_X(0) = I_{n\times n} \quad \text{(the identity matrix)},$$

by the exponent laws. Thus *Exp* maps radial lines $\mathbb{R} \cdot X$ through the origin in matrix space to (nonlinear) one-dimensional subgroups in $\mathrm{GL}(n, \mathbb{K})$ emanating from the identity element I. In Section 5.1 of LA I we saw that

$$\frac{d\gamma_X}{dt}(t) = X \cdot \gamma_X(t) \quad \text{(matrix product)}$$

for all t, and in particular the tangent vector at $I = \gamma_X(0)$ is $\gamma_X'(0) = X$.

EXERCISE 5.74. Compute the one-parameter groups $\gamma_X(t) = e^{tX}$ in closed form for the following matrices (evaluate the series sum).

$$\text{(a) } A_1 = \begin{pmatrix} -1 & 0 \\ 0 & -1 \end{pmatrix} \quad \text{(b) } A_2 = \begin{pmatrix} -1 & 0 \\ 0 & 1 \end{pmatrix} \quad \text{(c) } A_3 = \begin{pmatrix} 0 & 1 \\ -1 & 0 \end{pmatrix}$$

$$\text{(d) } A_4 = \begin{pmatrix} 0 & 1 \\ 1 & 0 \end{pmatrix} \qquad \text{(e) } A_5 = \begin{pmatrix} i & 1 \\ -1 & 0 \end{pmatrix}$$

where $i = \sqrt{-1}$.

---

[4] When we identify $\mathrm{M}(n, \mathbb{R}) \cong \mathbb{R}^{n^2}$ the matrix entries are actually *real-analytic* functions of the n^2 entries in A, with absolutely convergent multivariate Taylor series expansions at every base point.

The Logarithm Log(A) of a Matrix. Recall that $e^{A+B} = e^A \cdot e^B$ for commuting matrices. The following complex power series are convergent on the domains indicated:

$$\text{Exp}(z) = \sum_{k=0}^{\infty} \frac{1}{k!} z^k \qquad \text{for all } z \in \mathbb{C}$$

$$\text{Log}(z) = \sum_{k=1}^{\infty} \frac{(-1)^{k+1}}{k!} (z-1)^k \qquad \text{for } |z-1| < 1.$$

By grouping together even and odd powers in the series

$$\text{Exp}(iy) = 1 + iy - \frac{1}{2!}y^2 - i\frac{1}{3!}y^3 + \cdots$$

for real y, we can verify the basic computational identity

$$e^{x+iy} = e^x \cdot e^{iy} = e^x \cos(y) + ie^x \sin(y) \quad \text{for } x, y \in \mathbb{R}.$$

Then $e^{2\pi in} = 1$ for all n, so the complex exponential function $f(z) = e^z$ is periodic under translations by $\Delta z = 2\pi i$.

These series are norm convergent when applied to matrices of sufficiently small norm. The series Exp(X) converges for all X in any norm, while Log(A) converges when $\|A - I\| < 1$ for any norm such that $\|AB\| \leq \|A\| \cdot \|B\|$, for instance the operator norm $\|A\|_{op}$ introduced in Section 5.3 of LA I. By wrangling with absolute norm convergence of series one can verify that the matrix-valued versions of these series behave very much like their scalar-valued counterparts, with

$$\begin{aligned} &\text{Log}(\text{Exp}(X)) = X \quad \text{for } \|X\| < \log 2 \\ &\text{Exp}(\text{Log}(A)) = A \quad \text{for } \|A - I\| < 1. \end{aligned} \tag{5.18}$$

When X is a nilpotent matrix the series Exp(X) obviously reduces to a finite sum, and so does Log(A) when $A = I + N$ is unipotent, and these identities hold for all X and all $A \neq 0$.This becomes important in dealing with Jordan forms of complex matrices, which consist of Jordan blocks $B_k = \lambda_k I + N_k$ on the diagonal, where N_k is an elementary nilpotent matrix.

EXAMPLE 5.75. If λ is a nonzero complex number, we will show that

1. When $\lambda \neq 0$ the Jordan block

$$A = \lambda I + N = \begin{pmatrix} \lambda & 1 & 0 & \cdots & \cdots & 0 \\ 0 & \lambda & 1 & \cdots & \cdots & 0 \\ & 0 & \ddots & \ddots & & \vdots \\ \vdots & & & \ddots & 1 & 0 \\ & & & & \lambda & 1 \\ 0 & & \cdots & & 0 & \lambda \end{pmatrix}_{n \times n}$$

is the exponential of some complex matrix X.

2. The Jordan form can be used to conclude that every invertible complex matrix A is e^X for some X in M($n, \mathbb{C}$), so *Exp* is *surjective* from M($n, \mathbb{C}$) to GL($n, \mathbb{C}$).

3. There are no solutions in $\mathrm{M}(n, \mathbb{C})$ to the equation $\mathrm{Exp}(X) = 0_{n\times n}$. Nor are there any matrix solutions to $\mathrm{Exp}(X) = N$ when N is a nilpotent matrix.

DISCUSSION. In Part 1, we have

$$\begin{aligned}\lambda I + N &= \lambda \cdot \left(I + \frac{N}{\lambda}\right) = e^{\log(\lambda)} \cdot e^{\mathrm{Log}\left(I+\frac{N}{\lambda}\right)} \\ &= e^{\log(\lambda)} \cdot \mathrm{Exp}\left(\frac{N}{\lambda} - \frac{1}{2}\left(\frac{N}{\lambda}\right)^2 + \cdots + \frac{(-1)^n}{(n-1)} \cdot \left(\frac{N}{\lambda}\right)^{n-1}\right) \\ &= \mathrm{Exp}(\log(\lambda)I + (\cdots))\end{aligned}$$

so $(\lambda I + N)$ is in the range of *Exp*.

If $A \in \mathrm{GL}(n, \mathbb{C})$,some conjugate SAS^{-1} will be in Jordan form, consisting of Jordan blocks on the diagonal,

$$SAS^{-1} = \mathrm{diag}(B_1, \dots, B_r).$$

Each block is $B_k = \lambda_k I + N_k$ with N_k elementary nilpotent, and corresponds to an L_A-invariant subspace E_k in $\mathbb{C}^n$, with a particular choice of basis $\mathfrak{X}_k$. For the basis $\mathfrak{X} = \mathfrak{X}_1 \cup \cdots \cup \mathfrak{X}_r$ we have $[L_A]_{\mathfrak{X}\mathfrak{X}} = \mathrm{diag}(B_1, \dots, B_r)$. If $\dim(E_k) = m_k$, each $B_k = e^{C_k}$ for some $m_k \times m_k$ matrix C_k, and if $C = \mathrm{diag}(C_1, \dots, C_r)$ we have

$$e^C = \mathrm{diag}(e^{C_1}, \dots, e^{C_r}) = \mathrm{diag}(B_1, \dots, B_r) = SAS^{-1}.$$

Since $S^{-1}\mathrm{GL}(n, \mathbb{C})S = \mathrm{GL}(n, \mathbb{C})$, A is in the range of *Exp*.

As for Item 3, if $e^X = 0$ for some X we would have $I = e^X \cdot e^{-X} = 0$, which is impossible. If N is nilpotent and $N = e^X$ then $0 = N^d = e^{dX}$ and again, no such matrix $d \cdot X$ or X can exist. In particular, the claim proved in Part 1 cannot hold if $\lambda = 0$ (so A is an elementary nilpotent matrix). ○

REMARK 5.76. Any complex number $z \neq 0$ has a logarithm $w = \log(z)$ such that $e^w = z$; in fact there are infinitely many, given by

$$\log(z) = \log|z| + i\theta + 2\pi i n \qquad n \in \mathbb{Z},$$

where $z = |z|e^{i\theta}$ is the polar form of z. If $z = r + i0$ is real and negative, it cannot have a *real* logarithm because $e^t > 0$ for all real t.

If you re-examine the proof in Part 1 of 5.75 you will find that:

> *If λ is nonzero and real, the only obstacle to finding a real matrix such that* $\mathrm{Exp}(X) = \lambda I + N$ *arises when the logarithm* $\log(\lambda)$ *is nonreal, which happens for instance when $\lambda = -1$.*

The blocks in the Jordan form of any complex matrix A have the form $B = \lambda I$, or $B = \lambda I + N$, so the *only* obstacle to finding a real matrix X such that $e^X = A$ for a nonsingular matrix $A \in \mathrm{M}(n, \mathbb{R})$ is the presence of at least one non-real eigenvalue when A is viewed as a complex matrix, since that would produce a non-real constant term $\log(\lambda)$ in the solution. ○

EXAMPLE 5.77. Understanding the map $\mathrm{Exp} : \mathrm{M}(n.\mathbb{K}) \to \mathrm{GL}(n, \mathbb{K})$ for $\mathbb{K} = \mathbb{C}$ or $\mathbb{R}$ begins with finding all solutions of the matrix equation

$$\mathrm{Exp}(X) = e^X = I_{n\times n} \quad (X \in \mathrm{M}(n, \mathbb{K})),$$

as well as the equation $e^X = 0_{n\times n}$. The only solutions of $e^X = I$ in $\mathrm{M}(n,\mathbb{C})$ are

$$X = \mathrm{diag}(2\pi i k_1, \ldots, 2\pi k_r) \quad \text{for } k_1, \ldots, k_r \in \mathbb{Z}.$$

Real solutions (if any) can be read out of this by inspection.

DISCUSSION. We get the Jordan form of X by a similarity transformation $J(X) = SXS^{-1}$; it too is a solution because $SIS^{-1} = I$. But $J(X) = \mathrm{diag}(B_1, \ldots, B_r)$ consists of Jordan blocks, and $e^{J(X)} = \mathrm{diag}(e^{B_1}, \ldots, e^{B_r})$, so we are reduced to solving $e^B = I_{d\times d}$ for a Jordan block $B = \lambda I + N$ with $\lambda \neq 0$. Then we have

$$I = e^B = e^{\lambda}I \cdot e^N = e^{\lambda} \cdot \begin{pmatrix} 1 & \frac{1}{1!} & \frac{1}{2!} & \frac{1}{3!} & \cdots & \frac{1}{(d-1)!} \\ 0 & 1 & \frac{1}{1!} & \frac{1}{2!} & \cdots & \frac{1}{(d-2)!} \\ & & \ddots & & & \\ & & & \ddots & & \\ & & & & \ddots & \\ 0 & & & & & 1 \end{pmatrix}_{d\times d}.$$

This cannot happen unless $d = 1$ and each $\lambda = 2\pi i k$ for some $k \in \mathbb{Z}$. Thus the blocks B_j in a soution of $e^X = I$ are all 1×1 but the entries $2\pi k_j\sqrt{-1}$ may differ from cell to cell. The solutions are the diagonal matrices

$$X = \mathrm{diag}(2\pi i k_1, \ldots, 2\pi i k_r)$$

with $k_j \in \mathbb{Z}$. ○

EXERCISE 5.78. Use the JCF to prove that $\mathrm{Exp} : \mathrm{M}(n,\mathbb{R}) \to \mathrm{GL}(n,\mathbb{R})$ need not be surjective by showing that

$$A = \begin{pmatrix} -1 & 1 \\ 0 & -1 \end{pmatrix}$$

is not in the range of the exponential map.

Singularities of the Exponential Map on $\mathbf{M}(n,\mathbb{C})$. Next we show that the differential $d(\mathrm{Exp})_0$ of the exponential map from $\mathrm{M}(n,\mathbb{K})$ into $\mathrm{GL}(n,\mathbb{K}) \subseteq \mathbb{K}^{n^2}$ has nonzero Jacobian determinant at the origin in matrix space. Existence of a locally defined $\mathcal{C}^\infty$ logarithm *Log* that inverts *Exp* near the zero matrix is then an immediate consequence of the inverse mapping theorem 5.10.

PROPOSITION 5.79 (Local Inverses of *Exp*). *There are open neighborhoods U about the zero matrix in $\mathrm{M}(n,\mathbb{K})$, and V about the identity $I_{n\times n}$ in $\mathrm{GL}(n,\mathbb{K})$, such that* $\mathrm{Exp} : U \to V$ *is bijective and $\mathcal{C}^\infty$ in both directions. The locally defined inverse* $\mathrm{Log}(A)$ *is defined on* $V = \mathrm{Exp}(U)$, *with* $\mathrm{Log}(I_{n\times n}) = 0_{n\times n}$.

PROOF. Since $e^X = \sum_{k=0}^{\infty} X^k/k!$ is absolutely norm convergent it is easily verified that the matrix entries $(e^X)_{ij}$ are $\mathcal{C}^\infty$ functions of the entries x_{ij} in X, so *Exp* is a $\mathcal{C}^\infty$ map on matrix space. By convergence of the exponential series, simple norm estimates (as in Chapter 5 of LA I) show that

$$\mathrm{Exp}(X) = I + X + o(\|X\|) \quad \text{with} \quad \lim_{X\to 0} \frac{o(\|X\|)}{\|X\|} = 0,$$

so the differential $(d\text{Exp})$ at $X = 0$ is the identity map I. Thus $\det(d\text{Exp})_0 = 1$ and *Exp* is smoothly invertible near $X = 0$ in $\text{M}(n, \mathbb{K}) = \mathfrak{gl}(n, \mathbb{K})$ by Theorem 5.10 □

This is of course a local result about behavior of $\text{Exp} : \text{M}(n, \mathbb{K}) \to \text{GL}(n, \mathbb{K})$ near $X = 0$. But with further effort we can compute the differential of *Exp* at any base point in $\text{M}(n, \mathbb{K})$, see equation (5.22) below. Then $(d\text{Exp})_X$ is singular precisely at the base points X where $\det(d\text{Exp})_X = 0$, and finding these singular points becomes a straightforward problem in linear algebra.

In order to derive such formulas we need a few more basic facts about the maps *Exp, Ad*, and *ad* associated with a matrix Lie group G and its Lie algebra $\mathfrak{g}$, so we digress momentarily to derive a crucial result relating vectors X in the Lie algebra to one-parameter subgroups $\gamma_X(t) = \text{Exp}(tX)$ in a matrix Lie group G. The analytic estimates involved are related to the calculus proof that

$$\lim_{n\to\infty} \left(1 + \frac{t}{n}\right)^n = e^t$$

THEOREM 5.80. *Let G be a matrix Lie group and $\gamma(t)$ a $\mathcal{C}^\infty$ curve in G defined on some interval $|t| < r$. If $\gamma(0) = I$ and $\gamma'(0) = X$, so X is in the Lie algebra $\mathfrak{g} = \text{TG}_e$, then the one-parameter group*

$$\gamma_X(t) = \text{Exp}(tX) \text{ is equal to } \lim_{k\to\infty} \gamma\left(\frac{t}{k}\right)^k,$$

and remains within G for all $t \in \mathbb{R}$.

PROOF. The last part follows from the first because G is a group and we have $\gamma_X(s + t) = \gamma_X(s) \cdot \gamma_X(t)$. In fact, since $\gamma_X(-r, r) \subseteq G$ the exponent law $\gamma_X(s + t) = \gamma_X(s) \cdot \gamma_X(t)$ implies that $\gamma_X(-2r, 2r) = \gamma_X(-r, r)^2 \subseteq G$, etc., so $\gamma_X(\mathbb{R}) \subseteq G$.

Now choose $\delta > 0$ so that $\gamma(t) \in V$ (the neighborhood of I in our theorem) for $|t| < 2\delta$; then $Z(t) = \text{Log}(\gamma(t))$ is a $\mathcal{C}^\infty$ curve with with $Z(0) = 0$. By the chain rule

$$Z'(0) = (d\text{Exp})_0^{-1} \cdot \gamma'(0) = I \cdot X = X,$$

so by Taylor's formula

$$Z(t) = tX + o(t^2) \quad \text{as } t \to 0.$$

For fixed X, if we replace t by t/k for a fixed $k \in \mathbb{N}$, the last identity becomes

$$Z\left(\frac{t}{k}\right) = \frac{t}{k}X + o\left(\frac{t^2}{k^2}\right),$$

hence there exists $\epsilon > 0$ such that

$$kZ\left(\frac{t}{k}\right) = tX + o\left(\frac{1}{k}\right) \quad \text{for } |t| < \epsilon \text{ and } k \in \mathbb{N}.$$

Therefore

$$\gamma\left(\frac{t}{k}\right)^k = \left(\text{Exp}\, Z\left(\frac{t}{k}\right)\right)^k = \text{Exp}\left(kZ\left(\frac{t}{k}\right)\right) = \text{Exp}\left(tX + o\left(\frac{1}{k}\right)\right)$$

is in G for all k and $|t| < \epsilon$. Now let $k \to \infty$; by continuity of *Exp* we conclude that $e^{tX} \in G$ for $|t| < \epsilon$. □

EXERCISE 5.81. Whether $\mathbb{K} = \mathbb{R}$ or $\mathbb{C}$, prove that

$$\lim_{n\to\infty}\left(I + \frac{X}{n}\right)^n = e^X \quad \text{in } \mathrm{GL}(n, \mathbb{K})$$

for any $X \in \mathrm{M}(n, \mathbb{K})$.

HINT. The group $\mathrm{GL}(n, \mathbb{K})$ with Lie algebra $\mathfrak{gl}(n, \mathbb{K}) = \mathrm{M}(n, \mathbb{K})$ is an *open set* in matrix space, so it contains all points A near the identity I (recall Proposition 5.41 and Exercise 5.43 of LA I). Now apply Theorem 5.80 to the curve $\gamma(t) = I + tX$ for small $t \in \mathbb{R}$ in GL.

When $\mathbb{K} = \mathbb{C}$ an instructive alternative proof of the identity in Exercise 5.81, can be obtained using the JCF. When $X = \mathrm{diag}((a_1, \ldots, a_n)$ is diagonal the result follows immediately from the classical calculus result just mentioned. If X is diagonalizable, with $SXS^{-1} = D$, then

$$S\left(I + \frac{X}{n}\right)^n S^{-1} = \left(I + \frac{SXS^{-1}}{n}\right)^n \to Se^D S^{-1} \quad \text{as } n \to \infty.$$

Applying the inverse similarity transformation $B \to S^{-1}BS$ to both sides of this identity, we get

$$\left(I + \frac{X}{n}\right)^n \to S^{-1}e^D S = e^X \quad \text{as } n \to \infty.$$

We leave it as an exercise for the reader to carry out the final step, in which X is not diagonalizable over $\mathbb{C}$ and the JCF must be invoked.

Relation Between Exp : $\mathfrak{gl} \to$ GL and Its Restriction exp : $\mathfrak{g} \to$ G. Theorem 5.80 yields a precise description of the vectors $X \in \mathrm{M}(n, \mathbb{K})$ in the Lie algebra $\mathfrak{g}$ of a matrix Lie group G. The following corollary shows, for any such group, that *Exp* maps $\mathfrak{g}$ into G, so the restriction $\exp_G = \mathrm{Exp}|_{\mathfrak{g}}$ (usually written as "*exp*" with G and $\mathfrak{g}$ understood) is a well-defined map from $\mathfrak{g} \to G$ that can be regarded as the "intrinsic exponential map" on the Lie algebra of a matrix Lie group.

COROLLARY 5.82. *For $\mathbb{K} = \mathbb{R}$ or $\mathbb{C}$, let G be a matrix Lie group in* $\mathrm{GL}(n, \mathbb{K})$. *Then*

1. *The Lie algebra of G in* $\mathrm{M}(n, \mathbb{K})$ *is*

$$\mathfrak{g} = \{X \in \mathrm{M}(n, \mathbb{K}) : \mathrm{Exp}(tX) \in G \textit{ for all } t \in \mathbb{R}\}.$$

2. *The restricted exponential map* $\exp = \mathrm{Exp}|_{\mathfrak{g}}$ *sends $\mathfrak{g}$ into G, and since* $(d\mathrm{Exp})_0$ *is the identity map on* $\mathfrak{gl}(n, \mathbb{K})$ *the differential* $(d\exp)_0$ *of its restriction to $\mathfrak{g}$ is the identity map on $\mathfrak{g}$. Thus* $\exp : \mathfrak{g} \to G$ *is a locally invertible $\mathcal{C}^\infty$ map whose inverse is a $\mathcal{C}^\infty$ map with respect to the manifold structure on G.*

PROOF. The Lie algebra $\mathfrak{g}$ can be identified with the manifold $M = \mathbb{R}^m$ ($m = \dim(G)$), and by Theorem 5.80 *exp* is a $\mathcal{C}^\infty$ map from M into the manifold G of the same dimension. We have noted that the differential $(d\exp)_0$ at the zero element in $\mathbb{R}^m$ is the identity map (hence nonsingular), so by the inverse mapping Theorem 5.10 *exp* maps an open neighborhood of zero $U \subseteq M$ bijectively to an open neighborhood V of the identity element $e \in G$. In particular, the inverse map $\log = \big(\exp|_U\big)^{-1}$ is a $\mathcal{C}^\infty$ map with respect to the manifold structures in G and $M = \mathbb{R}^m$. □

Case Study: Nilpotent Lie Groups. In the next stages of Lie theory (beyond the scope of these *Notes*) the goal is to understand the structure of a Lie algebra $\mathfrak{g}$ and the various connected Lie groups associated with it. This begins with investigations of *Lie subalgebras* $\mathfrak{h} \subseteq \mathfrak{g}$, vector subspaces that are closed under formation of commutators $[X, Y]$, and so are Lie algebras in their own right. This could be complicated business, but there is one venue in which the whole theory of Lie groups takes a particularly simple form: the study of *nilpotent Lie groups* and Lie algebras. In this subsection we will give examples illustrating the particularly simple form many concepts take for these groups.

Every Lie algebra determines a sequence of subalgebras

$$\mathfrak{g} = \mathfrak{g}^{(1)} \supseteq \mathfrak{g}^{(2)} \supseteq \mathfrak{g}^{(3)} \supseteq \cdots$$

defined inductively taking linear combinations of iterated commutators

$$[X_1, [X_2, \dots, [X_r, X_{r+1}] \dots]]$$

of vectors in $\mathfrak{g}$:

$$\begin{aligned} \mathfrak{g}^{(1)} &= \mathfrak{g} \\ \mathfrak{g}^{(2)} &= [\mathfrak{g}, \mathfrak{g}] = \mathbb{K}\text{-span}\{[X, Y] : X, Y \in \mathfrak{g}\} \\ \mathfrak{g}^{(3)} &= [\mathfrak{g}, \mathfrak{g}^{(2)}] = \mathbb{K}\text{-span}\{[X, Y] : X \in \mathfrak{g}, Y \in \mathfrak{g}^{(2)}\} \\ &\vdots \\ \mathfrak{g}^{(k+1)} &= [\mathfrak{g}, \mathfrak{g}^{(k)}] = \mathbb{K}\text{-span}\{[X, Y] : X \in \mathfrak{g}, Y \in \mathfrak{g}^{(k)}\} \\ &\vdots \end{aligned} \tag{5.19}$$

DEFINITION 5.83. A Lie algebra over $\mathbb{K}$ is *nilpotent* if this commutator series dies after finitely many steps, with

$$\mathfrak{g} = \mathfrak{g}^{(1)} \supsetneqq \mathfrak{g}^{(2)} \supsetneqq \cdots \supsetneqq \mathfrak{g}^{(r)} \supsetneqq \mathfrak{g}^{(r+1)} = \{0\}.$$

We leave it as an Additional Exercise for Section 5.4 to show that the subspaces $[\mathfrak{g}, \mathfrak{g}^{(k)}]$ are decreasing and closed under formation of commutators $[A, B]$, so they are all Lie subalgebras in $\mathfrak{g}$; furthermore, $\text{ad}_X : \mathfrak{g} \to \mathfrak{g}$ is a nilpotent operator. It is possible to have $[\mathfrak{g}, \mathfrak{g}] = \mathfrak{g}$ at the first step, and then the terms in the commutator series are constant. (This is true for the Lie algebras $\mathfrak{g} = \mathfrak{sl}(n, \mathbb{R}), \mathfrak{so}(n), \mathfrak{u}(n)$ for $n \geq 2$, and many other classical groups.) Finally, we say that a connected Lie group G is *nilpotent* if its Lie algebra $\mathfrak{g}$ is nilpotent.

The *Heisenberg groups* H_n are nilpotent Lie groups whose algebraic properties underlie the Heisenberg uncertainty principle for n-particle systems in quantum mechanics. They can be modeled as matrix Lie groups with dimensions $\dim_{\mathbb{R}} H_n = 2n+1$, but their basic features are clearly suggested by what happens when $n = 1$ and $\dim_{\mathbb{R}}(G) = 3$, a case we now examine in detail.

EXAMPLE 5.84 (The Heisenberg Group H_3). The lowest order Heisenberg group is the matrix group

$$G = \left\{ \begin{pmatrix} 1 & x & z \\ 0 & 1 & y \\ 0 & 0 & 1 \end{pmatrix} : x, y, z \in \mathbb{R} \right\}.$$

Although G is embedded in a 9-dimensional matrix space its dimension as a Lie group is $\dim(G) = 3$ because its elements are parametrized by the variables x, y, z. We shall indicate elements of G by writing

$$(x, y, z) = \begin{pmatrix} 1 & x & z \\ 0 & 1 & y \\ 0 & 0 & 1 \end{pmatrix},$$

with $I = (0,0,0)$. The differentiable structure on G is determined by a single chart (U_α, x_α) with $U_\alpha =$ (all of G), $V_\alpha = x_\alpha(U_\alpha) = \mathbb{R}^3$, and chart map

$$x_\alpha \begin{pmatrix} 1 & x & z \\ 0 & 1 & y \\ 0 & 0 & 1 \end{pmatrix} = (x, y, z) \quad \text{in } \mathbb{R}^3.$$

The group multiplication law in G is an easily computed *polynomial* in the parameters x, y, z,

$$(x, y, z) \cdot (x', y', z') = (x + x', y + y', z + z' + xy')$$

We leave it as an exercise to show that the inverse of an element in G is also described by a polynomial, with $(x, y, z)^{-1} = (-x, -y, -z + xy)$.

In the Lie algebra $\mathfrak{g}$ we can take basis vectors

$$X = \begin{pmatrix} 0 & 1 & 0 \\ 0 & 0 & 0 \\ 0 & 0 & 0 \end{pmatrix} \quad Y = \begin{pmatrix} 0 & 0 & 0 \\ 0 & 0 & 1 \\ 0 & 0 & 0 \end{pmatrix} \quad Z = \begin{pmatrix} 0 & 0 & 1 \\ 0 & 0 & 0 \\ 0 & 0 & 0 \end{pmatrix},$$

which are the derivatives $\gamma'(0)$ of the $\mathcal{C}^\infty$ curves

$$\gamma_1(t) = (t, 0, 0) \quad \gamma_2(t) = (0, t, 0) \quad \gamma_3(t) = (0, 0, t)$$

in G. The one-parameter groups determined by these tangent vectors are

$$\gamma_X(t) = e^{tX} = I + tX + \frac{t^2}{2!}X^2 + \cdots = I + tX$$

since $X^2 = 0$; similarly

$$\gamma_Y(t) = I + tY = (0, t, 0) \quad \text{and} \quad \gamma_Z(t) = I + tZ = (0, 0, t).$$

From this we get the one nontrivial commutation relation $[X, Y] = Z$ between the basis vectors X, Y, Z in $\mathfrak{g}$; the general commutator $[A, B]$ can be found using bilinearity of the Lie bracket. We prove $[X, Y] = Z$ using the construction presented in Proposition 5.57.

DISCUSSION. Consider the product of one-parameter subgroups

$$f(s,t) = \gamma_X(s)\gamma_Y(t)\gamma_X(s)^{-1} = e^{sX}e^{tY}e^{-sX}$$

with $\gamma_X'(0) = X, \gamma_Y'(0) = Y$ in $\mathfrak{g}$. For any matrix Lie group, we have shown that the commutator $[X, Y]$ of elements in $\mathfrak{g}$ is given by

$$[X,Y] = \left.\frac{\partial^2 f}{\partial s \partial t}\right|_{s=t=0}.$$

Applying $\partial/\partial s$ we get

$$\frac{\partial f}{\partial s}(s,t) = Xe^{sX} \cdot e^{tY}e^{-sX} + e^{sX}e^{tY} \cdot (-Xe^{sX}).$$

Setting $s = 0$ this becomes $Xe^{tY} - e^{tY}X$; then applying $\partial/\partial t$ and setting $t = 0$ we get $XY - YX = [X, Y]$. But $f(s, t)$ can be written another way using the multiplication law in G,

$$f(s,t) = (s,0,0) \cdot (0,t,0) \cdot (-s,0,0) = (s,t,st) \cdot (-s,0,0) = (0,t,st)$$

from which we get

$$\frac{\partial f}{\partial s}(s,t) = (0,0,t) = tZ \quad \text{and} \quad \left.\frac{\partial^2 f}{\partial s \partial t}\right|_{s=t=0} = Z.$$

Hence $[X, Y] = Z$, $[Y,X] = -Z$, and all other structure constants are zero.

EXERCISE 5.85. Show that the exponential map $\mathrm{Exp} : \mathfrak{g} \to G$ has the form

$$\mathrm{Exp}(aX + bY + cZ) = \mathrm{Exp}\begin{pmatrix} 0 & a & c \\ 0 & 0 & b \\ 0 & 0 & 0 \end{pmatrix} = \begin{pmatrix} 1 & a & \left(a + \frac{1}{2}ab\right) \\ 0 & 1 & b \\ 0 & 0 & 1 \end{pmatrix},$$

for $a, b, c \in \mathbb{R}$. Note that $A^3 = 0$ for the strictly upper triangular matrices in $\mathfrak{g}$, so the exponential series is finite.

EXERCISE 5.86. Show that the exponential map $\mathrm{Exp} : \mathfrak{g} \to G$ has a globally defined $\mathcal{C}^\infty$ inverse $\mathrm{Log} : G \to \mathfrak{g}$ given by

$$\mathrm{Log}\begin{pmatrix} 1 & x & z \\ 0 & 1 & y \\ 0 & 0 & 1 \end{pmatrix} = \begin{pmatrix} 0 & x & \left(z - \frac{1}{2}xy\right) \\ 0 & 0 & y \\ 0 & 0 & 0 \end{pmatrix},$$

for $x, y, z \in \mathbb{R}$.

Finally we show how to compute the smooth vector field $\widetilde{H}$ on G determined by an arbitrary vector $H = aX + by + cZ$ in the Lie algebra $\mathfrak{g}$. With respect to the chart coordinates $x_\alpha(g) = (x, y, z)$ the basis vectors are

$$X = \left(\left.\frac{\partial}{\partial x}\right|_e\right), \dots, Z = \left(\left.\frac{\partial}{\partial z}\right|_e\right),$$

where $e = I$.

EXAMPLE 5.87. The left-invariant vector field $\widetilde{H}$ on G determined by a tangent vector $H = aX+bY+cZ$ can be described in terms of the chart coordinates and the fields of basis vectors

$$(\partial/\partial x), \dots, (\partial/\partial z)$$

they induce throughout the chart domain $U_\alpha = G$.

DISCUSSION. Regarding tangent vectors in $\mathfrak{g}$ as derivations on the local algebra $\mathcal{C}^\infty(e)$ at the identity in G, $\widetilde{H}$ becomes a partial differential operator on the space $\mathcal{C}^\infty(G)$ of smooth globally defined functions on G and we have

$$\begin{aligned}\widetilde{H}f(x,y,z) &= \langle \widetilde{H}_{(x,y,z)}, f\rangle = \langle (d\lambda_{(x,y,z)})_e H, f\rangle \\ &= \langle H, f\circ\lambda_{(x,y,z)}\rangle \qquad \text{(by definition of } (d\lambda_g)_e \text{ on } \mathfrak{g}) \\ &= \langle aX+bY+cZ\,,\ f\circ\lambda_{(x,y,z)}\rangle \\ &= a\cdot\frac{\partial}{\partial x}(f\circ\lambda_{(x,y,z)})(0,0,0) + \cdots + c\cdot\frac{\partial}{\partial z}(f\circ\lambda_{(x,y,z)})(0,0,0)\end{aligned}$$

By the multiplication law we obtain

$$(f\circ\lambda_{(x,y,z)})(x',y',z') = f(x+x', y+y', z+z'+xy').$$

Take partial derivatives $(\partial/\partial x'), \dots, (\partial/\partial z')$ of this expression and set $x' = y' = z' = 0$ to get

$$\begin{aligned}\widetilde{H}f(x,y,z) &= a\cdot\Big[\frac{\partial f}{\partial x}(x,y,z)\Big] + b\cdot\Big[\frac{\partial f}{\partial y}(x,y,z)\Big] + c\cdot\frac{\partial f}{\partial z}(x,y,z) \\ &= \Big\langle\Big(a\frac{\partial}{\partial x} + b\frac{\partial}{\partial y} + (bx+c)\frac{\partial}{\partial z}\Big)\Big|_{(x,y,z)}, f\Big\rangle.\end{aligned}$$

Conclusion: at any base point $g\in G$ the value of the left-invariant field $\widetilde{H}$ is

$$\widetilde{H}_g = a\Big(\frac{\partial}{\partial x}\Big|_g\Big) + b\Big(\frac{\partial}{\partial y}\Big|_g\Big) + (b+cx)\Big(\frac{\partial}{\partial z}\Big|_g\Big)$$

if $g = (x,y,z)$.

The same ideas apply for any Lie group, but when several charts are needed to cover G we would have to do this calculation for each chart. ○

The Maps *Exp*, *Ad*, *ad* Revisited. Almost everything we need to know about these maps for general matrix Lie groups follows from the corresponding properties of the general linear group $G = \mathrm{GL}(n,\mathbb{K})$ and its Lie algebra $\mathfrak{g} = \mathfrak{gl}(n,\mathbb{K})$ in which: "*Exp*" is the classical matrix-exponential map

$$\mathrm{Exp}(X) = e^X = \sum_{n=0}^{\infty}\frac{1}{n!}X^n \quad \text{for } X\in \mathrm{M}(n,\mathbb{K}).$$

For matrix Lie groups G,

- By definition, the map "*ad*" sends each $X\in\mathfrak{g}$ to the linear operator $\mathrm{ad}_X(Y) = [X,Y] = XY - YX$ on the Lie algebra. These are all linear *Lie derivations* on $\mathfrak{g}$ in the sense that

$$\mathrm{ad}_X([Y,Z]) = [\mathrm{ad}_X(Y), Z] + [Y, \mathrm{ad}_X(Z)].$$

- The map "*Ad*" sends each $g \in G$ to the differential
$$\mathrm{Ad}_g = (di_g)_e : \mathfrak{g} \to \mathfrak{g}$$
of the conjugation operator $i_g(x) = gxg^{-1}$. Since i_g is an automorphism of G, the differential turns out to be an *automorphism* of the Lie algebra, with
$$\mathrm{Ad}_g([X,Y]) = [\mathrm{Ad}_g(X), \mathrm{Ad}_g(Y)] \quad \text{for } X, Y \in \mathfrak{g}, g \in G.$$
The automorphisms of $\mathfrak{g}$ form a group $\mathrm{Aut}(\mathfrak{g})$ under composition of operators and the map $\mathrm{Ad} : g \to \mathrm{Ad}_g$ is a group homomorphism from $(G, \cdot) \to (\mathrm{Aut}(\mathfrak{g}), \circ)$. (Recall Exercise 5.59.)

These definitions apply verbatim to arbitrary Lie groups.

The first fact we need connects all three players in this saga, and pervades most discussion of Lie groups and their Lie algebras. The proof is based on Corollary 5.82, in which we showed that the restriction $\exp = \mathrm{Exp}|_{\mathfrak{g}}$ of the classical exponential map *Exp* to the Lie algebra $\mathfrak{g} \subseteq \mathrm{M}(n, \mathbb{K})$ of a matrix Lie group G turns out to be the "intrinsic" exponential map $\exp : \mathfrak{g} \to G$ associated with G. To state the result in full generality we consider a $\mathcal{C}^\infty$ homomorphism $\phi : G \to H$ between two matrix Lie groups: a $\mathcal{C}^\infty$ map that is also a group homomorphism, so that
$$\phi(e_G) = e_H \quad \text{and} \quad \phi(x \cdot_G y) = \phi(x) \cdot_H \phi(y) \quad \text{for all } x, y \in G.$$

THEOREM 5.88. *For any $\mathcal{C}^\infty$ homomorphism $\phi : G \to H$ between matrix Lie groups we have*
$$\phi(\exp_G(tX)) = \exp_H(t\,(d\phi)_e X) = e^{t(d\phi)_e X} \quad \textit{for } X \in \mathfrak{g}, t \in \mathbb{R}, \tag{5.20}$$
where $\exp_G$ *and* $\exp_H$ *are the exponential maps from* $\mathfrak{g} \to G$ *and* $\mathfrak{h} \to H$.

PROOF. For $X \in \mathfrak{g}$ consider the $\mathcal{C}^\infty$ curves (one-parameter subgroups in H)
$$\gamma_1(t) = \phi(\exp_G(tX)) \quad \text{and} \quad \gamma_2(t) = \exp_H(t\,(d\phi)_e X) = e^{t\,(d\phi)_e X},$$
which are group homomorphisms from $(\mathbb{R}, +)$ into H because
$$\exp\big((s+t)X\big) = \exp(sX) \cdot \exp(tX)$$
for both $\exp_G$ and $\exp_H$. By Corollary 5.82 the curve $\gamma(t) = \exp_G(tX)$ is $\mathcal{C}^\infty$ in G with $\gamma(0) = I$ and $\gamma'(0) = X \in \mathfrak{g}$. Viewing elements of $\mathfrak{h}$ as derivations on functions f in the local algebra $\mathcal{C}^\infty_H(e)$ at the identity in H, it follows from the definition of the differential $(d\phi)_e$ and Lemma 5.44 that
$$\begin{aligned}
\langle \gamma_2'(0), f\rangle &= \langle (d\phi)_e \gamma'(0), f\rangle = \langle (d\phi)_e X, f\rangle = \langle \gamma'(0), f \circ \phi\rangle \\
&= \frac{d}{dt}\{f(\phi(\gamma(t)))\}|_{t=0} \\
&= \frac{d}{dt}\{f(\phi(\exp_G(tX)))\}|_{t=0} \\
&= \left\langle \frac{d}{dt}\{(\phi(\exp_G(tX))\}|_{t=0}\,,\, f\right\rangle \\
&= \langle \gamma_1'(0), f\rangle
\end{aligned}$$

for all $f \in \mathcal{C}^\infty_H(e)$. Thus

$$\gamma_2'(0) = (d\phi)_e X = \gamma_1'(0).$$

The matrix Lie groups G, H need not live in the same matrix space. [5] Nevertheless, if $G \subseteq \mathrm{M}(n,\mathbb{R})$ and $H \subseteq \mathrm{M}(m,\mathbb{R})$ we may regard both groups $\gamma_1(t)$ and $\gamma_2(t)$ in H as one parameter-groups in matrix space $\mathrm{M}(m,\mathbb{R})$, with the same derivatives $\gamma_k'(0) = (d\phi)_e X$ at $t = 0$. Viewed this way, both groups $\gamma_k(t)$ are vector-valued solutions $X(t)$ to the differential equation

$$\frac{dX}{dt} = (d\phi)_e \cdot X(t) \quad \text{with initial condition } X(0) = I_{m\times m},$$

on the Euclidean space $\mathrm{M}(m,\mathbb{R}) \cong \mathbb{R}^{m^2}$

We now invoke a fundamental existence theorem in the theory of systems of m constant-coefficient ordinary differential equations. Recasting the system as a single differential equation involving vector-valued functions $f(t)$ with values in $\mathbb{R}^m$, the existence theorem takes the form:

THEOREM 5.89 (Fundamental Existence Theorem for ODE). *Given a matrix $B \in \mathrm{M}(m,\mathbb{R})$, the vector-valued solutions of the differential equation*

$$\frac{dX}{dt} = B \cdot X(t) \quad \textit{with initial value } X(0) = \mathbf{c}$$

are defined and $\mathcal{C}^\infty$ for all $t \in \mathbb{R}$. The solution is unique once the initial value is specified, and is given by

$$X(t) = e^{tB}(\mathbf{c}) \quad \textit{for } -\infty < t < \infty.$$

Returning to the proof of Theorem 5.88, we may take $B = [(d\phi)_e]$ and initial condition $X(0) = I_{m\times m}$ to conclude that

$$\gamma_1(t) = \gamma_2(t) = e^{t(d\phi)_e X} \cdot I_{n\times n} = \mathrm{Exp}(t(d\phi)_e X) \quad \text{for all } t,$$

so the identity (5.20) is verified. □

This yields a frequently invoked special case when we take $G = H$ and $\phi(g)$ a typical inner automorphism $\phi = i_{\exp_G(X)}$ for some $X \in \mathfrak{g}$, so that $(d\phi)_e = \mathrm{Ad}_{\exp_G(X)}$. Writing $\mathrm{ad}_X(Y) = [X, Y]$ as usual, we get:

COROLLARY 5.90. *If G is a matrix Lie group and $X, Y \in \mathfrak{g}$ we have*

$$\mathrm{Ad}_{\exp(tX)}(Y) = e^{t\,\mathrm{ad}_X}(Y) = \sum_{n=0}^{\infty} \frac{t^n}{n!}(\mathrm{ad}_X)^n(Y) \quad \textit{for all } t \in \mathbb{R}, \tag{5.21}$$

where "exp" is the restriction $\exp_G = \mathrm{Exp}|_{\mathfrak{g}}$.

PROOF. The first equality is immediate from Theorem 5.88; the second follows upon replacing $e^{t\,\mathrm{ad}(X)}$ by its power series expansion. □

[5]The surrounding matrix spaces are irrelevant. The result is actually true as stated for $\mathcal{C}^\infty$ homomorphisms between general Lie groups G, H.

This result is reflected in the following commutative diagram.

$$\begin{array}{ccc} \mathfrak{g} & \xrightarrow{\exp} & G \\ \text{ad} \downarrow & & \downarrow \text{Ad} \\ \text{Der}(\mathfrak{g}) & \xrightarrow{\exp} & \text{Aut}(G) \end{array}$$

where *Der*($\mathfrak{g}$) is the space of derivation on $\mathfrak{g}$.

LEMMA 5.91. *If G is a matrix Lie group then*

1. *The linear operators* $\text{Ad}_g = (di_g)_e$ *from* $\mathfrak{g} \to \mathfrak{g}$ *are Lie algebra homomorphisms, so that*
$$\text{Ad}_g([X,Y]) = [\text{Ad}_g(X), \text{Ad}_g(Y)] \quad \textit{for all } X, Y \textit{ in } \mathfrak{g}.$$
2. *The map* Ad *that sends* $g \to \text{Ad}_g$ *is a group homomorphism from G into the automorphism group* $\text{Aut}(\mathfrak{g})$ *of its Lie algebra.*

PROOF. For Part 2: $\text{Ad}_e = I$ (the identity map on $\mathfrak{g}$) and we have $i_{g_1g_2} = i_{g_1} \circ i_{g_2}$. Then by Exercise 5.59 we get $\text{Ad}_{g_1g_2} = \text{Ad}_{g_1} \circ \text{Ad}_{g_2}$ and (2.) follows. For Part 1 apply Lemma 5.58 in which we proved that

$$\text{Ad}_g(X) = (di_g)_e(X) = gXg^{-1} \quad \text{(matrix product)}.$$

We also know the Lie bracket $[\mathfrak{g}, \mathfrak{g}]$ in $\mathfrak{g}$ is the associative commutator $[X,Y] = XY - YX$ computed in the full matrix algebra $\text{M}(n, \mathbb{K})$. (Remember: XY and YX are *not* elements of $\mathfrak{g}$, but this commutator is back in $\mathfrak{g}$ owing to cancellations hidden from view in the proof of Proposition 5.57.) Now compute in $\text{M}(n, \mathbb{K})$: if $g \in G$ and $X, Y \in \mathfrak{g}$ we have

$$\begin{aligned} \text{Ad}_g[X,Y] &= g[X,Y]g^{-1} = g(XY - YX)g^{-1} \\ &= [gXg^{-1}, gYg^{-1}] \quad \text{(after some cancellations } gg^{-1} = e) \\ &= [\text{Ad}_g(X), \text{Ad}_g(Y)]. \end{aligned}$$

□

When $G = (\mathbb{R}, +)$ the exponential map is $\text{Exp}(x) = e^x$, but the calculation is not so straighforward for $G = (\mathbb{R}^n, +)$. First, our discussion of *Exp* was framed for matrix Lie groups, for which the usual formula $\text{Exp}(A) = e^A$ makes sense because G is embedded in some matrix space $\text{M}(n, \mathbb{R})$. For general Lie groups the definitions of the Lie algebra $\mathfrak{g}$ and exponential map $exp : \mathfrak{g} \to G$ are quite a bit more subtle.

So, to apply our previous discussion and define an exponential map to $G = \mathbb{R}^n$ we must first realize G as a matrix group. As an immediate consequence of equation (5.14), the bracket operation on the Lie algebra of any abelian matrix Lie group is trivial: $[X,Y] = 0$ for all $X, Y \in \mathfrak{g}$. Thus $\mathfrak{g}$ is just a copy of $\mathbb{R}^n$ where $n = \dim_{\mathbb{R}}(G)$. Also, as indicated in Example 5.26, $(\mathbb{R}^2, +)$ is isomorphic to the matrix Lie group

$$G = \left\{ \begin{pmatrix} 1 & 0 & t_1 \\ 0 & 1 & t_2 \\ 0 & 0 & 1 \end{pmatrix} : \mathbf{t} = (t_1, t_2) \in \mathbb{R}^2 \right\} \quad \text{in } \text{M}(3, \mathbb{R}),$$

whose exponential map we shall calculate in the following exercise. Similar calculations apply to any $\mathbb{R}^n$.

EXERCISE 5.92 (Calculating Exp for $(G, \cdot) \cong (\mathbb{R}^2, +)$). Define the matrices

$$X = \begin{pmatrix} 0 & 0 & 1 \\ 0 & 0 & 0 \\ 0 & 0 & 0 \end{pmatrix} \qquad Y = \begin{pmatrix} 0 & 0 & 0 \\ 0 & 0 & 1 \\ 0 & 0 & 0 \end{pmatrix},$$

so that a typical element in G is $I + t_1X + t_2Y$.

(a) Calculate the one-parameter groups $\gamma_X(t_1) = e^{t_1X}, \gamma_Y(t_2) = e^{t_2Y}$.

(b) Calculate the derivatives $\gamma_X'(0), \gamma_Y'(0)$ and explain why X, Y are basis vectors for the Lie algebra $\mathfrak{g} \subseteq \mathrm{M}(3, \mathbb{R})$.

(c) Verify that

$$\mathrm{Exp}(t_1X + t_2Y) = \mathrm{Exp}(t_1X) \cdot \mathrm{Exp}(t_2Y) \text{ is equal to } I + t_1X + t_2Y$$

for $t_1, t_2 \in \mathbb{R}$, so $exp_G = \mathrm{Exp}|_{\mathfrak{g}}$ maps $\mathfrak{g}$ bijectively onto the hyperplane $M = I + \mathbb{R}X + \mathbb{R}Y$.

HINT. The associative product XY is zero.

We now extend Proposition 5.79 to identify all the singularities of the exponential map Exp on $\mathrm{M}(n, \mathbb{K})$.

PROPOSITION 5.93. *The differential* $(d\mathrm{Exp})_X$ *of the exponential map at an arbitrary base point* X *in* $\mathrm{M}(n, \mathbb{K}) = \mathfrak{gl}(n, \mathbb{R})$ *is:*

$$(d\mathrm{Exp})_X = (d\lambda_{\mathrm{Exp}(X)})_I \circ \left(\frac{I - e^{-\mathrm{ad}_X}}{\mathrm{ad}_X}\right) = (d\lambda_{\mathrm{Exp}(X)})_I \circ f(\mathrm{ad}_X), \tag{5.22}$$

where $f(z)$ *is the analytic function of complex variable*

$$f(z) = \frac{1 - e^{-z}}{z} = 1 - \frac{1}{2!}z + \frac{1}{3!}z^2 - \dots \quad (f(0) = I \text{ by L'Hôpital}),$$

which converges for all $z \in \mathbb{C}$ *by the ratio test.*

REMARK 5.94. Before we take up the proof, this identity needs some interpretation. The first factor in (5.22) is the differential of a left-translation operator λ_g, $g = \mathrm{Exp}(X)$, which is always a nonsingular linear operator on $\mathfrak{gl}(n, \mathbb{K})$; as such it is irrelevant in deciding whether $(d\mathrm{Exp})_X$ is nonsingular, so we can ignore it.

In the second factor, $e^{-\mathrm{ad}_X}$ is a nonsingular linear operator on $\mathfrak{gl}(n, \mathbb{K})$, but the difference $I - e^{-\mathrm{ad}_X}$ might have determinant zero for certain choices of X. By Proposition 5.40 of LA I the matrix-valued series obtained by setting $z = \mathrm{ad}_X$ is absolutely norm convergent, so the factor $f(A)$ always makes sense.

If we write A for $f(\mathrm{ad}_X) : \mathfrak{gl} \to \mathfrak{gl}$ in (5.22), deciding whether $(d\mathrm{Exp})_X$ is singular for a given X comes down to deciding whether A is singular, which happens $\Leftrightarrow \lambda = 0$ is in the complex spectrum $\mathrm{sp}_{\mathbb{C}}(A)$. By the spectral mapping theorem (Lemma 6.12 of LA I if $\mathbb{K} = \mathbb{C}$, or Theorem 1.62 if $\mathbb{K} = \mathbb{R}$), we have

$$\mathrm{sp}\big(f(\mathrm{ad}_X)\big) = f\big(\mathrm{sp}(\mathrm{ad}_X)\big).$$

We conclude that the factor A is singular $\Leftrightarrow f(\lambda) = 0$ for some complex eigenvalue of $\mathrm{ad}_X : \mathfrak{g} \to \mathfrak{g}$. The geometric problem of finding the singlarities of

$\mathrm{Exp} : \mathfrak{gl}(n, \mathbb{K}) \to \mathrm{GL}(n, \mathbb{K})$ has now been reduced to a straightforward problem in linear algebra. ○

EXERCISE 5.95. If $A \in \mathrm{M}(n, \mathbb{R})$ is viewed as a complex matrix that happens to have all-real entries, it is quite possible that $\mathrm{sp}_{\mathbb{C}}(A) \supsetneqq \mathrm{sp}_{\mathbb{R}}(A)$. Explain why

$$A \text{ is singular } \Leftrightarrow \lambda = 0 \text{ is in } \mathrm{sp}_{\mathbb{R}}(A) \Leftrightarrow \lambda = 0 \text{ is in } \mathrm{sp}_{\mathbb{C}}(A).$$

EXERCISE 5.96. Explain why $\mathrm{Exp} : \mathrm{M}(n, \mathbb{K}) \to \mathrm{GL}(n, \mathbb{K})$ has a smooth local inverse near X unless ad_X has an eigenvalue of the form $\lambda = 2\pi n i$ for $n = \pm 1, \pm 2, \ldots$.

We return to the task of proving Proposition 5.93.

PROOF. For a smooth curve $X(t)$ in $\mathrm{M}(n, \mathbb{K})$ define the function

$$Y(s,t) = \mathrm{Exp}(-sX(t)) \cdot \frac{\partial}{\partial t}\{\mathrm{Exp}(sX(t))\} \quad \text{for } s, t \in \mathbb{R}.$$

Applying $\partial/\partial s$ (and abbreviating $X(t) = X$ for the moment), we get

$$\begin{aligned}\frac{\partial Y}{\partial s}(s,t) &= (-X)\mathrm{Exp}(-sX) \cdot \frac{\partial}{\partial t}\{\mathrm{Exp}(sX))\} \\ &\quad + \mathrm{Exp}(-sX)\frac{\partial}{\partial s}\left\{\frac{\partial}{\partial t}\{\mathrm{Exp}(sX)\}\right\}\end{aligned}$$

Now write $(***)$ as a placeholder for the first term, and interchange the partial derivatives in s and t in the second, to get

$$\begin{aligned}\frac{\partial Y}{\partial s} &= (***) + \mathrm{Exp}(-sX) \cdot \frac{\partial}{\partial t}\{X \cdot \mathrm{Exp}(sX)\} \\ &= (***) + \mathrm{Exp}(-sX) \cdot \left[\frac{dX}{dt}\mathrm{Exp}(sX) + X \cdot \frac{\partial}{\partial t}\{\mathrm{Exp}(sX)\}\right].\end{aligned}$$

The term

$$X \cdot \mathrm{Exp}(-sX)\frac{\partial}{\partial t}\{\mathrm{Exp}(sX)\}$$

appearing here cancels a similar term in $(***)$ that involves $(+sX)$, leaving

$$\begin{aligned}\frac{\partial Y}{\partial s} &= \mathrm{Exp}(-sX)\,\frac{dX}{dt}\,\mathrm{Exp}(sX) \\ &= \mathrm{Ad}_{\exp(-sX)}\left(\frac{dX}{dt}\right) = e^{-s\,\mathrm{ad}_X} \cdot \left(\frac{dX}{dt}\right) \quad \text{(by Lemma 5.58).}\end{aligned}$$

Since $Y(0,t) \equiv 0$, when $s = 1$ we see that

$$\mathrm{Exp}(-X)\frac{d}{dt}\{\mathrm{Exp}(X)\} = Y(1,t) = \int_0^1 \frac{\partial Y}{\partial s}(s,t)\,ds,$$

and since

$$\frac{\partial Y}{\partial s} = \mathrm{Exp}(-s\,\mathrm{ad}_X)\frac{dX}{dt} = \sum_{k=0}^{\infty} \frac{(-1)^k s^k}{k!}(\mathrm{ad}_X)^k \cdot \frac{dX}{dt}$$

we may integrate term-by-term from $s = 0$ to $s = 1$ to get

$$\begin{aligned}\operatorname{Exp}(-X(t))\frac{d}{dt}\{\operatorname{Exp}(X(t))\} &= \sum_{k=1}^{\infty}\frac{(-1)^k}{(k-1)!}\left(\operatorname{ad}_{X(t)}\right)^k\cdot\frac{dX}{dt}\\ &= f\left(\operatorname{ad}_{X(t)}\right)\cdot\frac{dX}{dt}.\end{aligned}$$

Multiplying both sides on the right by $\operatorname{Exp}(X(t))$ yields

$$\frac{d}{dt}\{\operatorname{Exp}(X(t))\} = f(\operatorname{ad}_{X(t)})\cdot\operatorname{Exp}(X(t))\frac{dX}{dt}\qquad\text{for all } t$$

(since $\operatorname{Exp}(X(t))$ commutes with its derivative and with dX/dt). This is true for all smooth curves with $X(0) = X$; taking $t = 0$ we get (5.22). (Recall that $(d\lambda_g)_e(H) = g\cdot H$ (matrix product) for all $g \in G, H \in \mathfrak{g}$.) □

As a historical note, the most commonly cited analytic proof of the CBH formula (discussed next) follows once equation (5.22) is in hand, see [**2**]; a different, almost completely algebraic proof that proceeds along the lines developed by Dynkin can be found in [**3**, **4**].

Connected Lie Groups. A matrix Lie group G becomes a metric space when equipped with the Euclidean distance function

$$d(A,B) = \left(\sum_{i,j=1}^{n}|A_{ij} - B_{ij}|^2\right)^{1/2}$$

it inherits from matrix space $\mathrm{M}(n,\mathbb{K}) \cong \mathbb{K}^{n^2}$. A set $A \subseteq G$ is an *open set* if, for every $a \in A$, all $x \in G$ sufficiently close to a are also in A.

DEFINITION 5.97 (Open Sets in G). A subset A in a metric space X is an *open set* if, for any $a \in A$ there is a radius $r(a) > 0$ such that $d(x,a) < r \Rightarrow x \in A$. It is a *closed set* if its complement $X \sim A$ is open.

That said, we can give a precise definition of what it means for a metric space X to be *connected*. This is an important concept because some results on Lie groups apply only to groups that are connected (or nearly so). In the other direction we have *discrete groups* such as the integers $(\mathbb{Z},+)$ whose points are isolated from one another; in this extreme *dis*-connected situation there is no geometry to discuss and we are in the realm of pure algebraic group theory.

DEFINITION 5.98. A metric space X *disconnected* if it it contains nonempty open subsets U, V such that $X = U \cup V$ and $U \cap V = \emptyset$. X is *connected* if no such partition of X into isolated open subsets is possible.

The metric space X itself is open, and by default so is the empty set because there are no elements in it for which the definition fails. In a metric space, saying "A is closed set" is the same as saying that you cannot escape from A by taking a limit $\lim_{n\to\infty} a_n$ of points $a_n \in A$. It is not hard to verify two basic properties of open sets in any metric space.

(i) Finite intersections $U_1 \cap \cdots \cap U_n$ of open sets are open.

(ii) An arbitrary union $\bigcup_{\alpha \in I} U_\alpha$ of open sets is open.

The matrix Lie groups we are concerned with are "locally Euclidean" by definition, and such spaces are connected if and only if they are "arcwise connected," which is easier to verify (and often obvious).

DEFINITION 5.99. A metric space X is *arcwise connected* if any two points $a, b \in X$ can be joined by a continuous curve $\gamma : [0,1] \to X$ such that $\gamma(0) = a$ and $\gamma(1) = b$. Of course the roles of a and b are interchangeable since the orientation-reversed curve $-\gamma(t) = \gamma(1-t)$ goes from b to a within X.

A little thought should convince you that a matrix Lie group G is arcwise connected if and only if every element $a \in G$ is arc-connected to the identity element $e \in G$.

LEMMA 5.100. *A connected matrix Lie group G is generated by any relatively open neighborhood U of the identity element, no matter how small, in the sense that*

$$G = \bigcup_{n \in \mathbb{N}} U^n \quad \text{where} \quad U^n = U \cdot \ldots \cdot U \quad (n \text{ terms}).$$

In essence, if G is a connected Lie group the nature of G close near the identity propagates to determine the nature of the whole group.

PROOF. (Sketch Proof.) The *relatively open sets in G* mentioned here are those of the form $A = G \cap U$ for some open set U in the full matrix space $\mathrm{M}(n, \mathbb{R})$, just as open subsets of a hypersurface S in $\mathbb{R}^n$ are the intersections of S with actual open subsets in $\mathbb{R}^n$. (They are also the open sets in G determined by the metric in matrix space, restricted to the subset G.)

The proof proceeds in several steps:

- $U^{-1} = \{u^{-1} : u \in U\}$ is also open, so $U \cap U^{-1}$ is a *symmetric* open neighborhood of e in G. Thus it will suffice to show that $G = \bigcup_{n=1}^{\infty} U^n$ for any open neighborhood of e such that $U = U^{-1}$.
- It follows that the union $G_0 = \bigcup_{n=1}^{\infty} U^n$ is closed under forming products $x \cdot y$ or inverses x^{-1} of its elements. Hence G_0 is the subgroup of G generated by U (the smallest subgroup containing the set U). We used the property $U^{-1} = U$ to insure that $x \in G_0 \Rightarrow x^{-1} \in G_0$.
- In matrix Lie groups the product $A \cdot B = \{ab : a \in A, b \in B\}$ of two open sets is again an open set, and so is the set of inverses $A^{-1} = \{a^{-1} : a \in A\}$. This is an immediate consequence of the the fact that the group multiplication operation $P : G \times G \to G$ is jointly continuous and the inversion operation $J : G \to G$ is bicontinuous.
- Thus all product sets U^n are open sets in G, as is their union G_0 which is now seen to be an open subgroup in the larger group G.

- For the endgame we observe that no *connected* matrix Lie group group can have a proper open subgroup, which forces the desired conclusion that $G_0 = G$. To see why, note that left translations $\lambda_g(x) = gx$ on G are bicontiuous maps, so they send open sets to open sets. If H is an open subgroup in G the cosets (translates) $gH = \lambda_g(H)$ are also open sets.

  We now assert that G is a union of *pairwise disjoint* translates $g_iH \neq g_jH$ (with $g_1 = e$), which split G into nonempty disjoint open sets

$$H_1 = eH = H \qquad \text{and} \qquad H_2 = \bigcup_{g_i \neq e} g_i H\,.$$

  Unless $G = H$ (in which case we're done), we get a contradiction to the hypothesized connectedness of G, and that will complete the proof of Lemma 5.100. The assertion follows immediately once we observe that

  In a matrix Lie group, left translates xH and yH of an open subgroup H are either identical ($xH = yH$ as sets) or are disjoint ($xH \cap yH = \emptyset$), so the distinct translates of H partition G into disjoint open pieces.

  In fact, if xH and yH have nontrivial intersection we must have $xh_1 = yh_2$ for some $h_1, h_2 \in H$, and then

$$yH = x(h_1 h_2^{-1})H = xH$$

  because H is a subgroup in G, $h_1 h_2^{-1}$ is in H, and $hH = H$ for any $h \in H$. So in this case $xH = yH$, and otherwise xH and yH are disjoint as claimed.

That completes the proof of Lemma 5.100. □

It is also worth noting (without going into the proof) that any Lie group G, connected or not, contains a largest connected open subgroup H and G is a disjoint union of its translates. H is referred to as the "connected component of the identity element in G." To mention a few examples: $G = (\mathbb{R}^n, +)$ and the circle group $G = (S^1, \cdot\,)$ are arcwise connected, hence connected, and so are the unitary groups $G = \mathrm{U}(n)$ with $n \geq 2$. But the real orthogonal groups $\mathrm{O}(n)$ for $n \geq 3$ have two connected components: the maximal connected open subgroup $H = \mathrm{SO}(n)$ of proper orthogonal transformations having $\det(A) = +1$, and a disjoint translate whose elements all have $\det(A) = -1$.

As we noted in Corollary 5.82, the restricted exponential map $\mathrm{Exp}|_{\mathfrak{g}}$ on a connected matrix group sends $\mathfrak{g}$ into G and its range includes an open neighborhood U of the identity in G, which in turn generates the entire connected group G.

The Campbell-Baker-Hausdorff Formula: Recovering G from $\mathfrak{g}$. The Campell-Baker-Hausdorff formula (CBH) was the culmination of efforts by several mathematicians at the beginning of the 20th century to show that when the product operation $x \cdot y$ near the identity in G is transfered via the exponential map to an operation $X * Y$ on $\mathfrak{g}$,

$$X * Y = \mathrm{Log}\big(\mathrm{Exp}(X) \cdot \mathrm{Exp}(Y)\big) \quad \text{for } X, Y \in \mathfrak{g},$$

the result can be written as an infinite series

$$\begin{aligned} X * Y = X + Y + \tfrac{1}{2}[X,Y] + \tfrac{1}{12}[X,[X,Y]] - \tfrac{1}{12}[Y,[X,Y]] \\ - \tfrac{1}{48}[Y,[X,[X,Y]]] - \tfrac{1}{48}[X,[Y,[X,Y]]] + \cdots \end{aligned}$$

that is absolutely norm convergent near zero in $\mathfrak{g}$, and whose terms involve only iterated Lie brackets. Baker, Campbell, and Hausdorff discussed recursive procedures for determining the series coefficents; it was only in 1947 that E.B. Dynkin found explicit combinatorial formulas for all terms. Thus the bracket operation in $\mathfrak{g}$ indeed determines the group structure of any connected Lie group G, up to a local isomoprhism.

Full proof of CBH would take us beyond the scope of these *Notes*, but can be found in [**2**, **5**, **6**]. One approach can be sketched as follows. The Taylor expansion of $\mathrm{Log}(z)$ about $z = 1 + i0$ is

$$\mathrm{Log}(z) = \sum_{n=1}^{\infty} \frac{(-1)^{n+1}}{n}(z-1)^n \quad \text{for } |z-1| < 1 \text{ in } \mathbb{C}.$$

This yields an absolutely norm convergent sum if we replace z by any complex matrix with operator norm $\|A\|_{op} < 1$.

By formally multiplying the series expansions for e^X and e^Y and solving for Z in $e^Z = e^X \cdot e^Y$ we get

$$\begin{aligned} Z = \mathrm{Log}(e^X e^Y) &= \sum_{n=1}^{\infty} \frac{(-1)^{n+1}}{n}\big(e^X e^Y - I\big)^n \\ &= \sum_{n>0} \frac{(-1)^{n+1}}{n}\left(\sum_{\substack{p_i+q_i>0 \\ 1\le i\le n}} \frac{X^{p_1}Y^{q_1}\cdots X^{p_n}Y^{q_n}}{\big(\sum_{i=1}^n p_i+q_i\big)\cdot \prod_{j=1}^n p_j!\,q_j!}\right). \end{aligned}$$

Dynkin's formula is obtained by showing that the expressions $X^{p_1}\cdots Y^{q_n}$ can be replaced by closely related expressions involving commutators

$$[X^{p_1}\cdots Y^{q_n}] = \big((\mathrm{ad}_X)^{p_1}(\mathrm{ad}_Y)^{q_1}\cdots(\mathrm{ad}_X)^{p_n}(\mathrm{ad}_Y)^{q_n-1}\big)\,Y,$$

where $\mathrm{ad}_A(B) = [A,B]$ for $A, B \in \mathfrak{g}$. Since $[A,A] = 0$ a term is zero if $q_n > 1$, and it is zero (by definition) if $q_n = 0$ and $p_n > 1$. Absolute convergence of the series near $(X,Y) = (0,0)$ would also have to be addressed.

THEOREM 5.101 (Campbell-Baker-Hausdorff Formula). *Let $\mathfrak{g}$ be the Lie algebra of a matrix Lie group over $\mathbb{R}$ and write $X * Y = \mathrm{Log}(e^X e^Y)$ for X, Y in $\mathfrak{g}$. On a sufficiently small neighborhood of the origin we have an absolutely norm* *convergent series expansion of the product operation in G:*

$$X * Y = \mathrm{Log}(e^X e^Y)$$

$$(5.23)\qquad = \sum_{n>0} \frac{(-1)^{n+1}}{n} \sum_{\substack{p_i+q_i>0\\ 1\le i\le n}} \frac{\big((\mathrm{ad}_X)^{p_1}(\mathrm{ad}_Y)^{q_1}\cdots(\mathrm{ad}_x)^{p_n}(\mathrm{ad}_Y)^{q_n-1}\big)\,Y}{\big(\sum_{i=1}^n p_i+q_i\big)\cdot\prod_{j=1}^n p_j!\,q_j!}.$$

where $\mathrm{ad}_A(B) = [A,B]$.

Observe that this formula is "universal," with the same terms no matter which Lie group we consider.

In many situations we only need to know that such a series expansion *exists*, but there are many instances in which we want to know the exact form of the terms, particularly for nilpotent Lie groups, for which the CBH formula has only finitely may nonzero terms. Then $X * Y$ is a multivariate *polynomial* in the variables X and Y.

5.5. The Lie Correspondence: Lie Groups vs Lie Algebras

Let G be a connected d-dimensional matrix Lie group. We have mentioned that the Lie algebra $\mathfrak{g}$, a purely algebraic object, encodes almost all the information needed to reconstruct G. The obstacles to a complete reconstruction lie in the fact that Lie groups are geometric objects (manifolds) as well as algebraic objects (groups), and both aspects are important. The Lie algebra provides a good handle on the algebraic aspects, but the methods of differential geometry are needed to deal with large-scale geometric issues.

For instance, we defined the Lie algebra as the tangent space $\mathfrak{g} = \mathrm{TG}_e$ at the identity element e, so vectors in $\mathfrak{g}$ are derivatives $X = \gamma'(0)$ along smooth curves in G as they pass through e. These are determined entirely by the nature of G in the immediate vicinity of e; the geometry of the group far from e has no influence on the outcome. By comparison, algbraic properties near e propagate throughout the group (at least if G is connected) because, as mentioned earlier, G is generated by any open neighborhood U about e, no matter how small, in the sense that $G = \bigcup_{n\in\mathbb{N}} U^n$.

Now consider what it means for Lie algebras or Lie groups to be "isomorphic."

DEFINITION 5.102 (Isomorphism of Lie Groups vs Lie algebras). For Lie algebras and matrix Lie groups we say that:

1. Two Lie algebras are *isomorphic*, written as $\mathfrak{g} \cong \mathfrak{h}$, if there is a linear bijection $\phi : \mathfrak{g} \to \mathfrak{h}$ that intertwines the Lie algebra structures, with
$$\phi([X,Y]_{\mathfrak{g}}) = \big[\phi(X), \phi(Y)\big]_{\mathfrak{h}} \quad \text{for } X, Y \in \mathfrak{g}.$$
This equivalence relation divides all Lie algebras into isomorphism classes, and all members of a single class are regarded as being "the same."
2. Two Lie groups are *differentiably isomorphic*, written $G \cong H$, if there is a bijection $\rho : G \to H$ that is: (i) a $\mathcal{C}^\infty$ diffeomorphism (ρ and its

inverse are both $\mathcal{C}^\infty$ maps), and (ii) ρ is a group isomorphism, so that

$$\rho(x \cdot_G y) = \rho(x) \cdot_H \rho(y) \quad \text{for } x, y \in G.$$

By time-honored abuse of notation we often say that two Lie groups are "isomorphic" when $G \cong H$. This condition is sensitive to the large-scale geometries of G and H, because of condition 2.

It is easy to see that $G \cong H$ implies $\mathfrak{g} \cong \mathfrak{h}$. The following simple example shows why the reverse implication can fail. It is quite possible for G and H to be "locally isomorphic," near their identity elements (which guarantees that $\mathfrak{g} \cong \mathfrak{h}$), but have quite different large scale geometries and are not globally diffeomorphic Lie groups, as in (2.).

EXAMPLE 5.103 (Locally Isomorphic Lie Groups). The real line $(\mathbb{R}, +)$ and the Euclidean spaces $(\mathbb{R}^n, +)$ are Lie groups whose manifold structures are determined by a single chart. The unit circle $(S^1, \cdot) = \{z \in \mathbb{C} : |z| = 1\}$ with complex multiplication as the group operation is also a Lie group, but two overlapping charts are required to define its manifold structure.

By the exponent laws, the complex exponential function

$$\phi(t) = e^{2\pi i t} = \cos(2\pi t) + i\sin(2\pi t)$$

is a periodic $\mathcal{C}^\infty$ map such that $\phi(t+n) = \phi(t)$ for all $n \in \mathbb{Z}$ that repeatedly wraps the line around the circle. It is also a group homomorphism from $(\mathbb{R}, +)$ onto $(S^1, \cdot\,)$ because

$$\phi(s+t) = \phi(s) \cdot \phi(t) \quad \phi(0) = 1 + i0$$

for all $s, t \in \mathbb{R}$. Although ϕ is not one-to-one it is one-to-one for points near zero in $\mathbb{R}$, for instance on the interval $I = \left(-\frac{1}{2}, \frac{1}{2}\right)$. Furthermore, for s, t sufficiently close to the origin that $s + t \in I$ (say, with $|s|, |t| < \frac{1}{4}$), ϕ behaves like a group *isomorphism* since it is one-to-one and $\phi(s+t) = \phi(s) \cdot \phi(t)$ for all such s, t.

In other words, ϕ is a "local isomorphism" between these Lie groups. This phenomenon is encountered quite often. We saw earlier that $\mathrm{SO}(n, \mathbb{C})$ and $\mathrm{O}(n, \mathbb{C})$ are locally isomorphic, as are the rotation group $\mathrm{SO}(3)$ on $\mathbb{R}^3$ and the full group $\mathrm{O}(3)$ of linear rigid motions on $\mathbb{R}^3$ (including reflections across planes through the origin as well as rotations). The case of $\mathbb{R}$ vs S^1 is more intersting since these groups are connected, and not related by inclusion – for details see Additional Exercises to Section 5.2, Exercise #3.

This concept turns out to be just what is needed to capture the true relation between connected (matrix) Lie groups and their Lie algebras. We start with a precise definition, after which we can at last state the Lie Correspondence between (matrix) Lie groups and their Lie algebras.

DEFINITION 5.104 (Local Isomorphisms). A *local homomorphism* between Lie groups, $\phi : G \to H$, is a $\mathcal{C}^\infty$ map defined on an open neighborhood U of the identity in G such that

$$\phi(x \cdot_G y) = \phi(x) \cdot_H \phi(y) \text{ in } H \text{ for all } x, y \in U \text{ such that } x \cdot_G y \in U,$$

or equivalently, $\phi(x \cdot_G y) = \phi(x) \cdot_H \phi(y)$ for all x, y near the identity in G. A *local isomorphism* is a bijective diffeomeorphism between open neigborhoods U, V of the identities in G and H such that

$$\phi : U \to V \quad \text{and} \quad \psi = \phi^{-1} : V \to U$$

are local homomorphisms.

This brings us to a fundamental result in Lie theory.

THEOREM 5.105 (The Lie Correspondence). *If G and H are connected matrix Lie groups with Lie algebras $\mathfrak{g}$ and $\mathfrak{h}$, the following statements are equivalent.*

1. *There is a diffeomorphism ϕ between open neighborhoods U, V of the identities in G, H (a bijective $\mathcal{C}^\infty$ map with $\mathcal{C}^\infty$ inverse) that is also a local isomorphism $\phi : G \to H$, so $\phi(x \cdot_G y) = \phi(x) \cdot_H \phi(y)$ if $x, y \in U$ and $x \cdot_G y \in U$.*
2. *There is a bijective linear isomorphism $d\phi : \mathfrak{g} \to \mathfrak{h}$ between the Lie algebras that intertwines the Lie bracket operations, with*

$$d\phi([X, Y]_\mathfrak{g}) = \left[d\phi(X), d\phi(Y)\right]_\mathfrak{h}$$

for all $X, Y \in \mathfrak{g}$.

In short, G and H are locally isomorphic connected Lie groups if and only if they have algebraically isomorphic Lie algebras.

EXERCISE 5.106. Verify that the map $\phi : (\mathbb{R}, +) \to (S^1, \cdot\,)$ given by

$$\phi(t) = e^{2\pi i t} = \cos(2\pi t) + i \sin(2\pi t)$$

is a local isomorphism between these abelian Lie groups, but is not a global isomorphism.

EXERCISE 5.107. Explain why $\mathrm{SO}(n, \mathbb{R})$ and $\mathrm{SL}(n, \mathbb{R})$ cannot be locally isomorphic Lie groups.

Further Development of Lie Theory. This is as far as we can go with Lie theory in this brief chapter. We conclude by indicating some of the next steps in a more complete account.

1. *Making the transition from matrix Lie groups to general Lie groups.* This begins by elaborating an alternative model of the Lie algebra, mentioned earlier, in which the Lie algebra is modeled as the space $\tilde{\mathfrak{g}} = \mathcal{L}(G)$ of smooth left-invariant vector fields $\widetilde{X}$ on G. By interpreting these as smooth first-order partial differential operators on the manifold G we can embed $\tilde{\mathfrak{g}}$ in a much larger structure Diff(G), the associative algebra of all smooth partial differential operators on G (of any finite order). This turns out to have many advantages.
2. *Coordinate systems in Lie groups.* Exponential coordinates and multiplicative coordinates (akin to rectilinear vs polar coordinates) can be defined, which allow computational flexibility in attacking problems.

3. *The correspondence: (Lie subgroups in G) $\leftrightarrow$ (Lie subalgebras in $\mathfrak{g}$).* Many geometric subtleties arise in defining what should be meant by a "Lie subgroup" of a Lie group G. This reflects similar problems that arise in differential geometry when one defines "submanifolds" of a differentiable manifold M. When "Lie subgroups" are properly interpreted there is a *bijective* correspondence that matches structural features in G with those in the Lie algebra $\mathfrak{g}$.

4. *Resolving the ambiguities in the Lie correspondence.* There is an elegant answer to the question

   > *Which connected Lie groups G have the same Lie algebras $\mathfrak{g}$, and what are the global relations between them?*

   Every Lie algebra $\mathfrak{g}$ determines a unique "simply connected covering group" $\widetilde{G}$, such that every connected Lie group G with $\mathfrak{g}$ as its Lie algebra is a simple quotient $G \cong \widetilde{G}/D$ by a discrete subgroup $D \subseteq \widetilde{G}$. The subgroup D encodes information of profound importance about the large-scale geometric structure of G. All connected Lie groups with the same Lie algebra $\mathfrak{g}$ are easily understood once the covering group $\widetilde{G}$ and the discrete subgroup D are known. The quotient map $\pi : \widetilde{G} \to \widetilde{G}/D = G$ is a surjective local isomorphism between connected Lie groups.

5. *Complete analysis of compact connected Lie groups.* Using the tools mentioned here a complete analysis can be made of the algebraic and geometric structure of all compact connected Lie groups, which include many of the symmetry groups that govern interactions of fundamental particles in nuclear physics, as well as other "secret symmetries of the universe."

Additional Exercises

Section 5.1. Matrix Groups and Implicit Function Theorem

1. TRUE/FALSE QUESTIONS ("True" if the statement is *always* true.)

   (a) The map

   $$\phi(t) = e^{2\pi i t} = \cos(2\pi t) + i\sin(2\pi t)$$

   from $\mathbb{R}$ into $\mathbb{C} \cong \mathbb{R}^2$ is a $\mathcal{C}^\infty$ map with $\mathrm{rk}(d\phi)_t = 1$ at all base points.

   (b) The set of matrices

   $$C_t = \begin{pmatrix} 1 & t \\ 0 & 1 \end{pmatrix} \qquad (t \in \mathbb{R}),$$

   when equipped with matrix multiplication $(\cdot)$, is a matrix group that is isomorphic to the additive group $(\mathbb{R}, +)$ of real numbers.

(c) The set of matrices

$$C_{\mathbf{x}} = \begin{pmatrix} 1 & 0 & 0 & x_1 \\ 0 & 1 & 0 & x_2 \\ 0 & 0 & 1 & x_3 \\ 0 & 0 & 0 & 1 \end{pmatrix} \qquad (\mathbf{x} \in \mathbb{R}^3),$$

equipped with matrix multiplication $(\cdot)$, is a matrix group that is isomorphic to the additive group $(\mathbb{R}^3, +)$.

(d) If we identify the complex plane $\mathbb{C}$ with $\mathbb{R}^2$, the complex exponential map

$$\phi(z) = e^z = e^{x+iy} = \cos(x) + i\sin(y),$$

for $z = x + iy \in \mathbb{C}$, becomes a $\mathcal{C}^\infty$ map from $\mathbb{R}^2 \to \mathbb{R}^2$ whose rank $\mathrm{rk}(d\phi)_z$ is equal to 2 at *every* base point $z \in \mathbb{C}$.

(e) If G is a matrix Lie group in $\mathrm{M}(n, \mathbb{R})$, then the "chart map"

$$x_\alpha : A \to (a_{11}, \dots, a_{in}; \dots ; a_{n1}, \dots, a_{nn}) \in \mathbb{R}^{n^2}$$

that reads off the entries a_{ij} in A imposes Cartesian coordinates on G that make it into a differentiable manifold.

2. If we delete the origin $0 + i0$ from the complex plane $\mathbb{C}$, complex multiplication $P(z, w) = z \cdot w$ still makes sense in the "punctured plane" $\mathbb{C}^* = \mathbb{C} \sim 0$, even though the $(+)$ operation does not (why?). Answer the following questions about the system $(C^*, \cdot)$. Cartesian coordinates are imposed on $\mathbb{C}^*$ by the globally defined chart map.

(a) Show that $(\mathbb{C}^*, \cdot)$ is a *group* because

- It is closed under formation of products $z \cdot w$ for $z, w \in \mathbb{C}^*$
- It has a multiplicative identity element e such that $e \cdot z = z$ for all $z \in \mathbb{C}^*$ for all $z \in \mathbb{C}^*$
- For every $z \in \mathbb{C}^*$ there is a multiplicative inverse $w = z^{-1}$ such that $z \cdot w = w \cdot z = e$

Find z^{-1} if $z = 4 - 5i$.

(b) If we let $\mathbb{R}^2_* = \mathbb{R}^2 \sim \{(0, 0)\}$, multiplication $z \cdot w$ and the inversion map $J(z) = 1/z$ on $\mathbb{C}^*$ become $\mathcal{C}^\infty$ maps from $\mathbb{R}^2_* \times \mathbb{R}^2_* \to \mathbb{R}^2_*$, or from $\mathbb{R}^2_* \to \mathbb{R}^2_*$, when described in the coordinates (x, y) on $\mathbb{R}^2_*$.

3. The 3-dimensional *Heisenberg group* is the set of upper triangular matrices G in $\mathrm{M}(3, \mathbb{R})$:

$$A(x, y, z) = \begin{pmatrix} 1 & x & z \\ 0 & 1 & y \\ 0 & 0 & 1 \end{pmatrix} \quad \text{for } x, y, z \in \mathbb{R}.$$

(a) Verify that G is closed under formation of matrix products and compute the entries x'', y'', z'' in the product of matrices

$$A(x'', y'', z'') = A(x, y, z) \cdot A(x', y', z').$$

and in the inverse matrix A^{-1}. (Since G obviously contains the identity matrix $I_{3\times 3}$, it is then a *group* under matrix multiplication.)

(b) Viewing matrices $A \in \mathrm{M}(3, \mathbb{R})$ as vectors in $\mathbb{R}^9$,

$$A = \big(a_{11}, \dots a_{13}, \dots, a_{31}, \dots, a_{33}\big),$$

how would you describe G as a level set $L_{\phi(\mathbf{x})=\mathbf{c}}$ in $\mathbb{R}^9$ for a suitably chosen smooth mapping $\phi : \mathbb{R}^9 \to \mathbb{R}^r$, and for what values of $r > 0$ and $\mathbf{c}$?

(c) Write out the $r \times 9$ Jacobian matrix of ϕ,

$$D\phi(A) = \left(\frac{\partial \phi_k}{\partial a_{jk}}\right),$$

at the identity element $I_{3\times 3}$ in G and at an arbitrary base point $A = A(x, y, z)$ in G.

(d) Is G a smooth hypersurface (differentiable manifold) in $\mathbb{R}^9$, and if so, of what dimension? Explain.

Section 5.2. Matrix Lie Groups

1. True/False Questions ("True" if the statement is *always* true.)

(a) If G is a matrix Lie group in $\mathrm{M}(n, \mathbb{C})$, it is not a complex Lie group unless $\dim_{\mathbb{R}}(G)$ is even.

(b) A matrix Lie group $G \subseteq \mathrm{M}(n, \mathbb{C})$ is a complex Lie group if its dimension $\dim_{\mathbb{R}}(G)$ is even.

(c) The additive complex numbers $(\mathbb{C}, +)$ can be realized as a martix group in $\mathrm{M}(2, \mathbb{C})$ because the set of matrices

$$G' = \left\{\begin{pmatrix} 1 & z \\ 1 & 0 \end{pmatrix} : z \in \mathbb{C}\right\}$$

equipped with matrix multiplication $(\cdot)$ is isomorphic to $(\mathbb{C}, +)$.

2. If X is in the tangent space $\mathfrak{g} = \mathrm{TM}_e$ of a matrix Lie group G, prove that the globally defined vector field with $\widetilde{X}_g = (d\lambda_g)_e X$ in TG_g actually satisfies the left-invariance condition

$$\widetilde{X}_{g\cdot p} = (d\lambda_g)_e(\widetilde{X}_p)$$

for all base points p and group elements g.

Hint. Tangent vectors $X_p, Y_p \in \mathrm{TM}_p$ are equal if they have the same action on the local algebra $\mathcal{C}^\infty(p)$.

The Circle Group S^1. The unit circle $S^1 = \{z \in \mathbb{C} : |z| = 1\}$ in the complex plane is a group under complex multiplication $(\cdot)$ since $|z \cdot w| = |z| \cdot |w| = 1$ and $|1/z| = 1$. It can also be viewed as the unit circle

$$K = \{\mathbf{x} \in \mathbb{R}^2 : \phi(\mathbf{x}) = x^2 + y^2 = 1\}.$$

By the IFT, this is a smooth 1-dimensional manifold in $\mathbb{R}^2$, but from the latter point of view, its group structure is less apparent. Carry out the following steps to show $G = (S^1, \cdot\,)$ is an abelian (commutative) Lie group.

3. For the circle group $G = (S^1, \cdot\,)$:

(a) Verify that the requirements of the IFT are satisfied for the level set $K = L_{\phi(x,y)=1}$ in $\mathbb{R}^2$, where $\phi(x, y) = x^2 + y^2$ maps $\mathbb{R}^2 \to \mathbb{R}$.

(b) We can parametrize K via the map

$$\gamma(t) = e^{2\pi it} = (\cos t, \sin t) = (x(t), y(t)) \quad \text{for } t \in \mathbb{R}.$$

The multiplication K inherits as a subset $S^1 \subseteq \mathbb{C}$ is given by the usual exponent law $\gamma(t) \cdot \gamma(t') = \gamma(t + t')$. If you only know the Cartesian coordinates $\gamma(t) = (x, y)$ and $\gamma(t') = (x', y')$, find the Cartesian coordinates of the product $\gamma(t) \cdot \gamma(t')$ in terms of x, y, x', y'.

4. Regarding $N = (0, 1)$ and $S = (0, -1)$ as the "north and south poles" of the circle K of Additional Exercise 3, define the stereographic projection maps $x_\alpha : U_\alpha \to \mathbb{R}$ and $x_\beta : U_\beta \to \mathbb{R}$ that send the arcs

$$U_\alpha = K \sim \{S\} \quad \text{and} \quad U_\beta = K \sim \{N\}$$

bijectively onto the tangent lines

$$L_N = \{(t, +1) : t \in \mathbb{R}\} \quad \text{and} \quad L_S = \{(s, -1) : s \in \mathbb{R}\}$$

passing through N and S respectively.

(a) Draw a picture and find explicit formulas for the maps $t = x_\alpha(x, y)$ and $s = x_\beta(x, y)$ from S^1 into $\mathbb{R}$.

(b) Compute the coordinate transition maps $x_\beta \circ x_\alpha^{-1}$ and $x_\alpha \circ x_\beta^{-1}$ and verify that they are $\mathcal{C}^\infty$ maps from $\mathbb{R} \to \mathbb{R}$.

Since these two $\mathcal{C}^\infty$-related charts cover the circle, they impose a differentiable structure on $K = S^1$.

COMPUTATIONS INVOLVING $\mathrm{O}(2, \mathbb{C})$. The following exercises examine properties of the complex orthogonal group $\mathrm{O}(2, \mathbb{C})$ consisting of matrices such that $AA^\mathrm{T} = I_{2\times 2}$. Identifying $\mathrm{M}(2, \mathbb{C}) \cong \mathbb{C}^4$ and dropping one redundant identity, we are left with 3 irredundant complex constraint identities involving the rows $R_j = (A_{j1}, A_{j2})$ of A,

$$\phi_1 = B(R_1, R_1) = 1 + i0, \quad \phi_2 = B(R_2, R_2) = 1 + i0, \quad \phi_3 = B(R_1, R_2) = 0,$$

so $\mathrm{O}(2, \mathbb{C})$ is the level set $L_{\phi(A)=(1,1,0)}$ in $\mathbb{C}^4$. Here $B : \mathbb{C}^2 \times \mathbb{C}^2 \to \mathbb{C}$ is the canonical nondegenerate symmetric bilinear form B on $\mathbb{C}^2$

$$B(\mathbf{z}, \mathbf{w}) = \mathbf{z}^\mathrm{T} I_{2\times 2} \mathbf{w} = z_1 w_1 + z_2 w_2$$

for $\mathbf{z} = (z_1, z_2), \mathbf{w} = (w_1, w_2)$.

5. Regarding $\mathrm{O}(2, \mathbb{C})$ as a subset in $\mathrm{M}(2, \mathbb{C}) \cong \mathbb{C}^4$,

(a) Prove that a matrix

$$A = \begin{pmatrix} a & b \\ c & d \end{pmatrix}$$

with complex entries is in $O(2, \mathbb{C})$ if and only if

$$d = a \qquad c = -b \qquad a^2 + b^2 = 1.$$

(b) Verify that all matrices in $O(2, \mathbb{C})$ commute. (You could use (a) for this. It is no longer true for $O(n, \mathbb{C})$ when $n \geq 3$).

(c) In terms of the constraint identities that determine them, explain why $\dim_{\mathbb{R}} O(2, \mathbb{C}) = \dim_{\mathbb{R}} SO(2, \mathbb{C}) = 2$.

6. Viewing $M(2, \mathbb{C}) \cong \mathbb{R}^8$, rewrite the complex constraints in Part (a) of the previous exercise as a system of real constraint identities $\phi(\mathbf{x}) = \mathbf{c}$. Then verify that a constant rank condition $\mathrm{rk}(d\phi)_A = r$ holds near every base point $A \in O(2, \mathbb{C})$, as required to apply the IFT.

HINT. Calculations like those in (a) were previously encountered in discussing Euler's theorem (LA I, Theorem 6.84); Exercises 6.86 and 6.88 in LA I might also provide some guidance).

7. Verify that the set of "real points"

$$K = SO(2, \mathbb{C}) \cap \big(M(2, \mathbb{R}) + i0\big)$$

in $SO(2, \mathbb{C})$ is the real-orthogonal group

$$SO(2) = \left\{ \begin{pmatrix} \cos\theta & -\sin\theta \\ \sin\theta & \cos\theta \end{pmatrix} : \theta \in \mathbb{R} \right\}.$$

(a) Explain why this group of matrices is isomorphic to the group $\mathrm{Rot}(2)$ of rotations $\mathcal{R}$ about the origin in $\mathbb{R}^2$, with composition of operators

$$R_{\theta_1} \circ R_{\theta_2} = R_{\theta_1 + \theta_2}$$

as its multiplication operation.

(b) Explain why K is a closed, bounded set of matrices in $M(2, \mathbb{R})$.

(c) Explain how K acquires the differentiable structure needed to make it a matrix Lie group.

8. $G = SL(n, \mathbb{R})$ is a hypersurface in $M(n, \mathbb{R})$ with $\dim_{\mathbb{R}}(G) = n^2 - 1$. Taking $n = 3$ and identifying matrices $A \in M(3, \mathbb{R})$ with vectors $\mathbf{x} = (a_{11}, a_{12}, a_{13}; \dots; a_{31}, a_{32}, a_{33}) \in \mathbb{R}^9$, $G = SL(3, \mathbb{R})$ is determined by the single constraint $\phi(A) = \det(A) = 1$ where $\phi : \mathbb{R}^9 \to \mathbb{R}$.

(a) Write out the Jacobian matrix $[(d\phi)_A]$ and verify that its rank is $= 1$ near any $A \in G$.

(b) Exhibit a partition $[1, 9] = J' \cup J$ as in the IFT such that the projection $(\pi_{J'}|_G)$ from $\mathbb{R}^{J'} \to \mathbb{R}^8$ yields a standard chart on $SL(3, \mathbb{R})$ near the identity element $I_{3\times 3}$.

(c) Does *every* choice $J' = \{1 \leq i_1 < \cdots < i_8 \leq 9\}$ of eight matrix entries (out of 9) yield a valid coordinate chart about the identity $I_{3\times 3}$ in G? If not, exhibit an explicit choice for which this fails.

9. For $G = \mathrm{SL}(2,\mathbb{R})$, an obvious map $G \to \mathbb{R}^4$

$$A = \begin{pmatrix} a & b \\ c & d \end{pmatrix} \to x_\alpha(A) = (a,b,c,d)$$

assigns "coordinates" to matrices in G but does not provide a "chart" for the standard $\mathcal{C}^\infty$ manifold structure on G because $\dim_{\mathbb{R}} G = 3$. Explain why we can obtain a valid coordinate chart (U_α, x_α) near the identity element $I_{2\times 2}$ by deleting one of the four variables from a, b, c, d. Identify at least one choice of the "redundant variable" that *does not* work.

10. Repeat Exercise 9, taking $A = \begin{pmatrix} 2 & 3 \\ 0 & 1/2 \end{pmatrix}$ as the base point in $\mathrm{SL}(2,\mathbb{R})$.

Section 5.3. Lie Algebra Structure in Tangent Spaces of Lie Groups

1. TRUE/FALSE QUESTIONS ("True" if the statement is *always* true.)

(a) If a matrix Lie group is abelian ($xy = yx$ for all x, y), the Lie bracket on $\mathfrak{g}$ is trivial, with $[X,Y] = 0$ for all $X, Y \in \mathfrak{g}$.

(b) If G is a matrix Lie group, every automorphism of its Lie algebra $\mathfrak{g}$ is the differential $\mathrm{Ad}_g = (di_g)_e$ of some conjugation operation $i_g(x) = gxg^{-1}$ ($g \in G$).

(c) For the abelian Lie group $G = (\mathbb{R}^n, +)$, every invertible linear operator on $\mathfrak{g}$ is an automorphism of the Lie algebra $\mathfrak{g}$.

(d) The set

$$\mathcal{T}_n = \{A \in \mathrm{M}(n,\mathbb{K}) : a_{ij} = 0 \text{ unless } i < j\}$$

of strictly upper triangular $n \times n$ matrices (over any ground field $\mathbb{K}$) becomes a nilpotent Lie algebra when equipped with the associative bracket $[A,B] = AB - BA$.

(e) If $\mathfrak{g}$ is a nilpotent Lie algebra, any Lie subalgebra $\mathfrak{h} \subseteq \mathfrak{g}$ is also nilpotent.

2. The Lie algebra of the general linear group $\mathrm{GL}(n,\mathbb{R})$ is all of matrix space $\mathfrak{gl}(n,\mathbb{R}) = \mathrm{M}(n,\mathbb{R})$. A basis is provided by the "matrix units" E_{ij} whose entries are zero except for a "1" in the (i,j) spot.

(a) For $n = 3$, give explicit formulas for the one-parameter subgroups $\gamma_{ij}(t) = \mathrm{Exp}(tE_{ij})$ by summing the exponential series. The outcome will depend on i, j. What are the corresponding tangent vectors $\gamma_{ij}'(0)$ in $\mathfrak{gl}(n,\mathbb{R})$?

(b) The Lie algebra $\mathfrak{sl}(n,\mathbb{R})$ of $\mathrm{SL}(n,\mathbb{R})$ has dimension

$$\dim_{\mathbb{R}} \mathfrak{sl}(n,\mathbb{R}) = n^2 - 1 < \dim_{\mathbb{R}} \mathfrak{gl}(n,\mathbb{R}) = n^2.$$

Find a way to select a basis $Y_1, \dots, Y_{n^2-1}$ for $\mathfrak{sl}(n,\mathbb{R})$ from the full set of matrix units E_{ij} in $\mathfrak{gl}(n,\mathbb{R}) = \mathrm{M}(n,\mathbb{R})$.

HINT. In Example 5.61, we characterized matrices A in $\mathfrak{sl}(n,\mathbb{R})$ in terms of their traces $\mathrm{Tr}(A) = a_{11} + \cdots + a_{nn}$.

(c) Describe the nonzero commutators $[Y_i, Y_j] = Y_iY_j - Y_jY_i$ for the basis in (b).

(d) Compute the one-parameter groups $\gamma_Y(t) = \text{Exp}(tY)$ for the basis vectors $Y = Y_k$ in $\mathfrak{sl}(n, \mathbb{R})$ constructed in Part (b).

3. If $X, Y \in \mathfrak{g}$ and the left-invariant vector fields $\widetilde{X}, \widetilde{Y}$ they determine on a matrix Lie group G are viewed as operators on $\mathcal{C}^\infty(G)$, explain why the vector field $[\widetilde{X}, \widetilde{Y}] = \widetilde{X} \circ \widetilde{Y} - \widetilde{Y} \circ \widetilde{Y}$ is left-invariant.

4. In Example 5.69, we considered the group Aff($\mathbb{R}$) of affine transformations $T_{a,b}(x) = ax + b$ on the real line.

(a) Determine the multiplication law $T_{a,b} \circ T_{a',b'} = T_{a'',b''}$ for composing affine transformations.

(b) Check that Aff($\mathbb{R}$) is a *group*, with

(i) IDENTITY ELEMENT I such that $I \circ T = T \circ I$.
(ii) INVERSES T^{-1} such that $T^{-1}T = I = TT^{-1}$.
(iii) ASSOCIATIVE LAW $T_1(T_2T_3) = (T_1T_2)T_3$,

for all $T_{a,b} \in \text{Aff}(\mathbb{R})$.

NOTE. That last rule is important because it allows us to write products $T_1 \cdots T_n$ without worrying about where to put the parentheses. ○

(c) Obviously $T_{1,0} = I$ (identity transformation on $\mathbb{R}$). Compute the parameters a', b' such that $T_{a',b'} = \left(T_{a,b}\right)^{-1}$.

(d) Verify that

$$\phi\begin{pmatrix} a & b \\ 0 & 1 \end{pmatrix} = T_{a,b} \quad (\text{where } a, b \in \mathbb{R}, a \neq 0)$$

is a bijective isomorphism between the group $(G, \cdot\,)$ of matrices in Example 5.69 and the group $(\text{Aff}(\mathbb{R}), \circ)$.

NOTE. The group Aff($\mathbb{R}$) is a differentiable manifold whose structure is determined by two charts defined on disjoint open sets (it is a Lie group with two connected components)

$$U_+ = \{T_{a,b} : a > 0,\ b \in \mathbb{R}\} \quad (\text{the identity component})$$
$$U_- = \{T_{a,b} : a < 0,\ b \in \mathbb{R}\} \quad (\text{of a subgroup}).$$

Parts (a) and (c) show that the group operations (matrix multiply) and (matrix inverse) become $\mathcal{C}^\infty$ maps when expressed in local chart coordinates, so Aff($\mathbb{R}$) is a Lie group. Strictly speaking, it is not a "matrix Lie group," but it is isomorphic to the *bona fide* matrix Lie group defined in (d). ○

5. Matrices A in the $ax + b$ group G are of two types:

(i) Those with $\det(A) > 0$, for which $T_{a,b} : \mathbb{R} \to \mathbb{R}$ preserves the order $x < y$ of points on the number line.
(ii) Those with $\det(A) < 0$, which reverse their order.

Thus G is a union $G = G^+ \cup G^-$ of disjoint, open subsets, according to the sign of $\det(A)$. The piece G^+ containing the identity element $I_{2\times 2}$ is a subgroup; its companion G^- is not. (Why?)

(a) Prove that the "one-parameter subgroups" $\gamma_1(s) = e^{sX}$ and $\gamma_2(t) = e^{tY}$ of Exercise 5.70 uniquely label each g in G^+ by a pair (s,t) in $\mathbb{R}^2$ by showing that every $g \in G^+$ has a unique factorization

$$g = g(s,t) = e^{sX} \cdot e^{tY} = \begin{pmatrix} e^s & 0 \\ 0 & 1 \end{pmatrix} \cdot \begin{pmatrix} 1 & t \\ 0 & 1 \end{pmatrix}.$$

(b) If an element of G is presented as $A = \begin{pmatrix} a & b \\ 0 & 1 \end{pmatrix}$, what are the values of the parameters (s,t) in Part (a)?

6. If $\mathfrak{g}$ is a finite-dimensional Lie algebra over $\mathbb{R}$, its complexification

$$\mathfrak{g}_{\mathbb{C}} = \mathfrak{g} + i\mathfrak{g} = \{X + iY : X, Y \in \mathfrak{g}\}$$

is a vector space over $\mathbb{C}$ if we define $(+)$ and $(\cdot)$ operations

$$(X + iY) + (X' + iY') = (X + X') + i(Y + Y')$$
$$(a + ib) \cdot (X + iY) = (aX - bY) + i(aY + bX)$$

(recall the discussion in Section 2.2) Verify that $\mathfrak{g}_{\mathbb{C}}$ becomes a Lie algebra over $\mathbb{C}$ if we define its bracket to be

$$[(X + iY), (X' + iY')] = \\ ([X, X'] - [Y, Y']) + i([X, Y'] + [Y, X']).$$

(Checking the Jacobi identity is the messy part.)

7. Verify that

$$\mathfrak{su}(2) = \{A \in \mathrm{M}(2, \mathbb{C}) : AA^{\mathrm{T}} = I\}$$

is the complexification of its real points

$$\mathfrak{so}(3) = \mathfrak{su}(2) \cap (\mathrm{M}(2, \mathbb{R}) + i0).$$

Section 5.4. The Exponential Map for Matrix Lie Groups

1. TRUE/FALSE QUESTIONS ("True" if the statement is *always* true).

(a) If $\gamma(0) = I$ and $\gamma'(0) = X \in \mathfrak{g}$ for a smooth matrix-valued curve $\gamma(t)$ that lies within a matrix Lie group $G \subseteq \mathrm{M}(n, \mathbb{R})$, then for some $r > 0$, $\gamma(t) = \gamma_X(t) = e^{tX}$ for $-r < t < r$.

(b) $\mathrm{M}(n, \mathbb{R})$ equipped with the associative bracket $[A, B] = AB - BA$ is the Lie algebra $\mathfrak{gl}$ of the general linear group $\mathrm{GL}(n, \mathbb{R})$.

(c) Every matrix $A \in \mathrm{M}(n, \mathbb{C})$ can be written as $D + N$ where D is a diagonal matrix, N is nilpotent, and the factors commute with $DN = ND$.

HINT. Think (JCF).

(d) Whenever a matrix $A \in \mathrm{M}(n, \mathbb{C})$ can be decomposed as a sum $D + N$ of *commuting* diagonal and nilpotent matrices, the D and N are unique.

(e) The Lie algebra of the unitary group $\mathrm{U}(n)$ is

$$\mathfrak{u}(n) = \{A \in \mathrm{M}(n, \mathbb{C}) : |\det(A)| = 1\}.$$

(f) Many curves $\gamma(t)$ in G that pass through the identity $e = I$ can determine the same tangent vector $X = \gamma'(0)$ in $\mathfrak{g} = \mathrm{TG}_e$ when $t = 0$, but the one-parameter group $\gamma_X(t) = e^{tX}$ is the *only* smooth curve in G such that $\gamma(0) = I$ and $\gamma'(0) = X$.

(g) For any matrix Lie group G, the union of all the curves traced out by the one-parameter subgroups $\gamma_X(t)$ with generators $X \in \mathfrak{g}$ contains an open neighborhood of the identity I in G.

2. Which of the following identities between Lie algebras are valid?

(a) $\mathfrak{u}(n) = \mathfrak{su}(n)$ in $\mathrm{M}(n, \mathbb{C})$

(b) $\mathfrak{o}(n) = \mathfrak{so}(n)$ in $\mathrm{M}(n, \mathbb{R})$

(c) $\mathfrak{o}(n, \mathbb{C}) = \mathfrak{so}(n, \mathbb{C})$ in $\mathrm{M}(n, \mathbb{C})$

3. If $A \in \mathrm{M}(n, \mathbb{R})$, the curve $X(t) = e^{tA}$ is a $\mathcal{C}^\infty$ curve that passes through the identity I when $t = 0$ and is a solution of the matrix-valued differential equation

$$\frac{dX}{dt}(t) = A \cdot X(t) \quad \text{for all } t \in \mathbb{R},$$

with initial value $X(0) = I_{n\times n}$ (identity matrix).

(a) Explain why $X(t)$ satisfies this differential equation.

(b) What is the initial value $X(0)$ when $t = 0$?

(c) What is its derivative $X'(0)$ when $t = 0$? At an arbitrary time t?

(d) Explain by example why the solution to this equation fails to be unique if the initial value has not been specified.

4. What happens if we consider *complex* matrices $A \in \mathrm{M}(n, \mathbb{C})$ and allow complex-valued solution curves $Z(t) = X(t) + iY(t)$ for the differential equation $dZ/dt = A \cdot Z(t)$ with initial value $Z(0) = I$? Is $Z(t) = \mathrm{Exp}(tA)$ still a solution to the initial value problem

$$\frac{dX}{dt}(t) = A \cdot X(t) \quad \text{with } X(0) = I_{n\times n}.$$

(a) Prove that the vector space $\mathfrak{g}^{(2)} = [\mathfrak{g}, \mathfrak{g}]$ is a Lie algebra in its own right by showing that Lie bracket $[A, B]$ of vectors in $\mathfrak{g}^{(2)}$ is always back in $\mathfrak{g}^{(2)}$.

HINT. The Lie bracket is bilinear, and $[X, Y] \in \mathfrak{g}$ for $X, Y \in \mathfrak{g}$ by definition of Lie algebra.

(b) Prove that the chain of commutator subspaces is decreasing, with $\mathfrak{g}^{(k+1)} \subseteq \mathfrak{g}^{(k)}$ for $k = 1, 2, \ldots$.

(c) Prove that $[\mathfrak{g}^{(k)}, \mathfrak{g}^{(k)}] \subseteq \mathfrak{g}^{(k)}$ for all k, so *every* commutator subspace $\mathfrak{g}^{(k)}$ is a Lie subalgebra of $\mathfrak{g}$.

5. In Example 5.84 we defined the 3-dimensional Heisenberg group as a group of matrices in $\mathrm{M}(3,\mathbb{R})$, and introduced a systematic way to label its elements by triples $(x, y, z) \in \mathbb{R}^3$. In this scheme the law of matrix multiplication took the form

$$(x, y, z) \cdot (x', y', z') = (x + x', y + y', z + z' + xy').$$

(a) Find the triple such that (x', y', z') is the inverse of (x, y, z) in this group.

(b) Compute the effect of an inner automorphism $i_g(a) = gag^{-1}$ when $g = (x, y, z)$ and $a = (x', y', z')$ in G.

THE KILLING FORM $B(X, Y) = \mathrm{Tr}(\mathrm{ad}_X \mathrm{ad}_Y)$ ON A LIE ALGEBRA.[6] This bilinear form on finite-dimensional Lie algebras turned out to be crucial in understanding their remarkable algebraic structures. The next example asks you to compute the matrix $B_{\mathfrak{X}}$ of the Killing form for the Lie algebra $\mathfrak{so}(3) \subseteq \mathrm{M}(3,\mathbb{R})$ of the group of rotations in 3-dimensional space, using a basis

$$\mathfrak{X} = \{X_1 = X, X_2 = Y, X_3 = Z\}$$

whose pattern of structure constants $[X_i, X_j] = \sum_{k=1}^{3} c_{ij}^{(k)} X_k$ is particularly simple:

$$[X, Y] = Z \qquad [Y, Z] = X \qquad [Z, X] = Y$$

These commutation relations were discussed in Example 5.84; for simplicity we have changed the labels on basis matrices from $\{\mathbf{I}, \mathbf{J}, \mathbf{K}\}$ used there to $\{X, Y, Z\}$. Computing the products $\mathrm{ad}_{X_i} \mathrm{ad}_{X_j}$ of operators on $\mathfrak{g}$ and their traces is less daunting than it might seem if approached with a few basic principles in mind.

- Computing the 3×3 matrices $[\mathrm{ad}_{X_i}]_{\mathfrak{X}}$ is straightforward.
- If we label $A_i = [\mathrm{ad}_{X_i}]_{\mathfrak{X}}$, the products $A_i A_j$ are the 3×3 matrices

  $$A_{ij} = [\mathrm{ad}_{X_i}]_{\mathfrak{X}} \cdot [\mathrm{ad}_{X_j}]_{\mathfrak{X}} = [\mathrm{ad}_{X_i} \mathrm{ad}_{X_j}]_{\mathfrak{X}},$$

  whose traces are the entries $B(X_i, X_j)$ in $[B]_{\mathfrak{X}}$.
- Since $B(X, Y)$ is symmetric (why?), we only need to compute the matrix entries $B(X_i, X_j) = \mathrm{Tr}(A_i A_j)$ for $1 \le i \le j \le 3$. these can be found by evaluating the coefficients in

  $$[X_i, [X_j, H]] = aX + bY + cZ$$

  for $1 \le i \le j \le 3$ and $H = X, Y, Z$.

For example, taking $X_i = X$ and $X_j = X$ or Y, it follows from the commutation relations for X, Y, Z that

$$[X, [X, X]] = 0 \quad (\text{since } [X, X] = 0)$$
$$[X, [X, Y]] = -Y \quad \text{and} \quad [X, [X, Z]] = -Z.$$

[6] W. Killing was a contemporary of Lie, who was one of the first to attempt a detailed structural analysis of the "simple" Lie algebras over $\mathbb{C}$, which include many of the complex classical groups. He missed some of the possibilities, later filled in by E. Cartan; nevertheless his pioneering work stands as a foundation of Lie theory.

Continuing these calculations we get

$$A_{11} = A_1 A_1 = \begin{pmatrix} 0 & 0 & 0 \\ 0 & -1 & 0 \\ 0 & 0 & -1 \end{pmatrix}$$

and $B(X,X) = B(X_1, X_1) = \mathrm{Tr}(A_1 A_1) = -2$. The other entries are determined similarly.

9. Verify that the matrix of the Killing form B for the real Lie algebra $\mathfrak{g} = \mathfrak{so}(3)$ of the group of rotations on $\mathbb{R}^3$ is

$$[B]_{\mathfrak{X}} = \begin{pmatrix} -2 & 0 & 0 \\ 0 & -2 & 0 \\ 0 & 0 & -2 \end{pmatrix}.$$

NOTE. The Killing form SO(3) is symmetric and strictly *negative definite*, with signature $(0,3)$ and $B(H,H) < 0$ for all $H \neq 0$ in $\mathfrak{g}$. Inner products on $\mathbb{R}^3$ are bilinear forms with signature $(3,0)$ and have $B(H,H) > 0$ for all nonzero H. ○

Section 5.5 and 5.6. The Exponential Map and Lie Correspondence

1. TRUE/FALSE QUESTIONS ("True" if the statement is *always* true.)

(a) The CHB formula for the multiplication operation in a matrix Lie group reduces to a *finite sum* if the Lie algebra $\mathfrak{g}$ is nilpotent.

(b) If A is a nonzero $n \times n$ complex matrix, solutions $X \in \mathrm{M}(n, \mathbb{C})$ to the matrix equation $e^X = A$ always exist.

(c) For $A \in \mathrm{M}(n, \mathbb{C})$ close enough to the identity matrix I, the matrix equation $e^X = A$ has a *unique* solution $X = \mathrm{Log}(A)$.

(d) If A is a nonzero $n \times n$ complex matrix and $X \in \mathrm{M}(n, \mathbb{C})$ is a solution $X \in \mathrm{M}(n, \mathbb{C})$ to the matrix equation $e^X = A$, all other solutions have the form $X + 2\pi n i I$ for $n \in \mathbb{Z}$.

(e) The exponential map $\mathrm{Exp} : \mathrm{M}(n, \mathbb{R}) \to \mathrm{GL}(n, \mathbb{R})$ maps the set

$$\mathcal{N}_n = \{A \in \mathrm{M}(n, \mathbb{K}) : A^k = 0 \text{ for some } k \in \mathbb{N}\}$$

of nilpotent $n \times n$ matrices into the set of unipotent $n \times n$ matrices

$$\mathcal{U}_n = \{A \in \mathrm{M}(n, \mathbb{K}) : A - I \text{ is nilpotent}\}.$$

2. List all *real* matrices X such that $e^X = I_{n \times n}$.

3. Find all *nilpotent* solutions $X \in \mathrm{M}(n, \mathbb{C})$ of the matrix identities

$$\text{(a) } \mathrm{Exp}(X) = \begin{pmatrix} -1 & 0 & 0 \\ 0 & -1 & 0 \\ 0 & 0 & -1 \end{pmatrix} \qquad \text{(b) } \mathrm{Exp}(X) = \begin{pmatrix} 0 & 1 & 0 & 0 \\ 0 & 0 & 1 & 0 \\ 0 & 0 & 0 & 1 \\ 0 & 0 & 0 & 0 \end{pmatrix}$$

HINT. In (a), there are no nilpotent solutions X (why?). In (b), you could use the (finite) series for $\mathrm{Log}(X)$ or work backward from the finite series for $\mathrm{Exp}(A) = X$. Remember that a finite sum of commuting nilpotent operators is nilpotent.

4. The following identities can be proved by working with the scalar power series for *Exp* and *Log*.

$$\text{(i)}\ \ \mathrm{Log}(\mathrm{Exp}(X)) = X \quad \text{for all } X \text{ with } \|X\| < \sqrt{2}$$

$$\text{(ii)}\ \ \mathrm{Exp}(\mathrm{Log}(A)) = A \quad \text{for all } A \text{ with } \|A - I\| < 1.$$

But if X is nilpotent (or A unipotent), the series for Exp and Log have only finitely many nonzero terms. Are these identities valid for *all* nilpotent X (or unipotent A)? Explain your reasoning.

5. If G and H are locally isomorphic matrix Lie groups, explain why their Lie algebras must be isomorphic (so $\mathfrak{g} \cong \mathfrak{h}$), starting from the definitions of $\mathfrak{g}, \mathfrak{h}$ as the tangent spaces $\mathrm{TG}_e, \mathrm{TH}_e$.

6. For the 3-dimensional Heisenberg group G discussed in Example 5.84, the product $A * B$ of two typical vectors

$$A = aX + bY + cZ \qquad B = a'X + b'Y + c'Z$$

is given by a *finite* sum.

(a) Explain why all iterated commutators

$$[H_1, [H_2, \dots, [H_{r-1}, H_r], \dots]] \qquad H_k \in \mathfrak{g}$$

of length $r \geq 3$ are all zero.

(b) Write out the full expansion of the Campbell-Hausdorff product $A*B$ for arbitrary A and B in the Lie algebra $\mathfrak{g}$ of the 3-dimensional Heisenberg group. (The CBH formula is a *finite* sum for all nilpotent Lie groups. This is one of their most congenial features.)

Bibliography

[1] Manfredo Perdigão do Carmo, *Riemannian geometry*, Mathematics: Theory & Applications, Birkhäuser Boston, Inc., Boston, MA, 1992. Translated from the second Portuguese edition by Francis Flaherty. MR1138207

[2] Melvin Hausner and Jacob T. Schwartz, *Lie groups; Lie algebras*, Gordon and Breach Science Publishers, New York-London-Paris, 1968. MR0235065

[3] G. Hochschild, *The structure of Lie groups*, Holden-Day, Inc., San Francisco-London-Amsterdam, 1965. MR0207883

[4] Nathan Jacobson, *Lie algebras*, Interscience Tracts in Pure and Applied Mathematics, No. 10, Interscience Publishers (a division of John Wiley & Sons), New York-London, 1962. MR0143793

[5] Wulf Rossmann, *Lie groups*, Oxford Graduate Texts in Mathematics, vol. 5, Oxford University Press, Oxford, 2002. An introduction through linear groups. MR1889121

[6] V. S. Varadarajan, *Lie groups, Lie algebras, and their representations*, Graduate Texts in Mathematics, vol. 102, Springer-Verlag, New York, 1984. Reprint of the 1974 edition. MR746308

Published Titles in This Series

30 **Frederick P. Greenleaf and Sophie Marques,** Linear Algebra II, 2020
29 **Frederick P. Greenleaf and Sophie Marques,** Linear Algebra I, 2019
28 **László Erdős and Horng-Tzer Yau,** A Dynamical Approach to Random Matrix Theory, 2017
27 **S. R. S. Varadhan,** Large Deviations, 2016
26 **Jerome K. Percus and Stephen Childress,** Mathematical Models in Developmental Biology, 2015
25 **Kurt O. Friedrichs,** Mathematical Methods of Electromagnetic Theory, 2014
24 **Christof Schütte and Marco Sarich,** Metastability and Markov State Models in Molecular Dynamics, 2013
23 **Jerome K. Percus,** Mathematical Methods in Immunology, 2011
22 **Frank C. Hoppensteadt,** Mathematical Methods for Analysis of a Complex Disease, 2011
21 **Frank C. Hoppensteadt,** Quasi-Static State Analysis of Differential, Difference, Integral, and Gradient Systems, 2010
20 **Pierpaolo Esposito, Nassif Ghoussoub, and Yujin Guo,** Mathematical Analysis of Partial Differential Equations Modeling Electrostatic MEMS, 2010
19 **Stephen Childress,** An Introduction to Theoretical Fluid Mechanics, 2009
18 **Percy Deift and Dimitri Gioev,** Random Matrix Theory: Invariant Ensembles and Universality, 2009
17 **Ping Zhang,** Wigner Measure and Semiclassical Limits of Nonlinear Schrödinger Equations, 2008
16 **S. R. S. Varadhan,** Stochastic Processes, 2007
15 **Emil Artin,** Algebra with Galois Theory, 2007
14 **Peter D. Lax,** Hyperbolic Partial Differential Equations, 2006
13 **Oliver Bühler,** A Brief Introduction to Classical, Statistical, and Quantum Mechanics, 2006
12 **Jürgen Moser and Eduard J. Zehnder,** Notes on Dynamical Systems, 2005
11 **V. S. Varadarajan,** Supersymmetry for Mathematicians: An Introduction, 2004
10 **Thierry Cazenave,** Semilinear Schrödinger Equations, 2003
9 **Andrew Majda,** Introduction to PDEs and Waves for the Atmosphere and Ocean, 2003
8 **Fedor Bogomolov and Tihomir Petrov,** Algebraic Curves and One-Dimensional Fields, 2002
7 **S. R. S. Varadhan,** Probability Theory, 2001
6 **Louis Nirenberg,** Topics in Nonlinear Functional Analysis, 2001
5 **Emmanuel Hebey,** Nonlinear Analysis on Manifolds: Sobolev Spaces and Inequalities, 2000
3 **Percy Deift,** Orthogonal Polynomials and Random Matrices: A Riemann-Hilbert Approach, 2000
2 **Jalal Shatah and Michael Struwe,** Geometric Wave Equations, 2000
1 **Qing Han and Fanghua Lin,** Elliptic Partial Differential Equations, Second Edition, 2011